New Vistas in
Electro-Nuclear
Physics

NATO ASI Series

Advanced Science Institutes Series

A series presenting the results of activities sponsored by the NATO Science Committee, which aims at the dissemination of advanced scientific and technological knowledge, with a view to strengthening links between scientific communities.

The series is published by an international board of publishers in conjunction with the NATO Scientific Affairs Division

A	**Life Sciences**	Plenum Publishing Corporation
B	**Physics**	New York and London
C	**Mathematical and Physical Sciences**	D. Reidel Publishing Company Dordrecht, Boston, and Lancaster
D	**Behavioral and Social Sciences**	Martinus Nijhoff Publishers
E	**Engineering and Materials Sciences**	The Hague, Boston, and Lancaster
F	**Computer and Systems Sciences**	Springer-Verlag
G	**Ecological Sciences**	Berlin, Heidelberg, New York, and Tokyo

Recent Volumes in this Series

Series B: Physics

New Vistas in Electro-Nuclear Physics

Edited by

E. L. Tomusiak and
H. S. Caplan
University of Saskatchewan
Saskatoon, Saskatchewan, Canada

and

E. T. Dressler
Pennsylvania State University
Ogontz Campus
Abington, Pennsylvania

Plenum Press
New York and London
Published in cooperation with NATO Scientific Affairs Division

Proceedings of a NATO Advanced Study Institute on
New Vistas in Electro-Nuclear Physics,
held August 22–September 4, 1985,
in Banff, Alberta, Canada

Library of Congress Cataloging in Publication Data

NATO Advanced Study Institute on New Vistas in Electro-Nuclear Physics (1985:
  Banff, Alta.)
  New vistas in electro-nuclear physics.

  (NATO ASI series. Series B, Physics; v. 142)
  "Proceedings of a NATO Advanced Study Institute on New Vistas in Electro-
Nuclear Physics, held August 22–September 4, 1985, in Banff, Alberta, Canada"
—T.p. verso.
  "Published in cooperation with NATO Scientific Affairs Division."
  Includes bibliographies and index.
  1. Nuclear structure—Congresses. 2. Electromagnetic interactions—Con-
gresses. 3. Scattering (Physics)—Congresses. 4. Quarks—Congresses. I.
Tomusiak, E. L. (Edward L.) II. Caplan, H. S. (Henry S.) III. Dresser, E. T. (Edward T.)
IV. North Atlantic Treaty Organization. Scientific Affairs Division. V. Title. VI.
Series.
QC793.3.S8N38   1985   539.7   86-18677
ISBN 978-1-4684-5202-0   ISBN 978-1-4684-5200-6 (eBook)
DOI 10.1007/978-1-4684-5200-6

PREFACE

The NATO Advanced Study Institute "New Vistas in Electro-Nuclear Physics" was held in Banff, Alberta, Canada from August 22 to September 4, 1985. This volume contains the lecture notes from that Institute.

The idea to organize this Institute coincided with the award of funding for a pulse stretcher ring at the University of Saskatchewan's Linear Accelerator Laboratory. This together with the high level of interest in electron accelerators worldwide convinced us that it was an appropriate time to discuss the physics to be learned with such machines. In particular that physics which requires high energy and/or high duty cycle accelerators for its extraction was intended to be the focus of the Institute. Thus the scope of the lectures was wide, with topics ranging from the structure of the trinucleons to quark models of nucleons, QCD, and QHD. The theme however was that we are just trying to understand the nucleus and that the electromagnetic probe can serve as a powerful tool in such a quest.

Lectures were generally held in the mornings and evenings. The afternoons were free, and most participants took advantage of the fine weather and stimulating environment of the Banff area. There was room in the programme for a number of workshops which were organized more or less spontaneously. Although we were not able to include the proceedings of the workshops in this volume, they were enthusiastically attended by the participants. The subjects and organizers of these workshops were Electric and Magnetic Resonances (B. Castel and L. Zamick), Two and Three-Body Nuclear Physics (F. Khanna), Polarized Electron Effects (T. W. Donnelly), Evidence for Quarks in Nuclei (B. Gibson), and Electromagnetic Interactions (J. D. Walecka). We take this opportunity to thank the organizers and speakers in these workshops. The Institute ended with the lecture Outlook and Open Questions by J. D. Walecka.

Primary funding for the Institute came from NATO through its Advanced Study Institute programme. We gratefully acknowledge additional financial support from NSERC (Natural Sciences and Engineering Research Council of Canada), TRIUMF, Atomic Energy of Canada Ltd., Los Alamos National Laboratory, Saskatchewan Accelerator Laboratory, and the University of Saskatchewan.

The Institute was organized by a committee consisting of E. L. Tomusiak, H. S. Caplan, and M. A. Preston. We received much help on the program and other organizational matters from our International Advisory Committee consisting of Profs. R. Woloshyn (TRIUMF), B. Goulard (Montreal), G. Brown (Stony Brook), B. Gibson (LANL), J. S. O'Connell (NBS), D. Drechsel (Mainz), P. Sauer (Hanover), J. H. Koch (NIKHEF), Sir D. Wilkinson (Sussex), J. M. Laget (Saclay), and B. Bosco (Firenze).

One of the most arduous tasks was the publication of this volume of the proceedings. Most credit goes to the typists Vicky Tomusiak and Arlene Vedress. However we managed to impose heavily on most of the computing staff at the Accelerator Laboratory. For answering our innumerable questions on troff, eqn, etc., and for getting the Laser-Writer operational with legal size paper we express our gratitude to Doug Murray, Glen Wright, and Sharon Pywell.

Saskatoon, Saskatchewan   Edward L. Tomusiak, Director
May, 1986   Henry S. Caplan, Co-Director
                                                 Melvin A. Preston, Co-Director
                                                 Edward T. Dressler, Co-Editor

CONTENTS

ELECTROWEAK INTERACTIONS WITH NUCLEI [Lectures 1 and 2]

QUANTUM HADRODYNAMICS (QHD) [Lectures 3 and 4]

J. D. Walecka

Institute of Theoretical Physics
Department of Physics, Stanford University
Stanford, California 94305

A. INTRODUCTION

The material in these first two lectures is taken from Refs. 1, 2, and 3. In this short time I clearly cannot work through everything as I would like to; however, all the algebra is done in detail starting from scratch in Ref. 1, which also covers an extensive list of topics. If you are a student who wants to learn about electron scattering from nuclei and electromagnetic interactions, I strongly urge you to send for a copy of these notes. They are still available from Argonne. I will give two lectures on Part 1 and two lectures on Part 2. Lecture I is meant as a general introduction to the subject of electroweak interactions with nuclei.

B. MOTIVATION

1. Why Nuclear Physics?

The atomic nucleus is a unique form of matter. It consists of many baryons in close proximity. All the forces of nature are present in the nucleus - strong, electromagnetic, weak, and gravitation (if we include neutron stars, which are nothing more than enormous nuclei held together by the gravitational attraction). Nuclei provide unique microscopic laboratories in which to test the structure of the fundamental interactions. In addition, the nuclear many-body problem is of intrinsic intellectual interest. Furthermore, most of the mass and energy in the visible universe comes from nuclei and nuclear reactions. If the goal of physics is to understand nature, then surely we must understand the nucleus. Finally, in sum, nuclear physics is the study of the *structure of matter*.

2. Why Electroweak Interactions?

The last decade and a half has seen the development of a unified theory of the electroweak interactions[4-6]. This is surely one of the great intellectual achievements of our era. The basis for this unification lies in a local gauge theory built on the symmetry structure $SU(2)_W \otimes U(1)_W$. It is essential to continue to put this theory to rigorous tests in all possible domains. In order to do this, we need intense sources of electroweak probes (e^-, ν_l, μ^-). Once the nature of the fundamental interaction is understood, the electroweak interaction provides a powerful tool for studying nuclear structure. We have a clean probe and we know what we measure.

3. Strong Interactions

If I can get ahead of myself, we will see in the next lecture that the current picture of the strong interactions is based on quarks and gluons as the underlying set of degrees of freedom. The theory of the strong interactions binding quarks into the observed hadrons, baryons and mesons, is quantum chromodynamics (QCD). It is a Yang-Mills non-abelian gauge theory based on a threefold local color symmetry - $SU(3)_c$ [7]. The symmetry structure of the "Standard Model" of the strong and electroweak interactions is thus $SU(3)_c \otimes SU(2)_W \otimes U(1)_W$. We will discuss in the next lecture how electroweak interactions with nuclei can be used to probe this assumed structure of the fundamental interactions. We will see how nuclei can be used to study the strong interaction, confinement aspects of QCD.

C. THE NUCLEAR MANY-BODY PROBLEM

1. Traditional Approach

Let me define the "traditional approach" to the nuclear many-body problem as developed in detail, for example, in Ref. 8. The underlying set of degrees of freedom is assumed to consist of structureless nucleons, the proton and neutron p and n. One starts from static two-body potentials fit to two-body scattering data. The non-relativistic, many-particle Schrödinger equation is then solved in some approximation, providing energy levels and wave functions for the system. One then constructs electroweak currents from the properties of free nucleons, and these are used to probe the system. This traditional approach to nuclear physics has had a great many successes; however, it is inadequate for a more detailed understanding of nuclear structure.

2. Relativistic Models of Nuclear Structure

A more appropriate set of degrees of freedom for the nucleus consists of the *hadrons,* the strongly interacting baryons and mesons. One can give many reasons for this. For example:

a) The long-range part of the Paris potential, probably the most accurate two-nucleon potential currently available, is derived from the exchange of mesons, predominantly $\pi(J^\pi,T) = (0^-,1)$, $\sigma(0^+,0)$, $\omega(1^-,0)$, and $\rho(1^-,1)$. We have *experimental proof* that the long-range part of this interaction is governed by meson exchange;

b) One of the most important recent advances in nuclear physics is the unambiguous identification through electron scattering (e,e′) of exchange currents in nuclei. These are additional currents present in the many-baryon system which arise from the flow of charged mesons between baryons.

Furthermore, an important goal of nuclear physics is to study the properties of nuclear matter under *extreme conditions*. This is crucial, for example, in:

a) Astrophysics, where one studies the properties of condensed stellar objects, and supernovae;

b) High-energy heavy-ion reactions, where high density and high temperature properties of nuclear matter are studied under laboratory conditions.

In any extended description of nuclear structure, it is important to incorporate *general principles* of physics, in particular
   1) Quantum mechanics;
   2) Special relativity;
   3) Causality.

The only consistent theoretical framework we have for dealing with such a relativistic, interacting many-particle system is *relativistic quantum field theory based on a local Lagrangian density*. I like to refer to such a field theory formulated in terms of the hadronic degrees of freedom as *quantum hadrodynamics (QHD)*.

3. Quarks and QCD

The theory of QCD is simple at short distances, where the theory is essentially free (asymptotic freedom). It is a complicated, strong-coupling theory at large distances (confinement)[7]. An important thesis of this set of lectures is the following:

4. The appropriate set of degrees of freedom depends on the distance scale *at which we probe the nucleus*.

D. ELECTRON SCATTERING

1. Why Electron Scattering?

Electron scattering (e,e′) provides a powerful tool for studying nuclear structure. First, the interaction is known; it is given by quantum electrodynamics (QED). Second, electrons provide a clean probe; we know what we measure. The electrons interact with the local electromagnetic charge and current densities in the target. Third, the interaction is relatively weak, of order α the fine structure constant, and one can make measurements without greatly disturbing the structure of the target.

In electron scattering one measures a macroscopic diffraction pattern in the laboratory, and this pattern is essentially the Fourier transform of the static and transition charge and current densities. Upon inversion of the Fourier transform, one has a determination of the detailed microscopic spatial distribution of the charge and current densities. This is a unique source of information about nuclear structure.

Electron scattering also provides a versatile probe. There is kinematic flexibility (discussed below). Furthermore, in addition to the Coulomb interaction with the charge density and the transverse-photon-exchange interaction with the convection current density, there is an interaction with the magnetization current density in the nucleus resulting from the intrinsic spin magnetization of the nucleons.

2. Lepton Kinematics in (e,e′)

Let us first concentrate on the lepton variables in the process (e,e′) as illustrated in Fig. 1. There are three free variables (k_1, k_2, θ) the initial and final wave numbers of the electrons and the scattering angle, or equivalently (ν, q^2, θ), where $\nu = q \cdot p/M_T = \varepsilon_1 - \varepsilon_2$ is the electron energy loss in the laboratory system, q^2 is the four-momentum transfer, and θ is the scattering angle. The first two combinations in the latter set are Lorentz invariants.

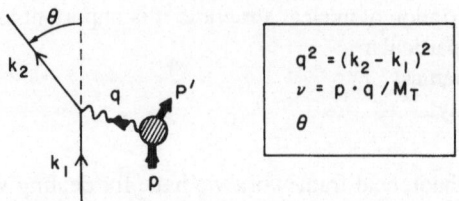

Fig. 1   Kinematics and lepton variables in semileptonic processes.

3. Response Tensor

The cross section for electron scattering can be written in the form $d\sigma \propto \eta_{\mu\nu}W_{\mu\nu}$. Here $\eta_{\mu\nu}$ is a known tensor constructed from the lepton variables. The target response is completely characterized in terms of the response tensor defined by

$$W_{\mu\nu} = (2\pi)^3 V \sum_i \sum_f <i|J_\nu(o)|f> <f|J_\mu(o)|i> (E)\delta^{(4)}(p'-p+q) \tag{1.1}$$

where $J_\mu(x) = (\vec{J}(x), i\rho(x))$ is the electromagnetic current density in the target. The average over initial nuclear states, and sum over final states simplifies this tensor, since it can only be constructed from the two independent four vectors p_μ and q_μ remaining in the problem. The coefficients in this expansion must be functions of the two Lorentz scalars q^2 and $q \cdot p$. In this fashion one is led directly to the basic electron scattering cross section, a result which I hope is familiar to all of you

$$\left[\frac{d^2\sigma}{d\Omega_2 d\varepsilon_2}\right]_{ee'}^{ERL} = \frac{\alpha^2 \cos^2\frac{\theta}{2}}{4\varepsilon_1^2 \sin^4\frac{\theta}{2}} \frac{1}{M_T}\left[W_2(q^2,q\cdot p) + 2W_1(q^2,q\cdot p)\tan^2\frac{\theta}{2}\right] \tag{1.2}$$

The extreme relativistic limit ERL shall mean that the lepton mass is neglected. The Mott cross section

$$\sigma_M = \frac{\alpha^2 \cos^2\frac{\theta}{2}}{4\varepsilon_1^2 \sin^4\frac{\theta}{2}} \tag{1.3}$$

is that for scattering a relativistic Dirac electron from a point charge. The two response surfaces W_1 and W_2 can be separated by keeping their arguments fixed and making a straight-line Rosenbluth plot against $\tan^2\theta/2$, or by working at $\theta = 180°$ where only W_1 contributes. The assumptions that go into the derivation of this result are general: one-photon exchange, Lorentz invariance, and current conservation.

For transitions to a discrete state of the target, one can perform the integration over final electron energies

$$\int \frac{d\varepsilon_2}{M_T} W_i(q^2,q\cdot p) = \omega_i(q^2)\, r \tag{1.4}$$

where the recoil factor is defined by

$$r^{-1} = 1 + \frac{2\varepsilon_1 \sin^2 \frac{\theta}{2}}{M_T} \, . \tag{1.5}$$

4. Multipole Analysis

Let us assume further that the nuclear transition densities are localized in space and that the initial and final nuclear states are eigenstates of angular momentum. A multipole decomposition of the electromagnetic interaction can be carried out with the result that[1],

$$2\omega_1 = \frac{4\pi}{2J_i+1} \sum_{J=1}^{\infty} \left[|<J_f\|\hat{T}_J^{mag}(q)\|J_i>|^2 + |<J_f\|\hat{T}_J^{el}(q)\|J_i>|^2 \right] \cdot \frac{E'}{M_T} \tag{1.6}$$

$$\omega_2 - \frac{q^2}{\bar{q}^2}\omega_1 = \frac{q^4}{\bar{q}^4} \cdot \frac{4\pi}{2J_i+1} \sum_{J=0}^{\infty} |<J_f\|\hat{M}_J(q)\|J_i>|^2 \cdot \frac{E'}{M_T} \, . \tag{1.7}$$

The multipole operators are defined by (here $q \equiv |\vec{q}|$)

$$\hat{M}_{JM}(q) = \int j_J(qx) Y_{JM}(\Omega_x) \hat{\rho}(\vec{x}) d\vec{x} \tag{1.8}$$

$$\hat{T}_{JM}^{el} = \frac{1}{q} \int \vec{\nabla} \times [j_J(qx)\vec{Y}_{JJ1}^{M}(\Omega_x)] \cdot \vec{J}(\vec{x}) d\vec{x} \, . \tag{1.9}$$

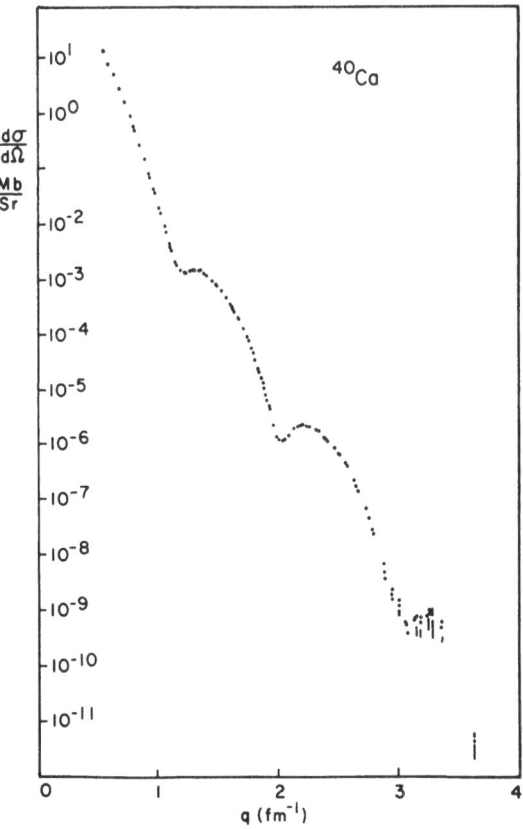

Fig. 2   Elastic (e,e) cross section for ^{40}Ca vs. momentum transfer.[9] The scattering here is from the charge distribution.

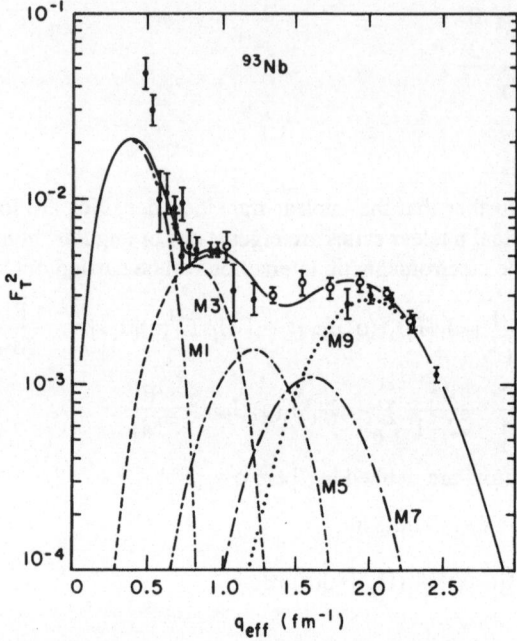

Fig. 3   Elastic transverse (e,e) form factor for ^{93}Nb.[10]
            The scattering here is from the magnetization distribution.

$$\hat{T}_{JM}^{mag} = \int [j_J(qx)\vec{Y}_{JJ1}^{M}(\Omega_x)] \cdot \vec{J}(\vec{x})d\vec{x} . \tag{1.10}$$

They are irreducible tensor operators in the nuclear Hilbert space. This decomposition
permits one to extract the M-dependence and angular momentum selection rules for the
nuclear matrix elements through the use of the Wigner-Eckart theorem. The longitudinal
multipoles

$$\hat{L}_{JM}(q) = \frac{i}{q} \int \left[\vec{\nabla}(j_J(qx)Y_{JM}(\Omega_x)) \right] \cdot \vec{J}(\vec{x})d\vec{x} \tag{1.11}$$

have been eliminated here through the use of current conservation.

5. Some Selected Examples

Figs. 2-5 show some selected examples of electron scattering results.

Fig. 2 shows the measured elastic monopole charge scattering cross section for $^{40}_{20}$Ca.[9]
Note the logarithmic scale of the left hand side of this figure! This is a state-of-the art
example of a measurement of the macroscopic diffraction pattern referred to above. The
high q data comes from Saclay.

Fig. 3 shows elastic magnetic scattering[10] from $^{93}_{41}$Nb whose shell-model
configuration is $(1g_{9/2})$. Note how all the magnetic multipoles show up as the momentum
transfer is increased (parity and time reversal limit one to odd magnetic multipoles in
elastic scattering). A measurement of the elastic magnetic cross section at all q allows
one to determine the spatial distribution of the ground-state magnetization, and hence of
the last valence proton in this nucleus. One can actually see that there is a little current
loop in the ground state of this nucleus! The high q data is from Bates.

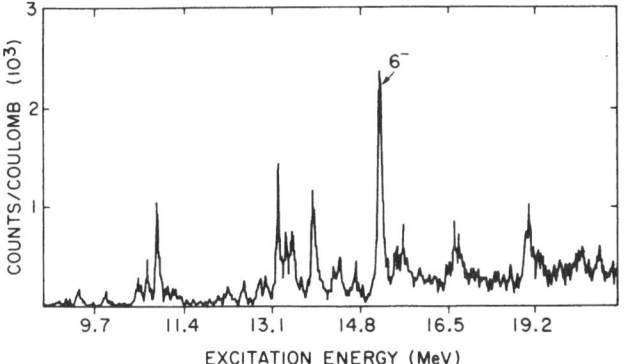

Fig. 4 ^{24}Mg (e,e') data taken at $\theta = 160°$ and q=2.13 fm^{-1}. From Ref 11.

Fig. 4 shows an inelastic spectrum from $^{24}_{12}$Mg taken at large angle and large momentum transfer[11]. Note how the state at 15.0 MeV dominates the spectrum. It can be identified as a stretched, high-spin magnetic excitation, in this case predominantly to the $(1d_{5/2})^{-1} (1f_{7/2})_{6^{-1}}$ configuration. (The large isovector magnetic moment of the nucleon implies isovector excitations dominate under these conditions.) Fig. 5 shows the inelastic form factor for this state, the area under the peak, as a function of momentum transfer. It exhibits a characteristic M6 multipolarity. The data was taken at Bates.

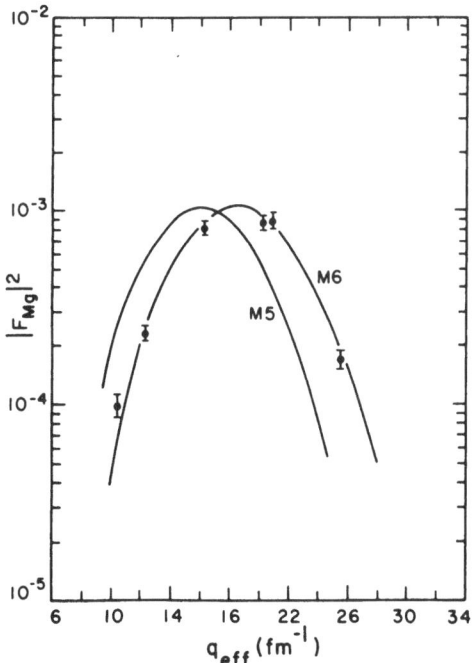

Fig. 5   Transverse form factor vs. q for the state at 15.0 MeV in Fig. 4, Ref 11.

E. SEMILEPTONIC WEAK INTERACTIONS

1. Effective Low-Energy Lagrangian

At energies small compared to the masses of the W and Z weak intermediate bosons (-100 GeV), semileptonic interactions in the Standard Model can be described by the following effective contact Lagrangian densities

$$L_{\text{eff}}^{(\pm)} = \frac{iG}{\sqrt{2}} \left\{ [\bar{e}\gamma_\lambda(1+\gamma_5)\nu_e + (e \leftrightarrow \mu)]J_\lambda^{(+)} \right.$$

$$\left. + [\bar{\nu}_e\gamma_\lambda(1+\gamma_5)e + (e \leftrightarrow \mu)]J_\lambda^{(-)} \right\} \qquad (1.12)$$

$$L_{\text{eff}}^{(\nu)} = \frac{iG}{\sqrt{2}} \left[\bar{\nu}_e\gamma_\lambda(1+\gamma_5)\nu_e + (e \leftrightarrow \mu) \right] J_\lambda^{(0)} \qquad (1.13)$$

$$L_{\text{eff}}^{(e)} = \frac{-iG}{\sqrt{2}} \left[\bar{e}\gamma_\lambda(1+\gamma_5)e - 4\sin^2\theta_w\bar{e}\gamma_\lambda e + (e \leftrightarrow \mu) \right] J_\lambda^{(0)} . \qquad (1.14)$$

Typical processes described by these Lagrangians are shown on the left above. The Fermi constant determined from μ-decay is (*)

$$G = \frac{1.02 \times 10^{-5}}{M^2} . \qquad (1.15)$$

2. Hadronic Currents

The hadronic current densities entering into these Lagrangians have the following general structure in the Standard Model

$$J_\mu = J_\mu + J_{\mu 5} \qquad \text{V-A} \qquad (1.16)$$

$$J_\mu^{(\pm)} = J_\mu^{V_1} \pm iJ_\mu^{V_2} \qquad \text{Isovectors} \qquad (1.17)$$

$$J_\mu^{\gamma} = J_\mu^{S} + J_\mu^{V_3} \qquad \text{E.M. Current} \qquad (1.18)$$

$$J_\mu^{(\pm)} = J_\mu^{V_1} \pm iJ_\mu^{V_2} \qquad \text{CVC} \qquad (1.19)$$

$$J_\mu^{(0)} = J_\mu^{V_3} - 2\sin^2\theta_w J_\mu^{\gamma} \qquad \text{WSG, Standard Model} . \qquad (1.20)$$

The first relation says that the weak hadronic current has both Lorentz vector and axial-vector parts. The second exhibits the isovector nature of the weak charge-raising and -lowering hadronic currents. The third exhibits the isospin structure of the electromagnetic current, and the fourth is the statement of the conserved vector current theory (CVC)[12]. It states that the vector part of the charge-raising and -lowering currents form an isovector triplet with the isovector part of the electromagnetic current. In consequence, the matrix elements of these operators can be related through the Wigner-Eckart theorem applied to isospin. The physical content of this relation is that the electromagnetic and weak interactions couple to different isospin components of the same current! One can make a strong argument that CVC provided the original insight into the unification of the electroweak interactions. The final relation is the form of the weak

_____

(*) Here we assume $G^{(\pm)} = G \cos\theta_c \equiv G$ (see later).

neutral current in the Standard Model. The third isovector component of the charge-changing current in Eq. 1.17, is mixed with the electromagnetic current of Eq. 1.18, through $\sin^2\theta_W$ where θ_W is the Weinberg angle.

3. Parity Violation

The contribution of the weak neutral current interaction of the electron in Eq. 1.14 is completely masked by electromagnetic interactions, unless one looks at an effect that necessarily involves the weak interaction. Parity violation, which is forbidden to all orders in the electromagnetic interaction, is such an effect. Consider the difference in cross section for the scattering of right- and left-handed longitudinally polarized electrons

$$(1.21)$$

which can only arise from parity violation. We calculate the difference in cross section arising from the interference of one-photon and Z° exchange (Eq. 1.14)

$$\gamma \qquad Z^\circ$$

$$(1.22)$$

An analysis in terms of response tensors leads to the following general result[1]

$$\overset{\leftarrow\ A\ \rightarrow}{\left[\frac{d\sigma_\uparrow - d\sigma_\downarrow}{d\sigma_\uparrow + d\sigma_\downarrow}\right] \cdot \left[W_2^\gamma \cos^2\frac{\theta}{2} + 2W_1^\gamma \sin^2\frac{\theta}{2}\right] = \frac{Gq^2}{4\pi\alpha\sqrt{2}}}$$

$$\times \left\{ b\left[W_2^{int}\cos^2\frac{\theta}{2} + 2W_1^{int}\sin^2\frac{\theta}{2}\right] - a\left[\left(\frac{2W_8^{int}}{M_T}\right)\sin\frac{\theta}{2}\left(q^2\cos^2\frac{\theta}{2} + \vec{q}^2\sin^2\frac{\theta}{2}\right)^{\frac{1}{2}}\right]\right\} .$$

This expression assumes Lorentz invariance and a V-A coupling; in the Standard Model the lepton couplings have the following form

$$a = -(1-4\sin^2\theta_W) \cong 0$$

$$b = -1 .$$

$$(1.24)$$

We also assume that the hadronic states have definite parity. (Parity admixtures in the nuclear states give a very different q^2 dependence, and, except perhaps at very low q^2, the above interference term dominates[13]). The response surfaces in this expression arise from the following combination of currents:

$$W_{1,2}^{int} \quad \text{from} \quad (J^{(0)}J^\gamma + J^\gamma J^{(0)})$$

$$W_8^{int} \quad \text{from} \quad (J_5^{(0)}J^\gamma + J^\gamma J_5^{(0)}) . \tag{1.25}$$

Note that Eq. 1.23 holds at all q^2 and $q \cdot p$.

4. Multipole Analysis

A multipole analysis can again be carried out just as in electron scattering. For example, the ERL cross section for neutrino (antineutrino) induced semileptonic weak processes takes the form[3]

$$\left(\frac{d\sigma}{d\Omega}\right)_{\nu(\bar{\nu})}^{ERL} = \frac{G^2 \varepsilon_2^2 \cos^2 \frac{\theta}{2}}{2\pi^2} r \cdot \frac{4\pi}{2J_i+1} \left\{ \sum_{J=0}^\infty |<J_f||\hat{M}_J(q) - \frac{q_0}{q} \hat{L}_J(q)||J_i>|^2 \right. \tag{1.26}$$

$$+ \left[\frac{q^2}{2\bar{q}^2} + \tan^2 \frac{\theta}{2} \right] \sum_{J=1}^\infty \left[|<J_f||\hat{T}_J^{mag}(q)||J_i>|^2 + |<J_f||\hat{T}_J^{el}(q)||J_i>|^2 \right]$$

$$\mp \tan\frac{\theta}{2} \left(\frac{q^2}{\bar{q}^2} + \tan^2\frac{\theta}{2}\right)^{\frac{1}{2}} \sum_{J=1}^\infty \left[2Re <J_f||\hat{T}_J^{mag}(q)||J_i><J_f||\hat{T}_J^{el}(q)||J_i>^* \right] \right\} .$$

The multipole operators appearing in this expression are computed with the weak current $J_\mu(x)$; hence each is a sum of two terms with opposite parities.

F. NUCLEAR CURRENTS

We proceed to discuss the construction of the nuclear current operators within the traditional approach.

1. Traditional Nuclear Many-Body Problem

As discussed above, the traditional approach starts from static two-body potentials and the non-relativistic Schrödinger equation[8]. We *define* the nuclear current density operator at the origin in second quantization as

$$\hat{J}_\mu(0) \equiv \sum_{\vec{k}\lambda\rho} \sum_{\vec{k}'\lambda'\rho'} c_{\vec{k}'\lambda'\rho'}^\dagger <\vec{k}'\lambda'\rho'|J_\mu(0)|\vec{k}\lambda\rho> c_{\vec{k}\lambda\rho}$$

$$\leftarrow \text{s.n.m.e.} \rightarrow . \tag{1.27}$$

This expression implies a *one-body* current. $c_{\vec{k}\lambda\rho}^\dagger$ and $c_{\vec{k}\lambda\rho}$ are the creation and destruction operators for plane-wave states with momentum $\vec{k}$, spin λ and isospin ρ. The single-nucleon matrix elements appearing in this expression are taken from those measured for a free nucleon

$$<p'|J_\mu(0)|p> = \frac{i}{V} \bar{u}(p') \left[F_1(q^2)\gamma_\mu + F_2(q^2)\sigma_{\mu\nu}q_\nu \right] u(p) . \tag{1.28}$$

The isospin dependence of each form factor is given by $F = (F^s + \tau_3 F^V)/2$. Given Eq. 1.27, the corresponding first-quantized densities can be immediately constructed[8]

$$\hat{J}_\mu(\vec{x}) = \sum_{j=1}^{A} [J_\mu(j)\delta^{(3)}(\vec{x}-\vec{x}_j)] .\tag{1.29}$$

They are given through order $1/M^2$ in Appendix A.

If one now selects some complete set of central-field single-particle states with quantum number $\alpha = (nljm_j;\text{\scriptsize{1/2}} m_t)$, then the nuclear multipole operators can be expanded as

$$\hat{T}_{JM_J,TM_T}(q) = \sum_\alpha \sum_\beta c_\alpha^\dagger <\alpha|T_{JM_J,TM_T}(q)|\beta>c_\beta .$$

$$\leftarrow \text{s.p.m.e.} \rightarrow\tag{1.30}$$

Where $c_\alpha^\dagger$ and c_α are the creation and destruction operators for the single-particle states and the single-particle matrix elements are computed from the appropriate multipoles of the current.

2. Many-Body Matrix Elements

Matrix elements of the nuclear multipole operators in Eq. 1.30 between *exact* nuclear states take the following form

$$<\Psi_f|\hat{T}_{JM_J,TM_T}(q)|\Psi_i> = \sum_\alpha \sum_\beta <\alpha|T_{JM_J;TM_T}(q)|\beta>\psi_{\alpha\beta}^{fi}$$

$$\leftarrow \text{s.p.m.e.} \rightarrow\tag{1.31}$$

$$\psi_{\alpha\beta}^{fi} \equiv <\Psi_f|c_\alpha^\dagger c_\beta|\Psi_i> \qquad ; \text{numerical coefficients} .\tag{1.32}$$

The transition matrix elements of the creation and destruction operators in Eq. 1.32 are simply numerical coefficients. These results are *exact* within this traditional approach. They are very simple and powerful. No matter how complicated a many-body shell-model calculation one carries out, for example, if it is carried out in some truncated space, then the transition matrix elements in Eq. 1.31 will simply involve a finite sum of terms running over the retained single-particle configurations.

The different q dependence of the single-particle matrix elements appearing in Eq. 1.31 now suggests an alternate procedure for carrying out a unified analysis of electroweak interactions with nuclei. One can use electron scattering (e,e′) to *determine* the numerical coefficients $\psi_{\alpha\beta}^{fi}$ within some appropriate basis, and then use these coefficients to predict the corresponding weak processes. We will give an example.

3. Weak Currents

The nuclear weak current operators can be constructed in an analogous fashion. Invariance principles dictate that the required single-nucleon matrix element of the charge-changing weak current must have the form[3]

$$<p'|J_\mu^{(-)}(0)|p> = \frac{i}{V} \bar{u}(p') [F_1\gamma_\mu + F_2\sigma_{\mu\nu}q_\nu + iF_S q_\mu$$

$$+ F_A\gamma_5\gamma_\mu - iF_P\gamma_5 q_\mu - F_T\gamma_5\sigma_{\mu\nu}q_\nu] \tau_- u(p) .\tag{1.33}$$

There are no second-class currents in the Standard Model and so $F_S = F_T = 0$. CVC implies that $F_{1,2} = F_{1,2}^V$ can be taken from electron scattering (e,e′), and the β-decay of the neutron yields $F_A(0) = -1.23 \pm 0.01$. The Goldberger-Treiman relation, together with pion-pole dominance of the matrix element of the divergence of the axial-vector current, yields the result

$$F_p = \frac{2MF_A(0)}{q^2 + m_\pi^2} \tag{1.34}$$

although this relation has never been accurately tested.

All of our calculations include a universal single-nucleon form factor of the dipole form $f_{s.n.} = [1 + q^2/(855 \text{ MeV})^2]^{(-2)}$ and a harmonic oscillator center-of-mass correction $f_{c.m.}$.[1]

G. SELECTED TOPICS

1. An Application

I will discuss one application of the above ideas carried out with Bill Donnelly[14]. Consider the following processes in the A=6 system

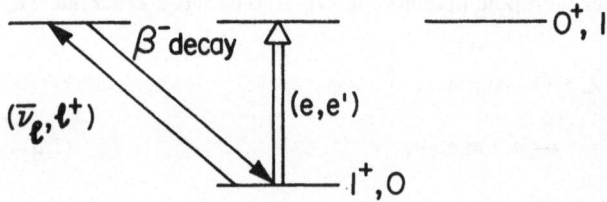

$$\begin{matrix} {}^6_2\text{He} & {}^6_3\text{Li} & {}^6_4\text{Be} \end{matrix} \tag{1.35}$$

as well as elastic magnetic electron scattering (e,e′) from ^{6}Li. We truncate to the p-shell and use a harmonic oscillator single-particle basis. The nuclear states must then take the following form

$$|1^+0\rangle = A|(1p_{3/2})^2 1^+0\rangle + B|(1p_{3/2} 1p_{1/2})1^+0\rangle + C|(1p_{1/2})^2 1^+0\rangle$$

$$|0^+1\rangle = D|(1p_{3/2})^2 0^+1\rangle + E|(1p_{1/2})^2 0^+1\rangle . \tag{1.36}$$

They are assumed normalized. The parameter set $\{A,B,C,D,E;b_{osc}\}$ is then determined from electron scattering. (*) Within this basis the multipole matrix elements must have the form

$$\frac{\langle 1^+0||\hat{T}^{mag}_{10}||1^+0\rangle}{\sqrt{3}(\frac{iq}{M})e^{-y}f_{s.n.}f_{c.m.}} \equiv p(y) = \alpha_e + \beta_e y \tag{1.37}$$

with a similar expression for the inelastic transition. The linear dependence on $y = (qb_{osc}/2)^2$ shown in Fig. 6 provides a check on the truncation procedure and allows an extraction of the numerical coefficients.

(*) Since we have the wavefunctions in this case, we have all the coefficients in Eq. 1.32.

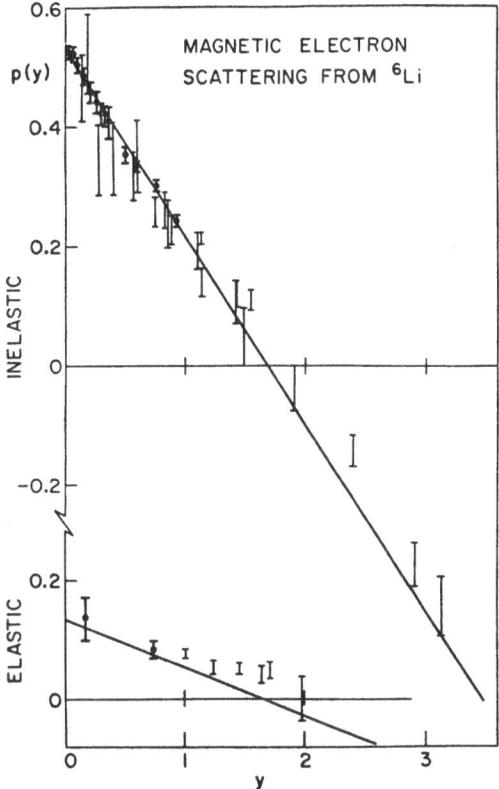

Fig. 6   Magnetic electron scattering from ^{6}Li in terms of p(y) Eq. 1.37. The straight
line is a minimum χ^2 fit to the accurate data at q$\leq$ 200 MeV (heavy error
bars)[14].

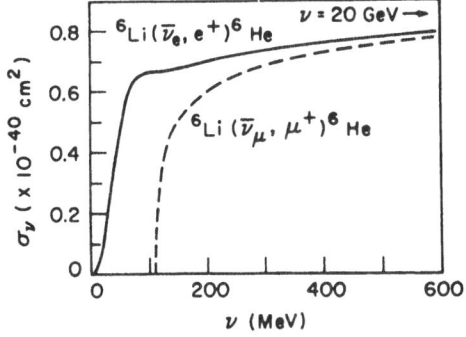

Fig. 7   Antineutrino cross section to ^{6}He$_{g.s.}$.[14]

The calculated and measured β-decay rates for ^{6}He are compared in Table 1.1. This is now an absolute calculation, and the agreement is to better than 5%. The predicted antineutrino cross sections are shown in Fig. 7.

Table 1.1: β- Decay Rate for ^{6}He (Ref. 14)

ω_β	$= 0.877 \pm 0.023 \text{ sec}^{-1}$	Theory
	$= 0.864 \pm 0.003 \text{ sec}^{-1}$	Experiment

This application illustrates how well we can do nuclear physics, in favorable cases, with the traditional approach.

2. Exchange Currents

We discuss next an extension of the traditional analysis to allow for additional *two-body* currents arising from the exchange of charged mesons between nucleons. Although many exchange current calculations exist, we focus on those of Dubach, Koch, and Donnelly (DDK) for concreteness[15]. These authors keep the static limit (leading terms in 1/M) of the time-ordered Feynman diagrams shown in Fig. 8. Each of these processes clearly represents an additional contribution to the current. If the two-nucleon potential is of the form $V = V_{neutral} + V_{OPEP}$ where the first term arises from neutral meson exchange (See Lectures III and IV) and the second is the one-pion-exchange potential, then the electromagnetic current of DDK is conserved. Furthermore, the threshold pion-electroproduction parts of the graphs satisfy the Kroll-Ruderman Theorem. The approach of DDK thus yields the correct form of the longest-range part of the exchange current. In this approach, the charge density operator is unmodified, and only the current density $\vec{J}(\vec{x})$ receives exchange current contributions, which are pure isovector.

Assume that ^{3}He can be described by a $(1s_{1/2})^{(-1)}$ harmonic oscillator shell-model configuration. The magnetic moment calculated with the inclusion of exchange currents $\mu = -2.078$ nm is now closer to the experimental value $\mu = -2.127$ nm than is the Schmidt value $\mu = -1.913$ nm, indicating we are in the right ballpark. The effect on the elastic magnetic electron scattering form factor at intermediate q is shown in Fig. 9. Here the oscillator parameter has been determined from elastic charge scattering. This figure illustrates *the marginal role of exchange currents in the traditional nuclear physics domain.*

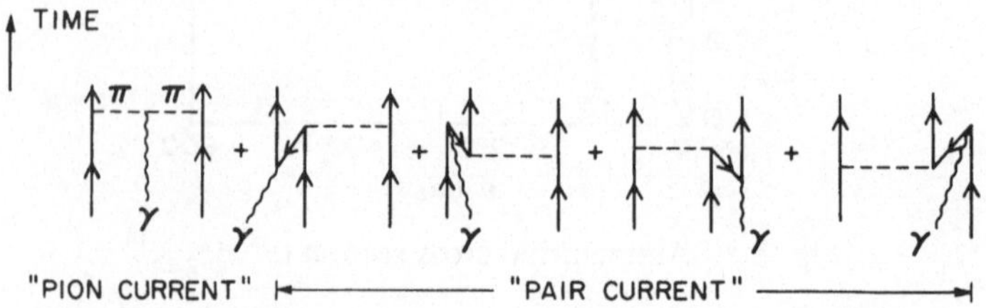

Fig. 8   Time-ordered Feynman diagrams retained in the one-pion-exchange current calculation of Ref 15.

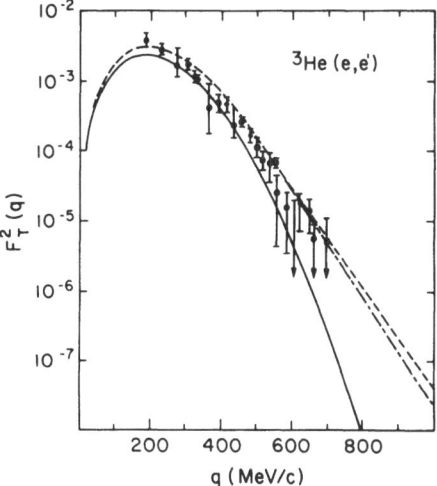

Fig. 9   Elastic transverse form factor for ^{3}He (e,e) with (dashed) and without (solid) one-pion-exchange currents[15].

Fig. 10 taken from Ref. 16, illustrates the state-of-the-art with respect to elastic magnetic electron scattering from ^{3}He. The dashed line shows the result obtained from the best three-body calculation done in the traditional approach (the three-body wave function is obtained by solving the Faddeev equations with potentials fit to the two-body data). The best three-body calculation in the traditional picture, clearly fails at high q^2. Also shown in Fig. 10 are two meson exchange current calculations[17-18], which include the pion exchange current discussed above, as well as other contributions. The

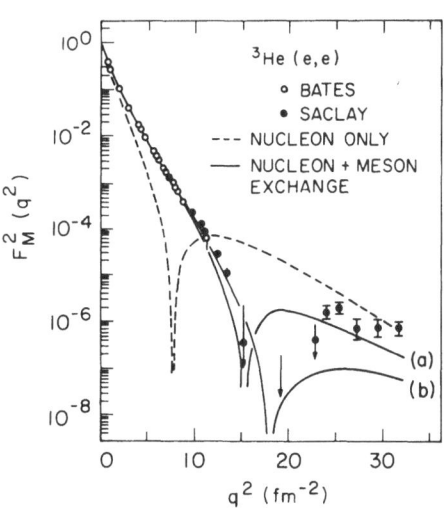

Fig. 10   Elastic magnetic form factor for ^{3}He (e,e) out to high q^2.[16] Two exchange current theories are shown[17b,18a].

difference between these two curves is a good measure of the present theoretical uncertainty. Note that the exchange current contribution is now a *dominant effect* at large q^2. This application *illustrates the need for QHD*.

3. Coincidence

Coincidence electron scattering experiments $(e,e'X)$ will form a major future thrust of the field of electromagnetic interactions with nuclei. An extensive discussion of the general phenomenological analysis of such processes is developed in Ref. 1, and we will hear a great deal more about this topic from the speakers at this school.

NEW VISTAS

A. SOME MODEL-INDEPENDENT RELATIONS

Let us use the previous analysis to derive some relations that are *independent of the details of nuclear structure*. They serve as tests of the Standard Model in the strong interaction, confinement regime appropriate to nuclear physics. The basic idea is to use the nucleus as a filter to isolate various pieces of the current. In particular, we shall use the isospin selection rules.

1. T=0 → T=1 Transitions

In this case only the isovector parts of the currents in Eqs. 1.16-1.20 can contribute, and the effective currents are given by

$$J_\mu^{(0)} \doteq J_\mu^{V_3}(1-2\sin^2\theta_W) + J_{\mu_s}^{V_3} \tag{2.1}$$

$$J_\mu^\gamma \doteq J_\mu^{V_3}. \tag{2.2}$$

With the aid of the general analysis in terms of response surfaces one can then derive relations between semileptonic electroweak cross sections[19,2].* For example,

$$\frac{d\sigma_{v_l v_l} - d\sigma_{\bar v_l \bar v_l}}{\left[d\sigma_{v_l l^-} - d\sigma_{\bar v_l l^+}\right]^{ERL}} = \frac{1}{2}(1-2\sin^2\theta_W). \tag{2.3}$$

The right hand side of this ratio of neutrino cross sections must be a constant depending only on the Weinberg angle. A better test of the unification of the electroweak interactions which also involves the electron scattering cross section is the relation

$$\frac{1}{2}\left[d\sigma_{v_l l^-} + d\sigma_{\bar v_l l^+}\right]^{ERL} - \left[d\sigma_{v_l v_l} + d\sigma_{\bar v_l \bar v_l}\right] = \frac{G^2 q^4}{4\pi^2\alpha^2}(\sin 2\theta_W)^2 \, d\sigma_{ee}^{ERL}. \tag{2.4}$$

2. T=0 → T=0 Transitions

In this case there can only be isoscalar contributions, and the effective currents satisfy the relation

---

* Such relations were anticipated by Weinberg[4].

$$J_\mu^{(0)} \doteq -2\sin^2\theta_W J_\mu^\gamma .$$ (2.5)

This leads to a direct proportionality between the cross sections for neutrino and electron scattering

$$d\sigma_{\underset{\bar{v}_l\bar{v}'_l}{v_l v'_l}} = \sin^4\theta_W \; \frac{G^2 q^4}{2\pi^2\alpha^2} \; d\sigma_{ee'}^{ERL} .$$ (2.6)

3. Discussion

Now these are truly *remarkable results!* They are independent of nuclear structure, and they hold at all q^2.* Thus they hold at wavelengths where only the gross features of nuclear structure are important, down to wavelengths where the baryon/meson degrees of freedom are crucial, and on down to distances where the quark/gluon degrees of freedom manifest themselves. They are based on a minimal set of assumptions: Lorentz invariance, the structure of the currents in Eqs. 1.16-1.20, isotopic spin invariance in the nuclear system, and the one-photon exchange analysis of (e,e').

Consider Eq. 2.6. It states that one should be able to lay the cross sections (appropriately scaled) on top of each other. They should be identical! And this holds at all q^2, and hence through all of the diffraction structure (seen, for example, in Fig. 2)! No matter how complicated an exchange current contribution one comes up with in electron scattering, for example, exactly the same contribution must be present in neutrino scattering through the weak neutral current. This, to me, provides *a true test of the unification of the electroweak interactions*.

4. Parity Violation - An Example

Consider elastic scattering from a 0^+ nucleus. In this case there can only be a charge monopole form factor, and the parity violation asymmetry

$$A \equiv (d\sigma_\uparrow - d\sigma_\downarrow)/(d\sigma_\uparrow + d\sigma_\downarrow)$$

in Eq. 1.23 takes the form[1]

$$A = \frac{Gq^2}{2\pi\alpha\sqrt{2}} \; b \left[Re \frac{F^{(0)}(q^2)}{F^{(\gamma)}(q^2)} \right] .$$ (2.7)

This relation assumes only Lorentz invariance and a V-A structure of the weak neutral current (WNC).** Now assume further that the ground state has isospin T=0 and use the result in Eq. 2.5. The electron scattering and WNC form factors must then be *identical*

$$F^{(0)}(q^2) = -2\sin^2\theta_W F^{(\gamma)}(q^2) .$$ (2.8)

This is again a truly *remarkable result!* All the comments made above in the discussion in Section 3 again apply. Eq. 2.8 is a relation which must hold between the charge and WNC form factors at all q^2. Substitution of Eq. 2.8 into Eq. 2.7, leads to

---

* Actually all q^2 and q·p.

** There is no contribution from parity admixtures in the ground state in this case with one photon exchange

$$A = \frac{Gq^2}{\pi\alpha\sqrt{2}} \sin^2\theta_W .$$ (2.9)

This result is originally due to Feinberg[20]. An experiment to measure A for ^{12}C is under-way at Bates. To me, the most important aspect to test is the linear dependence on q^2 in Eq. 2.9; we really want to test Eq. 2.8 in detail at all q^2. This again provides a true test of the unification of the electroweak interactions. Note that the parity-violation experiment, as hard as it is, is easier than a test of Eq. 2.6 which involves the cross section for neu-trino scattering.

In order to understand the simplicity of the above relations, we must examine the underlying structure of hadrons.

B. QUARKS AND QCD

There is now convincing evidence that hadrons are composed of a simpler substruc-ture, quarks, the lightest four of which are assigned the quantum numbers shown in Table II.1. Since the existence of quarks is *inferred*, it is useful to review the basic arguments for their existence.

Table II.1

Quark Quantum Numbers

Field/Particle	T	T_3	Q	B	S	C	Y=B+S+C
u	1/2	1/2	2/3	1/3	0	0	1/3
d	1/2	-1/2	-1/3	1/3	0	0	1/3
s	0	0	-1/3	1/3	-1	0	-2/3
c	0	0	2/3	1/3	0	1	4/3

1. Hadron Multipoles

If one assumes that the baryons are composed of three quarks (qqq) and the mesons of a quark-antiquark pair ($\bar{q}q$), and if the quarks are assigned the quantum numbers in Table II.1, then one obtains a concise description of the observed supermultiplets of hadrons and predictions for new ones.

2. Electroweak Currents

If one assumes that the electroweak currents are constructed from pointlike, Dirac quark fields with the quantum numbers in Table II.1, then a marvelously simple, accu-rate, and predictive description of these currents is obtained. The electromagnetic current, for example, is simply given by

$$J_\mu^\gamma = i\left[\frac{2}{3}(\bar{u}\gamma_\mu u + \bar{c}\gamma_\mu c) - \frac{1}{3}(\bar{d}\gamma_\mu d + \bar{s}\gamma_\mu s)\right] . \tag{2.10}$$

The charge-raising weak current is given by

$$J_\mu^{(+)} = i\bar{u}\gamma_\mu(1+\gamma_5)\,(d\cos\theta_c + s\sin\theta_c)$$
$$+ i\bar{c}\gamma_\mu(1+\gamma_5)\,(-d\sin\theta_c + s\cos\theta_c) . \tag{2.11}$$

The one complexity here is the occurrence of the Cabbibo angle θ_c, dictated by the empirically observed strangeness-changing weak interactions, which slightly mixes the quark fields appearing in this expressions. It disappears from the WNC in the Standard Model, which is given by

$$J_\mu^{(0)} = \frac{i}{2}\left[\bar{u}\gamma_\mu(1+\gamma_5)u + \bar{c}\gamma_\mu(1+\gamma_5)c - \bar{d}\gamma_\mu(1+\gamma_5)d - \bar{s}\gamma_\mu(1+\gamma_5)s\right]$$
$$- 2\sin^2\theta_W J_\mu^\gamma . \tag{2.12}$$

Indeed, it was the argument of Glashow et al.[6] that there should be no strangeness-changing WNC that led them to predict the existence of the c quark - a prediction later fully confirmed by experiment.

We shall define the *nuclear domain* by truncating the Hilbert space so that it contains only u and d quarks. We can still have any number of pairs of these quarks present (and hence any number of mesons) so that the states can still be very complicated. The quark field can then be written in the form of an isodoublet

$$\psi \doteq \begin{pmatrix} u \\ d \end{pmatrix} \qquad ; \text{nuclear domain} . \tag{2.13}$$

We further redefine the charge-changing Fermi constant by *

$$G^{(\pm)} \equiv G\cos\theta_c . \tag{2.14}$$

The currents in Eqs. 2.10 - 2.12 can then be recast in the form

$$J_\mu^\gamma = i\left[\bar{\psi}\gamma_\mu\frac{1}{2}\tau_3\psi + \frac{1}{6}\bar{\psi}\gamma_\mu\psi\right] \tag{2.15}$$

$$J_\mu^{(\pm)} = i\bar{\psi}\gamma_\mu(1+\gamma_5)\tau_\pm\psi \tag{2.16}$$

$$J_\mu^{(0)} = i\bar{\psi}\gamma_\mu(1+\gamma_5)\frac{1}{2}\tau_3\psi - 2\sin^2\theta_W J_\mu^\gamma . \tag{2.17}$$

These relations imply the previously assumed structure of the currents in Eqs. 1.16 - 1.20! We note that there is now a correction term to the WNC due to the $(\bar{s}s)$ and $(\bar{c}c)$ pairs in the nucleus

$$\delta J_\mu^{(0)} \equiv J_\mu^s = \frac{i}{2}\left[\bar{c}\gamma_\mu(1+\gamma_5)c - \bar{s}\gamma_\mu(1+\gamma_5)s\right] . \tag{2.18}$$

The contribution of this term, which takes us outside of the "nuclear domain" is expected to be small, but it is very difficult to estimate quantitatively.

---

* The factor of $\cos\theta_c$ can easily be incorporated in the previous Eqs. 2.3 - 2.4.

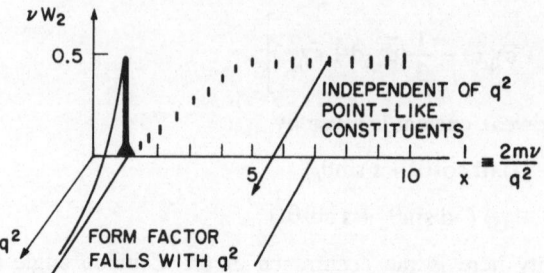

Fig. 11   Qualitative sketch of SLAC results on deep inelastic electron scattering from the nucleon.

3. Deep Inelastic Scattering (e,e′)

Dynamic evidence for the existence of a pointlike substructure in the nucleon is obtained from deep inelastic electron scattering (e,e′). The situation is illustrated qualitatively in Fig. 11, where we sketch the two-dimensional response surface νW_2 against q^2 and $1/x \equiv 2M\nu/q^2$. For elastic scattering* or inelastic scattering to discrete levels, there is a form factor which falls off with q^2. We have seen many examples of this in the nuclear case. At large values of

$$\omega \equiv \frac{1}{x} = \frac{2M\nu}{q^2} \tag{2.19}$$

(i.e., in the deep inelastic region) the cross section is observed to be *independent of* q^2. This is graphically illustrated by the SLAC data shown in Fig. 12.[21] The absence of a

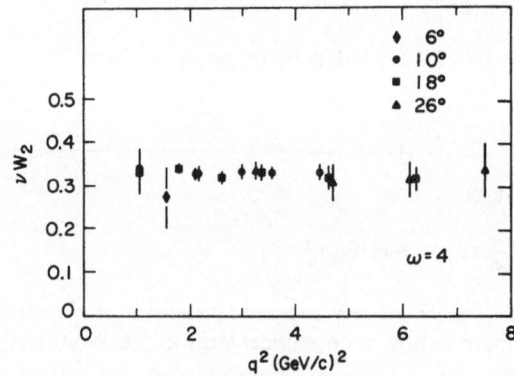

Fig. 12   Experimental demonstration that the response surface νW_2 for the proton is independent of q^2 at fixed $\omega = 1/x = 2 \, m\nu/q^2$ in deep inelastic scattering. Taken from Ref. 21.

---

* For elastic scattering from the nucleon, x=1.

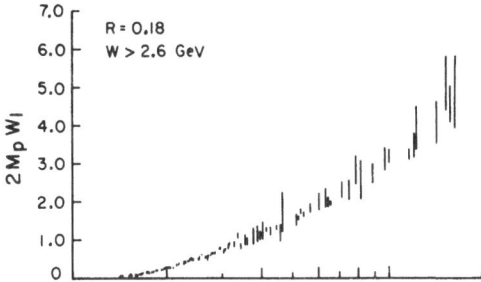

Fig. 13   Response surface $2mW_1$ for the proton as a function of the Bjorken scaling
variable $\omega = 1/x$.[21]

form factor here indicates that one is scattering from pointlike objects. The statement of
Bjorken scaling is that the two-dimensional response surface becomes a function of the
single variable in this region

$$\nu W_2(\nu, q^2) \rightarrow F_2(x) .\tag{2.20}$$

The structure functions for the proton determined by the SLAC experiments are shown in
Figs. 13 and 14.[21]

The quark-parton model (due to Feynman, and Bjorken and Paschos[22]) provides a
simple framework for evaluating the structure function in the scaling region

$$F_2(x) = \sum_N P(N) <\sum_i Q_i^2>_N xf_N(x) .\tag{2.21}$$

Here $P(N)$ is the probability that a very high momentum proton will have a structure consisting of N pointlike constituents ("partons"). The second factor is the sum of the
squares of the charges of the partons in this configuration. The quantity $f_N(x)$ is the

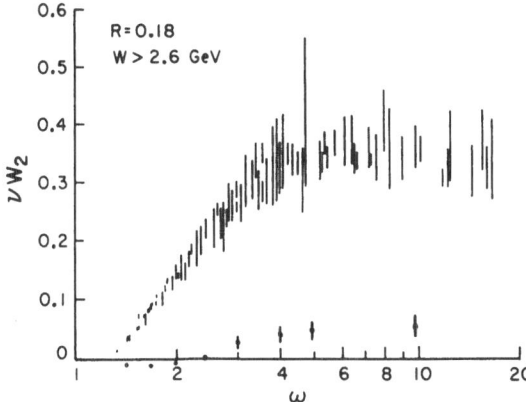

Fig. 14   Same as Fig. 13 for νW_2.[21] (The modification of this function in ^{56}Fe observed
by the EMC at CERN is shown at the bottom of this figure).

probability that a parton will carry a fraction x of the longitudinal momentum of the nucleon in this frame.

Also sketched in Fig. 14, for orientation, is the magnitude of the EMC effect (see e.g. Ref. 23), about which we shall hear a great deal at this school. It represents the difference in cross section/nucleon for $^{56}Fe(\mu,\mu')$ and $^{2}H(\mu,\mu')$. *The EMC effect is a clear and unambiguous demonstration of the modification of the quark structure of nucleons in the nucleus.*

C. QUANTUM CHROMODYNAMICS (QCD)

Quantum chromodynamics (QCD) is a relativistic quantum field theory of the strong interactions binding quarks into hadrons[7]. We cannot do justice to this subject in a few short lectures; however, the use of relativistic quantum field theory in the nuclear many-body problem, as well as the theory of QCD, are developed in detail in Ref. 24, which also contains an extensive list of references. If you are a student who wants to learn about these things, I urge you to start by reading this book.

1. Color

In addition to the "flavor" quantum numbers shown in Table II.1, quarks are given an additional intrinsic degree of freedom called "color", which takes three values i=R,G,B. This is analogous to isospin for the nucleons. The quark field then becomes

$$\psi = \begin{pmatrix} u \\ d \\ s \\ c \end{pmatrix} \rightarrow \begin{pmatrix} u_R & u_G & u_B \\ d_R & d_G & d_B \\ s_R & s_G & s_B \\ c_R & c_G & c_B \end{pmatrix} \equiv (\psi_R, \psi_G, \psi_B) \equiv \psi_i \ ; \quad i = R,G,B. \tag{2.22}$$

We introduce the three-component fields

$$\Psi \equiv \begin{pmatrix} \psi_R \\ \psi_G \\ \psi_B \end{pmatrix} . \tag{2.23}$$

This is actually a very compact notation; each field has many flavors, and each flavor is a 4-component Dirac field. The *electroweak currents are assumed to be independent of color.* They are written for example, as

$$\overline{\Psi}\gamma_\mu\Psi = \overline{\psi}_R\gamma_\mu\psi_R + \overline{\psi}_G\gamma_\mu\psi_G + \overline{\psi}_B\gamma_\mu\psi_B . \tag{2.24}$$

Thus each color field has identical electroweak couplings as indicated in Fig. 15.

2. QCD

QCD is a local gauge theory built on color. It is invariant under local unitary transformations on the three-component field in Eq. 2.23, and hence the theory possesses the local symmetry $SU(3)_C$. It is a non-abelian Yang-Mills theory. To construct such a theory one first introduces 8 massless gauge boson fields, the *gluon* fields, $A_\mu^a(x)$ with a=1,2.....,8; there is one for each generator. (These are the analogues of the photon field $A_\mu(x)$ in the abelian theory QED). The use of the covariant derivative (the analogue of (p-eA) in QED) then leads to the QCD Lagrangian density

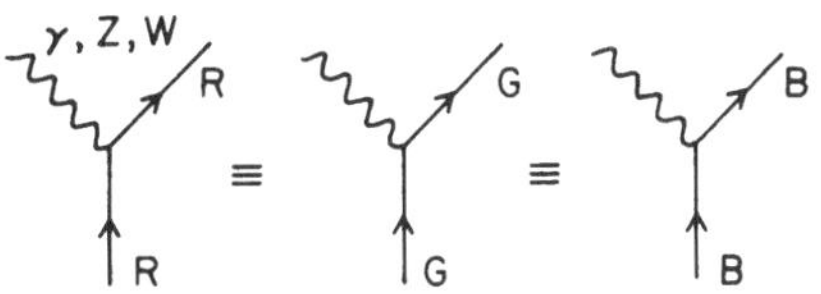

Fig. 15　Weak and electromagnetic quark couplings are independent of color.

$$L_{QCD} = -\frac{1}{4}F^{a}_{\mu\nu}F^{a}_{\mu\nu} - \overline{\Psi}\gamma_{\mu}\left[\frac{\partial}{\partial x_{\mu}} - \frac{i}{2}g\lambda^{a}A^{a}_{\mu}(x)\right]\Psi. \tag{2.25}$$

The SU(3) matrices λ_{a} appearing in the covariant derivative satisfy the algebra of the generators

$$\left[\tfrac{1}{2}\lambda^{a},\tfrac{1}{2}\lambda^{b}\right] = if^{abc}\tfrac{1}{2}\lambda^{c}. \tag{2.26}$$

The field tensor appearing in Eq. 2.25 must be more complicated than the Maxwell tensor of QED to maintain local gauge invariance. It is given by

$$F^{a}_{\mu\nu} = \frac{\partial}{\partial x_{\mu}}A^{a}_{\nu} - \frac{\partial}{\partial x_{\nu}}A^{a}_{\mu} + gf^{abc}A^{b}_{\mu}A^{c}_{\nu}. \tag{2.27}$$

The cubic and quartic cross terms in the square of this quantity imply that the QCD Lagrangian is intrinsically non-linear in the gluon fields. (Note that there are induced non-linearities in QED coming from electron loops; here, however, the Lagrangian itself is intrinsically non-linear.)

Since L_{QCD} is written in terms of the Ψ in Eq. 2.23, the strong color interactions are independent of flavor as illustrated in Fig. 16 (just multiply out the matrix product using the definition in Eq. 2.22). Eq. 2.25 has been written for massless quarks, but a mass term of the form

$$\delta L_{mass} = -\overline{\Psi}\underline{M}\Psi, \quad \underline{M} = \begin{bmatrix} M & & \\ & M & \\ & & M \end{bmatrix} \tag{2.28}$$

where $\underline{M}$ is the identity matrix with respect to color which evidently preserves local SU(3)$_{c}$ invariance.

The Feynman rules for QCD may now be derived, and they have the components

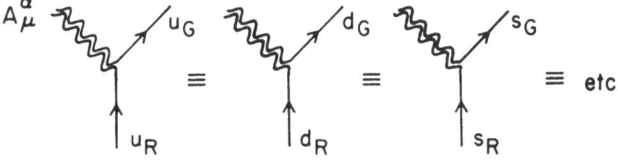

Fig. 16　Strong color interactions of the quarks are independent of flavor.

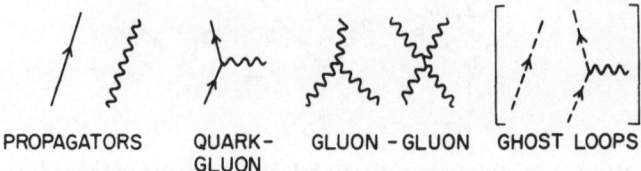

PROPAGATORS   QUARK-   GLUON - GLUON   GHOST LOOPS
                 GLUON

Fig. 17   Components of Feynman Rules for QCD.

illustrated in Fig. 17.* The ghost loops in the gluon amplitudes are a technical complexity; they are there to ensure that one generates the correct Lorentz invariant, gauge invariant, unitary S-matrix in this non-abelian gauge theory.

QCD has two remarkable properties[7]:

i) *Asymptotic Freedom.* At large momenta (small distances) the theory is essentially free. Due to the non-linear gluon couplings the charge is antishielded, and the renormalized coupling constant measured at low momentum transfers is *larger* than the effective coupling constant seen at high momenta, or short distances. This is in contrast to the situation in QED where the pairs arising from vacuum polarization shield the bare charge. When the effective coupling is small, one can do perturbation theory;

ii) *Confinement.* Quarks and color are confined to the interior of hadrons. We never see the underlying degrees of freedom in QCD as asymptotic free states in the laboratory! There is strong evidence from lattice gauge theory calculations, where QCD is solved on a finite space-time lattice, that confinement is indeed a dynamic property of QCD. Confinement is a manifestation of the strong-coupling aspect of QCD.

3. It is important to note that *in the Standard Model of the strong, electromagnetic, and weak interactions based on* $SU(3)_c \otimes SU(2)_W \otimes U(1)_W$, *the gluons have no weak or electromagnetic interactions.*

D. SLAC PARITY VIOLATION EXPERIMENT $^2H(\vec{e},e')$ [25]

Let us combine the results in these two lectures to get a simple expression for the parity violation asymmetry parameter measured in deep inelastic electron scattering from the deuteron $^2H(\vec{e},e')$. We use the model of the nucleon shown in Fig. 18 - we work in the nuclear domain and assume quark configurations p=(uud) and n=(udd); the states may have any number of additional gluons. The general expression for A in Eq. 1.23 may now be employed, with $\theta \to 0$ as in the SLAC experiments. For simplicity, it will be assumed that $\sin^2\theta_W = 1/4$ so that $\underline{a} = 0$. The quark-parton model in Eq. 2.21 then provides a simple expression for the required ratio of response functions

---

* They are given in detail in Ref. 24.

3 QUARKS: p = (uud)
              n = (udd)

⊕ ANY NUMBER OF GLUONS

Fig. 18   Model of the nucleon.

$$\left(\frac{vW_2^{int}}{vW_2^{\gamma}}\right)_{^2H} = \frac{<2\sum_i Q_i^{\gamma}Q_i^{(0)}>_{3-quark}^n + (n \leftrightarrow p)}{<\sum_i (Q_i^{\gamma})^2>_{3-quark}^n + (n \leftrightarrow p)} = \frac{4}{5}. \tag{2.29}$$

The n and p response surfaces are added incoherently since one is adding cross sections. Everything else cancels in this ratio, and only the indicated ratios of charges remain. They are immediately read off from Eq. 2.12 and Table II.1; the result in Eq. 2.29 is just 4/5. The asymmetry is then immediately given by the simple expression[1]

$$A_D = \frac{-Gq^2}{4\pi\alpha\sqrt{2}} \cdot \frac{4}{5}. \tag{2.30}$$

This result is compared with the SLAC data in Fig. 19.[25] The agreement is all that one could ask.

E. PICTURE OF THE NUCLEUS

On the basis of this discussion, we arrive at the picture of the nucleus illustrated in Fig. 20. We proceed to make some observations concerning this figure.

1. Confinement

The structure of confinement in a many-baryon system is an unsolved problem:
   a) QCD is simple at short distances, or high momenta;
   b) QCD is a complicated, strong-coupling theory at large distances - at the boundary of the confinement region;
   c) If QCD is to be a correct theory it must reproduce *nuclear physics;* in particular

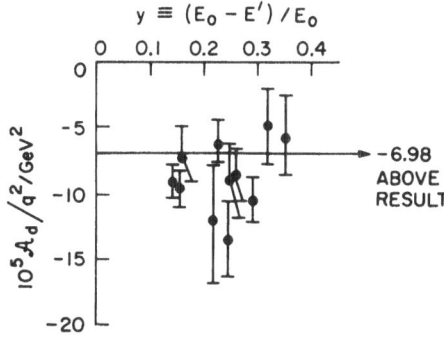

Fig. 19   Result in Eq. 2.30 compared with SLAC data for parity-violation asymmetry in deep inelastic $(\vec{e}, e')$ from ^2H.[25]

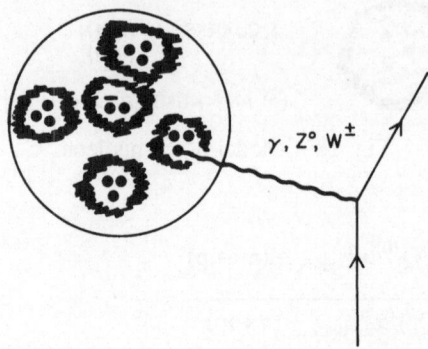

Fig. 20   Picture of the nucleus in the Standard Model.

it must reproduce:
   i) meson exchanges;
   ii) baryon dynamics in nuclei;
   iii) meson dynamics in nuclei.

2. Electroweak Interactions

The electroweak interactions couple to the quarks:
   a) The electroweak interactions see through the hadronic structure to the interior quark structure;
   b) The gluons providing the confinement are *absolutely neutral* to the electroweak interactions;
   c) The electroweak interactions are *color-blind*.

3. Simplicity of Previous Relations

We can now understand the simplicity of the previous relations, derived within the framework of the Standard Model, which are independent of the details of nuclear structure (Eqs. 2.3, 2.4, 2.6 and 2.9).
   a) No matter how complicated the nuclear structure in terms of baryons (qqq) and meson pairs ($\bar{q}q$);
   b) The electroweak interactions *see through* this structure to the interior quark structure of the hadrons;
   c) The electroweak currents are simple in terms of quarks;
   d) The various semileptonic electroweak processes then just probe different components of these basic quark currents.

We close these lectures with two conclusions that help to define the new vistas in electronuclear physics:
   1) The electroweak interactions see the quark structure of nuclei;
   2) Nuclear physics is the study of the strong-interaction aspects of QCD.

QUANTUM HADRODYNAMICS (QHD)

A. Background

These two lectures are based on a book "The Relativistic Nuclear Many-Body Problem" written with Brian Serot which will appear as volume 16 of the series *Advances in Nuclear Physics* edited by J. W. Negele and E. Vogt[24]. The chapter headings are

INTRODUCTION
RELATIVISTIC BARYONS
A SIMPLE MODEL

RELATIVISTIC HARTREE DESCRIPTION OF NUCLEI
QUANTUM HADRODYNAMICS (QHD)
THE DYNAMICAL QUANTUM VACUUM

CHARGED MESONS
RELATIVISTIC PION DYNAMICS
TWO-NUCLEON CORRELATIONS

ELECTROWEAK INTERACTIONS WITH NUCLEI
QUANTUM CHROMODYNAMICS (QCD)
SUMMARY

APPENDICES
    Notation and Conventions
    Dimensional Regularization
    Path-Integral Derivation of Feynman Rules
    The Feynman Rules in Local Gauge Theories

The purpose of the book is pedagogical. The aim is to instruct nuclear theorists on the use of relativistic quantum field theory as a basis for the description of the nuclear many-baryon system. The analysis is developed in detail in the book, and there is an extended list of topics. I urge students who are interested in any of these subjects, and want to learn, to read the book. In these lectures I will discuss a set of *selected topics* taken from this material.

One technical point. In the book the metric and conventions of Bjorken and Drell are employed[26]. Here I will use a metric such that $x_\nu = (\vec{x}, ix_0) = (\vec{x}, it)$ and $a \cdot b = \vec{a} \cdot \vec{b} - a_0 b_0$. The gamma matrices I will use are hermitian and satisfy $\gamma_\mu \gamma_\nu + \gamma_\nu \gamma_\mu = 2\delta_{\mu\nu}$. Also $\hbar = c = 1$. The conversion key between metrics is given in Table XII of Ref. 24.

B. Motivation

I would like to take some time to repeat the motivation for the present discussion.

Let me start by defining the "traditional non-relativistic many-body problem" as developed in detail, for example, in Ref. 8 (I will assume that you are familiar with the material in that reference.) In this approach a static two-body potential obtained from fits to two-body scattering and bound-state data is inserted in the many-particle Schrödinger equation and that equation is solved in some approximation to give energies and wave functions. The electroweak currents are then constructed from the properties of the free constituents and used to probe the static and transition densities of the many-particle system. Although this approach based on structureless nucleons interacting through static two-body potentials has had remarkable success in providing an understanding of the nucleus, it is clearly inadequate for a more detailed understanding of the nuclear system.

Evidently a more appropriate set of degrees of freedom for the nucleus consists of the *hadrons,* the strongly interacting mesons and baryons. Many arguments can be given for this. For example:

(1) The long-range part of the Paris two-nucleon potential, probably the most accurate currently available, is derived from the exchange of mesons, the most important being $\pi(J^\pi,T) = (0^-,1)$, $\sigma(0^+,0)$, $\omega(1^-,0)$, and $\rho(1^-,1)$. We have *experimental proof* that the long-range part of this interaction comes from meson exchange.

(2) There is now convincing proof from high-momentum transfer electron scattering (e,e') for the existence of *exchange currents* in nuclei. These are additional currents arising from the exchange of charged mesons between baryons in the many-body system.

(3) Meson factories daily study the production and interaction of these basic nuclear constituents.

Furthermore, a basic goal of nuclear physics is to describe nuclear matter under *extreme conditions.* For example, in astrophysics we need to understand the behavior of nuclear matter at high density and high temperature to describe condensed stellar objects and supernovae. The behavior of nuclear matter under extreme conditions is also studied in the laboratory in energetic heavy-ion reactions.

Any such attempt to describe the nucleus must incorporate *general principles* of physics, in particular
     1) Quantum mechanics
     2) Special relativity
     3) Causality.

The only consistent theoretical framework we have for describing such an interacting, relativistic, many-body system is *relativistic quantum field theory with a local Lagrangian density.* I like to refer to such a theory formulated in terms of *hadronic degrees of freedom* as "quantum hadrodynamics (QHD)".

We will also demand that the theory be *renormalizable.* This needs a little more justification. The essential reason is that we seek a consistent theoretical framework within which one can, in principle, calculate to arbitrary accuracy and compare with experiment. Our ignorance is summarized in terms of a minimal number of coupling constants and masses, which must be determined phenomenologically. The theory may, or may not, provide a correct description of nature. But that question can now be answered, as with any question in physics, by a detailed comparison between theory and experiment. In practice, the condition of renormalizability severely restricts the class of Lagrangians that we will consider. In addition, renormalizability makes one as insensitive as possible to the short-distance behavior of the theory, a feature that will become more important as we progress.

C. A Simple Model (QHD-I)

We focus the discussion in these lectures on the simple model that Brian and I refer to as QHD-I. It is formulated in terms of the following set of hadronic fields: a baryon field $\psi = \begin{bmatrix} p \\ n \end{bmatrix}$, a massive, neutral scalar meson field ϕ coupled to the scalar density $\bar{\psi}\psi$, and a massive, neutral vector meson field V_λ coupled to the baryon current $i\bar{\psi}\gamma_\lambda\psi$. The motivation for this choice is as follows:

(1) We want to study the bulk properties of nuclear matter. These fields give rise to the

smoothest possible average hadronic interactions (see the discussion of mean-field theory in Section C).

(2) Empirically one sees large Lorentz scalar and four-vector interactions in intermediate energy nucleon-nucleon scattering (see below).

(3) In the static limit with heavy baryons (which is *not* assumed), this theory gives rise to an effective baryon-baryon potential of the form

$$V_{static} = \frac{g_v^2}{4\pi} \frac{e^{-m_v r}}{r} - \frac{g_s^2}{4\pi} \frac{e^{-m_s r}}{r}. \tag{3.1}$$

This potential can exhibit a short-range repulsion and a long-range attraction, two dominant, qualitative features of the interaction between two nucleons.

We assume a Lagrangian density of the following form

$$L = -\frac{1}{4}F_{\mu\nu}F_{\mu\nu} - \frac{1}{2}m_v^2 V_\mu V_\mu - \frac{1}{2}[(\frac{\partial\phi}{\partial x_\mu})^2 + m_s^2\phi^2]$$

$$- \overline{\psi}[\gamma_\mu(\frac{\partial}{\partial x_\mu} - ig_v V_\mu) + (M - g_s\phi)]\psi \tag{3.2}$$

where the vector field tensor is defined by

$$F_{\mu\nu} \equiv \frac{\partial}{\partial x_\mu} V_\nu - \frac{\partial}{\partial x_\nu} V_\mu. \tag{3.3}$$

Hamilton's principle and Langrange's equations lead to the following field equations in QHD-I

$$\frac{\partial}{\partial x_\nu} F_{\mu\nu} + m_v^2 V_\mu = ig_v \overline{\psi}\gamma_\mu\psi \tag{3.4}$$

$$(\Box - m_s^2)\phi = -g_s\overline{\psi}\psi \tag{3.5}$$

$$[\gamma_\mu (\frac{\partial}{\partial x_\mu} - ig_v V_\mu) + (M - g_s\phi)]\psi = 0 \tag{3.6}$$

together with the adjoint of the last relation. The vector meson field Eqs. 3.4 are just Maxwell's equations with massive quanta and the *conserved baryon current*

$$B_\mu = i\overline{\psi}\gamma_\mu\psi \tag{3.7}$$

as source. Eq. 3.5 is the Klein-Gordon equation for the scalar field with the scalar baryon density as source. Eq. 3.6 is the Dirac equation for the baryon field with V_μ and ϕ included in a "minimal" fashion.

Recall that the energy-momentum tensor (stress tensor) in continuum mechanics is given in terms of the Lagrangian density by

$$T_{\mu\nu} \equiv L \delta_{\mu\nu} - \frac{\partial L}{\partial\left[\frac{\partial q}{\partial x_\mu}\right]} \frac{\partial q}{\partial x_\nu}. \tag{3.8}$$

For a *uniform system* at rest the expectation value of the stress tensor must take the form

$$< \hat{T}_{\mu\nu} > = P\delta_{\mu\nu} + (P + \epsilon)u_\mu u_\nu \tag{3.9}$$

where P is the pressure, ϵ is the energy density, and $u_\mu = (\vec{0}, i)$ is the four-velocity of the fluid. This relation allows us to identify the pressure and energy density. The Hamiltonian density, for example, from which the energy density is obtained, is given by

$$H = -T_{44} = \pi_q \dot{q} - L \tag{3.10}$$

where

$$\pi_q \equiv \frac{\partial L}{\partial \left[\dfrac{\partial q}{\partial t} \right]} \tag{3.11}$$

which corresponds to the canonical expression.

D. Mean Field Theory (MFT)

The coupling constants g_s and g_v are large, implying a strong- coupling theory. We have thus not made much progress by simply writing down a set of field equations unless a sensible starting approximation to their solution can be found. Fortunately one exists. Consider a uniform system of B baryons in a volume V, and imagine compressing it. As the baryon density $\rho = B/V$ gets large, so do the source terms on the righthand-side of the meson fields (Eqs. 3.4-3.5). It should then be a good approximation to replace the meson field by *classical fields* and their sources by expectations values (just as we deal with classical electromagnetic fields). Thus we replace

$$\hat{\phi} \to <\hat{\phi}> \equiv \phi_0$$
$$\hat{V}_\lambda \to <\hat{V}_\lambda> \equiv i\delta_{\lambda 4} V_0 . \tag{3.12}$$

The expectation value of the spatial part of the vector field vanishes by rotational invariance, and a crucial aspect of these relations is that the classical fields ϕ_0 and V_0 will be *constants independent of space and time* for uniform nuclear matter. The vector meson field (Eq. 3.4) for example, in this case reduces to

$$V_0 = \frac{g_v}{m_v^2} \rho_B . \tag{3.13}$$

Since the baryon current is conserved, the baryon number B, and hence the baryon density ρ_B for a uniform system, is a constant of the motion; the vector field V_0 is thus determined in terms of conserved quantities.

The MFT Lagrangian is obtained by substituting Eq. 3.12 into Eq. 3.2

$$L_{MFT} = \frac{1}{2} m_v^2 V_0^2 - \frac{1}{2} m_s^2 \phi_0^2 - \overline{\psi} \left[\gamma_\mu \frac{\partial}{\partial x_\mu} + \gamma_4 g_v V_0 + M^* \right] \psi . \tag{3.14}$$

Here the effective mass of the baryons, which plays a central role in the ensuing discussion, is defined by

$$M^* \equiv M - g_s \phi_0 . \tag{3.15}$$

The Dirac equation is linearized with the substitutions of Eq. 3.12, and it may be solved easily. We look for normal-mode solutions of the form $\psi = U(\vec{p}) \exp{(i\vec{p} \cdot \vec{x} - iEt)}$ with the result that

$$(\vec{\alpha} \cdot \vec{p} + \beta M^*) U(\vec{p}) = (E - g_v V_0) U (\vec{p}) . \tag{3.16}$$

The square of this relation yields the eigenvalue relation

$$E = g_v V_0 \pm (\vec{p}^2 + M^{*2})^{\frac{1}{2}} .$$ (3.17)

We note the important relation between the scalar and probability densities derived from Eq. 3.16

$$\bar{U}(\vec{p})U(\vec{p}) = \frac{M^*}{(\vec{p}^2 + M^{*2})^{1/2}} \, U^\dagger(\vec{p})U(\vec{p}) .$$ (3.18)

We choose to normalize to unit probability where $U^\dagger U = 1$.

The solutions to the Dirac equation provide a complete basis in which to expand the quantum field operator for the baryons. In the Schrödinger picture, where the operators are independent of time, the baryon field is given by

$$\hat{\psi}(\vec{x}) = \frac{1}{\sqrt{V}} \sum_{\vec{k}\lambda} \left[U(\vec{k}\lambda)A_{\vec{k}\lambda}e^{i\vec{k}\cdot\vec{x}} + V(-\vec{k}\lambda)B_{\vec{k}\lambda}^\dagger e^{-i\vec{k}\cdot\vec{x}} \right] .$$ (3.19)

We impose periodic boundary conditions in the volume V. The canonical (anti) commutation relations of the field operator imply that the normal-mode amplitudes satisfy the anticommutation relations

$$\{ A_{\vec{k}\lambda}, A_{\vec{k}'\lambda'}^\dagger \} = \delta_{\vec{k}\vec{k}'}\delta_{\lambda\lambda'}$$

$$\{ B_{\vec{k}\lambda}, B_{\vec{k}'\lambda'}^\dagger \} = \delta_{\vec{k}\vec{k}'}\delta_{\lambda\lambda'} .$$ (3.20)

They may thus be readily identified as the creation and destruction operators for the appropriate normal modes.

The Hamiltonian density follows immediately from Eqs. 3.4 and 3.10; it is obtained by inserting the field expansion and using the orthonormality of the Dirac wavefunctions, as well as the anticommutation relations of the creation and destruction operators. (The manipulations in this section are similar to those in Ref. 26.) The Hamiltonian density can be written in the form

$$\hat{H} = \hat{H}_{MFT} + \delta H$$ (3.21)

where the mean field result takes the form

$$\hat{H}_{MFT} = \frac{1}{2} m_s^2 \phi_0^2 - \frac{1}{2} m_v^2 V_0^2 + g_v V_0 \hat{\rho}_B +$$

$$+ \frac{1}{V} \sum_{\vec{k}\lambda} (\vec{k}^2 + M^{*2})^{\frac{1}{2}} (A_{\vec{k}\lambda}^\dagger A_{\vec{k}\lambda} + B_{\vec{k}\lambda}^\dagger B_{\vec{k}\lambda})$$ (3.22)

and the baryon density operator is given by

$$\hat{\rho}_B = \frac{1}{V} \sum_{\vec{k}\lambda} (A_{\vec{k}\lambda}^\dagger A_{\vec{k}\lambda} - B_{\vec{k}\lambda}^\dagger B_{\vec{k}\lambda}) .$$ (3.23)

The additional term in Eq. 3.21 is given by

$$\delta H = -\frac{1}{V} \sum_{\vec{k}\lambda} \left[(\vec{k}^2 + M^{*2})^{\frac{1}{2}} - (\vec{k}^2 + M^2)^{\frac{1}{2}} \right] .$$ (3.24)

Here the difference with respect to the vacuum has been taken in order to define the energy scale. The term in Eq. 3.24 arises from the normal ordering of the operators in Eq. 3.22; it is readily interpreted in Dirac hole theory as the zero-point energy of all of the filled negative energy states. The baryon density has also been defined by taking the difference with respect to the vacuum

$$\hat{\rho}_B \equiv \hat{\psi}^\dagger(\vec{x})\hat{\psi}(\vec{x}) - <0|\hat{\psi}^\dagger(\vec{x})\hat{\psi}(\vec{x})|0> \equiv :\hat{\psi}^\dagger(\vec{x})\hat{\psi}(\vec{x}):$$

$$\leftarrow\left[\frac{1}{V}\sum_{\vec{k}\lambda}1\right]\rightarrow \tag{3.25}$$

It is this normal-ordered current which is the true source of interaction, just as in quantum electrodynamics (QED).

We shall neglect the term δH in Eq. 3.21 for the present, returning to its role in the next lecture. The operators $\hat{H}_{MFT}$ and $\hat{\rho}_B$ in Eqs. 3.22-3.23 are now *diagonal and the model MFT has thus been solved exactly;* all the eigenstates and eigenvalues are known. We recall that MFT is expected to provide the correct equation of state in the high baryon density limit $\rho_B \rightarrow \infty$ in QHD-I.

D. Nuclear Matter

The ground state of the Hamiltonian in Eq. 3.22 for uniform nuclear matter, is evidently obtained by filling the momentum states up to a Fermi level k_F, with a spin-isospin degeneracy of γ for each state ($\gamma=4$ for nuclear matter composed of protons and neutrons with spin up and spin down). The equation of state can then be written in the following form

$$\epsilon(\rho_B;\phi_0) \equiv E/V$$

$$= \frac{g_v^2}{2m_v^2}\rho_B^2 + \frac{m_s^2}{2g_s^2}(M-M^*)^2 + \frac{\gamma}{(2\pi)^3}\int_0^{k_F}d\vec{k}(\vec{k}^2 + M^{*2})^{\frac{1}{2}} \tag{3.26}$$

$$P(\rho_B;\phi_0) = \frac{g_v^2}{2m_v^2}\rho_B^2 - \frac{m_s^2}{2g_s^2}(M-M^*)^2 + \frac{1}{3}\frac{\gamma}{(2\pi)^3}\int_0^{k_F}d\vec{k}\frac{\vec{k}^2}{(\vec{k}^2+M^{*2})^{1/2}}. \tag{3.27}$$

The baryon density is given by

$$\rho_B = \frac{\gamma}{(2\pi)^3}\int_0^{k_F}d\vec{k} = \frac{\gamma}{6\pi^2}k_F^3. \tag{3.28}$$

The first term in Eq. 3.26 is the vector meson interaction where Eq. 3.13 has been used to eliminate the vector field. The second term is the scalar meson mass term, written in terms of M^* (Eq. 3.15). The final term is the relativistic energy of a Fermi gas of baryons of mass M^*.

The scalar field ϕ_0 appearing in these relations remains to be determined; this is most directly done with the aid of *thermodynamic* arguments. At a fixed V and B the system will minimize its energy. Thus

$$\left[\frac{\partial E}{\partial \phi_0}\right]_{V,B} = 0. \tag{3.29}$$

This implies

$$\phi_0 = \frac{g_s}{m_s^2}\rho_s \tag{3.30}$$

where the scalar density is defined by

$$\rho_s \equiv \frac{\gamma}{(2\pi)^3} \int_0^{k_F} d\vec{k} \, \frac{M^*}{(\vec{k}^2 + M^{*2})^{1/2}} \, . \qquad (3.31)$$

This is a self-consistency equation for ϕ_0 (or equivalently for $M^* = M - g_s\phi_0$) which must be solved at each density. Eq. 3.30 is recognized as nothing more than the scalar meson field (Eq. 3.5) in MFT. The pressure in Eq. 3.27 is obtained from Eqs. 3.9-3.10; alternatively, one can also derive it from the thermodynamic relation $P = - (\partial E/\partial V)_B$.

There are two parameters in this MFT of nuclear matter (only the ratio of meson coupling constants to masses enter). We choose to determine these parameters by fitting the observed binding energy and density of nuclear matter (Fig. 21)

$$c_s^2 \equiv g_s^2 \, (M^2/m_s^2) = 267.1$$

$$c_v^2 \equiv g_v^2 \, (M^2/m_v^2) = 195.7 \, . \qquad (3.32)$$

There are several comments concerning these results:

(1) The mechanism for nuclear saturation in this model (Fig. 21) is the repulsion between like-baryons and the damping of the scalar meson attraction with increasing density.

(2) Saturation is here entirely a relativistic effect. A Hartree-Fock variational calculation with the static potential of Eq. 3.1 shows the corresponding non-relativistic many-body system is *unstable against collapse* (see Ref. 8).

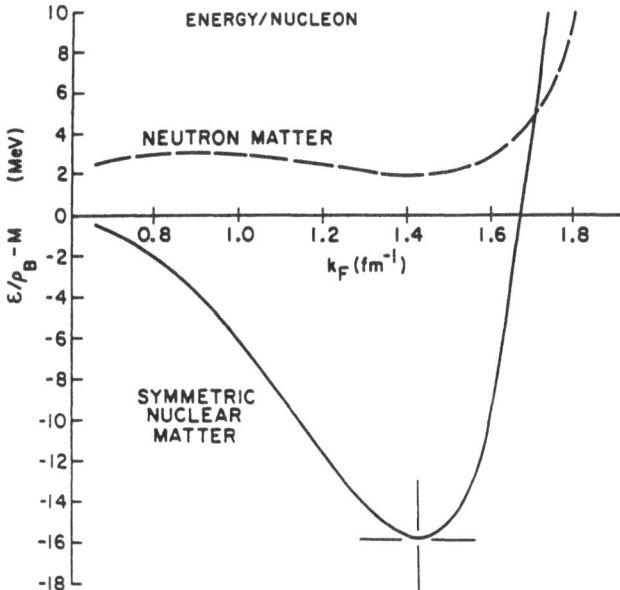

Fig. 21   Saturation curve for nuclear matter. These results are calculated in the relativistic mean-field theory with baryons and neutral scalar and vector mesons
(QHD-I). The coupling constants Eq. 3.32 are chosen to fit the value and position of the minimum. The prediction for neutron matter ($\gamma = 2$) is also
shown[24].

(3) The solution to the self-consistency equation for M^* as a function of density is shown in Fig. 22. Note that at nuclear matter saturation density $M^*/M = 0.56$, and we clearly have a *new energy scale* in this model problem; the scalar meson field energy is of the same order as the nucleon mass itself (see the right-hand-side of Fig. 22).

(4) We note that while the scalar meson density $\overline{\psi}\psi$ is the simplest thing one can write down relativistically, its *non-relativistic limit is complicated,* since (c.f. Eq. 3.18)

$$\frac{M^*}{(\overline{p}^2 + M^{*2})^{1/2}} = 1 - \frac{1}{2}\frac{\overline{p}^2}{M^{*2}} + \frac{3}{8}\frac{\overline{p}^4}{M^{*4}} + \cdots . \tag{3.33}$$

The non-relativistic limit is an infinite series of velocity-dependent interactions!

With the determination of the coupling constants in Eq. 3.32, all other properties of nuclear and neutron matter are now predicted. The results for neutron matter are obtained by simply setting $\gamma=2$ in the previous equations. Neutron matter is unbound (Fig. 21). The equation of state P vs. ε for neutron matter is shown in Fig. 23. Note the approach to the causal limit $P = \varepsilon$, where the thermodynamic speed of sound is equal to the velocity of light, from below. There is a phase separation in this model, similar to the liquid-gas phase transition in the Van der Waal's equation of state; the properties of the two phases are determined by a Maxwell construction.

The neutron matter equation of state may be inserted in the Tolman, Oppenheimer, Volkoff equations for a spherically symmetric metric in general relativity, and the neutron star mass $M/M_\odot$ (in units of the solar mass) plotted against the central density (Fig. 24). We find a maximum neutron star mass of $(M/M_\odot)_{max} = 2.57$. The present equation of state is about as "stiff" as one can get and still be consistent with causality and the observed saturation properties of nuclear matter.

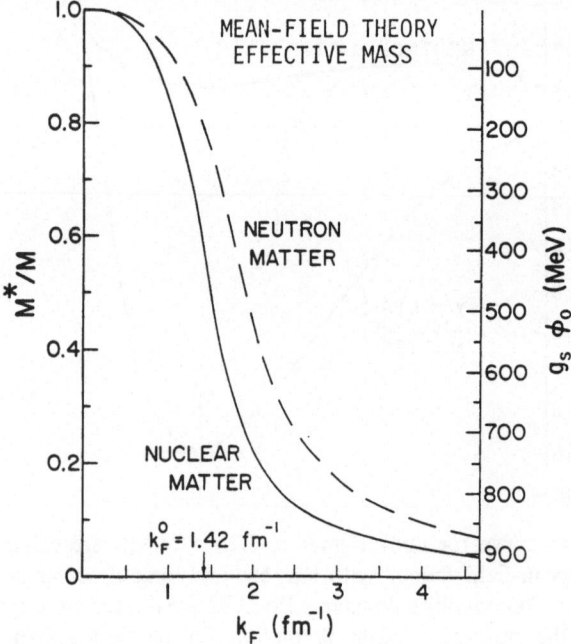

Fig. 22   Effective mass as a function of density for nuclear ($\gamma = 4$) and neutron ($\gamma = 2$) matter based on Fig. 21.[24]

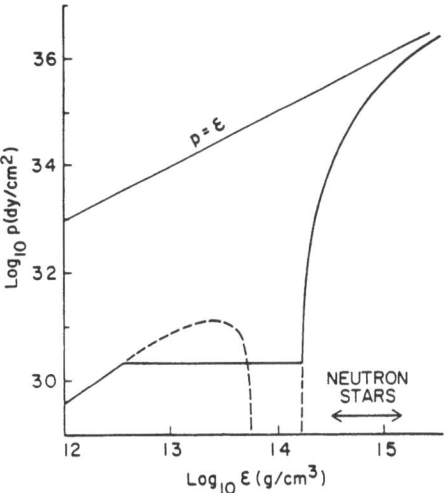

Fig. 23   Predicted equation of state for neutron matter at all densities based on Fig. 21. A Maxwell construction is used to determine the equilibrium curve in the region of the phase transition. The density regime relevant for neutron stars is also indicated[24].

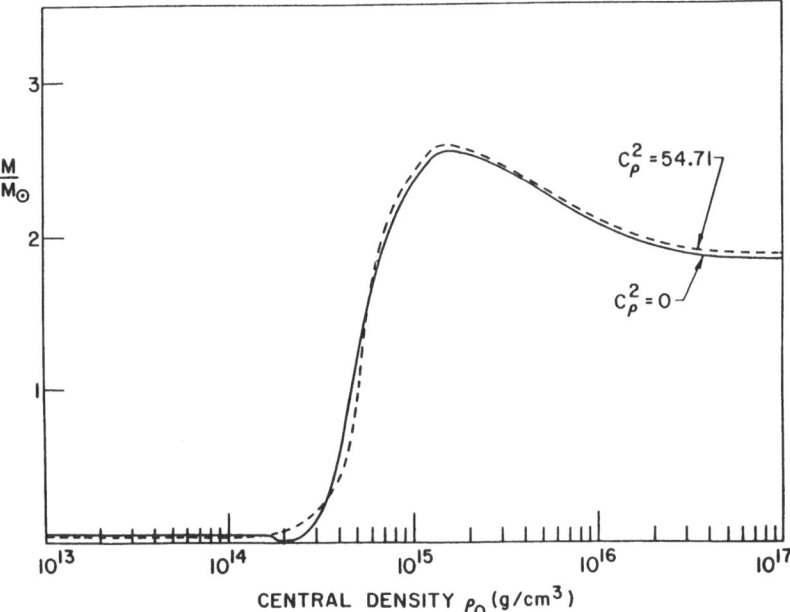

Fig. 24   Calculated neutron star mass in units of the solar mass $M_\odot$ as a function of the central density based on Fig. 23 (solid curve)[24].

E. Finite Temperature

Since the MFT Hamiltonian and baryon densities in Eqs. 3.22-3.23 are diagonal, we can immediately calculate any properties of nuclear matter. One important application is the behavior at finite temperature. We first review some basic results of statistical mechanics[8].

Imagine the uniform system of nuclear matter in a volume V, in contact with a heat bath at temperature T and a particle bath of baryons at a chemical potential μ. The thermodynamic potential at this system is given by

$$\Omega(\mu,V,T) = -\frac{1}{\beta} \, ln \, Z_G \tag{3.34}$$

where the grand partition function is defined by

$$Z_G \equiv Tr\left\{ e^{-\beta(\hat{H}-\mu\hat{B})} \right\} . \tag{3.35}$$

The Trace involves a sum over all diagonal elements in the Hilbert space. Here we use the customary notation $\beta \equiv 1/k_BT$. Knowledge of the thermodynamic potential provides a complete description of the equilibrium properties of the system for we have the thermodynamic relations

$$\Omega = -PV \tag{3.36}$$

$$d\Omega = -SdT - PdV - Bd\mu \tag{3.37}$$

and hence the entropy S, the pressure P, and the baryon number B can be determined by partial differentiation. (The pressure P can be immediately determined from Ω itself through Eq. 3.36.)

In the MFT the Hamiltonian and baryon number operators obtained from Eqs. 3.22-3.23 are diagonal, and we can immediately compute the Trace in the basis of eigenstates of the number operators

$$A^\dagger_{\vec{k}\lambda}A_{\vec{k}\lambda} \mid n_{\vec{k}\lambda} > = n_{\vec{k}\lambda} \mid n_{\vec{k}\lambda} > \tag{3.38}$$

$$B^\dagger_{\vec{k}\lambda}B_{\vec{k}\lambda} \mid \bar{n}_{\vec{k}\lambda} > = \bar{n}_{\vec{k}\lambda} \mid \bar{n}_{\vec{k}\lambda} > . \tag{3.39}$$

The calculation proceeds exactly as that for a non-interacting Fermi gas[8].

The vector field V_0 appearing in the thermodynamic potential may be determined by taking a thermal average of the vector meson field equations. For a uniform system we have the result (compare Eq. 3.13)

$$m_v^2V_0 = g_v <<\hat{\rho}_B>> \equiv g_v\rho_B(\mu,V,T; \phi_0,V_0) . \tag{3.40}$$

The dependence of V_0 is explicit in the Hamiltonian, and it follows immediately that the partial derivative of the thermodynamic potential with respect to V_0 now vanishes

$$\left[\frac{\partial\Omega}{\partial V_0} \right]_{\mu,V,T; \, \phi_0} = 0 . \tag{3.41}$$

The scalar field ϕ_0 may be determined by using the thermodynamic argument that a system at fixed μ,V,T will minimize its thermodynamic potential. Thus

$$\left[\frac{\partial\Omega}{\partial\phi_0} \right]_{\mu,V,T} = \left[\frac{\partial\Omega}{\partial\phi_0} \right]_{\mu,V,T; \, V_0} = 0 . \tag{3.42}$$

The dependence on ϕ_0 is again explicit in the Hamiltonian (recall $M^* = M - g_s\phi_0$) and Eq.3.42 then yields (compare Eq. 3.30)

$$m_s^2\phi_0 = g_s\rho_s(\mu,V,T; \phi_0,V_0) .\tag{3.43}$$

As usual, we now fix ρ_B, that is, we adjust the chemical potential μ until a given baryon density is obtained. Eq. 3.40 then implies that the vector field is determined by

$$V_0 = \frac{g_v}{m_v^2} \rho_B .\tag{3.44}$$

The equation of state now becomes

$$\varepsilon(\rho_B,T) \equiv E/V$$

$$= \frac{g_v^2}{2m_v^2} \rho_B^2 + \frac{m_s^2}{2g_s^2} (M-M^*)^2 + \frac{\gamma}{(2\pi)^3} \int d\vec{k}(\vec{k}^2 + M^{*2})^{\frac{1}{2}} (n_k + \bar{n}_k)\tag{3.45}$$

$$P(\rho_B,T) = \frac{g_v^2}{2m_v^2} \rho_B^2 - \frac{m_s^2}{2g_s^2} (M-M^*)^2 + \frac{1}{3} \frac{\gamma}{(2\pi)^3} \int d\vec{k} \frac{\vec{k}^2}{(\vec{k}^2+M^{*2})^{1/2}} (n_k+\bar{n}_k)\tag{3.46}$$

where the baryon density is given by

$$\rho_B = \frac{\gamma}{(2\pi)^3} \int d\vec{k} (n_k - \bar{n}_k) .\tag{3.47}$$

The self-consistency equation for $\phi_0 = (M-M^*)/g_s$ is

$$\phi_0 = \frac{g_s}{m_s^2} \rho_s = \frac{g_s}{m_s^2} \frac{\gamma}{(2\pi)^3} \int d\vec{k} \cdot \frac{M^*}{(\vec{k}^2+M^{*2})^{1/2}} (n_k+\bar{n}_k) .\tag{3.48}$$

Note the crucial difference in signs in Eqs. 3.47 and 3.48. The thermal distribution functions appearing in these expressions are defined by

$$n_k \equiv \frac{1}{e^{\beta(E_k^*-\mu^*)} + 1}\tag{3.49}$$

$$\bar{n}_k \equiv \frac{1}{e^{\beta(E_k^*+\mu^*)} + 1}\tag{3.50}$$

where

$$E_k^* \equiv (\vec{k}^2 + M^{*2})^{\frac{1}{2}}\tag{3.51}$$

$$\mu^* \equiv \mu - g_vV_0 = \mu - \frac{g_v^2}{m_v^2} \rho_B .\tag{3.52}$$

The last equality follows since V_0 has now been determined by Eq. 3.44.

As $T \to 0$, the baryon distribution becomes a step function and the antibaryon contribution vanishes; this reproduces the previous results at $T = 0$ (Eqs. 3.26-3.31). As $T \to \infty$, pairs are produced, the self-consistent mass goes to zero, and the energy density and pressure are given by

$$\varepsilon = \frac{7\pi^2\gamma}{120} (k_BT)^4\tag{3.53}$$

$$P = \frac{1}{3}\,\varepsilon.$$ (3.54)

These relations are analogous to a *black-body* spectrum and equation of state.

Eqs. 3.45-3.52 are readily solved numerically at all T, and the resulting isotherms for neutron matter, that is, the constant temperature cuts of the surface of the equation of state are shown in Fig. 25. There is a phase transition (c.f. Fig. 23). In the region of the phase transition *Gibb's criteria* for phase equilibrium are satisfied

$$P_1 = P_2$$

$$\mu_1 = \mu_2$$

$$T = \text{constant}.$$ (3.55)

This is accomplished by plotting P against μ at fixed T and finding where the curve crosses itself. One sees a critical region and a critical temperature above which the phase transition disappears. The isotherms *terminate* as the energy density is decreased. There is a finite, limiting value of the energy density as the baryon density goes to zero; it is just the black-body energy density and pressure.

The solution to the self-consistency relation for the baryon mass at vanishing baryon density (and hence, from Eq. 3.47, at vanishing baryon chemical potential) is shown in Fig. 26. The physics of this result is as follows: pairs can be produced. This does not change ρ_B, but it will change ρ_s. Increasing ρ_s decreases M^*, which makes it easier to produce pairs. The solution to the resulting self-consistency relation is shown in this figure. The interesting feature of this curve is the abrupt vanishing of the baryon mass

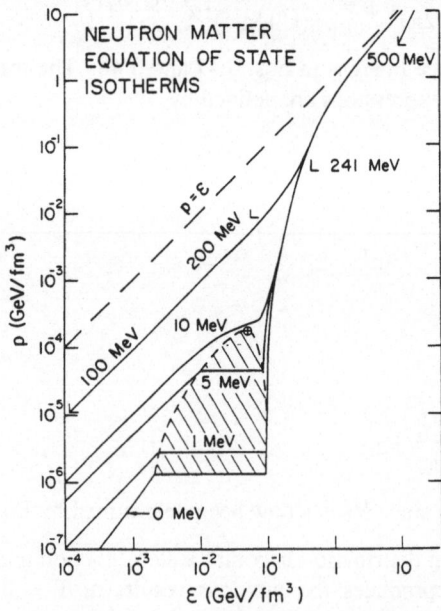

Fig. 25   Isotherms of the neutron matter equation of state at finite temperature, as calcu-
          lated in the mean-field theory of QHD-I. The curves are labeled by the value
          of $k_B T$, and the left-hand end-point of an isotherm corresponds to zero baryon
          density. The shaded area shows the region of phase separation, and the critical
          point at a temperature $k_B T_c = 9.1 \pm 0.2$ MeV is indicated by $\oplus$.[24]

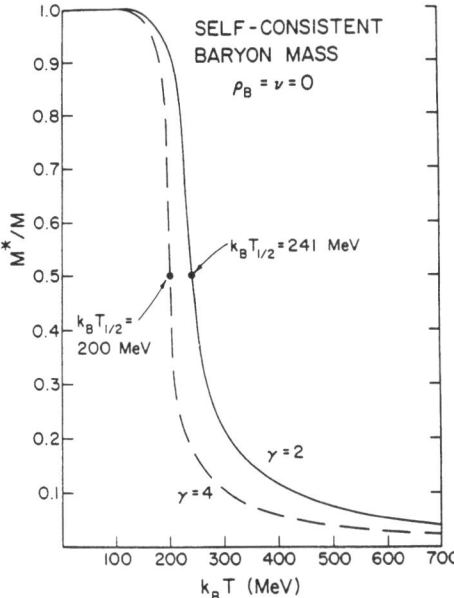

Fig. 26   Self-consistent nucleon mass as a function of temperature at vanishing baryon density. Results are indicated both for neutron matter ($\gamma = 2$ - based on Fig. 25) and nuclear matter ($\gamma = 4$).[24]

for $k_BT \ll M$. At high temperature the baryons are massless. As the temperature is lowered they acquire a mass due to the self-consistent freezing out of the vacuum pairs; they then retain this mass down to $T = 0$.

F. Finite Nuclei

In order to describe finite nuclear systems, the condensed meson fields $\phi_0(|\vec{x}|)$ and $V_0(|\vec{x}|)$ can be given a spatial dependence. There will be corresponding additional gradient terms in the Lagrangian density. We concentrate here on the static properties of nuclei, and assume the fields are still independent of time. When the classical, condensed meson fields and sources have a spatial dependence (here assumed spherically symmetric with $r = |\vec{x}|$), the field Eqs. 3.3-3.6, become

$$(\nabla^2 - m_v^2)V_0 = -g_v\rho_B(r) \tag{3.56}$$

$$(\nabla^2 - m_s^2)\phi_0 = -g_s\rho_s(r) \tag{3.57}$$

$$\left[\frac{1}{i}\vec{\alpha}\cdot\vec{\nabla} + g_vV_0(r) + \beta(M-g_s\phi_0(r))\right]\psi = i\frac{\partial\psi}{\partial t}. \tag{3.58}$$

It will now be assumed that the baryons move in well-defined single-particle orbitals characterized by a set of single-particle quantum numbers κ, and that the levels are filled up to some value F. The local source terms in the meson field equations are then evaluated by summing the Dirac densities over the occupied orbitals

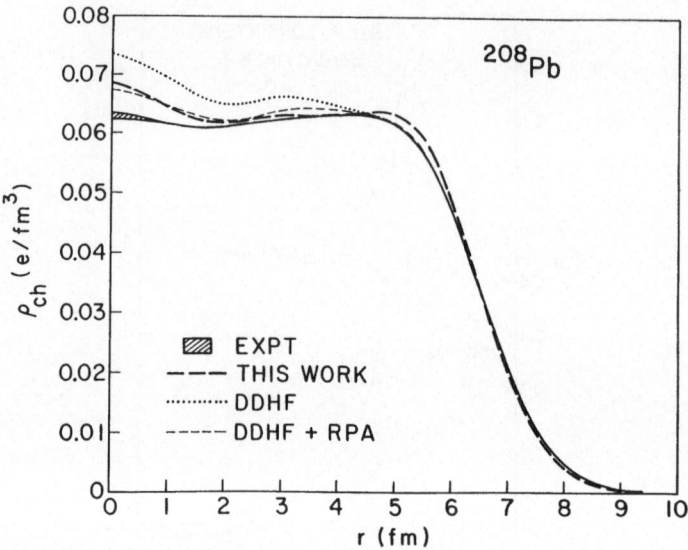

Fig. 27   Charge density for ^{208}Pb. The solid curve and shaded area represent the fit to
experimental data. Relativistic Hartree results are indicated by the long dashed
lines.[27,24]

$$\rho_B(r) = \sum_{\kappa}^{F} U_{\kappa}^{\dagger}(r)U_{\kappa}(r) \tag{3.59}$$

$$\rho_s(r) = \sum_{\kappa}^{F} \overline{U}_{\kappa}(r)U_{\kappa}(r) . \tag{3.60}$$

In order to determine these densities, it is necessary to solve the Dirac equation in the
fields determined by these densities. The equations are evidently coupled and non-linear.
Fortunately, iteration procedures converge rapidly. These equations have been solved by
Horowitz and Serot and our discussion is based on their work[27,24]. Horowitz and Serot
also include a condensed, neutral rho field $\rho_0^0(r)$ coupled to the isovector density
$ig_\rho \overline{\psi}\gamma_\mu\tau_3\psi/2$.* This field is non-zero if $N \neq Z$. The Coulomb potential $A_0(r)$ is also
retained.

There are four parameters in this relativistic Hartree theory of finite nuclei
$\{g_s, g_v, g_\rho, m_s\}$. Horowitz and Serot choose to fit the nuclear matter values of E/B, k_F,
and a_4 (symmetry energy). One length scale is required, and the mean-square charge
radius of ^{40}Ca is also fit. The masses $m_v \equiv m_\omega$ and m_ρ are taken from experiment.

The resulting nuclear charge densities are illustrated in Figs. 27-29. They are compared with experimental results obtained from elastic electron scattering (e,e). The central density of ^{208}Pb defines the value of k_F for nuclear matter (Fig. 27) and the height is
thus fit. The mean-square charge radius of ^{40}Ca is also fit (Fig. 28). The charge density

---

*   Charged mesons are included in QHD-II[24].

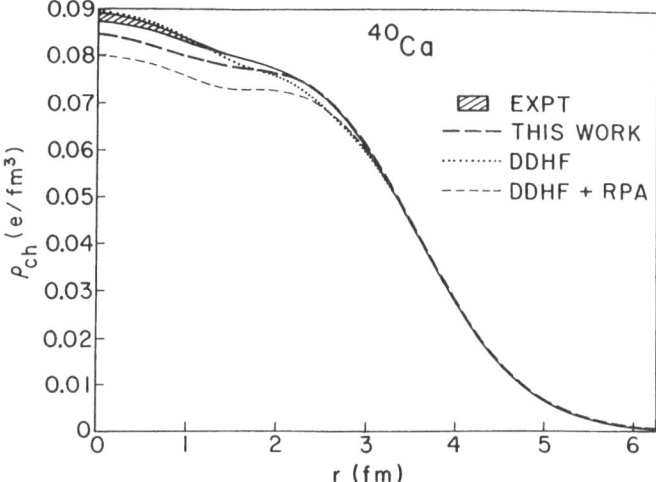

Fig. 28   Same as Fig. 27 for ^{40}Ca.

of ^{16}O is then obtained for free (Fig. 29).

One gets something else for free. Fig. 30 shows the the calculated Hartree single-particle spectrum for the occupied orbitals in ^{208}Pb. One sees all the shell closures of the nuclear shell model. The physics behind this result is that of a Dirac particle moving in spatially varying fields $\phi_0(r)$ and $V_0(r)$. Recall the analogous situation in an atom where the spin-orbit interaction arises from Dirac electrons moving in the spatially varying Coulomb potential $A_0(r)$. Whereas the binding energy of nuclear matter arises from a *cancellation* of large contributions from the scalar and vector fields, the spin-orbit interaction receives *additive* contributions from them. The spin-orbit interaction is clearly of the right sign and magnitude (recall that the fields themselves are very large!), and since it is a sum of the two effects, it is relatively stable against improved

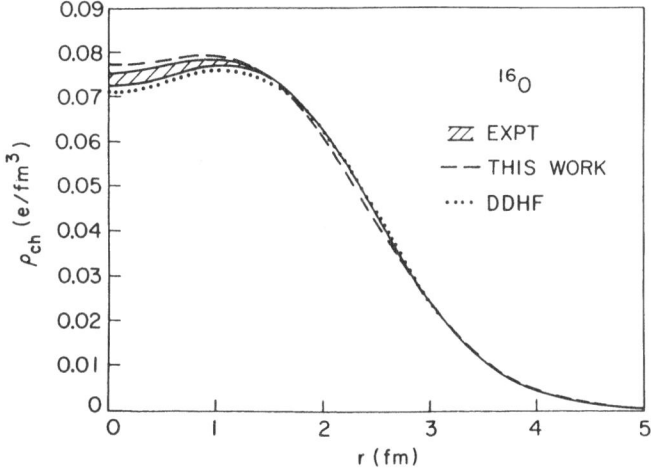

Fig. 29   Same as Fig. 27 for ^{16}O.

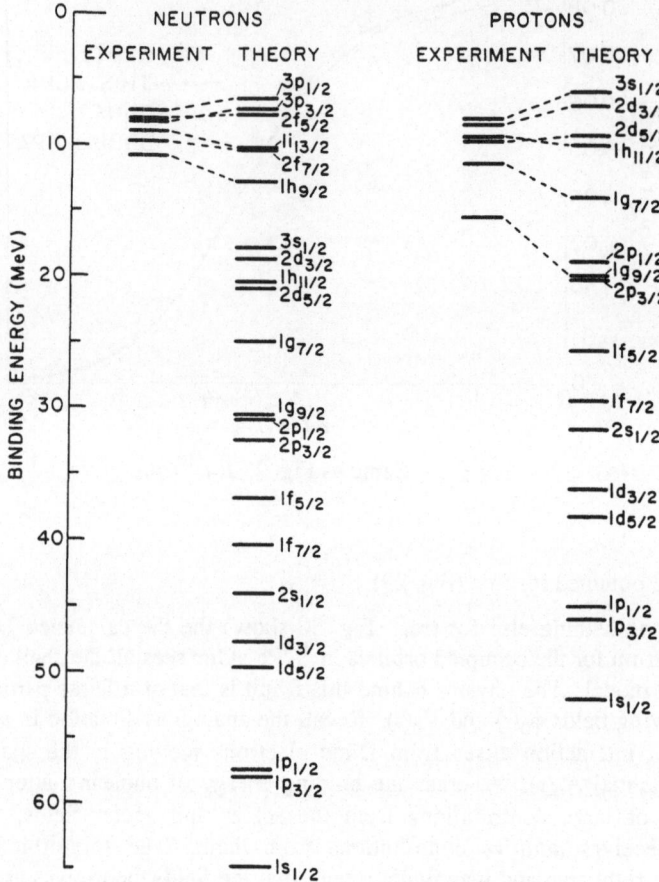

Fig. 30   Predicted spectrum for occupied levels in ^{208}Pb. Experimental levels are from neighboring nuclei[27,24].

approximations. *One thus derives the nuclear shell model by fitting only a few bulk properties of nuclear matter.*

One gets even more for free. The preceding analysis should also be applicable to nucleons above the Fermi surface, that is, to nucleons scattered by the nucleus. Now QHD-1 is too simple to describe the detailed spin dependence of the free N-N scattering amplitude. Let us then agree to compromise and take the scattering amplitude

$$f_{NN} = f_s 1^{(1)} \cdot 1^{(2)} + f_v \gamma_\mu^{(1)} \cdot \gamma_\mu^{(2)} + \dots \tag{3.61}$$

from experiment. (There are five independent amplitudes in this expression). It is an empirical fact that when written this way in terms of Lorentz-invariant combinations, this scattering amplitude is *predominantly scalar and vector!* The previously calculated Hartree densities may now be used to construct an optical potential $U \propto \rho \otimes f_{NN}$, and the Dirac equation solved for the scattering of a baryon by this potential. This approach to

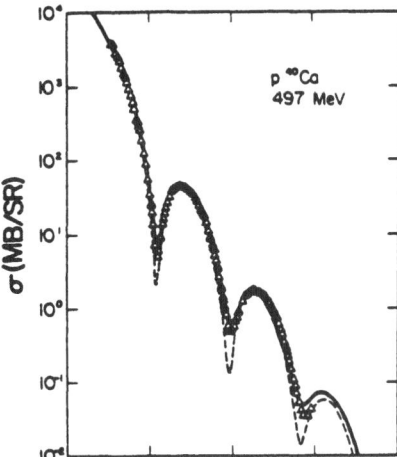

Fig. 31   Calculated cross section for $p + {}^{40}Ca$ at $T_L = 497$ MeV using the Dirac impulse
approximation (RIA- solid curve).

nucleon-nucleus scattering goes by the name of the "Relativistic Impulse Approximation
(RIA);" it is developed in Refs. 28 and 29. The results are illustrated in Figs. 31-33.
Fig.31 compares the calculated and experimental differential cross section for protons on
${}^{40}Ca$ at an incident energy of 497 MeV. The experiments were carried out at LAMPF.
Fig. 32 compares the predicted and measured polarization, and Fig. 33 shows similar
results for the spin-rotation function. The agreement is striking. The underlying physics
here is exactly the same source of spin dependence that leads to the existence of the
nuclear shell model described above.

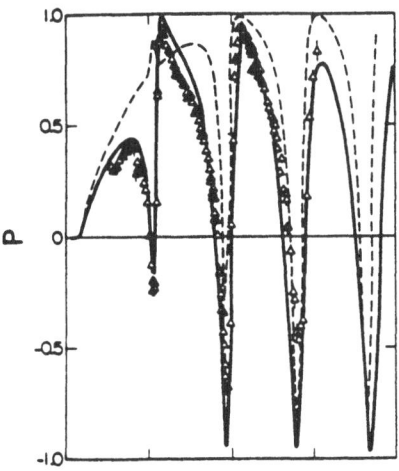

Fig. 32   Calculated analyzing power for $p + {}^{40}Ca$ at $T_L = 497$ MeV using the Dirac
impulse approximation (RIA- solid curve).

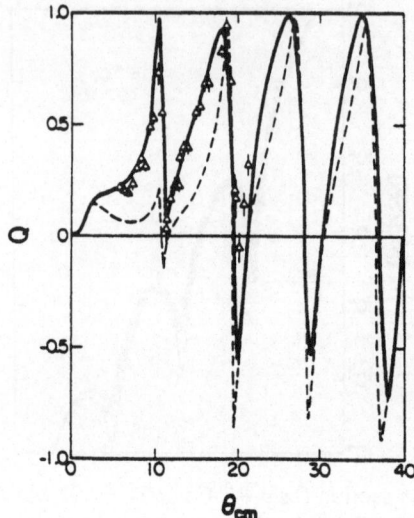

Fig. 33   Calculated spin rotation function for $p + {}^{40}Ca$ at $T_L = 497$ MeV using the Dirac
impulse approximation (RIA- solid curve). (Figs. 31-33 prepared by Professor
B. C. Clark. From Refs. 28, 24.

QUANTUM CHROMODYNAMICS (QCD)

We have seen in the previous lecture that the phenomenology of the mean field
theory built on quantum hadrodynamics is quite successful. The original motivation,
however, was to develop a model field theory in which we could, in principle, calculate
to arbitrary accuracy and compare with experiment. Let us then start this lecture by
returning to the nuclear matter problem and examining corrections to the MFT in QHD-I.

A. Quantum Hadrodynamics QHD-1

The content of the relativistic many-body theory can be summarized in terms of a set
of Feynman rules for the Green's functions[8]. The baryon Green's function, for example,
is defined by

$$iG_{\alpha\beta}(\vec{x}_1 t_1, \vec{x}_2 t_2) \equiv < \Psi \mid P[\hat{\psi}_\alpha(\vec{x}_1 t_1), \hat{\bar{\psi}}_\beta(\vec{x}_2 t_2)] \mid \Psi>$$

$$= \int \frac{d^4k}{(2\pi)^4} e^{ik\cdot(x_1 - x_2)} iG_{\alpha\beta}(k) . \qquad (4.1)$$

The time-ordered product includes a factor of (-1) for the interchange of the fermion
operators. The Green's functions allow us to calculate the expectation values of products
of field operators. In fact, the baryon contribution to $\hat{T}_{\mu\nu}$ can be calculated from the
Green's function in Eq. 4.1.

The *Feynman rules* for the nth-order contribution to iG(k) are:

(1) Draw all topologically distinct, connected diagrams

(2) Vertices

$$ig_s \cdot 1 \qquad\qquad\qquad -g_v \cdot \gamma_\mu \qquad\qquad\qquad\qquad (4.2)$$

(3) Propagators

$$k \quad \frac{1}{i} \frac{1}{k^2 + m_s^2} \qquad ;\text{scalar meson} \qquad\qquad\qquad (4.3)$$

$$k \quad \frac{1}{i} \frac{1}{k^2 + m_v^2} \left[\delta_{\mu\nu} + \frac{k_\mu k_\nu}{m_v^2} \right] \qquad ;\text{vector meson} \qquad\qquad (4.4)$$

$$p \quad \frac{1}{i} \left[\frac{1}{i p\!\!\!/ + M} + 2\pi i (i p\!\!\!/ - M)\, \delta\, (p^2 + M^2)\, \theta\, (p_0)\, \theta\, (k_F - |\vec{p}|) \right] \qquad ;\text{baryon} \qquad (4.5)$$

(4) Conserve four-momentum at each vertex

(5) Integrate $\int d^4 q/(2\pi)^4$ for each independent internal line

(6) Dirac matrix product along fermion lines

(7) Factor $(-1)^F$ for closed fermion loops

(8) If a particle line closes on itself, include a factor $\exp(i k_0 \eta)$ where $\eta \to 0^+$.

The masses in the propagators carry a small negative imaginary part to give the proper Feynman singularities. The term $k_\mu k_\nu$ in the vector meson propagator goes out in any S-matrix element since the vector meson couples to the conserved baryon current. The proof is identical to that which demonstrates the vanishing contribution of the gauge-dependent parts of the photon propagator in QED[26]. In fact, the theory is analogous to massive QED with an additional scalar interaction; it is renormalizable.

It is the extra contribution in the baryon propagator, present at finite baryon density, which complicates the finite-density relativistic nuclear many-body problem. The role of this extra term is to move a finite number of poles from the fourth to the first quadrant in the complex frequency plane so that when one evaluates expectation values by closing contours in the upper-half p_0 plane, there will be contributions from the occupied single-particle states (Fig. 34). We note that when the frequency contours are so-closed, one *cannot avoid picking up the contribution of the negative-frequency poles in the second*

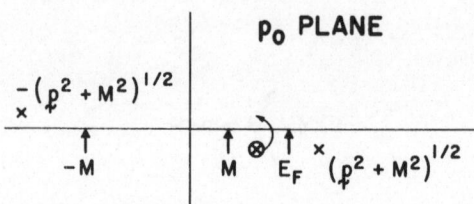

Fig. 34   Poles of the baryon propagator in the complex frequency plane. Here
$E_F = (\vec{k}_F{}^2 + M^2)^{1/2}$.

quadrant. These contributions are an essential feature of the relativistic many-body problem. They are absent in the non-relativistic many-body problem where these antiparticle contributions are pushed off to infinity and ignored[8].

As one application, consider the Relativistic Hartree Approximation (RHA) to nuclear matter. It is a self-consistent, one-baryon-loop calculation. It is done in detail two different ways in Ref. 24 - by summing diagrams and by using path integrals. Here we simply outline it. The RHA is defined to be the self-consistent summation of tad-poles. This statement may be summarized diagrammatically in terms of Dyson's equation for the baryon Green's function

$$\text{(4.6)}$$

Self-consistency enters through the use of this same Green's function to compute the tad-pole loops. Once determined, this Green's function can be used to compute the expecta-tion value of $\hat{T}_{\mu\nu}$. Tadpole contributions to the meson propagators are also retained in the RHA. The baryon loop calculation diverges, and counter-terms must be added to the Lagrangian to renormalize the theory. Since the theory is renormalizable, these counter-terms will contain powers of the scalar field only up through the fourth order. Thus we add

$$\delta L_{CTC} = \sum_{n=1}^{4} \frac{C_n}{n!}\, \phi^n \qquad (4.7)$$

to the Lagrangian density of QHD-I. The counter-terms are fixed in the *vacuum* sector, and our renormalization prescription (chosen to minimize many-body forces in nuclei) is to cancel the one-baryon-loop contributions to the appropriate scalar meson amplitudes at $q_i = 0$. In principle, there could be finite cubic and quartic scalar meson interactions remaining at these points after renormalization.

With the energy density defined by $\varepsilon \equiv E/V$, the previous MFT result is modified by a correction term

$$\varepsilon_{RHA} = \varepsilon_{MFT} + \Delta\varepsilon_{VF} \, . \tag{4.8}$$

The "vacuum fluctuation" correction provides a proper evaluation of the previous result in Eq. 3.24

$$\Delta\varepsilon_{VF} = \delta H - \delta L_{CTC} \, . \tag{4.9}$$

It is given by the expression[24]

$$\Delta\varepsilon_{VF} = \frac{-\gamma}{16\pi^2} \left[M^{*4} ln\frac{M^*}{M} + M^3(M-M^*) - \frac{7}{2}M^2(M-M^*)^2 \right.$$

$$\left. + \frac{13}{3} M(M-M^*)^3 - \frac{25}{12} (M-M^*)^4 \right] \, . \tag{4.10}$$

The modification of the MFT equation of state is shown in Fig. 35.[24] Note that the MFT result remains correct at high density. This provides a partial justification of our initial derivation of the MFT in Lecture III. (The effect on the finite temperature equation of state is discussed in Ref. 30). The modification of the MFT binding energy curve is shown in Fig. 36.[24] The term $\Delta\varepsilon_{VF}$ is evidently a small shift on the new energy scale in the nuclear problem (i.e. several hundred MeV); however, it is important for a quantitative description of the saturation properties of nuclear matter in this theory. We note that this additional contribution is completely *absent* in the non-relativistic many-body

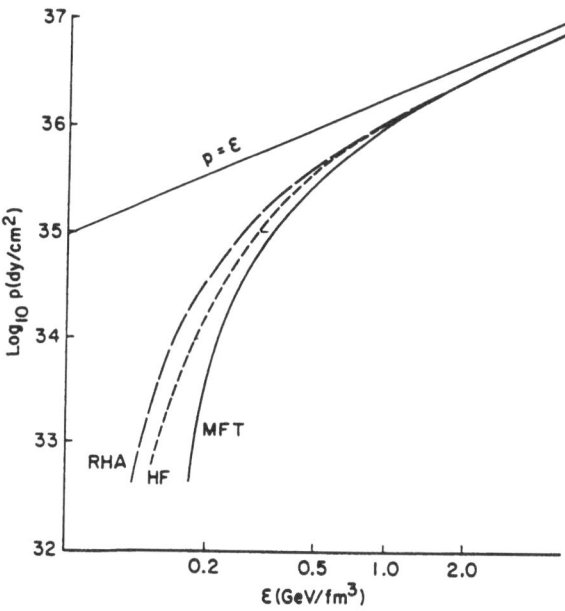

Fig. 35   Nuclear matter equation of state. The mean-field theory (MFT) results are shown as the solid line. The relativistic Hartree approximation (RHA), which includes vacuum fluctuations, produces the long-dash line[24].

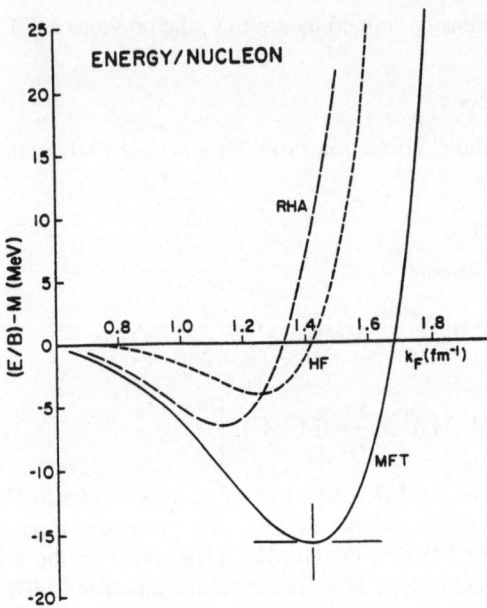

Fig. 36   Energy/nucleon in nuclear matter. The curves are calculated and labeled as in Fig. 35.[24]

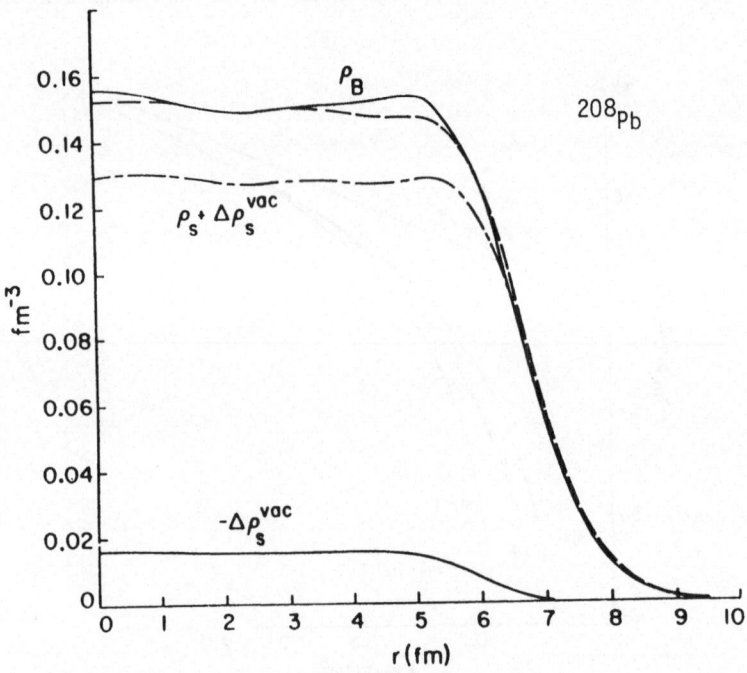

Fig. 37   Density profiles in ^{208}Pb. The total Hartree baryon density of Fig. 27 is shown by the solid curve, and the corresponding result including vacuum fluctuations is given by the dashed curve. Also shown are the total scalar density $\rho_s + \Delta\rho_s$ and (minus) the vacuum fluctuation correction $-\Delta\rho_s$. All curves are "point" densities that do not include single-nucleon form factors. (From Ref. 24).

problem; it is inherently a relativistic effect. If $\Delta\varepsilon_{VF}$ is included in the energy density in a Thomas-Fermi approximation, then the modification of the Hartree charge and scalar densities for ^{208}Pb is illustrated in Fig. 37.[24] The change in densities is small, but it is important if one seeks to investigate small effects, such as the neutron contribution to the charge density.

B. Electroweak Interactions

We have seen that QHD provides a relativistic framework for describing nuclear structure. The static Hartree properties have been determined from the interactions with neutral mesons (QHD-I). We would like to probe this structure with electron scattering (e,e′). In order to have a realistic description of this process, however, it is necessary to enlarge our theoretical framework to include charged mesons (QHD-II). There is an *internal* electromagnetic structure of the nucleon in QHD-II coming, for example, from the following process

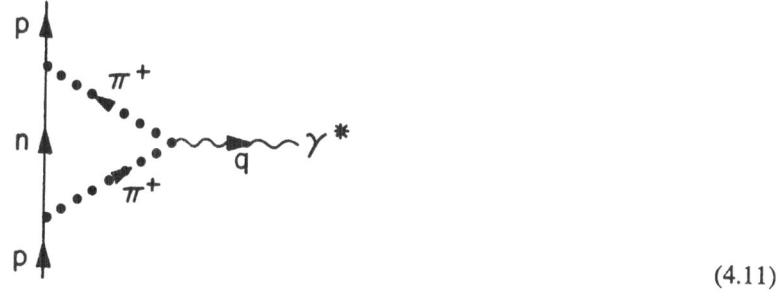

$$(4.11)$$

Now one can show on general grounds that the electromagnetic form factors of the nucleon have a spectral representation

$$F(q^2) = \frac{1}{\pi} \int_{(2m_\pi)^2}^{\infty} d\sigma^2 \, \frac{\rho(\sigma^2)}{\sigma^2 + q^2} \, . \qquad (4.12)$$

The lowest-mass part of the spectral weight function $\rho\,(\sigma^2)$ comes from the lightest-mass intermediate state in the absorptive part (which consists of two pions as illustrated in Eq. 4.11). A simple Fourier transform of the static limit of Eq. 4.12 shows that the lowest-mass meson spectrum gives rise to the longest-range part of the ground-state charge and current density of the nucleon.

Let us make the simplifying assumption that the two-pion contribution dominates the spectral weight function everywhere. The resulting magnetic properties of the nucleon calculated under this assumption are then equivalent to those calculated directly from the Feynman diagram in Eq. 4.11; they are shown in Table IV.1[24].

A detailed calculation of the internal electromagnetic properties of a free nucleon is as yet prohibitive in QHD-II, but these arguments indicate that we may hope to at least be in the right ballpark.

In order to probe the Hartree structure of nuclei, we shall for now be content to use an effective current, which includes this internal electromagnetic structure. We first assume, consistent with observations over the range of momentum transfers of interest to nuclear physics, that there is a common single-nucleon form factor. We thus define an

Table IV.1

Magnetic Properties of the Nucleon Calculated from the
Two-pion Contribution in QHD-II [Ref. 24] (*)

	$\frac{1}{2}(\lambda_p+\lambda_n)$	$\frac{1}{2}(\lambda_p-\lambda_n)$	$(<r^2>_{2p})^{\frac{1}{2}}$
Theory	0	1.60	0.49 fm
Experiment	-0.06	1.853	0.8 fm

(*) This calculation gives $F_{2n}(q^2) = -F_{2p}(q^2)$

effective Moller potential

$$\frac{1}{q^2} f_{s.n.}(q^2) . \tag{4.13}$$

We next define an *effective electromagnetic current operator* which is to be used in lowest order in QHD-I to describe electromagnetic interactions. We write

$$\hat{J}_\mu^\gamma(x) = i\bar{\psi}\gamma_\mu Q\psi + \frac{\partial}{\partial x_\nu} \frac{1}{2M} (\bar{\psi}\sigma_{\mu\nu}\underline{\lambda}\psi) \tag{4.14}$$

where the charge and anomalous moments are defined by

$$Q \equiv \frac{1}{2} (1+\tau_3)$$

$$\underline{\lambda} \equiv \frac{1}{2} (1+\tau_3)\lambda_p + \frac{1}{2} (1-\tau_3)\lambda_n . \tag{4.15}$$

Although oversimplified, this current does have the following essential features to recommend it:

    (1) It is local
    (2) It is covariant
    (3) It is conserved in QHD-I.

$$\frac{\partial}{\partial x_\mu} \hat{J}_\mu^\gamma(x) = 0 \tag{4.16}$$

    (4) It gives the correct result for a free nucleon.

There is nothing that now intrinsically limits theoretical calculations to low q^2 in the nuclear case.

As one application, we show the calculations of Serot [31] for elastic magnetic scattering from ^{209}Bi. The valence proton is in a $(1h_{9/2})$ orbital, and the upper and lower radial components of the Dirac-Hartree wave function are shown in Fig. 38. The transverse elastic form factor is shown in Fig. 39, along with the non-relativistic "Schrödinger-equivalent" result[31]. The experimental value of the magnetic moment is $\mu = 4.10$ nm which lies between the Dirac value $\mu = 5.07$ nm and the Schmidt value $\mu = 2.64$ nm. The enhancement of the moment, and of the intermediate-q magnetic form factor, can be traced to the increased convection current $\vec{p}/M^*$ in the Dirac case.

An effective weak current can be similarly constructed. We define the nuclear axial-vector current by

$$\hat{J}_{\mu 5}^{(\pm)}(x) = \left[\delta_{\mu\nu} - \frac{1}{\Box - m_\pi^2} \frac{\partial}{\partial x_\mu} \frac{\partial}{\partial x_\nu} \right] \hat{j}_{\nu 5}^{(\pm)}(x) \tag{4.17}$$

where the basic axial current is given by

$$\hat{j}_{\mu 5}^{(\pm)}(x) = i\overline{\psi}\gamma_5\gamma_\mu\omega^\pm \psi \tag{4.18}$$

and the weak charge by

$$\omega^{(\pm)} \equiv F_A(0)\tau_\pm . \tag{4.20}$$

As justification, we note that the axial-vector current in Eq. 4.17 has the following properties:

(1) It is covariant
(2) It satisfies PCAC (it is partially-conserved)

$$\frac{\partial}{\partial x_\mu} \hat{J}_{\mu 5} = O\ (m_\pi^2) \tag{4.21}$$

(3) It gives the correct result for a free nucleon.

The full electroweak currents take the form

$$\hat{J}_\mu^{(\pm)} = \hat{J}_\mu^{(\pm)} + \hat{J}_{\mu 5}^{(\pm)} \tag{4.22}$$

$$\hat{J}_\mu^{(0)} = \hat{J}_\mu^{V_3} - 2 \sin^2\theta_w\ \hat{j}_\mu^\gamma . \tag{4.23}$$

We now have a model relativistic nuclear many-body theory where the electroweak hadronic currents have all the general properties assumed in Lecture II.

C. Quantum Chromodynamics (QCD)

Quantum chromodynamics (QCD) is a theory of the strong interactions binding quarks into hadrons. It is a Yang-Mills non-abelian gauge theory built on color. We reviewed the basic properties of QCD in Lecture II and wrote the Lagrangian in Eq. 2.52. In Lecture II we also discussed the two remarkable properties of QCD: asymptotic freedom and confinement. QCD appears to be the underlying theory of the strong interactions; it is certainly a very successful theory of the high-momentum, short-distance microscopic world.

We are evidently faced with a problem. How are we to reconcile QHD with QCD? There are various possibilities, for example:

(1) One can assume a separation radius R in coordinate space for the hadrons. QHD can then be used outside of this radius, and asymptotically free QCD inside of this radius.

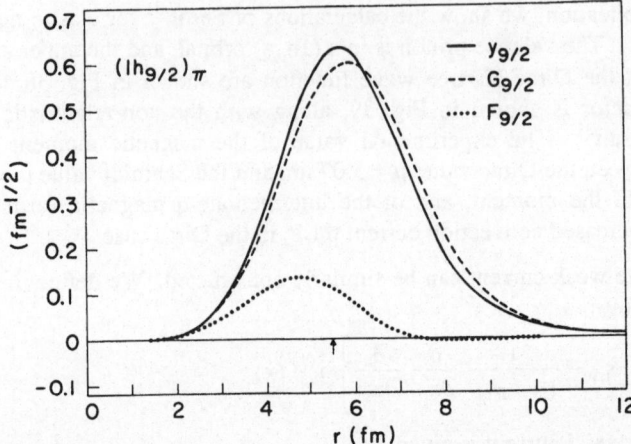

Fig. 38   Valence proton wave functions for ^{209}Bi. The relativistic large (G) and small
(F) components are indicated along with the "Schrödinger-equivalent" reduced
wave function (y). This 1hg/$_2$ proton state is bound by 2.06 MeV in both cases.
The arrow indicates the empirical rms radius of the core nucleus.[31,24]

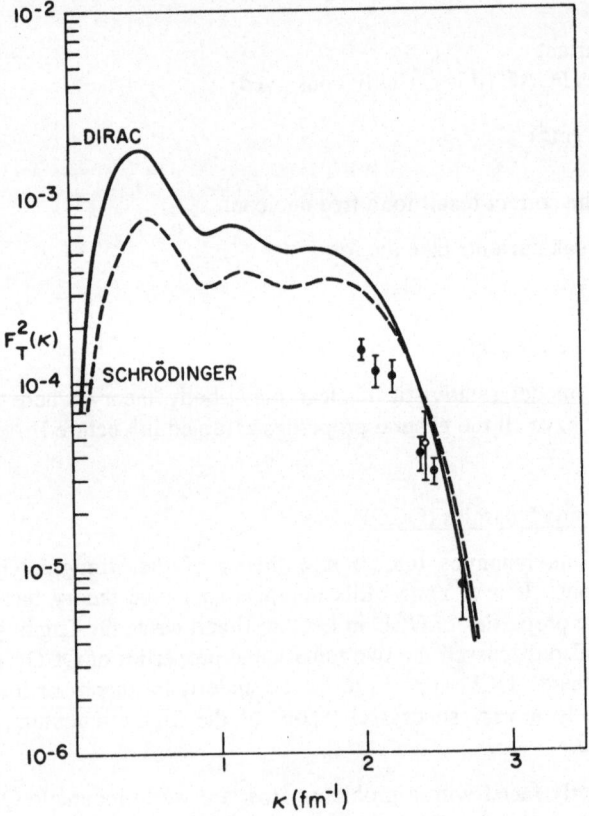

Fig. 39   Transverse elastic form factor of ^{209}Bi. Both relativistic and nonrelativistic
results are shown as functions of the momentum transfer $\kappa = q$ (From
Refs. 31,24).

This is the basis for the bag model, and extended bag models, of hadrons.

(2) One can assume a separation in momentum space. Observed hadrons can be used to evaluate the contribution of nearby singularities in dispersion relations and spectral representations, and the far-off contributions can be evaluated from asymptotically free QCD.

(3) One can imagine that one has two different models for two distinct phases of nuclear matter

> QHD (solved in mean field theory) for a baryon/meson phase
> QCD (solved as asymptotically free) for a quark/gluon phase

It is this last approach that I would like to pursue.

D. Phase Diagram of Nuclear Matter

Let us combine the material in these lectures to carry out a very simple, model calculation of the phase diagram of nuclear matter.

For the *baryon/meson* phase, we use QHD-I solved in the MFT in Lecture III. The equation of state at all temperatures and densities is given by Eqs. 3.45-3.52.

For the *quark/gluon* phase, we use QCD with the following simplifying approximations:

(1) We restrict the discussion to the *nuclear domain* where one works in that sector of the Hilbert space containing only u and d quarks. Any number of pairs of these objects may be present, however, so the states can still be very complicated. In the nuclear domain the quark field reduces to the expression given in Eq. 2.13. The u and d quarks will be assumed massless.

(2) The confinement property is modeled by assuming that it costs a positive energy/volume to create a "vacuum bubble" into which the quarks and gluons are inserted.

$$\left[\frac{E}{V}\right]_{vac} = +b \qquad\qquad (4.24)$$

This is the essence of the M.I.T. bag model. Here, however, only the volume energy plays a role, and one need not model the complicated surface region of the hadrons.

(3) Consistent with asymptotic freedom, we assume non-interacting quarks and gluons with degeneracy factors given by

$$\gamma_Q = (3 \text{ colors}) \quad x \quad (2 \text{ flavors}) \quad x \quad (2 \text{ spins}) = 12$$

$$\gamma_G = (8 \text{ colors}) \quad x \quad (2 \text{ helicities}) \qquad\qquad = 16 \tag{4.25}$$

Note the eight gluons are massless, and like the photon they have two helicity states.

The equation of state at all temperatures and densities follows immediately from elementary statistical mechanics[8]

$$\varepsilon \equiv \frac{E}{V} = b + \frac{\gamma_Q}{(2\pi)^3} \int k d\vec{k}(n_k + \bar{n}_k) + \frac{\gamma_G}{(2\pi)^3} \int \frac{k d\vec{k}}{e^{\beta k} - 1} \tag{4.26}$$

$$P = -b + \frac{1}{3} \left\{ \frac{\gamma_Q}{(2\pi)^3} \int k d\vec{k}(n_k + \bar{n}_k) + \frac{\gamma_G}{(2\pi)^3} \int \frac{k d\vec{k}}{e^{\beta k} - 1} \right\} . \tag{4.27}$$

Since quarks carry baryon number 1/3, the baryon density is given by

$$\rho_B = \frac{1}{3} \frac{\gamma_Q}{(2\pi)^3} \int d\vec{k}(n_k - \bar{n}_k) . \tag{4.28}$$

The thermal distribution functions appearing in these expressions are defined by

$$n_k \equiv \frac{1}{e^{\beta(k - \mu/3)} + 1} \tag{4.29}$$

$$\bar{n}_k \equiv \frac{1}{e^{\beta(k + \mu/3)} + 1} . \tag{4.30}$$

Here μ is the baryon chemical potential. This equation of state is simpler than that discussed in Lecture III since there is no self-consistency equation to be solved. We can, in fact, immediately derive the following analytic results for the quark/gluon phase of nuclear matter in this model:

(1) A linear combination of Eqs. 4.26 and 4.27 yields the equation of state at all T and ρ_B

$$3 (P+b) = \varepsilon - b . \tag{4.31}$$

(2) At finite baryon density $\rho_B \equiv 2k_F^3/3\pi^2$ and zero temperature $T = 0$,

$$3 (P+b) = \varepsilon - b = \frac{3}{2\pi^2} k_F^4 . \tag{4.32}$$

Here the Fermi pressure of the quarks keep the vacuum bubble from collapsing.

(3) At finite temperature $T \neq 0$ and zero baryon density $\rho_B = \mu = 0$

$$3 (P+b) = \varepsilon - b = \frac{37}{30} \pi^2 (k_B T)^4 . \tag{4.33}$$

(4) At finite temperature $T \neq 0$, zero baryon density $\rho_B = 0$, and *zero pressure P=0*

$$b = \frac{37}{90} \pi^2 (k_B T_0)^4 . \tag{4.34}$$

Above this temperature, the thermal pressure of the quarks, antiquarks, and gluons causes the bubble to expand.

We now have two different models for two different phases of nuclear matter. To combine these results, we again appeal to *Gibb's criteria* for phase equilibrium given in Eq. 3.55. When these conditions are satisfied, the two phases can coexist in equilibrium. (Elsewhere, for fixed V,T,B, it is the phase with the lowest Helmholtz free energy that is stable.)

There is one parameter b in this model equation of state. We *arbitrarily choose*

$$R \equiv 3 \, (2\pi^2 b)^{\frac{1}{4}} \equiv 1.2 \, M \, . \tag{4.35}$$

Quark/gluon matter then saturates well above nuclear matter (Fig. 40); we ensure that observed nuclear matter is in the baryon/meson phase.

The resulting isotherms for nuclear matter are shown in Fig. 41. Let us follow one of them: at low density, it is the baryon/meson phase described in Lecture III which is the stable form. The pressure can then be increased until a value is reached where the quark/gluon phase begins to form and the two phases coexist in equilibrium. Additional pressure then converts the system entirely to the quark/gluon phase and moves the system up along the equation of state curve given in Eq. 4.31.

The vapor pressure curve of nuclear matter, that is the pressure at which the two phases are in equilibrium, is plotted against 1/T in Fig. 42. At high temperature and high pressure, the equilibrium phase is always quark/gluon. At high temperature there is a limiting pressure (corresponding to zero baryon density); it is just the "black-body" result in Eq. 4.33.

This is a very simple model, but it does have several essential features to recommend it:

    1) It provides a completely relativistic calculation of the phase diagram.
    2) The QHD description of the baryon/meson phase is consistent with most

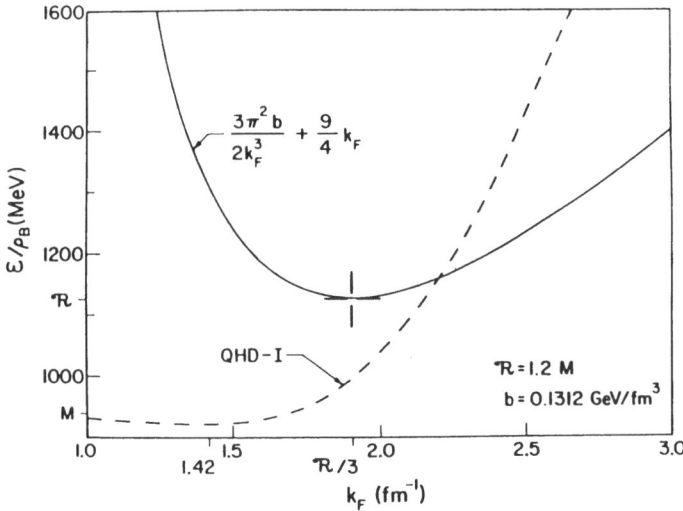

Fig. 40   Saturation curve at T = 0 for nuclear matter. The solid curve denotes the quark-gluon result in the present model, with R defined by Eq. 4.35. The baryon density is parametrized by k_F through $\rho_B = 2k_F^3/3\pi^2$. The corresponding result for observed nuclear matter is indicated by the dashed curve. A value of R = 1.2M is used for illustration.

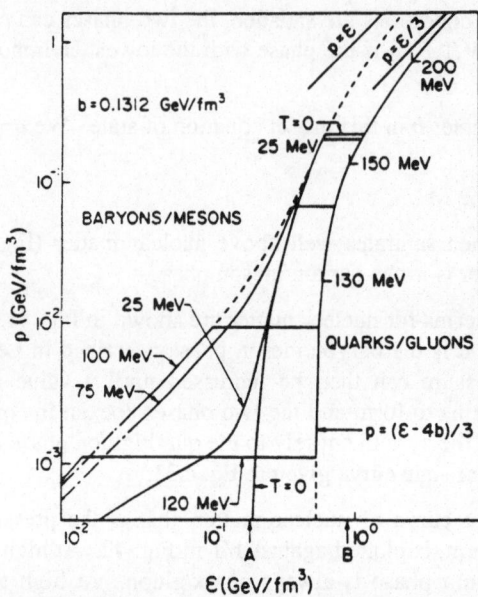

Fig. 41   Equation of state isotherms for nuclear matter for the indicated value of $k_B T$ calculated as described in the text. Equilibrium between the baryon/meson and quark/gluon phases exists along the horizontal segments. The arrows "A" and "B" indicate the energy density at nuclear matter saturation and at the center of the most massive neutron star in Fig. 24 respectively. The left-hand end-points of the higher temperature curves correspond to zero baryon density.

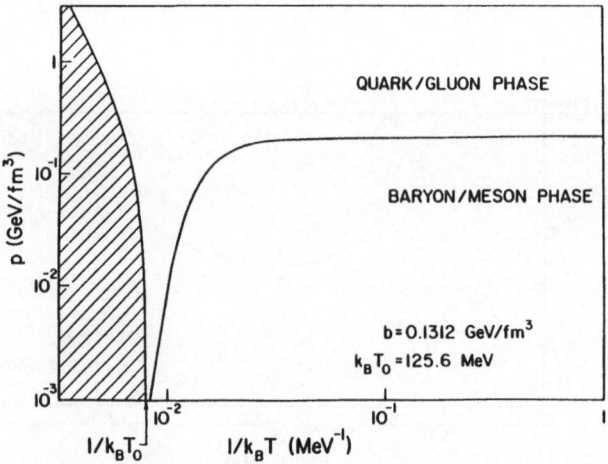

Fig. 42   Phase diagram for nuclear matter based on Fig. 41. The equilibrium vapor pressure is plotted against $1/k_B T$. The boundary of the shaded region is given by Eq. (4.33) and represents the minimum obtainable pressure at the indicated temperature.

observed properties of real nuclei.

3) The description of the quark/gluon phase is consistent with asymptotic free-dom.

4) The statistical mechanics has been done exactly.

Some important references for this phase transition are Refs. 32-35.

The search for this phase transition in high-energy heavy-ion reactions, the signal for its formation, the conditions under which statistical equilibrium is attained, and indeed, whether a phase transition exists at all in more sophisticated calculations are all current and challenging problems in nuclear physics.

E. Summary

Let me try to summarize these lectures.

Quantum hadrodynamics (QHD) is a relativistic quantum field theory of nuclear structure based on hadronic degrees of freedom; it provides a theoretical basis for the relativistic nuclear many-body problem. It correlates and explains many features of nuclear structure, including

1) Nuclear densities
2) The shell model
3) Intermediate energy (p,p) scattering.

QHD is evidently an approximation to quantum chromodynamics (QCD). QHD is in some sense "correct" at large distances.

Within the framework of QHD there are many outstanding problems, for example:

(1) Pions are a problem. The strong spin and isospin dependence of the coupling of pions to nucleons implies that pion exchange largely averages out in the bulk properties of nuclear matter. However, a consistent description of the small s-wave pion interaction, of pion-nucleon resonances (the nucleon isobars), and of nuclear matter has yet to be obtained in QHD. Charged mesons, in general, are difficult to handle.

(2) While the single-particle spectrum in nuclei is in the right ballpark, close inspection shows the level spacings are too large near the Fermi surface. A self-consistent relativistic Hartree-Fock calculation for finite nuclei should be carried out.

(3) QHD is a strong-coupling theory at short distances. One would like to have an exact solution to the theory with which to compare in this regime. A lattice calculation here is an interesting possibility.

(4) The big question is, can one derive QHD from QCD? Can one demonstrate the existence of condensed scalar and vector fields ϕ_0 and V_0 in the many-baryon system starting from QCD?

Quantum chromodynamics (QCD) is the underlying theory of the strong interactions binding quarks and gluons into hadrons. It is formulated in terms of a completely new set of degrees of freedom for the nuclear problem. The theory is simple at short distances (asymptotic freedom), and is successful at describing short-distance phenomena.

At large distances QCD is a complicated strong-coupling theory (confinement). To be a correct theory, QCD must reproduce nuclear structure including

1) Meson exchanges
2) Baryon dynamics in nuclei

    3) Meson dynamics in nuclei.

It will be a long time before one will be able to describe these phenomena starting from the Lagrangian of QCD. The following is my favorite example in this regard: imagine being given the Lagrangian of QED, the underlying theory of atomic structure, and being asked to predict the phenomenon of superconductivity!

    As for the *future*, in my opinion the most exciting possibility is to look for interesting, qualitatively new nuclear phenomena based on this new underlying set of degrees of freedom, quarks and gluons. The phase transition we discussed is just one example. There will undoubtedly be new collective phenomena, and new macroscopic color configurations in the many-baryon system. Where should one look? What are their properties? It is an exciting time in nuclear physics.

    What one will most likely do in the future is compute nuclear properties as accurately as possible in both QHD and QCD and then compare the calculations with each other and with experiment in order to study
    1) Quarks in nuclei
    2) The inadequacies of QHD
    3) Approximations to QCD.
It is a long, tough program, and it will take a lot of hard work; however, the physics payoff should be high.

    I would like to close with three quotations which I find very interesting and thought-provoking:

    "We have been doing nuclear physics for 50 years without quarks, why do we need them now?" (H. L. Anderson, LAMPF II Workshop (1983))

This is actually a profound question for nuclear physics and nuclear physicists. I ask you to think about it very carefully.

    "The single most important *practical* application of the advances in particle physics may be the revolution in our picture of the nucleus." (R. R. Wilson, private communication (1984))

And finally, an appropriate new definition of the field,

    "Nuclear physics is the study of the strong interaction aspects of QCD." (N. Isgur, CEBAF Workshop (1984)).

APPENDIX A - NUCLEAR CURRENTS

    The traditional procedure discussed in the text in Eqs. 1.27-1.29 yields matrix elements of the one-body nuclear electromagnetic current densities of the following form

$$\vec{J}(\vec{x})_{fi} = \vec{J}_c(\vec{x})_{fi} + \vec{\nabla} \times \vec{\mu}(\vec{x})_{fi}$$

$$\rho(\vec{x})_{fi} = \rho_0(\vec{x})_{fi} + \vec{\nabla} \cdot \vec{s}(\vec{x})_{fi} + \nabla^2 \phi(\vec{x})_{fi} \tag{A.1}$$

The operators appearing in these expressions are given in first quantization by[1]

$$\hat{\rho}_0(\vec{x}) = \sum_{j=1}^{A} e(j) \, \delta^{(3)} \, (\vec{x} - \vec{x}_j)$$

$$\hat{\vec{J}}_c(\vec{x}) = \sum_{j=1}^{A} e(j) \frac{1}{M} [\vec{p}(j), \delta^{(3)}(\vec{x}-\vec{x}_j)]_{sym}$$

$$\hat{\vec{\mu}}(\vec{x}) = \sum_{j=1}^{A} \mu(j) \frac{1}{2M} \vec{\sigma}(j) \, \delta^{(3)}(\vec{x}-\vec{x}_j)$$

$$\hat{\vec{s}}(\vec{x}) = \sum_{j=1}^{A} s(j) \frac{1}{4M^2} \vec{\sigma}(j) \times [\vec{p}(j), \delta^{(3)}(\vec{x}-\vec{x}_j)]_{sym}$$

$$\phi(\vec{x}) = \sum_{j=1}^{A} s(j) \frac{1}{8M^2} \delta^{(3)}(\vec{x}-\vec{x}_j) . \tag{A.2}$$

These results are obtained from an expansion carried through order $1/M^2$. The nucleon charges and moments appearing in these expressions are defined by

$$e = \frac{1}{2}(1 + \tau_3)$$

$$\lambda = \frac{1}{2}(1 + \tau_3)\lambda_p + \frac{1}{2}(1 - \tau_3)\lambda_n$$

$$\mu = e + \lambda$$

$$s = e + 2\lambda \tag{A.3}$$

The first two of Eqs. A.2 are the familiar charge and convection current densities of non-relativistic quantum mechanics.

REFERENCES

[1] J. D. Walecka, "Electron Scattering," Lectures given at Argonne National Laboratory, Argonne, Illinois, ANL-83-50 (1983).

[2] J. D. Walecka, "Electroweak Interactions with Nuclei," in Proc. Conf. on Intersections Between Particle and Nuclear Physics, Steamboat Springs, Colorado, May, 1984, ed. R. Mischke, A.I.P. Conf. Proc. No. **123**, A.I.P. (N.Y.) p.1.

[3] J. D. Walecka, "Semi-Leptonic Weak Interactions in Nuclei," in Proc. "Weak-Interactions - 1977," ed. D. B. Lichtenberg, A.I.P. Conf. Proc. No. **37**, A.I.P. (N.Y.) p. 125.

[4] S. Weinberg, Phys. Rev. Lett. **19**, 1264 (1967); Phys. Rev. **D5**, 1412 (1972).

[5] A. Salam and J. C. Ward, Phys. Lett. **13**, 168 (1964).

[6] S. L. Glashow et al., Phys. Rev. **D2**, 1285 (1970).

[7] F. Wilczek, Ann. Rev. of Nucl. Sci. **32**, 177 (1982).

[8] A. L. Fetter and J. D. Walecka, *Quantum Theory of Many-Particle Systems*, McGraw-Hill, New York (1971).

[9] I. Sick et al., Lect. Notes in Physics **108**, Springer, Berlin (1979).

[10] R. C. York and G. A. Peterson, Phys. Rev. **C19**, 574 (1979).

[11] H. Zarek et al., Phys. Rev. Lett. **38**, 750 (1977).

[12] R. P. Feynman and M. Gell-Mann, Phys. Rev. **109**, 193 (1958).

[13] B. D. Serot, Nucl. Phys. **A322**, 408 (1979).

[14] T. W. Donnelly and J. D. Walecka, Phys. Lett. **44B**, 330 (1973).

[15] J. Dubach, J. H. Koch and T. W. Donnelly, Nucl. Phys. **A271**, 279 (1976).

[16] J. M. Cavedon et al., Phys. Rev. Lett. **49**, 986 (1982).

[17] D. Riska, Nucl. Phys. **A350**, 227 (1980).

[18] E. Hadjimichael, B. Goulard and R. Bornais, Phys. Rev. **C27**, 831 (1983).

[19]   J. D. Walecka, "Neutrino Interactions with Nuclei," in Proc. Second LAMPF II Workshop, eds. H. A. Thiessen *et al.*, LA-9572-C, Vol. II. Los Alamos National Laboratory, Los Alamos, New Mexico (1982) p. 560.
[20]   G. Feinberg, Phys. Rev. **D12**, 3575 (1975).
[21]   J. I. Friedman and H. W. Kendall, Ann. Rev. Nucl. Sci. **22**, 203 (1972).
[22]   J. D. Bjorken and E. A. Paschos, Phys. Rev. **185**, 1975 (1969).
[23]   R. L. Jaffe, Phys. Rev. Lett. **50**, 228 (1983).
[24]   B. D. Serot and J. D. Walecka, "The Relativistic Nuclear Many-Body Problem," *Advances in Nuclear Physics* eds. J. W. Negele and E. Vogt, Vol. **16**, (in press).
[25]   C. Y. Prescott *et al.*, Phys. Lett. **77B**, 347 (1978); **84B**, 524 (1979).
[26]   James D. Bjorken and S. D. Drell, *Relativistic Quantum Mechanics,* McGraw-Hill (New York) 1964; *Relativistic Quantum Fields,* McGraw Hill (New York) 1965.
[27]   C. J. Horowitz and B. D. Serot, Nucl. Phys. **A368**, 503 (1981).
[28]   B. C. Clark, S. Hama, R. L. Mercer, L. Ray and B. D. Serot, Phys. Rev. Lett. **50**, 1644 (1983); B. C. Clark *et al.*, Phys. Rev. **C28**, 1421 (1983).
[29]   J. A. McNeil, J. R. Shepard and S. J. Wallace, Phys. Rev. Lett. **50**, 1439, 1443 (1983).
[30]   R. A. Freedman, Phys. Lett. **71B**, 369 (1977).
[31]   B. D. Serot, Phys. Lett. **107B**, 263 (1981).
[32]   J. C. Collins and M. J. Perry, Phys. Rev. Lett. **34**, 1353 (1975).
[33]   G. Baym and S. A. Chin, Phys. Lett. **62B**, 241 (1976).
[34]   S. A. Chin, Phys. Lett. **78B**, 552 (1978).
[35]   J. Kuti *et al.*, Phys. Lett. **95B**, 75 (1980); **98B**, 199 (1981).

ELECTRON SCATTERING FROM NUCLEI AT SEVERAL GeV

R. G. Arnold

The American University   and   Stanford Linear Accelerator Center
Washington, DC 20062   Stanford, CA 94305
USA   USA

1. INTRODUCTION

A basic goal of physics is to uncover and then explain the layers in the structure of matter. In the last decade evidence has accumulated from experiments at GeV energies that nucleons and mesons are composed of a new substructure containing charged point-like quarks interacting by gluon exchange. This is a peculiar substructure. The quarks and gluons carry a new quantum number called color, but they have never been observed isolated in the laboratory. The objects observed in real experiments appear to be combinations of quarks bound into color neutral particles or color singlets. Quantum chromo-dynamics (QCD) is the gauge theory of colored quarks and gluons invented to describe this substructure. A central task of modern nuclear physics is to understand how the structure and interactions of nucleons arise from the interactions among quarks.

Nuclear physics is today in a rather peculiar state. On the one hand, we are confident that the quark substructure exists and that ultimately all of nuclear physics must be understood to be governed by quark dynamics. On the other hand, much of what we see in nuclear experiments can be fairly well understood using nucleon and meson degrees of freedom. So the question arises: where do we see quarks in nuclei?

What people often mean by that questions is: where is perturbative QCD applicable to describe nuclear structure? The interactions between quarks have the odd feature that at large distances (of the order of nucleon size) they become very strong and complex, while at short distances they become weaker and things get simpler. People are attracted to start with the simpler regime where useful results can be calculated in perturbation expansions. The basic idea of perturbative QCD is that at some sufficiently high energy or momentum transfer (how high is a big open question) in a particular reaction, the primary interaction is a hard scattering process involving a few quarks and gluons. The dynamics of that hard process determines the shape (versus energy, momentum, or angle) of the overall cross section, while the magnitude of the cross section is determined by the relative probability for having a particular configuration of quarks and gluons in the initial and final states. One way to 'see' quarks in nuclei is to find kinematic regions where the shape of the cross sections are determined from interactions among a few quarks at short distance.

Ideally, we would like to conduct quark-quark experiments to measure quark interactions. Since that is not possible, we must alter somewhat our old standards for hard experimental evidence and work by inference from many pieces of indirect evidence and be guided by strong theoretical prejudices. Two examples of perturbative QCD predictions are for scaling in deep inelastic lepton scattering, and power law fall off for hadronic form factors. Experimental verification of these predictions using nuclear targets would provide strong evidence for the role of quarks in nuclei.

These lectures are primarily about experimental results, limitations, and hopes and plans for new experiments. My goal is to give an impression of where we stand in a few cases in the process of gathering evidence for or against perturbative QCD. The experiments I describe are electron scattering measurements from nucleon and nuclear targets using electrons in the GeV energy range. The electromagnetic probe at high energies is one of the best ways to study quark structure of nuclei. The electromagnetic interaction is clean and the kinematic range available in feasible experiments offers resolution for the probe that ranges from nuclear size down to subnucleon size.

Someday if it is clear that perturbative QCD is applicable in a certain kinematic region in some reaction, then we can venture to the next level of question such as:

   a) What is the perturbative structure of the hadrons, i.e., what are the quark-gluon wave functions?

   b) Can we use nuclei to learn about features of quark dynamics not available in individual hadrons? Eventually nuclear physicists will have to face the more difficult problem, how to understand the long range part of the nuclear interaction in terms of nonperturbative quark dynamics. This is clearly a very difficult task, but that is where we are heading.

These lectures are divided into two parts. The first describes elastic scattering and is limited to experiments on the proton, neutron and the deuteron. The second describes inelastic scattering, with an emphasis on deep-inelastic lepton scattering from nuclei and the recently discovered distortion of quark momentum distributions in nuclei, called the EMC effect.

2. ELASTIC SCATTERING

Elastic electron scattering from nucleons and nuclei is one of the basic reactions for learning about the structure of hadronic matter. Elastic experiments at low energy revealed that the proton and nuclei are not pointlike. Early models of the nucleon pictured a core surrounded by a meson cloud. Following the discovery of the quarks, nucleons were imagined to be made of three pointlike valence quarks accompanied by a cloud of quark-antiquark pairs. The fractionally charged quarks were found to carry only about half the momentum of the nucleons, but they carry all the charge. An electron scatters from a nucleon or a nucleus by interacting with the electromagnetic current of the quarks, and, therefore, elastic (and inelastic) electron experiments probe the distribution of quarks in the target.

Perturbative QCD makes definite predictions for elastic form factors. These predictions began with a series of dimensional scaling laws[1] for the Q^2 dependence of exclusive scattering processes based upon quark counting and the scale invariance of the quark-quark interaction at short distances. The cross section for elastic electron scattering at momentum transfer Q^2 is proportional to the probability for transferring a collection of quarks with momentum P_i in the initial state to $P_i + Q$ in the final state, leaving the target bound as in Fig. 1. The transfer of momentum from the electron to all the quarks in the

(a)

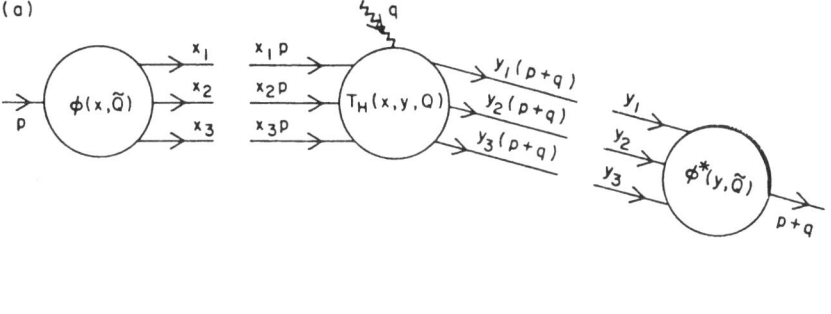

(b)

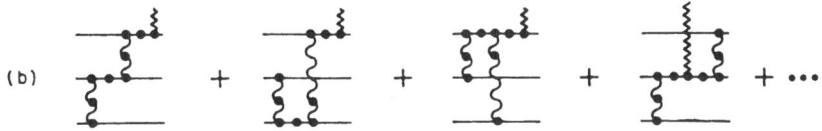

Fig. 1   Diagram for elastic electron-nucleon scattering in lowest order perturbative
QCD. a) Factorization of the nucleon form factor at large Q^2 into parts
depending on the initial and final wave functions ϕ and the factor T_H containing the hard scattering between the three valence quarks, b) leading order hard
scattering diagrams.

target takes place by gluon exchange between the quarks. At large Q^2 the contribution to
elastic scattering from components of the target wave function containing extra quark-
antiquark pairs is predicted to fall away faster with Q^2 than the contribution from scattering on the valence quarks alone. The form factor falls with a power of Q^2 determined by
the number of valence quarks in the target. The more quarks there are whose momentum
must be changed to $P_i + Q$, the faster the form factor falls with increasing Q^2. The
dimensional scaling laws give only the shape of the form factors. The absolute value is
governed by the relative probability of finding particular quark momentum distributions
in the target wave function, and eventually that also must be calculated.

The first step is to determine if and where the perturbative QCD predictions for
hadron form factors are correct. The next is to deduce quantitative information about the
wave functions of the target from the form factor data. Let us look at the evidence.

2.1 The Proton

The cross section for elastic electron scattering from a nucleon can be written

$$\sigma = \sigma_M \left[\frac{G_E^2 + \tau G_M^2}{1 + \tau} + 2\tau G_M^2 \tan^2(\theta/2) \right] \tag{1}$$

where $\tau = Q^2/4m^2$ and G_E and G_M are the electric and magnetic form factors. They are in
turn related to the Dirac and Pauli form factors by

$$G_E = F_1 - \tau F_2 \tag{2}$$

$$G_M = F_1 + F_2 . \tag{3}$$

At some sufficiently large Q^2 where the electron scatters predominantly from the
three valence quarks in the nucleon, as in Fig. 1, elastic scattering can be viewed as
electron-quark scattering followed by two hard quark-quark scatterings.

The dimensional scaling prediction[2] for the nucleon form factor is

$$F_1 = \frac{C_1}{Q^4} \tag{4}$$

$$F_2 = \frac{C_2}{Q^6} . \tag{5}$$

The Dirac form factor F_1 decreases with a power $1/Q^4$ because there are two off-shell quark propagators in Fig. 1 each falling with $1/Q^2$. The Pauli form factor F_2 falls with one additional power of Q^2 because it involves an extra helicity flip of one of the quarks. The normalization constants C_1 and C_2 are determined by the nucleon wave function.

If we consider only the valence quarks, the nucleon spin-isospin wave function can be written schematically as

$$|p> = \frac{1}{\sqrt{18}} \qquad (2|u\uparrow u\uparrow d\downarrow> + 2|u\uparrow d\downarrow u\uparrow> + 2|d\downarrow u\uparrow u\uparrow> + \text{antisymmetric parts})$$

$$|n> = u \rightarrow d . \tag{6}$$

The constants C_1 and C_2 would be determined by a sum over the squares of the charges of the struck quarks weighted by the probability for the struck quark to absorb momentum Q in the initial state and remain bound in the final state. In the dimensional scaling predictions the constants C_1 and C_2 are either not given or a portion of them is estimated from a guess at the relative contribution of various quark charges. We will see below some predictions for the ratio of the neutron and proton from factors using these ideas.

A complete prediction for the form factors requires knowledge of the nucleon quark wave functions. The first explicit perturbative QCD calculation of nucleon form factors were given by Brodsky and Lepage[3]. They evaluated the contributions from the four first order QCD hard scattering diagrams as in Fig. 1. For large Q^2 where the contribution from F_2 has died away and G_M is determined by F_1, they predict

$$G_{Mp} = \frac{\alpha_s^2(Q^2)}{Q^4} \ \phi(Q^2) . \tag{7}$$

The factor $1/Q^4$ follows from the assumption that the perturbative expansion is valid and the scattering is from the three valence quarks only. The two powers of $\alpha_s(Q^2)$ arise from the strong coupling at the quark-gluon vertices. The factor $\phi(Q^2)$ contains the dependence on the initial and final wave functions.

This first order QCD prediction suggests that the form factor G_M, or F_1, may not fall exactly like $1/Q^4$, but may have an additional Q^2 dependence due to the change of the coupling constant α_s with Q^2. The QCD coupling constant is not really constant. It is predicted to vary with Q^2 like $1/\ln (Q^2/\Lambda_{QCD}^2)$, where Λ_{QCD} is the only parameter of QCD. If elastic electron-proton scattering at measurable Q^2 is determined by first order QCD processes, and if uncertainties from lack of knowledge of the wave functions are small, perhaps we could observe the change of the coupling constant $\alpha_s(Q^2)$ with Q^2, if accurate data over a range of Q^2 were available.

Brodsky and Lepage[3] made predictions for G_{Mp} using some models for nucleon wave functions of the type $\phi \sim (x_1 x_2 x_3)^n$ where x_i is the fraction of the nucleon momentum carried by one of the valence quarks, and n is a parameter. The models they used are symmetric about $x_1, x_2, x_3 = 0.3$, and ϕ falls off to zero when any x_i gets near one. It seems reasonable that the nucleon momentum would be more or less shared equally by all the valence quarks. The Brodsky-Lepage predictions for G_{Mp} along with the previous data are shown in Fig. 2. The data are consistent with a nearly constant value for Q^4G_{Mp} for Q^2 greater than 10 $(GeV/c)^2$, although the precision of the data above 10 $(GeV/c)^2$ is not good enough to rule out a substantial deviation from $1/Q^4$ behavior. The absolute normalization of the theoretical predictions is not determined; only the shape versus Q^2 is calculated. The curves in Fig. 2 are arbitrarily normalized to the data at Q^2 around 8 $(GeV/c)^2$.

Recently there has been a huge debate[4] about the applicability of perturbative QCD to exclusive processes in the Q^2 range attainable in real experiments. Isgur and Llewellyn-Smith[5] have argued that the proton form factor will get large contributions up to very large Q^2 from processes involving soft components of the wave functions, components containing quark-antiquark pairs in addition to the valence quarks. For proton wave functions symmetric in $x_1x_2x_3$, there are sharp cancellations between certain terms in Fig. 1 that suppress the first order QCD contributions to G_{Mp}. The only way for processes in Fig. 1 to be important at low Q^2 is for the wave functions to be asymmetric in $x_1x_2x_3$.

A series of theoretical investigations[6,7] of the nucleon form factors have recently appeared using the Brodsky-Lepage perturbative QCD approach, but with wave functions that are asymmetric in momentum space. Cherniak and Zhitnitsky[6], have employed a QCD sum rule technique to determine asymmetric wave functions. They evaluate G_{Mp} and G_{Mn} and find that the lowest order QCD terms are large at low Q^2 and their results are close to the data.

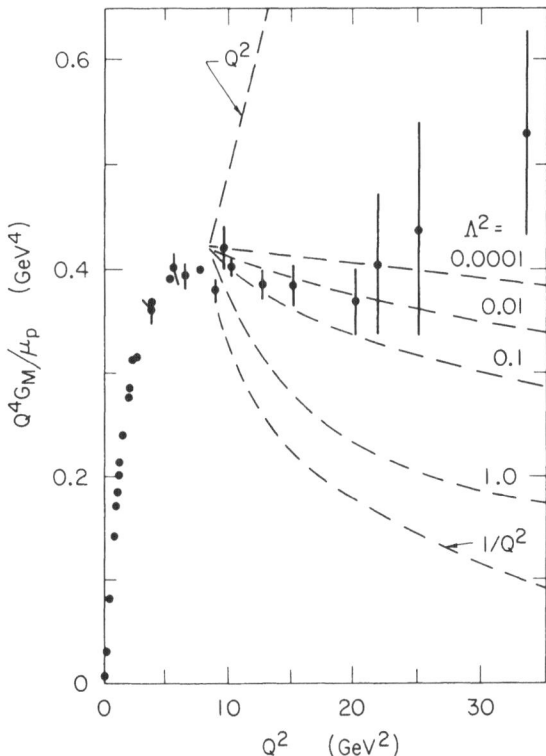

Fig. 2   Previous world data for the proton magnetic form factor G_{Mp} times Q^4 plotted versus momentum transfer Q^2. The curves are perturbative QCD predictions from Ref. 3.

Before describing the results of a recent experiment let me restate the physics questions at hand. The origianl motivation for a new measurement of ep elastic scattering, stimulated by the Brodsky-Lepage calculation in Fig. 2, was:

A) to verify that $Q^4 G_{Mp}$ is nearly constant at large Q^2 to establish where perturbative QCD becomes applicable;

B) to look for small variations of the data from $1/Q^4$ that might indicate a change of $\alpha_S(Q^2)$ with Q^2.

These goals assume that the theoretical uncertainty from lack of knowledge of the wave functions is small. We now think that the questions are more complicated because of the Isgur and Llewellyn-Smith argument, and that the proton wave function plays a crucial role in the interpretation of such data. This has the disadvantage that what was once viewed as a potentially clean test of QCD may not be so clean. However, it has the advantage that such data may be more important for establishing a rather surprising result, that the valence quarks in a nucleon do not share the nucleon momentum more or less equally. An asymmetry in the momentum distribution, if it exists, will have profound effects in other areas of nucleon and nuclear structure physics.

The experiment[8] I describe briefly here was performed recently at SLAC. The data taking was completed in May 1984, and results are nearly ready for publication. This experiment used the high energy, high intensity electron beam from the SLAC linac and the facilities of End Station A to measure ep elastic scattering in a single arm spectrometer.

The major equipment needed for this experiment was already in existence at SLAC. With some improvements to the detectors and the target, and by making a long run in one experiment dedicated to measuring ep elastic scattering, it was possible to significantly improve the quality of the data out to Q^2 of 31 $(GeV/c)^2$. The major improvement was the substitution of a new set of multiwire proportional chambers in place of the old scintillator hodoscopes used for track measurements in the 8 GeV/c spectrometer. The new chambers made it possible to measure low elastic cross sections in the presence of large random background. A nitrogen filled gas Cherenkov counter and a lead glass array were used for triggering and for identifying electrons. A schematic of the detector package is

GAS CHERENKOV COUNTER   WIRE CHAMBERS   LEAD GLASS SHOWER COUNTER

SCINTILLATORS

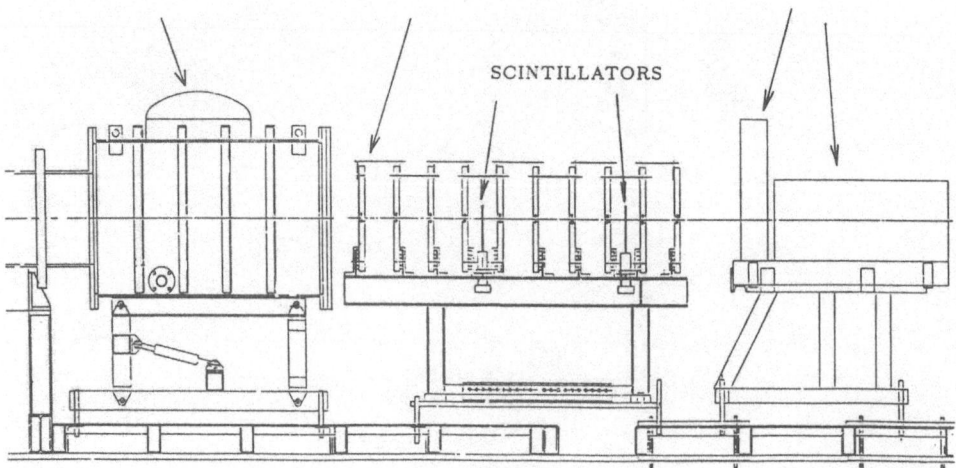

Fig. 3   Electron detectors in the SLAC 8 GeV/c spectrometer used in experiments E136[8] and E139[43].

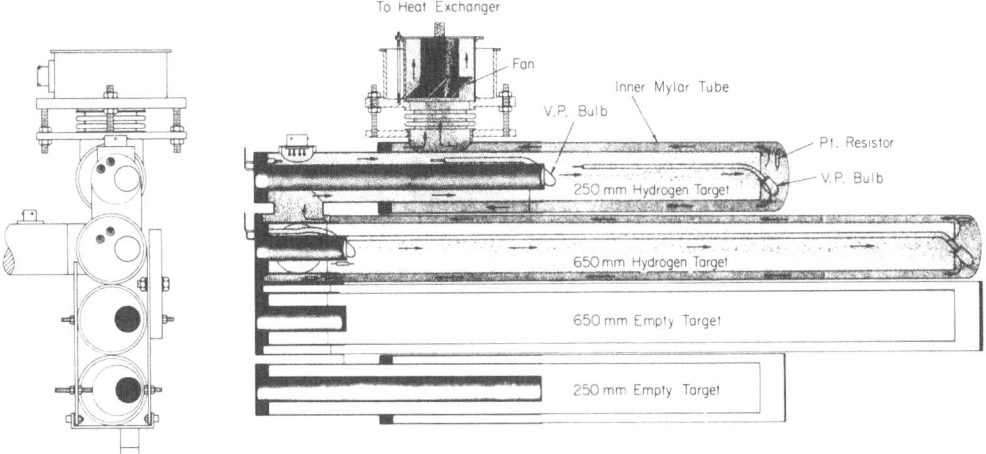

Fig. 4   Liquid hydrogen target system[9] used in SLAC experiment E136 on electron-proton elastic scattering[8].

shown in Fig. 3. This detector system was required to pick out a signal of high energy elastic electrons at the rate of a few counts per day under conditions where there were as many as five or six hit wires per plane per beam pulse from soft room background.

Another reason for the success of this experiment was the availability at SLAC of the expertise and technology for constructing long liquid hydrogen targets able to withstand the high beam power[9]. The targets in this experiment were cylindrical in shape, as shown in Fig. 4, with the axis along the incident beam. Two lengths were used, one 25 cm and one 65 cm long, as well as empty targets for measuring scattering from the target end caps. The long target was designed with tungsten shields to obscure the end caps from view by the spectrometer and eliminate end cap background. The high powered electron beam deposits up to 250 watts in the liquid hydrogen. To keep the liquid cool and to prevent local boiling along the beam path, the liquid was circulated in a closed loop by a large submerged fan through the cells and to a heat exchanger at the rate of 2 m/sec..

Three spectra of the counts versus missing mass for scattered electrons are shown in Fig. 5. The elastic peak and its radiative tail are the large signal in the region $0.7 \leq W^2 \leq 1.5$ GeV^2. The background from end caps and slit scattering was negligible. The signal decreased rapidly with increasing Q^2. The spectrum in Fig. 5b contains 20 events in the elastic peak at $Q^2 = 31.2$ (GeV/c)^2 and was obtained in approximately 20 days of running the full SLAC beam into 65 cm of liquid hydrogen. This indicates the realistic upper limit on the Q^2 range attainable with the present, and perhaps ever, available experimental apparatus.

Preliminary results for the magnetic form factor G_{Mp} scaled by Q^4 are shown in Fig. 6, with error bars artificially increased pending final determination of the cross sections. The new data are in good agreement with previous results at low Q^2, and the precision is significantly improved in the range above $Q^2 = 15$ (GeV/c)^2. This is in rough agreement with the prediction from QCD shown in Fig. 2 for values of Λ_{QCD} in the range of 50 to 200 MeV. The new data have a shape similar to the theoretical curves from Cherniak and Zhitnitsky[6] in Fig. 7. The deviation from pure $1/Q^4$ behavior in the theoretical result is primarily due to the change in $\alpha_s(Q^2)$ with Q^2.

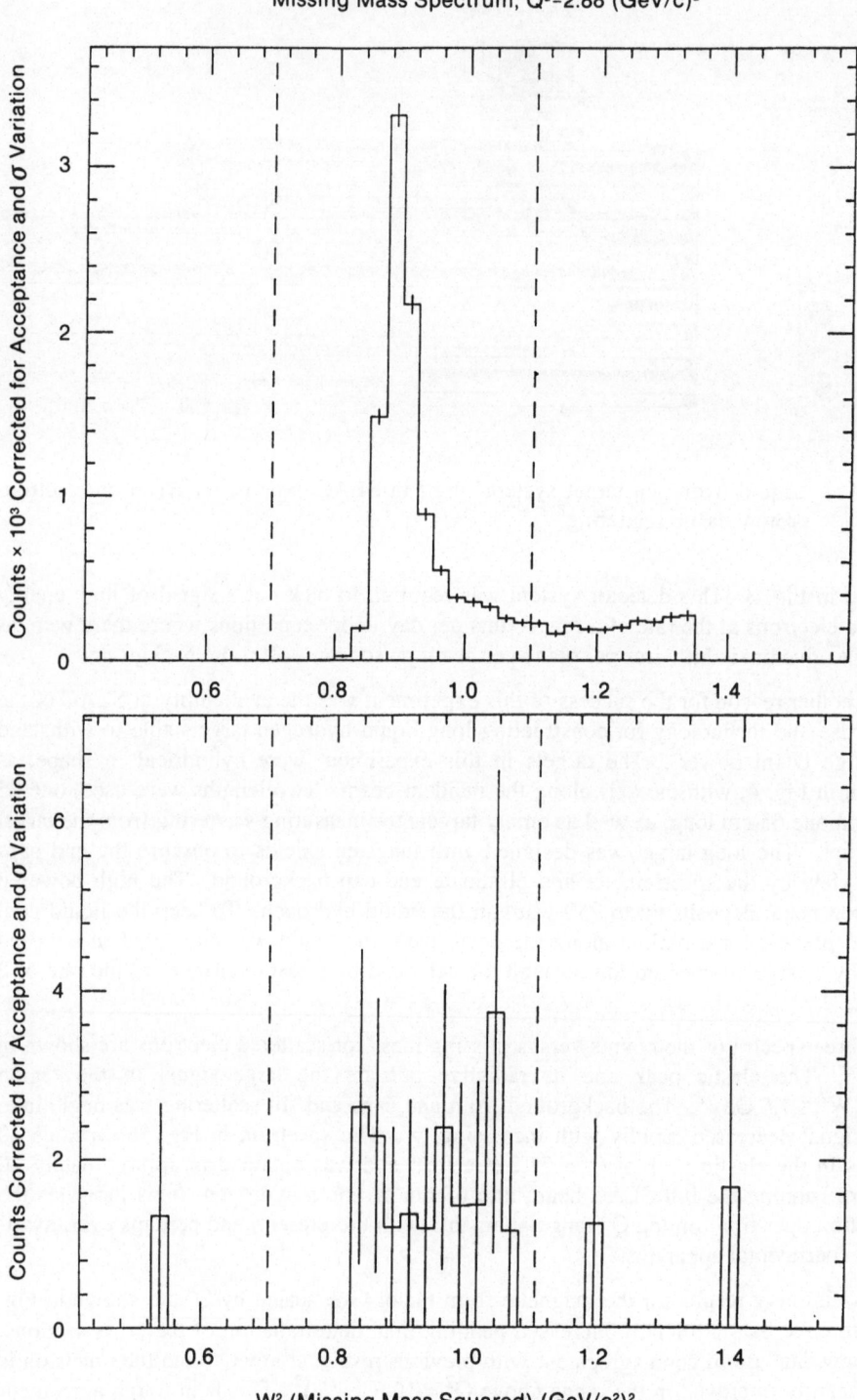

Fig. 5   Spectra of electrons from ep elastic scattering in SLAC experiment E136[8].
versus missing mass: Upper $Q^2 = 2.8$; Lower $Q^2 = 31.2$ $(GeV/c)^2$.

It is premature at this time to make statements about quantitative tests of QCD or try to extract Λ_{QCD} from such data, pending the outcome of the debate over the applicability of perturbative QCD at this range of Q^2. It is however clear that this more precise measurement of G_{Mp} over a large range of Q^2 in one experiment will provide important constraints on our knowledge of proton structure.

2.2 The Neutron

Electron scattering from the neutron can in principle be used to measure the neutron form factors G_{En} and G_{Mn} from which we can learn about the quark structure of the neutron. Unfortunately, there are no free neutron targets and measurements of neutron structure must be done using deuterium targets. During the 1950's and 1960's several measurements[10] were made of electron-deuteron quasi elastic scattering from which neutron form factors were extracted. The function G_{En} is small, and may be zero, but it is poorly determined from the data. The results for G_{Mn} are better known, but the old data[11] ended with large errors at $Q^2 = 3.7$ $(GeV/c)^2$. The previous measurements were usually compared to predictions from vector dominance models and were used to limit the number and the mass of mesons contributing to the virtual photon interaction with the nucleon[12].

Following the discovery of quarks and the development of QCD, the neutron form factors were re-examined as a possible place to test the theory. The cross section for electron-neutron scattering is given by Eq. 1. It is useful to look at the theoretical prediction for the ratio of neutron and proton cross sections, especially for comparison with experiments where deuterium is the target. For small values of the scattering angle, the term proportional to $\tan^2(\theta/2)$ can be neglected and we can write

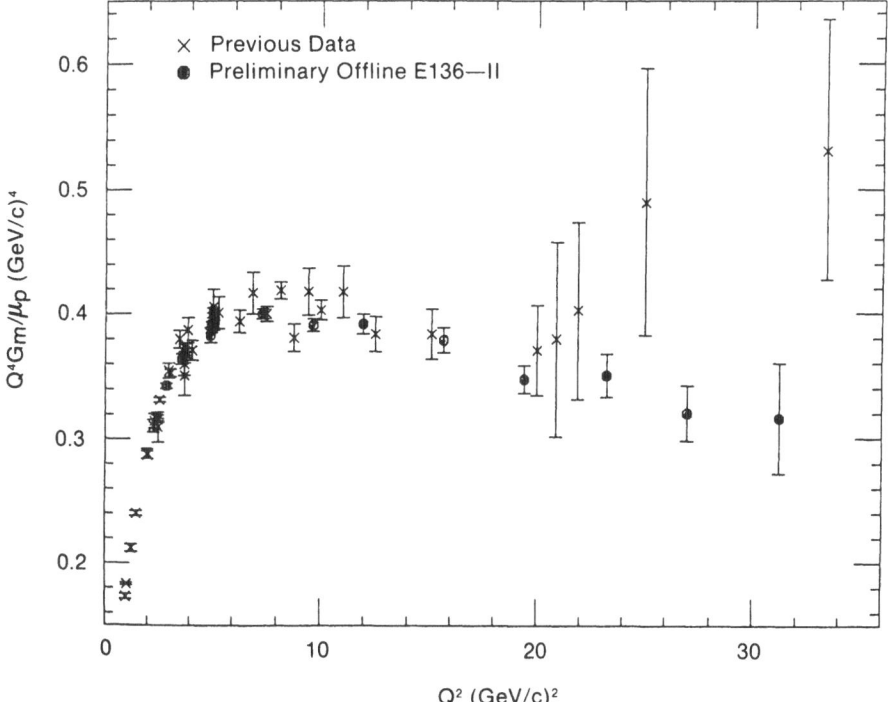

Fig. 6   Preliminary results for G_{Mp} from SLAC experiment E136[8].

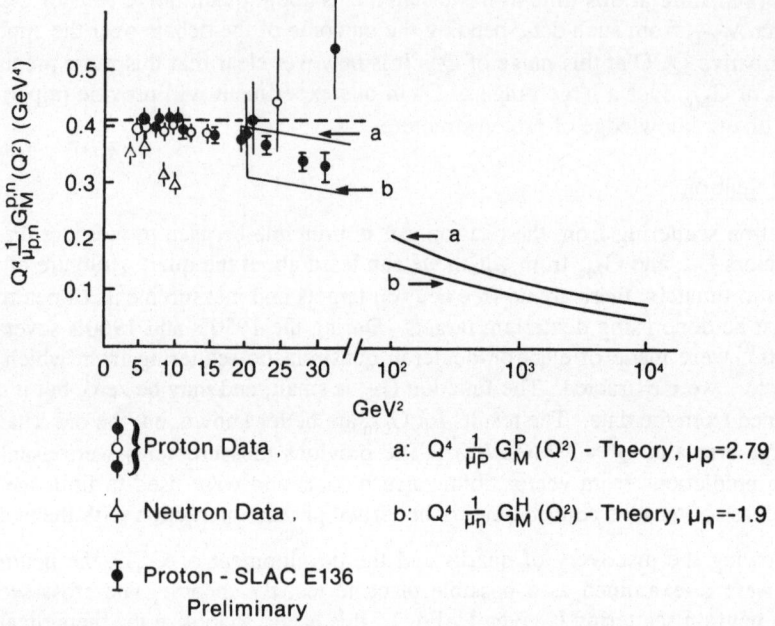

Fig. 7   Proton and neutron form factors. The curves are perturbative QCD results
from Chernyak and Zhitnitsky (Figure from Ref. 6). Preliminary results from
E136 in Fig. 6 are also plotted.

$$\frac{\sigma_n}{\sigma_p} = \frac{G_{En}^2 + \tau G_{Mn}^2}{G_{Ep}^2 + \tau G_{Mp}^2} \tag{8}$$

or

$$\frac{\sigma_n}{\sigma_p} = \frac{F_{1n}^2 + \tau F_{2n}^2}{F_{1p}^2 + \tau F_{2p}^2} . \tag{9}$$

From the dimensional scaling arguments the form factors F_1 and F_2 are predicted to fall
with powers of Q^2 as in Eq. 2. There are several possible theoretical predictions for the
ratio σ_n/σ_p depending upon details of the nucleon wave functions that determine the constants C_1 and C_2.

Case (a) The neutron wave function (and presumably the proton wave function also)
is spatially symmetric, so the neutron is neutral everywhere in space. In that case[2]

$$F_{1n} = 0$$

$$F_{2n} = \frac{C_2}{Q^6} \tag{10}$$

and

$$\frac{\sigma_n}{\sigma_p} \rightarrow \left[\frac{C_{2n}}{C_{1p}}\right]^2 \frac{4m}{Q^2} \tag{11}$$

and the ratio σ_n/σ_p falls with Q^2.

Case (b) The neutron wave function is asymmetric and $F_{1n} \neq 0$. The ratio σ_n/σ_p becomes a constant.

$$\frac{\sigma_n}{\sigma_p} \rightarrow \left[\frac{C_{1n}}{C_{1p}}\right]^2 . \tag{12}$$

If we use a model of the spin-isospin nucleon wave function with valence quarks only, represented schematically by Eq. 6, it is possible to compute the relative size of constants C_{1n} and C_{1p} from a sum over the squares of the charges on the struck quarks. If the quark wave function is asymmetric in space, and, therefore, also in momentum, it is possible that one or the other type of quark is favored to be near $x = 1$ and have a relatively large probability for absorbing a virtual photon and recoiling bound in the final state. If the struck quark has the same flavor as the nucleon, i.e., if it is most likely that the virtual photon connects to an up quark in the proton and a down quark in the neutron, then

$$\left[\frac{C_{1n}}{C_{1p}}\right]^2 = \frac{1}{4} . \tag{13}$$

Alternatively, it might be more important for the struck quark to have the same helicity as the nucleon, because contributions from quarks of opposite helicity would involve an extra spin flip and be suppressed by an extra power of Q^2. This would give

$$\left[\frac{C_{1n}}{C_{1p}}\right]^2 = \frac{3}{7} = 0.428 . \tag{14}$$

Therefore, it might be possible, if data were available at Q^2 large enough for the dimensional scaling ideas to be applicable, that a simple ratio of neutron and proton cross sections would give hints about the symmetry properties of the nucleon wave functions.

This was the situation in 1980 when it became apparent that a new measurement of the neutron magnetic form factor was needed. Using the facilities at SLAC, the method of quasi elastic scattering could be pushed to higher Q^2 and the precision of the data at lower Q^2 could be significantly improved. A new measurement was made[13], and I describe here the experiment and results briefly.

The method relies on high statistics measurement of quasi elastic scattering from deuterium in conjunction with ep elastic and inelastic scattering to measure ep structure functions for subtraction from the ed data. Measurements were made at a fixed scattering angle of 10 degrees using the SLAC 20 GeV/c spectrometer. Data were taken with incident beam energies ranging up to 20 GeV at five settings corresponding to electron-neutron elastic scattering at Q^2 of 2.5, 4, 6, 8, and 10 $(GeV/c)^2$.

The neutron from factors were extracted using a few assumptions about the cross sections, and using various deuteron wave function models for smearing the proton data. The primary assumption is that the single arm inelastic ed cross section as a function of the missing mass W^2 is assumed to be of the form

$$\sigma_d(W^2) = <\sigma_p^{el}> + <\sigma_n^{el}> + <\sigma_p^{in}> + <\sigma_n^{in}> \tag{15}$$

where σ_p^{el} and σ_n^{el} are the elastic proton and neutron cross sections, and the brackets indicate these are smeared over W^2 due to the Fermi motion. The terms σ_p^{in} and σ_n^{in} represent the inelastic cross sections, which also get smeared. Another assumption is that the elastic form factors for bound nucleons are the same as for free onshell nucleons and depend only on Q^2. This is expected to be a good approximation for the loosely bound nucleons near the quasi elastic peak, the region of the data that determines the results.

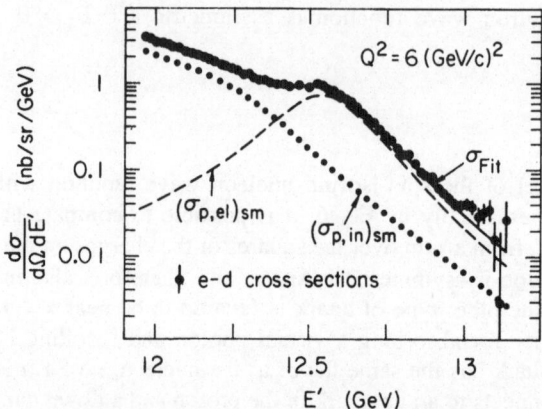

Fig. 8   The cross section for ed scattering at incident energy E = 15.742 GeV versus
           scattered electron energy E' at 10 degrees[13]. The dashed (dotted) curve is the
           smeared elastic (inelastic) ep cross section. The solid curve is the result of a fit
           of the other curves to the data as described in the text.

Since there is no experimental information on the inelastic cross sections for the free neutron, we assume that the shape of σ_n^{in} versus W^2 is the same as for the proton σ_p^{in}.

To extract σ_n^{el} from data on $\sigma_d(W^2)$, we use our measured proton elastic and inelastic
cross sections and models for deuteron smearing to construct smeared cross sections.
Then the ratio of neutron and proton elastic cross sections is determined from fits to the
deuteron quasi elastic data using the function

$$\sigma_d(W^2) = \left[1 + \frac{\sigma_n^{el}}{\sigma_p^{el}} \right] <\sigma_p^{el}> + C_{in}<\sigma_p^{in}> . \tag{16}$$

The ratio $\sigma_n^{el}/\sigma_p^{el}$ and the inelastic normalization constant C_{in} are parameters of the fit.

An example of the data and the contributions to the fit at $Q^2 = 6$ (GeV/c)^2 are shown
in Fig. 8. The main features of the method are illustrated here. The spectra are composed of a broad quasi elastic peak on a smooth background from smeared inelastic
scattering. The total cross section near the peak is mostly due to the smeared elastic
cross sections. There is some variation in the shapes of the smeared proton spectra in the
tails of the quasi elastic peak when different deuteron models are used. The fit parameters $\sigma_n^{el}/\sigma_p^{el}$ are primarily constrained by the data points near the quasi elastic peak and
are not much affected by the tails of the peak.

This experiment has pushed the single arm quasi elastic method for neutron form factor determination to the practical upper limit in Q^2 set mainly by the small neutron cross
section compared to the proton elastic and the inelastic background. At higher Q^2 the
quasi elastic peak disappears into the inelastic background.

The results from this experiment are displayed in Figs. 9 and 10. The experimental
errors include both statistical errors and systematic errors from model dependence of the
analysis procedure. Fig. 9 shows the form factor, essentially G_{Mn} in this case, multiplied
by the power of Q^2 from the dimensional scaling prediction, along with similar data for
the pion, proton and the light nuclei. On the log scale of this plot the neutron data

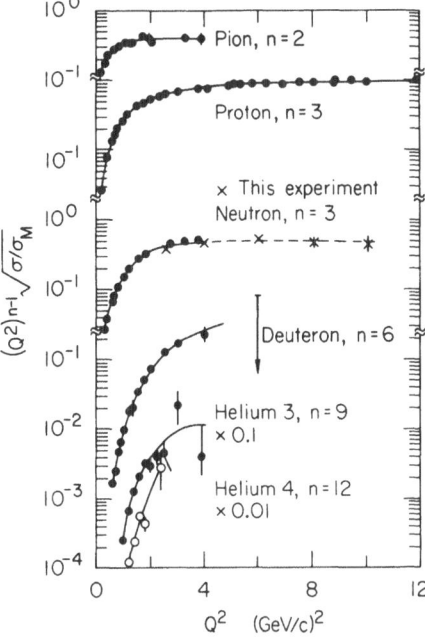

Fig. 9   Quark dimensional scaling of the elastic form factors of the pion, nucleons, and three light nuclei. The power n is the number of elementary quark constituents.

become nearly constant versus Q^2 as it should if G_{Mn} falls like $1/Q^4$.

A close examination of Fig. 9 shows that $Q^4 G_{Mn}$ is not constant versus Q^2, and is slightly different from the proton curve. This can be seen more clearly in Fig. 10 where the ratio σ_n/σ_p is plotted. The new data, with much reduced errors, overlap with the old

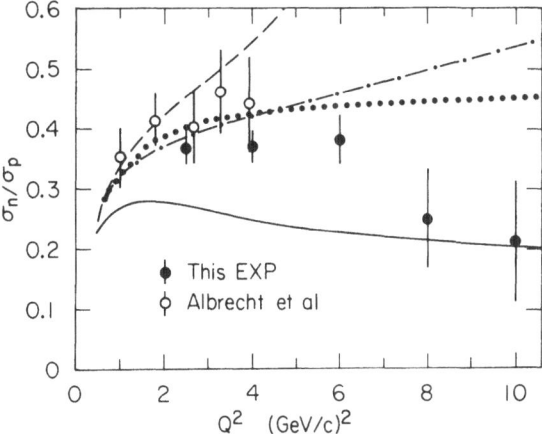

Fig. 10   The ratio σ_n/σ_p as a function of Q^2. Data are from Refs. 11 and 13. The dashed and solid curves are vector-dominance models from Ref. 12. The dotted curve is form factor scaling Eq. 18 with $G_{En} = 0$. The dashed-dotted curve is the dipole law for G_{Mn} with $G_{En} = 0$ and σ_p from experiment.

data at a value around 0.37 out to $Q^2 = 4$ (GeV/c)^2, and then the ratio descends to a value around 0.21 ± 0.10 at $Q^2 = 10$ (GeV/c)^2. These results are in stark contrast to the vector dominance models and naive form factor scaling. They are also consistent with the prediction that σ_n/σ_p falls with Q^2, as it would if the quark wave functions are symmetric and $F_{1n} = 0$ (Case (a) above). It is also possible that the data are approaching a constant value near 0.2 at high Q^2. This would be consistent with the prediction that $\sigma_n/\sigma_p \rightarrow 1/4$ when the quark wave function is asymmetric and the struck quark has the same flavor as the nucleon. The data are inconsistent with the value $\sigma_n/\sigma_p \rightarrow 3/7 = 0.429$ predicted when the struck quark has the same helicity as the nucleon.

The neutron form factors have also been calculated by Cherniak and Zhitnitsky[6] and their results are displayed in Fig. 7. The new data do not extend out to large Q^2 where the theoretical prediction is given. However, a straightline extrapolation of the neutron data to higher Q^2 would give a curve falling faster with Q^2 than the QCD prediction. Perhaps this is an indication that up to $Q^2 = 10$ (GeV/c)^2 there are still large contributions from soft components of the wave function that have not yet died away.

More data for the neutron at higher Q^2 would be very important for testing these ideas. Unfortunately, the single arm quasi elastic technique does not work at higher Q^2. One possible method for reducing the background that has been used in previous experiments at lower Q^2 is to detect the recoil neutron in coincidence with the scattered electron. This requires high energy (E > 10 GeV) to reach high Q^2 and high duty factor to make neutron detection feasible.

2.3 The Deuteron

The deuteron has three electromagnetic form factors - the charge, quadrupole and magnetic - G_C, G_Q, and G_M. The cross section for electron deuteron elastic scattering has the form

$$\sigma_d(Q^2) = \sigma_M [A(Q^2) + B(Q^2)\tan^2(\theta/2)] . \tag{17}$$

The structure function $A(Q^2)$ measured at forward angles is a combination of the squares of all three form factors. The $B(Q^2)$ function depends on G_M only, and it can be extracted from the cross section by measuring at backward angles. The previous data[14] at the highest Q^2 is a measurement of $A(Q^2)$ out to 4 (GeV/c)^2. The previous data[15] for $B(Q^2)$ extend only up to $Q^2 = 1$ (GeV/c)^2.

There are many ways to view deuteron structure depending upon your stating point in physics. Traditionally the deuteron form factors are calculated in the nonrelativistic impulse approximation as the sum of scattering from the moving neutron and proton. It is expected that the nonrelativistic impulse approximation does not contain the whole story, and that there will be modifications at high Q^2 from scattering on meson exchange currents or from relativisitic effects. In the framework of the traditional models the data can be viewed either as a test of the product of deuteron wave functions and nucleon form factors, or as a search for modifications to the simple impulse picture from higher order effects.

On the other hand, if you are interested in looking for quarks in a nucleus, then you would compare the data to the dimensional scaling predictions and look for power law behavior. One of the crucial tests for the applicability of the dimensional scaling ideas is that the form factor must fall smoothly with increasing Q^2, and there can be no diffraction features. Such features would be determined by relatively long range properties of the nucleon interaction (of the order of the nucleon size), not by the hard scattering off the valence quarks interacting at short distance.

Unfortunately, by a conspiracy of nature, the beautiful diffractive shapes predicted for the individual form factors G_C, G_Q, and G_M in most traditional models are completely merged into a smooth curve when squared and added together in $A(Q^2)$. Therefore, the crucial tests of models from the location and size of diffractive features are not possible from data on $A(Q^2)$. We need separate experimental determination of the individual form factors over a range of Q^2. The G_C and G_Q can only be extracted if deuteron polarization in either the initial or final state is measured. The G_M form factor can be measured directly by scattering electrons at backward angles. The existing data for $A(Q^2)$ and $B(Q^2)$ and three dramatically different predictions for $B(Q^2)$ are shown in Fig. 11.

Motivated by the desire to measure the deuteron and the other nuclear form factors out to large momentum transfer, we proposed[16] that a new electron injector be added to the SLAC linac to produce high current beams in the energy range 0.5 to about 5 GeV. SLAC works well when producing beams from 5 to 20 GeV, but the beam intensity and stability decrease at lower energy. The nuclear physics experiments we have in mind require high intensity beams at lower energy because the electron scattering cross sections are low. The new injector has been constructed, and has operated for about one year. It inserts electrons into the linac at a point 1/6 of the length from the downstream end, and it produces an electron beam with intensity a factor of 50 to 100 times as high as that available from the full linac when operated below 5 GeV. This beam makes possible a whole new range of electromagnetic nuclear physics experiments.

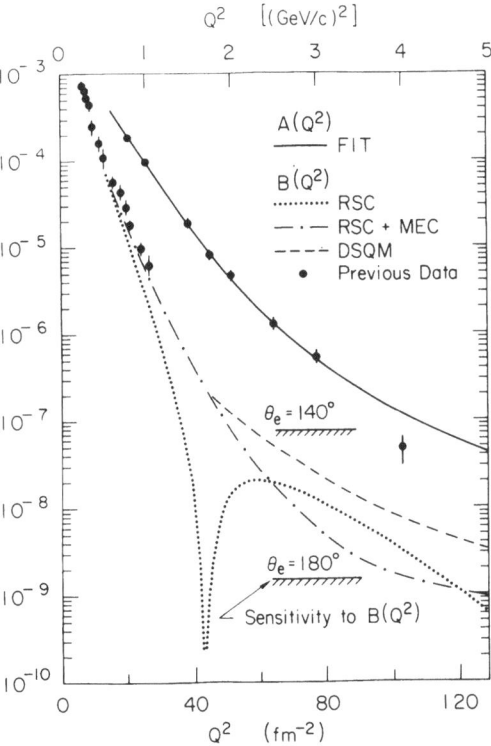

Fig. 11   Deuteron form factors $A(Q^2)$ and $B(Q^2)$. The theoretical curves for $B(Q^2)$ are: RSC- impulse approximation using Reid soft core wave functions; RSC+MEC-Reid soft core plus meson exchange currents; DSQM-dimensional scaling quark model[2] arbitrarily normalized at $Q^2 = 1.75$ (GeV/c)^2. The sensitivity for measurement of $B(Q^2)$ at large scattering angles is indicated.

The new injector is the centerpiece of a new research program[17] at SLAC, called Nuclear Physics at SLAC (NPAS) that uses the facilities of End Station A for a program of experiments. To date two experiments[18,19] have taken data. The first measured inclusive inelastic electron scattering from a series of nuclei[18], and the second measured ed elastic and inelastic scattering with the electrons detected at angles around 180 degrees[19]. The deuteron experiment was performed in a specially constructed double arm spectrometer system in which the scattered electrons were detected in coincidence with the recoiling nuclei. The data taking for the first phase ended in July 1984, and analysis is in progress.

The new ed elastic data points are at Q^2 of 1.2, 1.5, 1.75, 2.0, 2.25, and 2.5 $(GeV/c)^2$. The last two data points deviate substantially above a smooth curve through the points at lower Q^2. These data perhaps show evidence for a diffraction feature, but it is not possible with only a few data points to see the complete shape of this new feature. The experiment has recently been approved for additional running next year; this will be extremely important for uncovering the shape of $B(Q^2)$ above $Q^2 = 2$ $(GeV/c)^2$. If a diffraction feature exists, we will learn much about the short range nucleon interaction from its precise location and shape. The existence of a diffraction feature would force us to look to Q^2 higher than 2 $(GeV/c)^2$ for the region where perturbative QCD models are appropriate. This would be a big step forward in our study of the question: Where do we see quarks in nuclei?

2.4 Future Elastic Experiments

It is not difficult in a few minutes to draw up a list of elastic form factors we would dearly love to measure. Eventually we should look forward to the day when we can make a comparison in a consistent theoretical framework to the complete set of form factor data for the nucleons, the $A = 2$ and $A = 3$ nuclei out to some large Q^2. Data for the deuteron $B(Q^2)$ is coming, but many pieces of information are still missing. High on the list would be a separation of G_C and G_Q of deuterium, and more precise measurements of G_{En}. The proton electric form factor G_{Ep} is also not known[20] above $Q^2 = 4$ $(GeV/c)^2$, and the errors are large above $Q^2 = 1$ $(GeV/c)^2$. The magnetic form factors[21] for ^3He and the charge and magnetic form factors[22] for ^3H are not yet measured above $Q^2 = 1$ $(GeV/c)^2$.

I will confine this discussion to a description of a hypothetical experiment designed to measure G_{Ep} at large Q^2. The purpose is to illustrate both the present poor state of our knowledge of this form factor and the experimental limitations which make this information difficult to get.

The previous data[20] for G_{Ep} are shown in Fig. 12. The highest Q^2 points were obtained from Rosenbluth separations using data from Bonn, Desy and SLAC taken in the 1960's. Some improvement could be made to the precision attainable in this technique by careful attention to the experiment design and by using data with small statistical error measured in only one spectrometer. It is unlikely, however, that the Rosenbluth method can yield significant results for G_{Ep} at Q^2 much above 4 $(GeV/c)^2$. The problem is that G_{Ep} makes only a small contribution to the elastic cross section. If we use the form-factor-scaling approximation

$$G_{Ep}(Q^2) = \frac{1}{\mu_p} G_{Mp}(Q^2) \qquad (18)$$

as a guide, the term τG_{Mp}^2 in Eq. 1 dominates over the term G_{Ep}^2 above $Q^2 = 4$ $(GeV/c)^2$, and the contribution to the cross section from G_{Ep} becomes negligible even at small scattering angles.

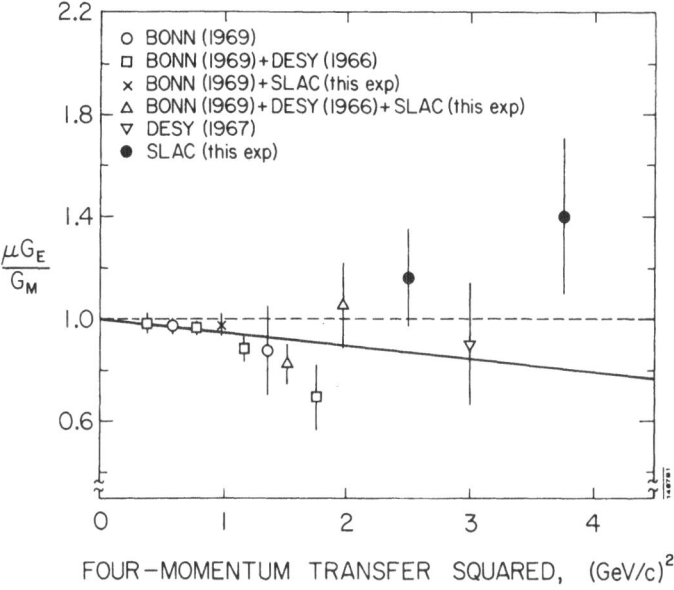

Fig. 12   The world data for the proton form factor G_{Ep} (Fig. from Ref. 20).

The only hope is to play a trick and somehow amplify the contribution of G_{Ep} compared to G_{Mp} in some measureable quantity. One possible trick would be to use polarized electrons and measure the polarization of either the initial or final state protons. The following paragraphs describe the rough outlines of an experiment designed to determine G_{Ep} by measuring the polarization transfer in elastic scattering of polarized electrons. The basis of this method is described in more detail in Ref. 23.

Consider an experiment in which longitudinally polarized electrons are scattered from unpolarized nucleons in a scattering geometry as in Fig. 13. Only longitudinal incident electron polarization is important at high energy, and parity conservation and time reversal invariance limit the possible recoil polarization components to lie in the

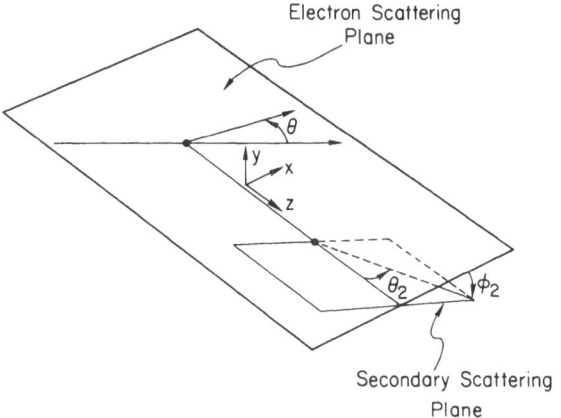

Fig. 13   Electron scattering geometry and the second scattering coordinate system for analysis of recoil polarization.

scattering plane along either the x or z direction in Fig. 13. The recoil polarization component p_x could be sensed by measuring the ϕ_2 dependence of some cross section in an analyzing reaction. The component p_z would not generate any ϕ_2 dependence. In principle, it is not necessary to detect the scattered electrons, but in practice that would help to identify recoil nucleons originating from elastic scattering.

The recoil nucleon polarizations p_x and p_z are given[23] by

$$I_0 p_x = -2[\tau(1+\tau)]^{\frac{1}{2}} G_M G_E \tan(\theta/2) \tag{19}$$

$$I_0 p_z = \frac{E+E'}{M} [\tau(1+\tau)]^{\frac{1}{2}} G_M^2 \tan^2(\theta/2) \tag{20}$$

where $I_0 = \sigma/\sigma_M(1+\tau)$. The most interesting polarization component is p_x. This observable is proportional to a product of G_E and G_M rather than the sum of squares with G_M enhanced by kinematic factors as in the cross section. This is precisely the trick needed to enhance the visibility of G_E. In the case of the neutron, where G_{En} is very small, perhaps even zero, measurement of p_x could provide the crucial leverage needed to pry that precious quantity from experimental data.

The best candidate for an analyzing reaction is proton-proton elastic scattering. An analyzer-detector would be needed to measure the up-down asymmetry in the scattering of recoiling protons on hydrogen. (Actually the full ϕ_2 distribution could be measured.) The cross section and analyzing power for pp elastic scattering are fairly large and have been measured[24] with uncertainties of 5% to 10% of their value in the kinematic range of interest.

If there are no spurious asymmetries caused by the apparatus, the raw asymmetry in the analyzing reaction is given by

$$\varepsilon = \frac{U-D}{U+D} = a p_x A_y \tag{21}$$

where a is the incident beam polarization, and A_y is the analyzing power for pp scattering. A primary task for the experimenters will be to ensure that the unwanted asymmetries are reduced by various techniques (such as reversing the beam polarization, measuring with unpolarized beam, etc.) to the point where statistical errors dominate. Then the error in the measured p_x will be proportional to the total number of analyzed counts

$$\frac{\Delta\varepsilon}{\varepsilon} = \frac{1}{\sqrt{U+D}} . \tag{22}$$

To illustrate the sensitivity to G_{Ep} in a measurement of polarization transfer, consider the following hypothetical experiment based on realistic extrapolations of the facilities at SLAC.

1) Beam current - 20 to 30 mA peak at 180 hz, which gives 10 μA average
2) Beam polarization[25] - a = 0.9
3) Proton primary target - 30 cm liquid hydrogen
4) Electron spectrometer - SLAC 8 GeV/c spectrometer, solid angle $\Delta\Omega = 0.9$ msr
5) Recoil Spectrometer - must be built, for example, a Q-Q-D-D system with reverse bend to minimize precession of the recoil spin, solid angle $\Delta\Omega = 3$ msr, maximum momentum 4 GeV/c
6) Analyzer - 1 to 2 m long cylinder of liquid hydrogen, efficiency 0.03 to 0.05

7) Electron detectors - standard chambers, Cherenkov, lead-glass
8) Recoil proton detectors - wire chambers plus scintillators.

If the electron scattering angle is around 20 degrees, then we expect p_x to be about 0.1. Using the conditions listed above, we can estimate the running time required to achieve counting statistics for analyzed events sufficient to measure an experimental asymmetry with statistical error of 10%. The results are summarized in Table I. If it is possible to reduce the systematic errors in such a measurement to less than 10% by careful design and procedures, then it appears feasible to obtain measurements of G_{Ep} perhaps up to $Q^2 = 5$ $(GeV/c)^2$ in reasonable counting time. The polarization transfer method seems, therefore, to offer a competitive alternative to the Rosenbluth method for measuring G_{Ep} to higher Q^2 than is presently known. It does not seem feasible to extend this technique to the very high Q^2 region where G_{Mp} is known. This situation is another example of the existence of an experimental horizon beyond which we may never venture.

Table I Hypothetical $\vec{e}p$ Recoil Polarization Experiment
($\Theta_e = 20°$, Expect $p_x \approx 0.1$, Analyser eff ≈ 0.03)

	Electrons		Recoil			
Q^2	E_i	E_f	Θ_p	p_p	N analyzed/ day	days for raw assym. stat. error $\dfrac{\Delta\varepsilon}{\varepsilon} = 0.1$
$(GeV/c)^2$		GeV	deg.	GeV/c		
2	4.64	3.57	42.6	1.77	2.8×10^5	0.3
4	6.92	4.79	34.1	2.92	4.2×10^4	3
6	8.83	5.63	28.6	4.03	1.0×10^4	23

3. INELASTIC SCATTERING

3.1 Introduction

Data on high energy inelastic lepton scattering from the proton and nuclear targets, since the first experiments in the late 1960's at SLAC, provides a major source of evidence for the existence of quarks and provides a primary motivation for QCD. After more than 15 years of experiments using electrons, muons, and neutrinos in a wide variety of experimental arrangements, there now exists a large body of deep-inelastic data and theoretical interpretation[26]. In this section I will focus on the description of a surprising discovery[27] made in 1983 in some data taken at CERN, subsequently called the EMC effect after the European Muon Collaboration which did the experiment. The EMC experiment used an iron target for deep-inelastic scattering of high energy muons in an attempt to measure the QCD scale parameter Λ_{QCD}. The assumption was that iron nuclei could be used to provide a high density target full of quarks, and that the quarks in iron would look the same as those in free nucleons. Much to the surprise of the physics world, this assumption turned out to be wrong.

The EMC effect is apparently caused by a modification of the momentum distribution of the quarks in nucleons embedded in heavy nuclei compared to the momentum distribution in the (nearly free) nucleons in deuterium. The existence of this effect opens up a window into the behavior of quarks in the nuclear medium. This discovery has generated much excitement and activity, and the story is not yet fully written[28]. I will describe the present data, indicate some of the suggested theoretical interpretations, briefly describe some future experiments that may help settle some of the unsolved questions. The entire subject is another chapter in answer to the question: Where do we see quarks in nuclei?

3.2 Overview of Deep-Inelastic Results

Before describing the EMC discovery, we first need to review some of the results from the huge body of deep-inelastic scattering (DIS) data. Kinematics and cross sections are described in many places[26] and only a brief outlining is given here. The kinematic variables for charged lepton scattering (i.e., electrons or muons) are sketched in Fig. 14. The cross section for inelastic scattering with detection of the scattered leptons only is given by

FEYNMAN DIAGRAM OF SINGLE PHOTON EXCHANGE

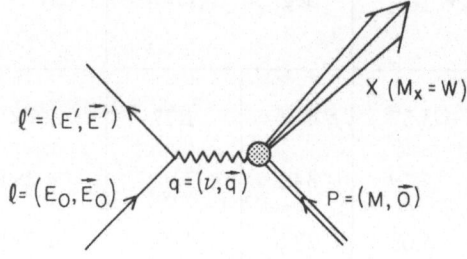

LABORATORY SCHEMATIC

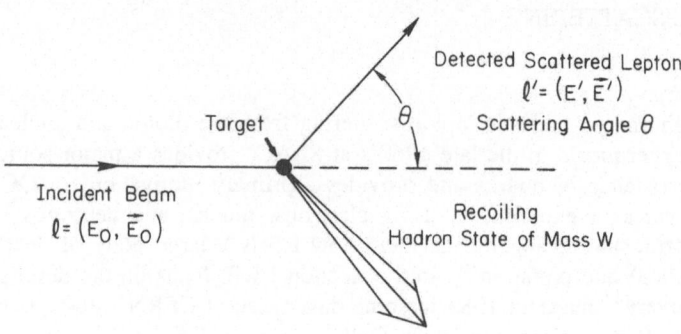

Fig. 14   Kinematic variables for lepton nucleon scattering.

$$\frac{d\sigma}{d\Omega dE'} = \sigma_M[W_2(\nu,Q^2) + 2W_1(\nu,Q^2)\tan^2(\theta/2)] \tag{23}$$

$$= \Gamma[\sigma_T(\nu,Q^2) + \epsilon\sigma_L(\nu,Q^2)] \tag{24}$$

where Γ is the flux of transverse virtual photons and

$$\epsilon = [1 + 2(1 + \nu^2/Q^2)\tan^2(\theta/2)]^{-1} \tag{25}$$

is the degree of longitudinal polarization of the virtual photon.

The following is a short list of some of the primary results and interpretations of deep inelastic scattering:

3.2.1 <u>Scaling (SLAC 1968)</u> The nucleon inelastic structure functions $\nu W_2(\nu,Q^2)$ and $2mW_1(\nu,Q^2)$ become a function of a single variable $x = Q^2/2m\nu$ at large Q^2 and ν. (See Fig. 15.)

$$\nu W_2(\nu,Q^2) \rightarrow F_2(x) \tag{26}$$

$$2mW_1(\nu,Q^2) \rightarrow F_1(x) . \tag{27}$$

The discovery of this pheomenon in the early days at SLAC prompted the development of the parton model[29], which pictured the nucleon at large Q^2 to look like a collection of nearly free pointlike charged constituents. The scaling variable $0 \leq x \leq 1$ is the momentum fraction of the nucleon carried by the struck constituent. The structure functions reduce to a universal function of x independent of the probe resolution $\sim 1/Q^2$ because the constituents are pointlike; they have no internal structure which would produce a form factor dependent on Q^2. Scaling was observed to be good to 20% or better over a wide kinematic range where the cross sections vary by orders of magnitude. The scaling of deep-inelastic structure functions provided the primary motivation for the development of QCD with scale invariant interaction between quarks at short distances (asymptotic freedom).

3.2.2 <u>Quarks have spin one-half.</u> The ratio of longitudinal and transverse cross sections $R = \sigma_L/\sigma_T$ was measured[30] to be small in the scaling region, which indicates that the quark-partons are spin one-half particles. The experimental results for R have relatively large errors because that ratio is small and difficult to measure. The best data are

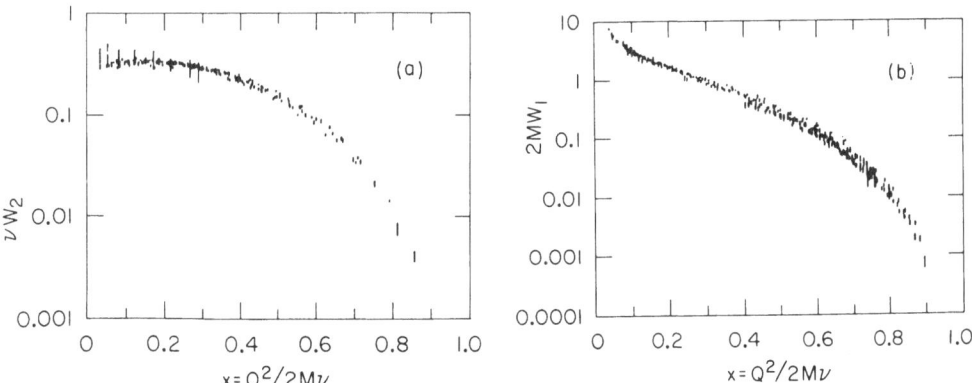

Fig. 15   Deep inelastic structure functions a) νW_2 and b) $2mW_1$ for the proton with $1 < Q^2 < 14$ (GeV/c)^2 and $W^2 > 4$ (GeV)^2. (Fig. from R. Taylor[26]).

electron scattering results from SLAC taken over a range of x and Q^2 on hydrogen and deuterium targets. The variations between data at different kinematic points and for different targets are about the same size as the experimental errors. A number often quoted is the average over all the SLAC data $<R> = 0.20 \pm 0.10$, including statistical and systematic errors [30].

For nearly massless spin-one-half quarks, the helicity is conserved in the quark-photon interaction. A spin-one virtual photon in a transverse interaction can flip the quark spin and conserve helicity, but a spin-zero photon in a longitudinal interaction does not conserve helicity. Hence, for spin-one-half quarks the transverse interaction dominates; $F_2 = 2xF_1$ and $R \rightarrow 0$ at large Q^2. In less naive models R is predicted to be finite due to effects from the finite quark mass and transverse momentum[31]. In second order QCD, R is predicted to be nonzero due to gluon radiation at the quark-photon vertex[32]. However, all these effects are expected to be small and they decrease with increasing Q^2. Therefore, the facts that scaling is observed experimentally with only small deviations and that R is small, are interpreted to mean that deep-inelastic scattering takes place from individual pointlike spin-one-half quarks.

3.2.3 Half the nucleon momentum is missing from the quarks; the rest is carried by uncharged matter (glue).

The F_2 structure function can be interpreted as the momentum distribution function for the charged constituents. In the parton model this can be expressed as

$$F_2(x) = \sum_i z_i^2 x f_i(x) \tag{28}$$

where z_i is the charge of the constituents and $f_i(x)$ is the probability for quark i to have momentum fractions x. With up and down quarks ($z_i = +2/3, -1/3$), the model gives

$$\int_0^1 dx F_2^N(x) = <z_i^2> = \frac{5}{18} . \tag{29}$$

The data gives

$$\int_{0.2}^{0.8} dx F_2^N(x) = 0.13 \sim \frac{1}{2} \times \frac{5}{18} . \tag{30}$$

Therefore at least half of the nucleon momentum is carried by matter not sensed by scattering charged particles.

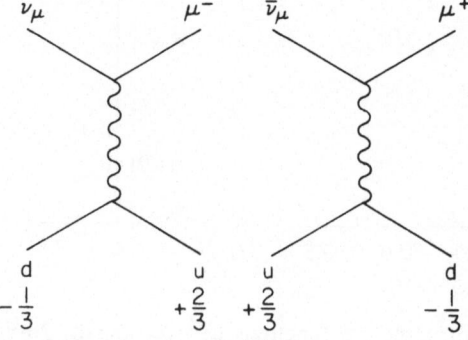

Fig. 16   Diagrams for neutrino scattering by charged current interactions.

3.2.4 Neutrino scattering gives detailed flavor composition of the quark momentum distributions.

The neutrino scattering from up and down quarks is shown in Fig. 16. Parity non-conservation of the weak interaction introduces a third structure function F_3. The neutrino and anti neutrino structure functions each get contributions from a different combination of quarks in the target. The proton structure functions are

$$F_2^{\nu p} = 2x(\bar{u} + d + s + \bar{c})$$

$$F_2^{\bar{\nu} p} = 2x(u + \bar{d} + \bar{s} + c)$$

$$xF_3^{\nu p} = 2x(-\bar{u} + d + s - \bar{c})$$

$$xF_3^{\bar{\nu} p} = 2x(u - d - \bar{s} + c)$$

where u,d,s,c ($\bar{u},\bar{d},\bar{s},\bar{c}$) are the distributions of up, down, strange and charmed quarks (anti quarks). By taking sums and differences of neutrino and anti neutrino cross sections on hydrogen and deuterium, the individual quark and anti quark distributions can be deduced. The gluon distributions can also be determined indirectly by comparison of the

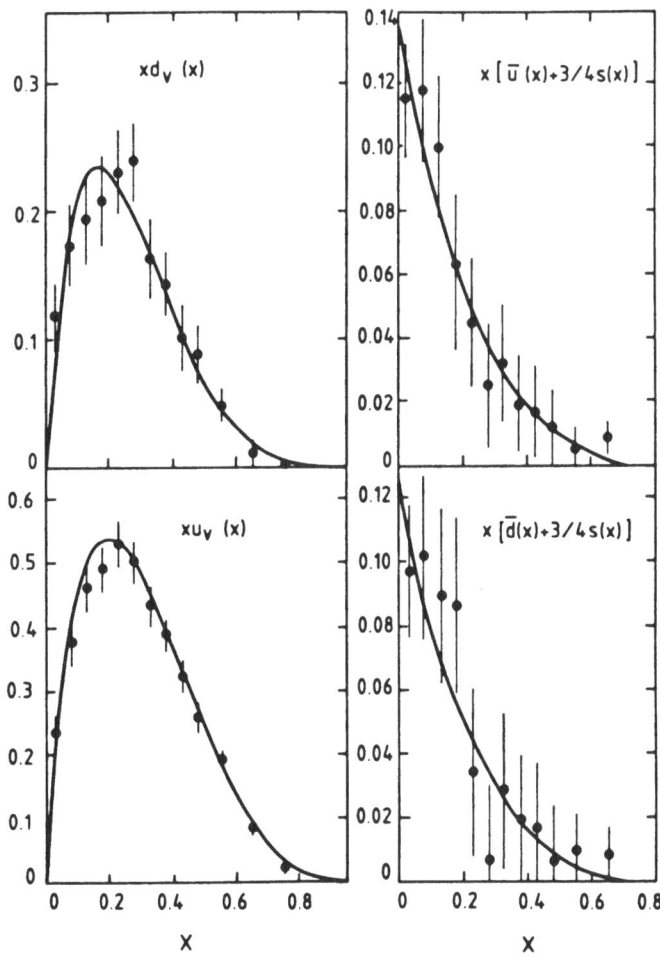

Fig. 17 Valence- and ocean-quark momentum distributions of the proton at $\langle Q^2 \rangle \approx 8(\text{GeV}/c)^2$ from neutrino scattering[34].

data to QCD models[33]. Some results[34] for quark momentum distributions in the proton
are shown in Fig. 17. These measurements indicate that there are no anti quarks above x
= 0.3. For x larger than 0.3, the nucleon momentum is carried by the valence quarks.
This fact will be important in the interpretation of the EMC effect.

3.2.5 <u>Scaling violations</u> The observed scaling of the deep-inelastic structure functions
is not perfect. The theory of QCD predicts that the functions will not scale perfectly with
Q^2. These deviations from perfect scaling are expected to contain quantitative information about the strength of the strong coupling constant. This can be seen schematically in
Fig. 18. A virtual photon at a certain Q^2 will have some probability for connecting to a
quark with momentum fraction x. As the Q^2 is increased, and the photon resolution
decreased, there is some increased chance that the quark will have shed momentum by
emitting a gluon just before the photon interacts. The gluons emitted can in turn create
more soft quark-anti quark pairs and make self interactions. The overall result is an
apparent shift of momentum from quarks at a particular x to quarks (and glue) at lower x
as Q^2 is increased. If the deviation from perfect scaling could be observed over a large
enough range of x and Q^2, it might be used, so goes the hope, to measure the scale
parameter Λ_{QCD}. The desire to extract Λ_{QCD} from scaling violations in DIS has been a
primary motivation for a huge international effort in this field for the last decade. It was

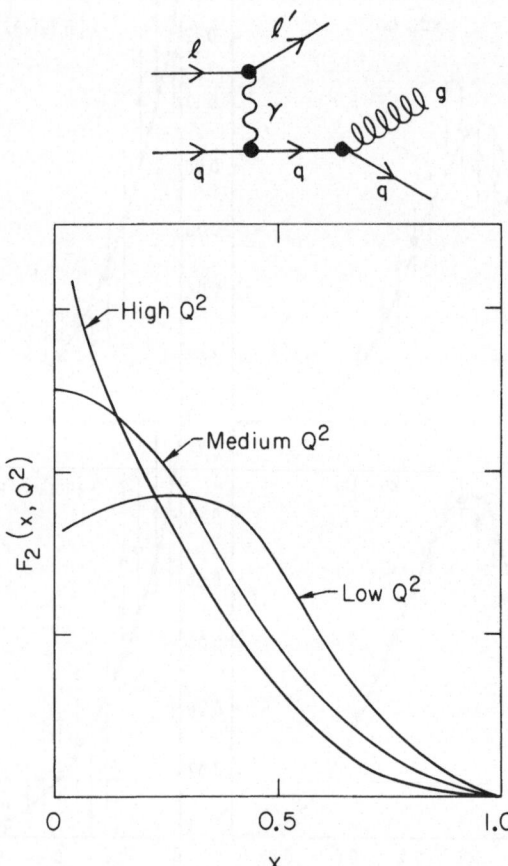

Fig. 18   Quark-gluon interactions in deep inelastic scattering: a) bremsstrahlung of a
            gluon from a quark, b) the qualitative behavior of $F_2(x,Q^2)$ at various Q^2 due
            to gluon radiation.

an experiment of this kind using high energy muons to scatter from iron and deuterium targets that lead to the EMC discovery.

The list of experimental results from DIS goes on to include measurements of the heavy quark production (strange and charmed quarks) from nucleons[35], measurement of parity violation from electroweak interference[36], measurements of nucleon spin structure functions[37], and measurements of the particle distributions in the final state[38], to mention a few topics. These experiments in some cases may also contain evidence for modification of quark momentum distributions in nuclei, and they will provide areas for future study. However, there is not time here to cover all these subjects, so we go on now to look at the evidence for the EMC effect in the inclusive cross sections.

3.3 Prelude to the EMC Discovery

In experiments with muon and neutrino beams, it has been necessary in many cases to use heavy nuclei for targets, rather than hydrogen or deuterium, simply to get the luminosity large enough to measure significant numbers of events. The targets used have included[35], calcium carbonate[39], and neon[40] (in bubble chambers). The assumption made in interpreting the results from heavy nuclear targets was that the only modification to the quark momentum distributions arises from the Fermi motion of the bound nucleons. The effect on the nuclear structure functions was thought to be small for x below about 0.6 because the free nucleon structure functions are slowly varying in that region. (See the data for F_2 for the proton in Fig. 15). Corrections for the small change in shape for F_2 or F_1 below x = 0.6 could be calculated using Fermi smearing models.

For x larger than 0.6, where the structure functions fall steeply with increasing x, the Fermi smearing was expected to be a significant effect. Nuclear structure functions (per nucleon) were expected to be larger at high x than for free nucleons due to the smearing in from lower x. The nuclear structure functions are finite for x > 1, beyond the physical

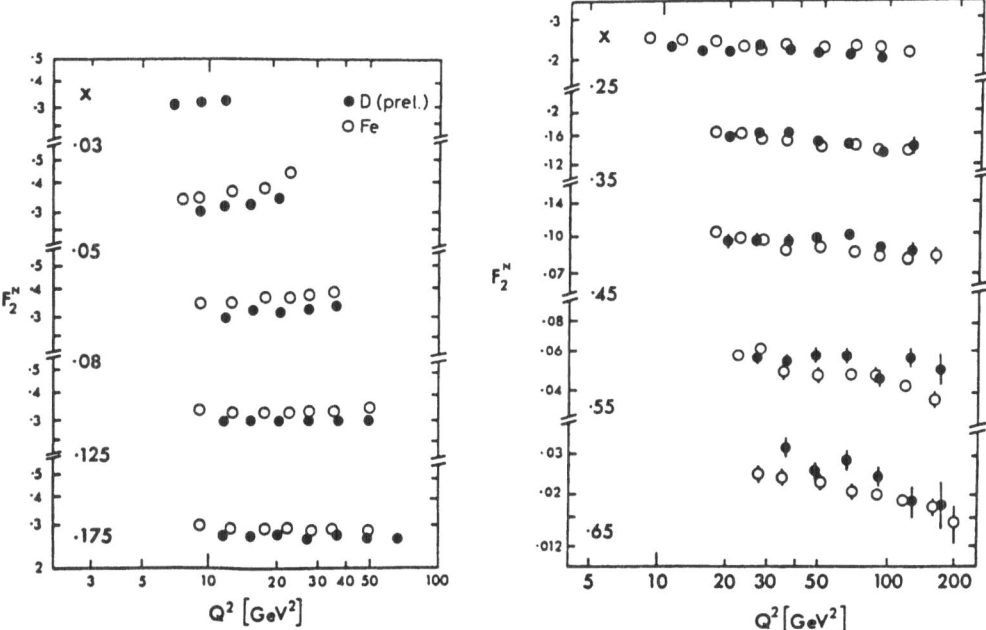

Fig. 19   Comparison of F_2(Fe) and F_2(d) from muon scattering by the EMC[27] (Fig. from F. Dydak[26]).

range for free nucleons, because of Fermi smearing. For x > 0.6 the cross sections decrease rapidly. The event samples at high x are small in most experiments, and the corrections for Fermi motion affected the interpretation of only a small portion of the available data.

3.4 The EMC Discovery and Confirmation

The data for F_2 from muon scattering on iron and deuterium as measured[27] by the EMC are shown in Fig. 19. These data show the characteristic deviation from perfect scaling; F_2 increases with Q^2 at low x and decreases with Q^2 at high x. The figure also shows the systematic difference in the results from iron and deuterium that attracted so much attention. In Fig. 20 is plotted the ratio of F_2 for iron and deuterium from the EMC. There is a ±15% variation from unity in this ratio in the region $0.05 \leq x \leq 0.65$, and the deviation is way outside the experimental errors.

Quickly following the EMC discovery, the effect was confirmed[41] in electron scattering data from experiments performed a decade ago at SLAC. By using a combination of archeological methods and sound detective work, the SLAC experimenters were able to locate old data tapes and log books and reconstruct results for electron scattering on the iron and aluminum end caps of their liquid hydrogen and deuterium target cells. The results are shown in Fig. 21, together with the data from EMC and several other previous experiments at lower Q^2 and low x. The SLAC data agrees with the EMC data for x > 0.2. The EMC data show a ratio significantly larger than one for x below 0.2, in disagreement with the data from an experiment at lower Q^2. Also shown in Fig. 21 is a theoretical curve calculated for the ratio of the cross sections assuming the only effect on the ratio is due to Fermi smearing of the single nucleon structure functions[42]. The data are completely different in shape from the Fermi model curve. The deviations from unity

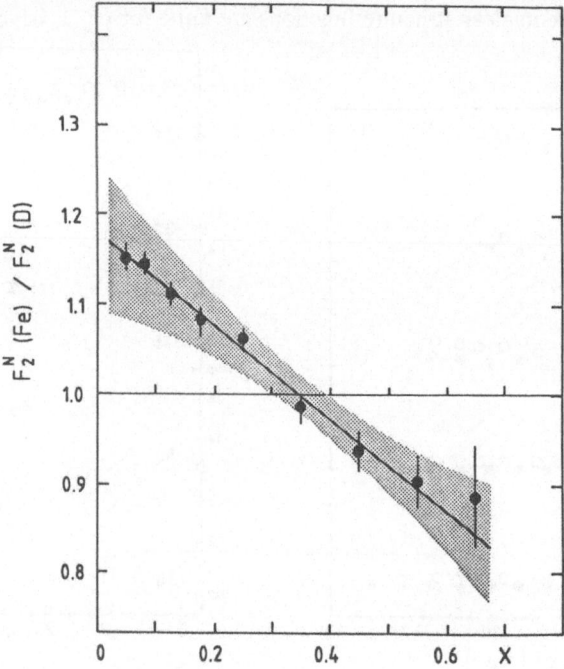

Fig. 20 The ratio of the structure functions F_2 for iron and deuterium from muon scattering by the EMC[27].

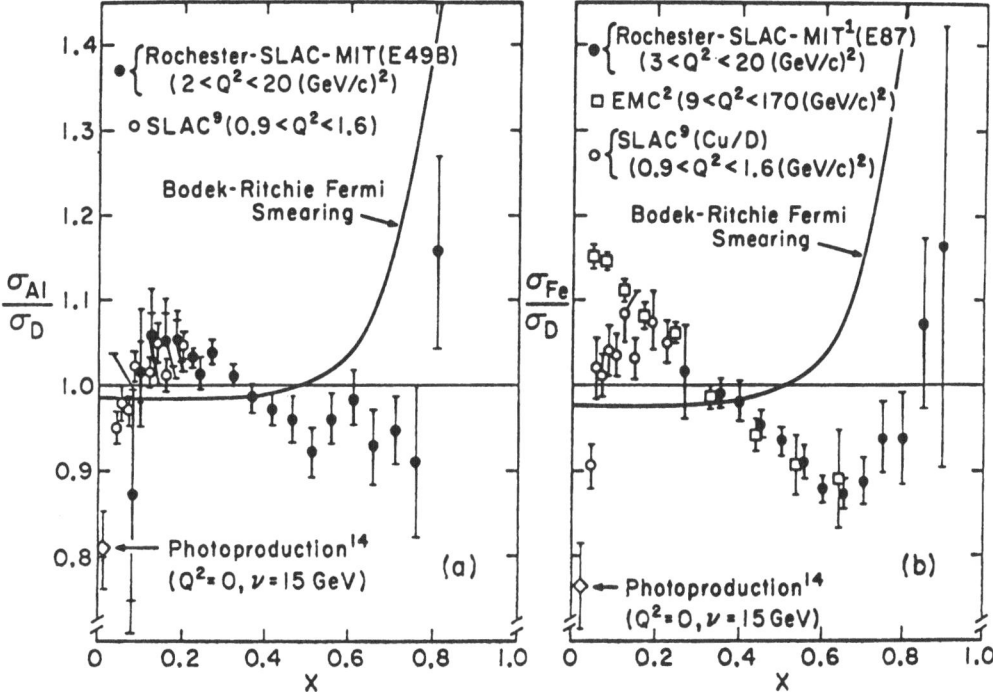

Fig. 21   Confirmation of the EMC effect in old electron scattering data from SLAC[41].

in the ratio of cross sections are similar in shape for iron and aluminum, with a hint of an increased size in iron.

This was the situation in early 1983. At that time we were in the middle of the measurements of ep elastic scattering, described in the previous section, when the experiment was temporarily suspended to repair a broken target. Rather than continue measuring ep elastic, we elected to build a completely different target containing a series of heavy nuclei and switch to a measurement of deep-inelastic scattering.

3.5 The A-Dependence of Deep-Inelastic Scattering

The primary objectives of the new experiment[43], known as Experiment E139 at SLAC, were to confirm the EMC discovery in an experiment dedicated to measuring the effect with a minimum of systematic errors, and to measure the A-dependence by scattering from a series of targets spanning the periodic table. The scope of the experiment was set by: the beam energy maximum of 21.5 GeV into SLAC End Station A, the momentum and angle range of the 8 GeV/c spectrometer, and the constraint that the experiment had to fit into less than one month of beam time. These conditions fixed the range in x and Q^2 that could be reached, shown in Fig. 22.

The targets chosen were selected on the criteria that they be readily available, sturdy materials that could be easily fabricated, and would not melt in the high powered electron beam. Because we suspected that the effect was caused by some distortion of the nucleons from the presence of near neighbors, we wanted to include ^4He in the targets. That nucleus has an anomalously large central density due to the tight binding of the four-nucleon system. If the distortion is caused by overlap of bound nucleons, then ^4He

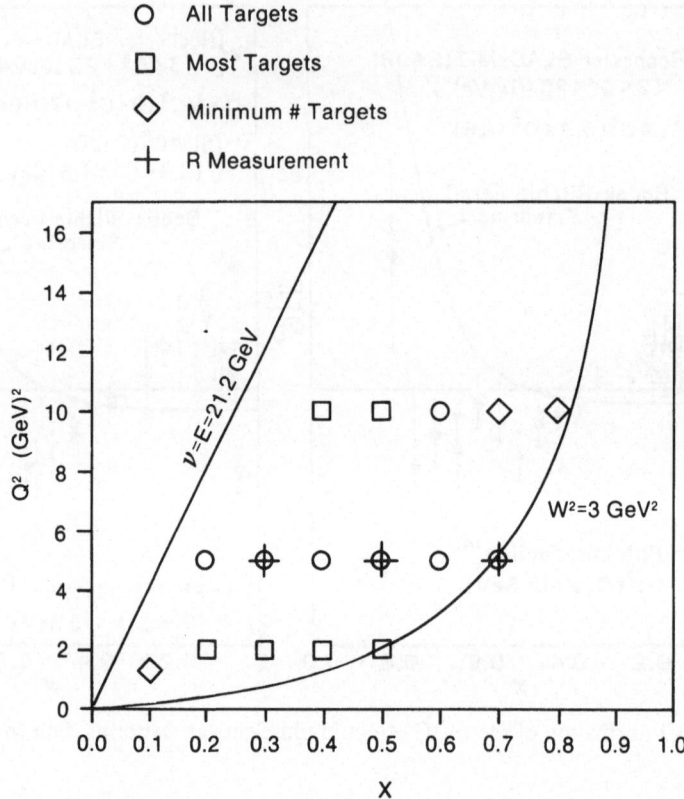

Fig. 22   Kinematic range covered by SLAC experiment E139 on deep inelastic scattering from nuclei[43].

might show a large effect. The rest of the targets, in addition to hydrogen and deuterium
for calibration, were chosen to conveniently span the periodic table about equally spaced
in log A. The solid targets were natural isotopes of Be, C, Al, Ca, Fe, Ag and Au. Targets of 2% radiation thickness were used up to Fe. For Al and above, there were targets
of 6% and 12% radiation thickness. Targets of a given radiation length decrease in
grams per cm^2 with increasing A, and the thicker targets were used for the higher A
nuclei to get larger counting rates. Targets of the same material with two thicknesses
were measured and results compared to check that we understood the portion of the radiative corrections proportional to the external radiators. In the final results, the data from
the 12% radiators were not used.

A primary aim was to cover as wide a range in x as possible at a value of Q^2
sufficiently large to be clearly in the deep-inelastic scaling region. The largest x span is
for the points at $Q^2 = 5$ $(GeV/c)^2$. A second goal was to look for possible Q^2 variation at
a value of x where the EMC effect was large. The data at x = 0.6 span the range 5
$\leq Q^2 \leq 15$ $(GeV/c)^2$. The experiment was designed to measure the ratio of scattering on
deuterium compared to the other targets with a minimum of systematic error in the ratio.
Targets were interchanged frequently at a given kinematic setting and running conditions
were chosen to keep all corrections for variations in acceptance, efficiency, pion contam-
ination, and dead time small. The overall systematic error in the ratio σ_A/σ_d was 1% to

2%. A big advantage for this experiment over that by the EMC is that the acceptance of the 8 GeV/c spectrometer does not change significantly when different targets are used and when the spectrometer is operated at different angles and momenta.

The measured cross sections were corrected for radiative effects, and the results for heavy nuclei were adjusted to give ratios of structure functions per nucleon for nuclei with equal number of neutrons and protons. The results are shown in Fig. 23. In Fig. 23a the results for iron at various Q^2 are shown, along with the EMC data. This plot, and similar ones for the other nuclei, not shown, indicate that there is no significant variation in the ratio of cross sections as a function of Q^2 in the range of 2 to 15 $(GeV/c)^2$. The cross section ratios scale, which we take as evidence that the scattering takes place from individual quarks. Since no Q^2 variation is observed, the rest of the data in Figs. 23b-d are averaged over Q^2.

These plots show clearly how the distortion of the nuclear structure functions increases with A. The new data agree within errors with the results for Fe and Al from the previous SLAC experiments. They also agree with the EMC data for x > 0.3, but they deviate from the EMC date at lower x. The E139 data for σ_A/σ_d are never significantly larger than unity. This fact has important consequences for some of the interpretation.

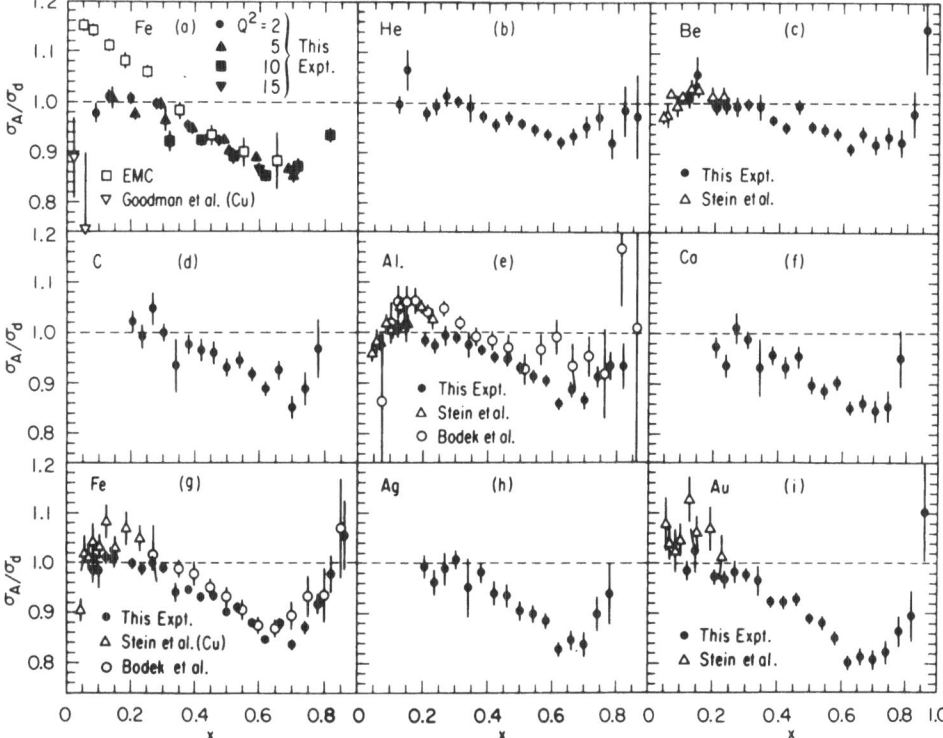

Fig. 23   (a) E139 results[43] for σ_{Fe}/σ_d as a function of x for various values of Q^2 along with muon data from Refs. 27 and 45. (b)-(i) σ_A/σ_d averaged over Q^2 as a function of x for various nuclei, and electron data from Ref. 44. Only statistical errors are shown.

Another way to look at the A-dependence is to plot the results at fixed x versus log A, as in Fig. 24. These and similar plots for other values of x show that the effect increases proportional to log A with a slope that varies with x. The ratio σ_A/σ_d deviates from unity linearly with the average nuclear density $\rho(A)$

$$\frac{\sigma_A}{\sigma_d} = a[1 + b\rho(A)] \tag{32}$$

where a and b are parameters that vary with x. The results for ^{4}He do not deviate significantly outside the experimental errors from the straight lines through the data for larger nuclei.

While the E139 experiment did confirm the existence of the EMC effect, and clearly measured its A-dependence, a problem still remains in the disagreement between the EMC data and the rest of the world's data for iron at low x. This disagreement can be seen clearly in Figs. 23a and 25, where the EMC data continues rising at low x, while the other data (within errors) are near or below unity. This difference could possibly be due to some unknown systematic error in the experiments, in which case more independent measurements may eventually clear up the problem. It could also be due to physics effects.

One suggestion is that these data disagree because the experiments were done at different Q^2. The EMC data start at $Q^2 = 9$ (GeV/c)^2 at the lowest x and span a Q^2 range up to 50 (GeV/c)^2 at x = 0.2 (see Fig. 19), whereas the E139 data for x ≤ 0.2 are more measured at 2 and 5 (GeV/c)^2. The data from Stein et al.[44] are all at Q^2 below 2 (GeV/c)^2. The structure functions at low x and low Q^2 are suspected to be complicated functions of x, Q^2 and A due to the competition between deep-inelastic scattering on the target quarks and shadowing of the interacting virtual photons. It is conceivable that the deviations we observe are caused by such phenomena, in which case more data at low x over a range of

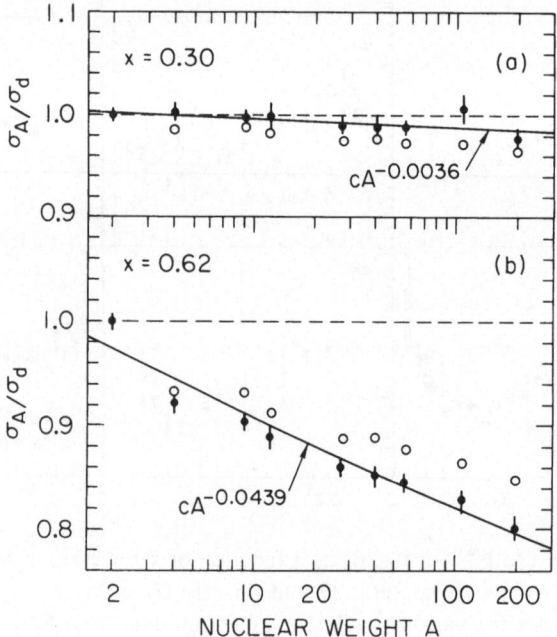

Fig. 24   E139 results[43] for Q^2 averaged ratios σ_A/σ_d versus log A for two values of x. The solid line is a fit to the form $\sigma_A/\sigma_d = cA^\alpha$. The errors are statistical only.

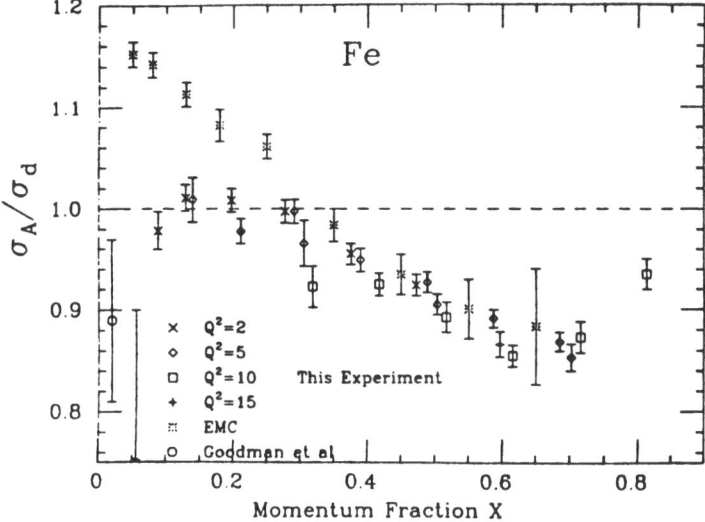

Fig. 25   E139 results[43] for σ_{Fe}/σ_d as a function of x for various values of Q^2 along with muon data from Refs. 27 and 45.

Q^2 and A will be important for mapping out this physics. The evidence for Q^2 variation of the cross sections within the data from a single experiment is not strong, however. Neither the EMC data in Fig. 19 nor the E139 data in Fig. 23a show large variation with Q^2 over the range measured in each experiment. It is also difficult to explain why at x = 0.1 and 0.15 there should be such a large jump between the E139 data at $Q^2 = 2$ and 5 $(GeV/c)^2$ and the EMC data which begin at $Q^2 = 9$ $(GeV/c)^2$.

Another possibility, which I will explore in more detail in the following paragraphs, is that the interpretation of the cross section data in terms of structure functions F_2 may be confused because the value of $R = \sigma_L/\sigma_T$ needed to make the extraction depends upon A. To see how this works, let us look at how the structure functions F_1 and F_2 are obtained from the data.

In Fig. 26 is plotted a schematic of the deep-inelastic cross sections from Eq. 24 versus the kinematic variable ε, which parameterizes the polarization of the virtual photon flux. Two numbers are required to specify the deep-inelastic cross section at a given x, Q^2, and ε. These can be either F_1 and F_2 or they can be $d^2\sigma/d\Omega dE'$ and the ratio

$$R = \frac{\sigma_L}{\sigma_T} = \frac{F_2}{2xF_1}\left[1 + \frac{4m^2x^2}{Q^2}\right] - 1 .\tag{33}$$

The longitudinal and transverse cross sections σ_L and σ_T are related to the structure functions F_1 and F_2 by

$$F_1(\nu,Q^2) = \frac{k}{4\pi^2\alpha}\,\sigma_T(\nu,Q^2)\tag{34}$$

$$F_2(\nu,Q^2) = \frac{k}{4\pi^2\alpha}\left[\frac{Q^2}{Q^2+\nu^2}\right]\left[\sigma_T(\nu,Q^2) + \sigma_L(\nu,Q^2)\right]\tag{35}$$

where

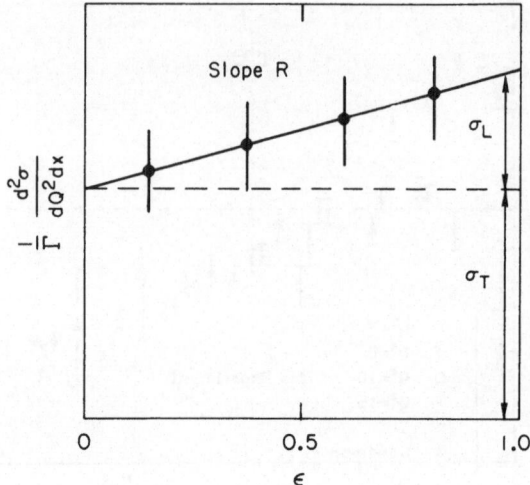

Fig. 26   Schematic of the deep inelastic cross section versus the virtual photon polarization parameter ε.

$$k = \frac{W^2 - M^2}{2M}.$$ (36)

To make an extraction of F_1 and F_2 from the data at fixed x and Q^2, it is necessary to
measure the cross section for various values of ε (essentially by measuring at various
scattering angles) and determine F_1 from the intercept at ε = 0 and F_2 from the intercept
at ε = 1.

We saw in the previous section, the ratio R = σ_L/σ_T in the deep-inelastic region is
measured (with large errors) to be small and nearly the same for hydrogen and deuterium. The basis of the quark model interpretation of the DIS data is that at high Q^2 the
scaling of the data has established the applicability of perturbative QCD, and we can
assume that the cross section is determined by interaction of the virtual photon with individual quarks. It should not matter if those quarks are in free or bound nucleons. The
value of R for a heavy nucleus should be the same as for hydrogen or deuterium.

A possible confusion could arise between the EMC data and the rest of the world's
data if R depends upon A and if we are not careful when comparing the results measured
at different values of ε. The EMC data were taken in a spectrometer that measured
muons scattered at forward angles where the parameter ε is near one. In this case an
extraction of F_2 from the cross sections involves a small extrapolation to ε = 1 that is
insensitive to uncertainties in the slope versus ε. The E139 data, on the other hand, are
measured at various ε in the range 0.4 ≤ ε ≤ 0.9. Extraction of F_2 from the cross section
data would be more sensitive to uncertainties in R. In any case, the ratio of cross section
σ_A/σ_d at all ε should be the same as the ratio of F_{2A}/F_{2d} at ε = 1, no matter what the true
value of R is, so long as it is the same in heavy nuclei as it is in deuterium. While there
may be a small slope in the individual σ_A and σ_d versus ε, from higher order QCD
effects, the ratio σ_A/σ_d should be constant versus ε.

There is some evidence[46] from E139 data that R depends upon A. This can be seen
in Fig. 27 where the ratios σ_{Fe}/σ_d for various values of x and Q^2 are plotted versus ε. If
we suspect that physics can cause the ratios to depend separately on x, Q^2 and ε, we

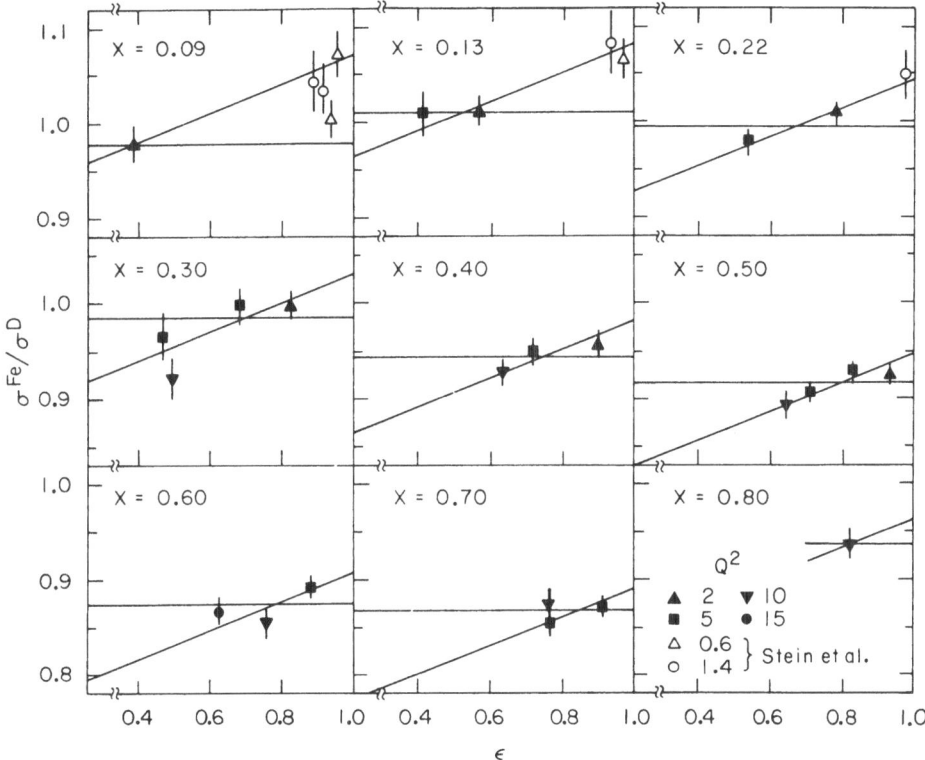

Fig. 27   E139 results[43,49] for σ_{Fe}/σ_d at various x and Q^2 values versus the virtual photon polarization parameter ε. Also shown are daata from a Cu target from Ref. 44. The error bars are statistical only.

should be careful about comparing data at different kinematic points. In this experiment, data were taken at three values of x = 0.3, 0.5, and 0.7 at fixed $Q^2 = 5$ $(GeV/c)^2$ for two values of ε. These data points, represented by square symbols in Fig. 27, were fit to a straight line versus ε, assumed to have the same slope for all x. The results of the fit

$$\frac{d(\sigma_{Fe}/\sigma_d)}{d\varepsilon} = 0.15 \pm 0.11 \tag{34}$$

are the sloped lines in Fig. 27. The other data points at other Q^2 from E139 and from the previous experiment by Stein et al.[44], not included in the fit, are also plotted for comparison. All these data seem to show a slope versus ε that corresponds to a larger value for R in iron than in deuterium. The slope does not change significantly with x or Q^2, but the data are sparse and the errors are large.

Fig. 28 shows the results for F_{2Fe}/F_{2d} extracted from the E139 cross sections at various ε assuming the slope in Eq. 34. The E139 data for F_{2Fe}/F_{2d} are closer to the EMC data at lower x than are the ratios of cross sections, but the F_2 ratios have large uncertainties from the extrapolation to $\varepsilon = 1$. While these data are not precise nor extensive enough to be conclusive, they do suggest a possible explanation for the discrepancy between the experiments.

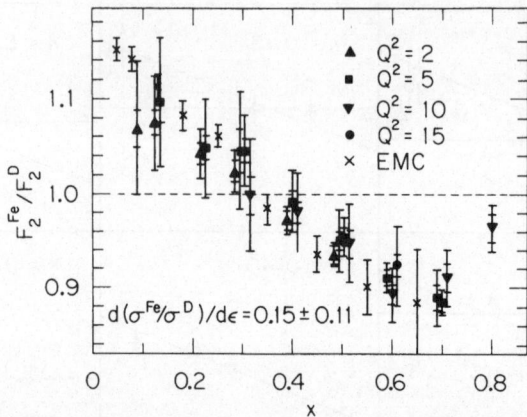

Fig. 28 E139 results[46] for the ratio of structure functions F_2 for iron and deuterium extracted at $\varepsilon = 1$ from cross section ratios σ_{Fe}/σ_d at various ε using the slope $d(\sigma_{Fe}/\sigma_d)/d\varepsilon = 0.15 \pm 0.10$. The inner error bar is the statistical error, while the outer bar indicates the additional uncertainty from the extrapolation to $\varepsilon = 1$. Also shown are the EMC data[27].

A brief summary of the experimental situation for the EMC effect is as follows:

1) The structure functions for deuterium and heavy nuclei (iron) scale for $Q^2 > 2$ $(GeV/c)^2$, which is interpreted as evidence for scattering from individual quarks.

2) σ_A/σ_d is not constant versus x. The deviation is large (10% to 20%) and negative for x > 0.3. Since there are only valence quarks at x > 0.3, this implies a shift of momentum away from valence quarks in a nucleus at x > 0.3 toward either: the glue, the quark-anti quark pairs in the ocean at low x, or the valence quarks at lower x (and also some to quarks at x > 1). Perhaps momentum is shifted in all these directions.

3) The effect increases with the log of A or linearly with the average nuclear density.

4) The ratio $R = \sigma_L/\sigma_T$ might be A-dependent.

5) The experimental situation for x < 0.2 is unclear, and more measurements are needed.

3.6 Theoretical Explanations of the EMC Effect

Since the discovery of the EMC effect, there has been a veritable explosion of theoretical activity in the world attempting to explain the phenomenon[28]. There is not time or space here to describe this work adequately. I will limit this to a brief description of the general outlines of some of the models. More details can be found in the references.

The proposed explanations for this effect fall into two broad classes: those that give an explanation within the framework of traditional nuclear physics and use models with nucleons, virtual mesons, and deltas; and those that use quarks as the relevant degrees of freedom.

Consider first the quark based models. The main problem is to understand how momentum gets shifted away from valence quarks at x > 0.3.

3.6.1 <u>Nucleons swell up in nuclei</u> In one subclass of the quark models [47], it is proposed that nucleons expand in volume when they are bound in nuclei. That means the valence quarks are confined to a larger volume, and there is a shift of momentum away from the valence quarks onto the glue, which does the confining, and perhaps also onto the quark-anti quark pairs. These models predict an A-dependence because the amount of swelling depends on how deeply the nucleons are buried in the nucleus. Lighter nuclei have a larger fraction of their nucleons in the skin where the swelling is less.

Presumably if nucleons are swollen, this would lead to consequences in other areas of physics. The electromagnetic form factors of fat nucleons would be different than for free ones, and this might show up as a modification to the cross section in quasi elastic electron scattering[48], for example. Several experiments have measured separations of longitudinal and transverse cross sections in quasi elastic scattering with results that do not agree with interpretation using free nucleon form factors. More experiments are needed in this area.

3.6.2 <u>Nucleons sometimes overlap in nuclei</u> Another subclass of explanations[49,50] suggest that somehow nucleons, which are pictured as clusters of quarks, can occasionally interpenetrate each other and allow their quarks to intermingle. This would produce a redistribution of momentum among the quarks and glue in the overlapped clusters. The valence quarks would effectively be confined to a larger volume than in single nucleons and they would lose momentum on average. Some of these models based on nucleon overlap are also called rescaling models. They are so named because the change in nuclear structure with A looks similar to a change in scale predicted by QCD for the evolution of the single nucleon structure functions with increasing Q^2.

A special feature of some of the cluster models[50] that distinguishes them from many of the other models is that they predict modification of the quark momentum distributions at very large x. The size and shape of nuclear structure functions for x > 1 is due entirely to nuclear binding. If multi-nucleon clusters are formed by close collisions of bound nucleons, then $F_2(x)$ at x > 1 would get contributions from quarks which carry some fraction of the momentum of the multi-nucleon clusters, plus quarks in single nucleons that are Fermi smeared up to x > 1. Data for nuclear structure functions at x > 1 might help distinguish between the competing models for the EMC effect. Some data exists[51], but more is needed. Unfortunately, the cross sections fall rapidly at high x and high Q^2 and such data is difficult to obtain.

3.6.3 <u>Nuclei are semi transparent to quarks</u> In one class of quark models, the nucleus is pictured as a collection of quark bubbles that can overlap to allow quarks to tunnel throughout the nucleus[52]. This would explain the EMC effect by the redistribution of momentum away from quarks that are allowed to move in a large volume. In these models the effect is predicted to increase with some power of the nuclear radius.

3.6.4 <u>More pions in nuclei</u> There are many papers that offer explanations in terms of traditional nuclear physics[53]. The main contenders in this class are models which explain the momentum distortion by an increase in the number of virtual pions per nucleon in heavy nuclei. The momentum distributions of these pions are soft and they enhance the structure functions at low x as well as rob momentum from the nucleons, which carry the momentum at high x. The main problem is to build a model which gives the desired effect with a number of extra pions consistent with other experimental constraints.

3.7 <u>Future Inelastic Experiments</u>

There are many outstanding problems in nucleon and nuclear substructure that will be addressed by more experiments in the near future. The discrepancy between the EMC data and the other experiments at low x will be explored by two new muon scattering experiments. The EMC apparatus at CERN is being upgraded by a new collaboration, and they plan to take data with high statistics on deuterium and various heavy nuclei[54]. The experiment design is being modified to decrease the absolute systematic error and the systematic error in ratios of deuterium to other targets.

Another experiment[55] is being prepared at Fermilab to use the new high energy muon beam from the Tevatron when it arrives. This experiment will be aimed primarily at studying the final state using a powerful series of detectors to identify and measure the momentum of particles emerging in high multiplicity events. These two high energy muon experiments will provide important new data from nuclear targets that will help unpack the physics at low x.

There is presently underway at SLAC a new experiment[56] E140 to measure R in deep inelastic scattering from hydrogen, deuterium, iron, and gold over a similar kinematic range as for E139. The aim is to reduce the errors on the absolute measurements of R to approximately ± 0.04 or better, and measure the ratio of R in deuterium to that in Fe and Au with even better precision. This data will provide an important test of QCD. Most of the present data for R is too high (but with large errors) to agree with QCD, which predicts R to be generally less than 0.10. If the new data confirm absolute values of R around 0.2 to 0.4 with reduced errors, that would present a problem for QCD. It is also possible that E140 will find that R depends on nuclear weight A. That would be completely contradictory to the present interpretation of the deep inelastic process as scattering from individual quarks.

The data from E140 has the potential to change our way of thinking about the quark structure of nucleons and nuclei. It has been suggested[57] that *all* of the scale breaking in deep-inelastic scattering can be attributed to contributions from coherent scattering processes that have form factors falling with powers of Q^2 rather than to variations in $\alpha_s(Q^2)$ from scattering on single quarks. These processes, sometimes called higher twist or power law suppressed terms, arise when the virtual photon connects to subgroups of quarks in the target. The proton, for example, may appear to the virtual photon as a superposition of components containing three valence quarks, one quark and a diquark pair, three quarks plus quark-antiquark pairs, etc.. The total structure function would be a sum of contributions from scattering on individual quarks, on diquark pairs, and on soft quark-antiquark pairs. The relative size of the contribution from each component would be weighted by the probability for finding each component times its form factor. Individual quarks have no form factors, whereas diquarks have a form factor proportional to $1/Q^2$. The chance for finding each component in the wave function would be a complicated function of x and Q^2.

If the proton wave function contains significant amounts of diquark components, they might show up as an enhancement to R. A diquark pair would be coupled to spin-one or spin-zero and would produce a large longitudinal cross section. A nucleon embedded in a heavy nucleus might contain a larger fraction of diquarks. If R for protons and deuterium is found to be larger than QCD predicts and is enhanced in heavy nuclei, then the present consensus about the interpretation of deep inelastic scattering will have to be revised.

More measurements of longitudinal and transverse separation in the quasi elastic and threshold inelastic region above x = 1 in nuclei are also being planned[58]. These could give evidence for or against the swelling and overlap of nucleons in nuclei.

ACKNOWLEDGEMENTS

This work was supported by the Department of Energy contract DE-AC03-76SF00515, and by the National Science Foundation under Grant No. PHY8410549.

REFERENCES

[1] S. Brodsky, G. Farrar, Phys. Rev. **D11,** 1309 (1975).

[2] S. Brodsky, B. Chertok, Phys. Rev. **D14,** 3003 (1976).

[3] S. Brodsky, G. Lepage, Phys. Rev. **D22,** 2157 (1980).

[4] Physics Today, August 1985, p. 17.

[5] N. Isgur, C. Llewellyn Smith, Phys. Rev. Letters **52,** 1080 (1984).

[6] V. I. Cherniak, I. R. Zhitnitsky, Nucl. Phys. **B246,** 52 (1984).

[7] A. Andrikopoulou, J. Phys. G: Nucl. Phys. **11,** 21 (1985).

[8] R. Arnold *et al.,* "A Proposal for Measurement of Electron-Proton Cross Sections at Large Momentum Transfer", SLAC Experiment E136 (1980).

[9] J. W. Mark, SLAC PUB 3169 (1983), Cryogenic Engineering Conference and International Cryogenic Materials Conference, Colorado Springs Colorado, August 11-19 (1983).

[10] W. Bartel *et al.,* Nucl. Phys. **B58,** 429 (1973); K. Hanson *et al.,* Phys. Rev. **D8,** 753 (1973).

[11] R. J. Budnitz *et al.,* Phys. Rev. **173,** 1357 (1968); W. Albrecht *et al.,* Phys. Rev. **26B,** 642 (1968).

[12] G. Höhler *et al.,* Nucl. Phys. **B114,** 505 (1976); S. Blatnik, N. Zovko, Acta Phys. Austriaca **39,** 62 (1974).

[13] S. Rock *et al.,* Phys. Rev. Letters **49,** 1139 (1982).

[14] R. Arnold *et al.,* Phys. Rev. Letters **35,** 776 (1975).

[15] S. Auffret *et al.,* Phys. Rev. Letters **54,** 649 (1985).

[16] R. Arnold *et al.,* "A Proposal to Build a New Injector at SLAC for a Program of Research in Experimental Nuclear Physics", SLAC (1982).

[17] R. Arnold, "NPAS - A Program of Nuclear Physics at SLAC", CEBAF Summer Workshop, Newport News, Virginia, June 25-29, 1984.

[18] D. Day *et al.,* "Proposal for Inclusive Electron Scattering from Nuclei", NPAS experiment NE3 (1984).

[19] R. Arnold *et al.,* "A Proposal to Measure Electron Scattering from Deuterium at Large Momentum Transfer at 180 Degrees", NPAS Experiment NE4 (1984).

[20] J. Litt *et al.,* Phys. Rev. Letters **31B,** 40 (1970).

[21] J. M. Cavedon *et al.,* Phys. Rev. Letters **49,** 986 (1982).

[22] F. P. Juster *et al.,* Phys. Rev. Letters **55,** 2261 (1985).

[23] R. Arnold, C. Carlson, F. Gross, Phys Rev. **C23,** 363 (1981); W. Donnelly, MIT Print CTP, 1254 (1985), Submitted to Annals of Phys.

[24] M. G. Albrow *et al.,* Nucl. Phys. **B23,** (1970); D. Miller *et al.,* Phys. Rev. **D16,** 2016 (1977); P. R. Bevington *et al.,* Phys. Rev. Letters **41,** 384 (1978).

[25] Based on a high polarization laser driven polarized source under development at SLAC, C. K. Sinclair, private communication.

[26] R. Taylor, "Inelastic Electron-Nucleon Scattering Experiments", Proceedings International Symposium on Lepton Photon Interactions, Stanford, August 21-27, (1976); J. Drees, "Deep Inelastic Scattering", Lectures at the 1980 CERN School of Physics; J. Drees, Proceedings 1981 International Symposium on Lepton and Photon Interactions at High Energy, Bonn August 24-29 (1981); F. Eisele, Journal de Physique **C3,** 337 (1982); F. Dydak, Proceedings 1983 International Lepton/Photon Symposium, Cornell University, Ithica, N.Y. (1983); W. D. Nowak, CERN Print PHE 84-10, submitted to Fortschritte der Physik.

[27] J. J. Aubert *et al.,* Phys. Letters **123B,** 275 (1983).

[28] N. N. Nikolaev, "EMC Effect and Quark Degrees of Freedom in Nuclei: Facts and Fancy", Oxfort Print TP-58/84, Invited Talk at VII International Seminar on Problems of High Energy Physics-Multiquark Interactions and Quantum Chromo-dynamics, Dubna USSR, 19-23 June 1984; R. R. Norton, "The Experimental Status of the EMC Effect", Rutherford Print RAL-85-054, Invited talk at Topical Seminar on Few and Many Quark Systems, San Marino, Italy 25-29 March 1985; E. L. Berger, "Interpretations of the Nuclear Dependence of Deep Inelastic Lepton Scattering", Argonne Print ANL-HEP-PR-85-70, and Proceedings of the Topical Seminar, San Marino (1985); H. J. Pirner, "Deep Inelastic Lepton-Nucleus Scattering", Proceedings Int. School of Nuclear Physics Erice, to be published. Progress in Particle and Nuclear Physics (1984).

[29] R. P. Feynman, Photon-Hadron Interactions, W. A. Benjamin Inc., Reading Mass. (1972).

[30] A. Bodek *et al.,* Phys. Rev. **D20,** 1471 (1979); M. D. Mestayer *et al.,* Phys. Rev. **D27,** 285 (1983).

[31] R. Barbieri *et al.,* Nucl. Phys. **B117,** 50 (1976); A. DeRujula *et al.,* Ann. Phys. **103,** 315 (1977).

[32] A. Buras *et al.,* Nucl. Phys. **B131,** 308 (1977), **132,** 249 (1978); L. F. Abbott *et al.,* Phys. Letters **88B,** 157 (1979).

[33] H. Abramowicz *et al.,* Z. Phys. **C17,** 283 (1983).

[34] Neutrino data from the BEBC (WA25) collaboration, figure from F. Dydak, Ref. 26.

[35] See review by F. Eisele, Ref. 26.

[36] C. Y. Prescott *et al.,* Phys. Lett. **77B,** 347 (1978).

[37] V. W. Hughes, J. Kuti, Ann. Rev. Nucl. Part. Sci. **33,** 611 (1983).

[38] H. E. Montgomery, Proceedings 10th International Symposium on Lepton-Photon Interactions, Bonn (1981), p. 508; N. Schmitz, ibid, p. 527; F. Dydak in Ref. 26.

[39] F. Bergsma *et al.,* Phys. Letters **123B,** 269 (1983); M. Jonker *et al.,* Phys. Letters **128B,** 117 (1983).

[40] M. A. Parker *et al.*, Nucl. Phys, **B232**, 1 (1984); A. E. Asratyan *et al.*, Moscow ITEP print 115, submitted to Sov. J. Nucl. Phys. (1985).

[41] A. Bodek *et al.*, Phys. Rev. Letters **50**, 1431; **54**, 534 (1983).

[42] A. Bodek, J. Ritchie, Phys. Rev. **D23**, 1070 (1981); **D24**, 1400 (1981).

[43] R. Arnold *et al.*, Phys. Rev. Letters **52**, 727 (1984).

[44] S. Stein *et al.*, Phys. Rev. **D12**, 1884 (1975).

[45] M. S. Goodman *et al.*, Phys. Rev. Letters **47**, 293 (1981).

[46] S. Rock, 22nd International Conference on High Energy Physics, Leipzig , East Germany, July 19-25 (1984); J. Gomez, SLAC PUB 3552 (1985).

[47] L. S. Celenza, A. Rosenthal, C. M. Shakin, Phys. Rev. Letters **53**, 892 (1984).

[48] P. J. Mulders, Phys. Rev. **54**, 2560 (1985).

[49] R. L. Jaffe *et al.*, Phys. Letters **134B**, 449 (1984); C. E. Carlson, T. J. Havens, Phys. Rev. Letters **51**, 261 (1983); F. E. Close *et al.*, Phys. Letters **129B**, 346 (1983).

[50] H. Pirner, J. Vary, Phys. Rev. Letters **46**, 1376 (1981).

[51] P. Bosted *et al.*, Phys. Rev. Letters **49**, 1380 (1982).

[52] O. Nachtmann, H. J. Pirner, Z. Phys. **C21**, 277 (1984).

[53] C. H. Llewellyn Smith, Phys. Rev. Letters **128B**, 107 (1983); M. Erickson, A. W. Thomas, Phys. Rev. Letters **128B**, 112 (1983); E. L. Berger, F. Coester, Phys. Rev. **D32**, 1071 (1985).

[54] D. Allasia *et al.*, CERN Proposal SPSC/P210, "Detailed Measurements of Structure Functions from Nucleons and Nuclei", February 1985.

[55] Fermilab experiment E665, T. B. Kirk, (FNAL), V. Eckardt (MPI) spokesmen; D. F. Geesaman, M. C. Green, Argonne print PHY-4622-ME-85 (1985).

[56] R. Arnold *et al.*, SLAC Proposal E140, "Measurement of the x, Q^2, and A-Dependence of R", February 1985.

[57] L. F. Abbott, R. M. Barnett, Annals Phys. **125**, 276 (1980).

[58] Z. E. Meziani, *et al.*, "Proposal to Measure Transverse and Longitudinal Response Functions for Several Nuclei at Momentum Transfer Near $Q^2 = 1$ (GeV/c)^2, NPAS experiment NE9 (1985).

QUARKS AND NUCLEAR PHYSICS

C. E. Carlson

Physics Department
College of William and Mary
Williamsburg, VA 23185 USA

INTRODUCTION

There has been much interest in QCD (Quantum Chromodynamics) among nuclear physicists as well as particle physicists. This is as it should be. QCD is (we believe!) the correct underlying theory for nuclear interactions. Eventually one must be able to use the right theory to explain things which are pure mystery in a phenomenological theory.

Solving QCD is hard. In these lectures we will look at something restricted: QCD treated using perturbation theory, sometimes to be abbreviated PQCD. This should be valid for processes taking place at very short distances or high momentum transfers. Our general motivation will be threefold: to see what limits traditional nuclear physics must approach as the kinematics becomes more extreme; to see where the boundary lies between the "strong interaction region of QCD" and the "asymptotically free" region of QCD, where PQCD is valid; and to get some quark results that are different from what could be got with classical nuclear physics.

The lectures will be divided into four sections. The first and last sections will be two applications of PQCD using just the elementary properties of that theory. The two subjects will be the deuteron form factors and the N–Δ transition. There will also be some data-based discussion of what momenta or energies are needed for PQCD to work, and one application where we will require a more detailed study of PQCD, namely the nucleon electromagnetic form factors.

Before jumping into the deuteron form factor exposition, we will continue with some introductory remarks about how PQCD is used.

The QCD Lagrangian is

$$L_{QCD} = -\frac{1}{4} F_{\mu\nu}{}^a F^{\mu\nu,a} + \overline{\psi}\, \gamma^\mu \,(i\partial_\mu - \frac{1}{2}\, g\, \lambda_a\, A_\mu^{a})\psi$$

The notation is very compact[1]. The quark fields ψ have 3 unwritten indices, a Dirac index, a flavor index, and a color index; λ_a, a=1, ...8, is a set of 3 x 3 matrices operating on the color indices; A_μ^a represents the eight gluon fields; g is the coupling parameter (coupling constant); and

$$F_{\mu\nu}^a = \partial_\mu A_\nu^a - \partial_\nu A_\mu^a - g f^{abc} A_\mu^b A_\nu^c \,.$$

We will consider only up and down quarks and treat them as massless.

A salient feature of the above Lagrangian is that the quark-gluon coupling has the same space-spin structure as QED - vector bosons coupled to spin-1/2 fermions - allowing us to carry over much that we are already familiar with.

We will catalog three mainsprings of the elementary applications of PQCD.

1. Hadronic helicity conservation[2]. Helicity, usually denoted λ, is spin along the direction of motion. The assumptions we make are that the quark mass and binding effects are small, and that the valence quarks dominate the dynamics of the hadrons. These allow us to think when a hadron is moving fast in some direction that its quarks are moving in the same direction, and,

$$\lambda_h = \sum_{q \in h} \lambda_q \,.$$

Thus, quark helicity conservation implies hadron helicity conservation. To see that quark helicities are conserved, write the amplitude for some diagram like Fig. 1 where the quarks are massless and the incoming and outgoing ones are on shell. Use a discrete operator like γ_5 [$\gamma_5 u_\lambda (p) = \text{sgn}(\lambda) u_\lambda (p)$ for massless quarks of helicity 1] to show the diagram is zero unless $\lambda = \lambda'$, incoming helicity equals outgoing helicity. No general statement is possible about the helicities of the intermediate off-shell quarks.

2. Dimensional analysis[3]. This is sometimes useful when the quark mass is strictly zero.

3. Actual calculations of some lowest order diagrams, sometimes piece by piece, can be done to get $O(m/Q)$ terms, where m is some mass scale[4].

Onto the deuteron form factors!

DEUTERON ELECTROMAGNETIC FORM FACTORS

We will derive[5] the asymptotic Q^2 dependence of the form factors and also demonstrate that two of them have a fixed relation at high Q^2. First, however, we must define our notation.

Kinematics and Introduction

Elastic electron-deuteron scattering is given by

$$\frac{d\sigma}{d\Omega} = \frac{d\sigma}{d\Omega}\bigg|_{NS} (A + B \tan^2 \frac{\theta}{2}) \,.$$

"NS" stands for "no structure"; θ is the electron scattering angle in the lab. The quantities A and B are functions only of Q^2, the squared four-momentum transfer to the

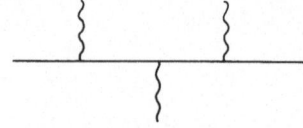

Fig. 1   A quark line with several gluons attached.

deuteron, and in terms of the charge magnetic, and quadrupole form factors we have

$$A(Q^2) = G_c^2 + \frac{2}{3}\,\eta\,G_M^2 + \frac{8}{9}\,\eta^2\,G_Q^2$$

$$B(Q^2) = \frac{4}{3}\,\eta\,(1+\eta)\,G_M^2$$

with

$$\eta = Q^2/4M_d^2 \;.$$

We consider a high Q^2 virtual photon hitting the deuteron and can work either in the Breit or brickwall frame (where the deuteron enters along some direction, absorbs the photon and exits in precisely the opposite direction but with the same magnitude 3-momentum) or in the infinite momentum frame (or IMF, where the deuteron enters moving infinitely fast in a chosen direction, and the photon enters sideways to it).

To get the Q^2 dependence of the form factors at high Q^2, it suffices to consider the deuteron as a collection of parallel-moving constituents, six quarks in QCD, as in Fig. 2(a). One of the quarks absorbs the virtual photon of momentum q, with $Q^2 = -q^2 > 0$. To rebind the deuteron the momentum must be shared equally among the six quarks. We deal with a Q high enough to be much greater than the mean Fermi momentum of the quarks. The Fermi momentum distribution will determine how much deviation from equal sharing of momentum is allowed and so sets the scale of normalization but does not determine the asymptotic dependence of the amplitude of Q^2.

The Q^2 dependence and also the spin dependence are got from analyzing this diagram, along with the other approximately 860,000 lowest order diagrams that contribute to $\gamma^* + d \rightarrow d$.

The results of high Q^2 are that

$$G_C = \frac{2}{3}\,\eta\,G_Q = \frac{Q^2}{6M_d^2}\,G_Q$$

and that

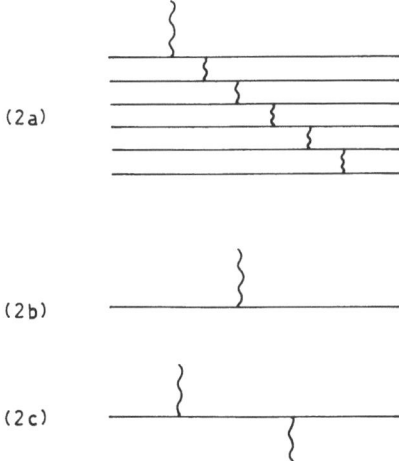

(2a)

(2b)

(2c)

Fig. 2   (a) One of the many lowest order diagrams for the deuteron form factor. (b) and (c) Pieces of the previous diagram.

$$G_C \sim 1/Q^{10}$$
$$G_M \sim m^2/Q^{12}$$
$$G_Q \sim m^2/Q^{12}$$

where m is some mass scale. We proceed to derive those.

The G_C–G_Q Relation[5]

The result will follow just from hadron helicity conservation. It is best to start with helicity amplitudes

$$G^I_{\lambda\lambda'} = <d'\lambda' \mid \varepsilon^\mu_I \cdot J^\gamma_\mu \mid d\lambda>$$

where J^γ is the electromagnetic current operator, d and d' are deuteron momenta, λ and λ' are deuteron helicities, and the index I on the photon polarization vector can be either T for transverse of L for longitudinal. There are 3 independent helicity amplitudes (others can be got by time reversal or space inversion), and in the Breit frame they are G^L_{00}, G^L_{+-}, and G^T_{+0}.

The first step is easy. In the limit that quark mass and binding effects are neglected, we have from hadron helicity conservation

$$G^L_{+-} = 0 = G^T_{+0}$$
$$G^L_{00} \neq 0 .$$

The second step is to relate the helicity amplitudes to the multipole form factors. The standard decomposition of the current gives[6]

$$G^I_{\lambda\lambda'} = <d'\lambda' \mid \varepsilon_I \cdot J^\gamma \mid d\lambda>$$

$$= \varepsilon^\mu_I \{-G_1(\xi'^* \cdot \xi)(d+d')_\mu + G_3 \frac{\xi \cdot q \xi'^* \cdot q}{2M_d^2}(d+d')_\mu$$

$$- G_M [\xi_\mu (\xi'^* \cdot q) - \xi'^*_\mu (\xi \cdot q)]\}$$

with

$$G_C = G_1 + \frac{2}{3}\eta G_Q$$
$$G_Q = G_1 - G_M + (1+\eta)G_3 ,$$

and ξ and ξ' are the initial and final deuteron polarization vectors. At high Q,

$$-\frac{1}{Q} G^L_{00} = G_C + \frac{4}{3}\eta G_Q ,$$

$$\frac{1}{Q} G^L_{+-} = G_C - \frac{2}{3}\eta G_Q ,$$

and

$$-\frac{1}{Q} G^T_{+0} = \sqrt{\eta} G_M .$$

That G^L_{+-} is zero but G^L_{00} not, leads to the quoted result

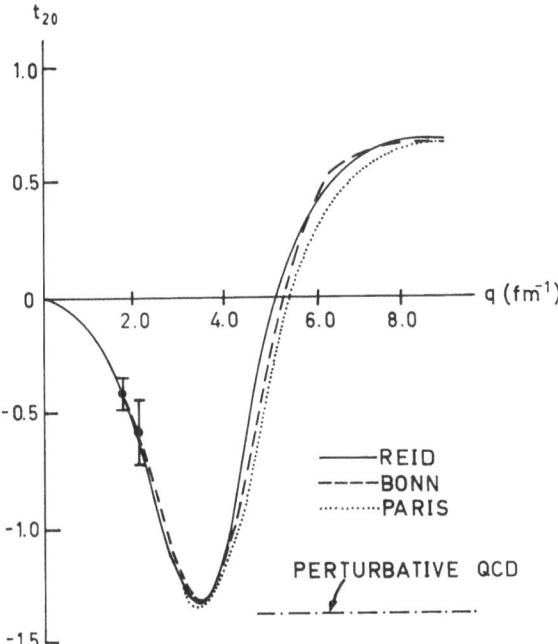

Fig. 3   The deuteron tensor polarization t_{20} as a function of q. The figure comes by courtesy of Roy Holt.

$$G_C = \frac{2}{3} \eta \, G_Q \, .$$

The contrast of this QCD prediction with traditional nuclear physics is dramatic for the tensor polarization, in particular for t_{20}, as may be seen in Fig. 3. If $x \equiv \frac{2}{3} \eta \, G_Q/G_C$, then[7]

$$t_{20} = \sqrt{2} \, (x(x+2))/(1+2x^2)$$

and the PQCD result is $-\sqrt{2}$ which is even the opposite sign from the traditional nuclear physics result for $Q \geq 5 \text{ fm}^{-1} \approx 1\text{GeV}^{-1}$. It will be very interesting to see where the data falls.

There remains the important question of whether the QCD result is valid at those low Q^2. We will defer this question, while stating we can find reason to hope that the result may work toward the edge of Fig. 3, near $Q \approx 2$ GeV.

We will further analyse Fig. 2a and find its Q^2 dependence and also in more detail the spin dependence. Fig. 2a is just a sample. Remember that there are lots of diagrams; many will be zero, and for those that are not we can show that the Q^2 and spin dependence is the same as Fig. (2a). Calculating the normalization is harder, but may soon be possible by teaching the computer to enumerate and evaluate the diagrams[8].

Two pieces of Fig. 2a are shown in Figs. 2b and 2c. They may be described by some simple rules.

The One-Gluon Rule (Fig. 2b) This part of the larger diagram either is proportional to Q and conserves quark helicity or is proportional to m (a mass scale) and flips quark helicity according to whether the absorbed gluon is transverse (T) or longitudinal (L), respectively.

The spin dependence of this rule can easily be verified in the Breit frame. That the helicity conserving possibility is proportional to Q requires only dimensional analysis; the result for the helicity flip is consistent with its being zero for m→0. Similar comments pertain to the next two rules.

The Two Gluon (or Gluon-Photon) Rule (Fig. 2c) If one gluon is absorbed and one emitted, the largest amplitude is constant in Q^2 (this includes the quark propagator but not the gluon propagators) and is the case where one gluon is transverse and the other longitudinal, and quark helicity is conserved. If both gluons are longitudinal, the amplitude is $O(m/Q)$ with quark helicity flipped, and if both gluons are transverse, the amplitude is zero.

The Transverse Gluon Rule Two quark lines connected by a transverse gluon have opposite helicity. This follows because the helicity direction of an absorbed transverse gluon is the same as the helicity direction of the quark that absorbs it; for an emitted transverse gluon the directions are opposite. There is no helicity correlation for quark lines connected by a longitudinal gluon.

These rules can be applied in Fig. 4, where Fig. 2a is reproduced three times with helicity labellings to correspond to G_{00}^L, G_{+-}^L, and G_{+0}^T. Remembering that a $1/Q^2$ from each gluon propagator is not already included in the rules gives

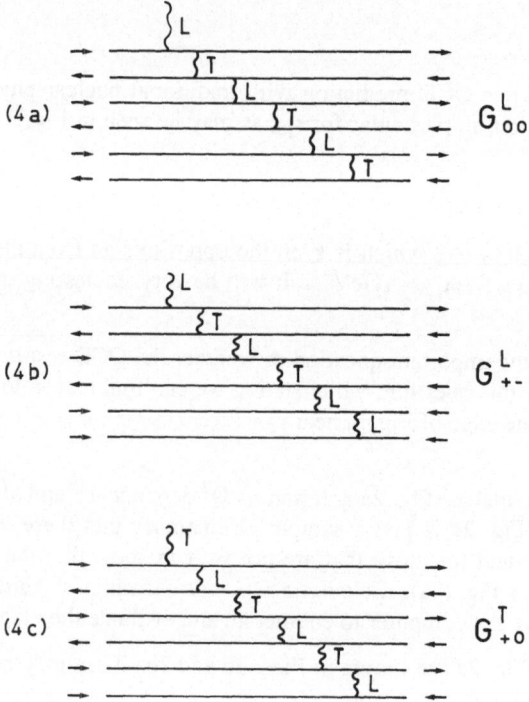

Fig. 4   Three versions of Fig. (2a), specifically labelled for the three independent helicity amplitudes.

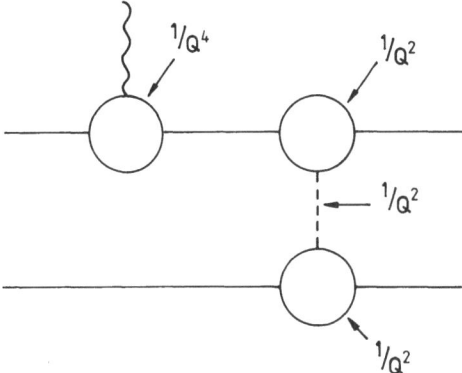

Fig. 5   The deuteron form factor in the "neutron-proton model".

$(1/Q)G_{00}^L \sim 1/Q^{10}$,

$(1/Q)G_{+-}^L \sim m^2/Q^{12}$,

and

$(1/Q)G_{+0}^T \sim m/Q^{11}$,

and this combined with the relations between the helicity amplitudes and multipole form factors leads to the stated results for the latter. Also note that asymptotically[3,9]

$$\sqrt{A(Q^2)} \sim 1/Q^{10} .$$

The result for $A(Q^2)$ was first presented in the context of the quark model. However, one could do a similar analysis for the "neutron-proton model" of the deuteron, as depicted in Fig. 5. The difference is that the nucleons are not elementary so that there are form factors at the vertices. With a $1/Q^4$ at the electromagnetic vertex, a $1/Q^2$ (so chosen because that is standard in nuclear physics calculation) at each meson-nucleon-nucleon vertex, and a $1/Q^2$ from the meson propagator, we get

$$\sqrt{A(Q^2)} \sim 1/Q^{10} ,$$

the same as the QCD result[10]. However, the spin structure and the G_C/G_Q relation will not be the same because the spins and couplings of the mesons exchanged are not the same.

The data[11] for $\sqrt{A(Q^2)}$ are plotted vs. Q^2 in Fig. 6. The data as a whole do not fall like $1/Q^{10}$, but the last three points may. This would suggest $Q^2 \sim 4\ GeV^2$ as a benchmark for the onset of asymptotic behavior.

Summary

Perturbative QCD predicts that $\sqrt{A(Q^2)}$ falls like $1/Q^{10}$ asymptotically, but this prediction is not unique to QCD.

PQCD also predicts that $G_C = \frac{2}{3}\eta\ G_Q$, and this function is reasonably unique. (It could be obtained in more traditional nuclear physics by making some form factors fall much faster than others, so that all that survived at high Q^2 were vector mesons with simple vector couplings. The analysis is then the same as PQCD. The result does not

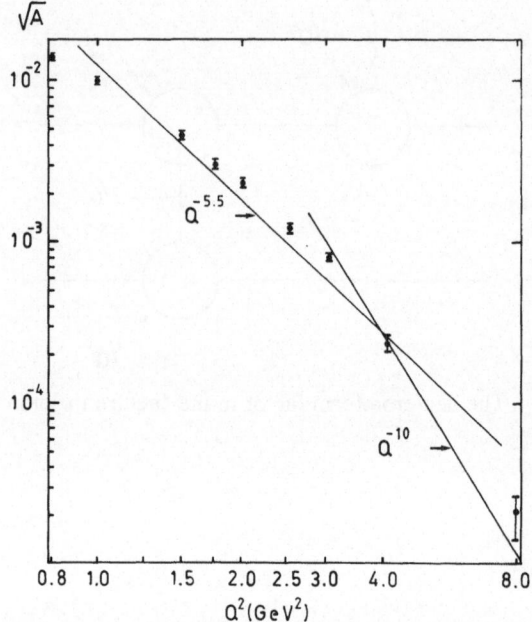

Fig. 6   The measured deuteron form factor $A^{1/2}$ plotted vs. Q^2.

depend on having 6 fermions in one case and 2 in the other. I have never seen the form factors so arranged but want to note the possibility).

Asymptotic behavior may begin at $Q^2 \sim 4$ GeV^2. More discussion of this will come.

Finally, experimental measurements of G_C/G_Q and G_M are important for understanding the deuteron and nuclear binding and should - and surely will - be measured independently of anything we may say.

NECESSARY ENERGY SCALE FOR PQCD TO WORK: EVIDENCE FROM DATA

We will make some comments based on data as to when PQCD may begin working. The data to be studied will fall into two categories, namely predictions for $d\sigma/dt$ in two body hadronic reactions $A+B \rightarrow C+D$ and predictions for the Q^2 dependence of electromagnetic form factors.

A typical lowest order diagram for a typical 2 body process, $\pi N \rightarrow \pi N$, is shown in Fig. 7a. Four gluons are needed to completely connect the diagram and share out the transferred momentum so that the sets of quarks in the final hadrons have parallel momenta. Then fix the scattering angle so there is only one energy scale, say s (the squared c.m. energy). The amplitude may then be analyzed either dimensionally or by rules like the ones we used for the deuteron form factors to learn[12]

$$ m \sim \frac{1}{s^3} f(\theta) $$

or that

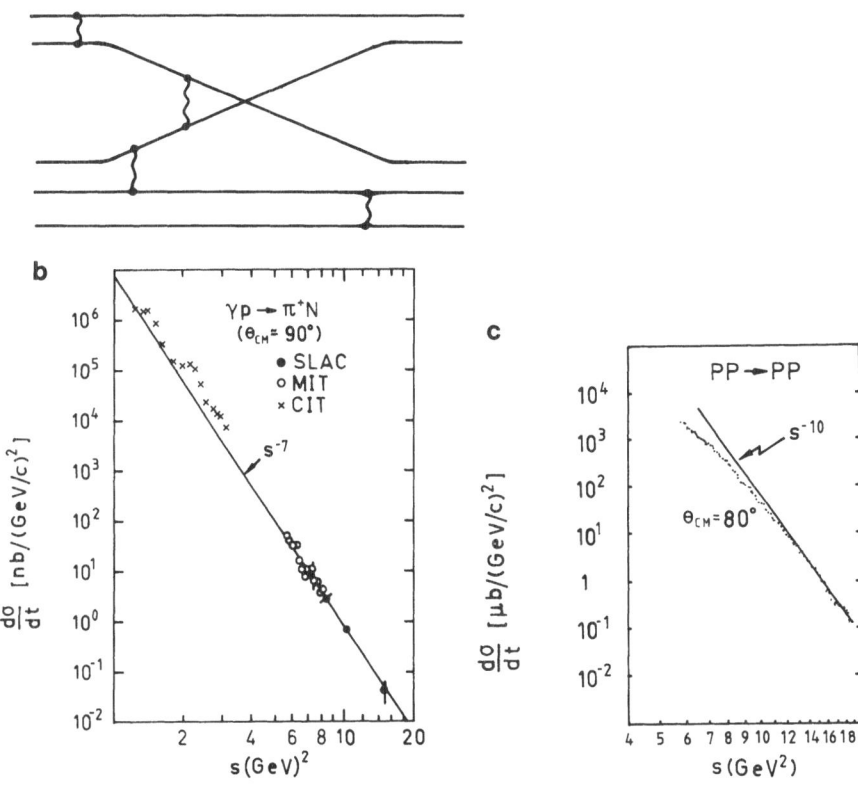

Fig. 7   (a) One lowest order completely connected diagram for πN→πN. (b) Plot of
dσ/dt vs. s for γp→π⁺n. (c) The same for pp elastic scattering. [(b) and (c)
come by courtesy of Stanley Brodsky.]

$$\frac{d\sigma}{dt} \sim \frac{1}{s^2} \ |m|^2 \sim \frac{1}{s^8}$$

at fixed angle.

For general A+B→C+D,

$$\frac{d\sigma}{dt} \sim s^{2-N_A-N_B-N_C-N_D}$$

where N_A is the number of quarks or elementary field in A, etc. Thus[21],

$$\frac{d\sigma}{dt} \sim \quad s^{-10} \quad pp \to pp$$

$$s^{-7} \quad \gamma p \to \pi^+ n .$$

Fig. 7 plots $\frac{d\sigma}{dt}$ vs. s for several processes. One can see that the QCD prediction
works above some minimum s which is about 10 GeV² for pp → pp but just 2 GeV² for
γp → π⁺n.

Regarding form factors, the PQCD result for the helicity conserving form factors is a Q^2 fall-off just given by the gluon propagators (Fig. 2a will do as an example); that is,

$$F(Q^2) \sim \left[\frac{1}{Q^2}\right]^{N-1}$$

for an N-quark system.

For the proton, G_M is the helicity conserving form factors, and $Q^4 G_M$ - which should be constant[4] at high Q^2 - is plotted vs. Q^2 in Fig. 8. At 5 GeV^2 (the first data point with an error bar indicated is at 5 GeV^2) and beyond the curve is substantially flat. Lines that the should follow if G_{Mp} fell like $1/Q^2$ or $1/Q^6$ are shown on the figure and are definitely far from the data.

So it seems that the power law behavior predicted by perturbative QCD is followed by the data for, as a benchmark, Q^2 or s of about 5 GeV^2.

PROTON FORM FACTOR NORMALIZATION

In the proton form factor calculation (and in other processes too), we should not be satisfied just to obtain and check the Q^2 dependence, but we should also try to obtain the normalization and see how it compares to the data. This requires that all the lower order perturbation theory diagrams be calculated and that the proton wave function be known. The first is quite possible; we will display the answer below. The proton wave function is not known and not at present calculable. (We should, however, mention that QCD sum rules allow determining some moments of the proton wave function which can be used to obtain a model wave function, and that lattice gauge theories may eventually give an answer. Some progress is being made on the pion wave function using lattice gauge theory.) Not knowing the wave function changes the question we must ask: Does there exist *some* wave function that fits the data? The wave function of course should fit the data without egregious side problems such as excessive quark transverse momenta or where terms neglected in calculating QCD perturbatively exceed both the PQCD

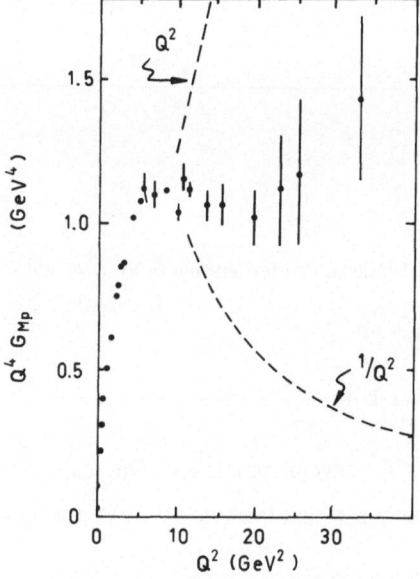

Fig. 8   The proton form factor data plotted vs. Q^2.

calculation and the data. We believe a suitable wave function can be found. We shall proceed to calculate (or show the result of calculating) the Feynman diagrams, analyze the results of choosing a simple and then a more general wave function, and then discuss some "soft contributions" omitted in PQCD.

Calculation of T_H

We shall concentrate on the 3 quark part of the wave function; Fock components with more constituents require more gluon exchanges to give all the constituents parallel momenta in the final state and so their contribution to the form factor falls faster with Q^2. Also we shall work in an infinite momentum frame where the entering proton is moving along the z-axis and shall use momentum components say for quark i;

$$\vec{k}_{iT} = (k_i^1, k_i^2)$$

and

$$x_i = k_i^+/p^+$$

where $k_i^+ = k_i^0 + k_i^3$ and p is the proton momentum. For the three quarks in the initial proton

$$\sum k_{iT} = 0$$

and

$$\sum x_i = 1 .$$

The proton form factor can be represented by Fig. 9a or by

$$G_M = \int [dx] [dy] \, \phi(x,Q) \, T_H(x,y,Q) \, \phi(y,Q)$$

where

$$\phi(x,Q) = \int^Q [dk_T] \, \psi \, (x,k_T) .$$

We use

$$[dx] = \Pi \, dx_i \, \delta(1-\Sigma x_j)$$

and

$$[dk_T] = \Pi \, \frac{d^2 k_{iT}}{16\pi^3} \cdot 16\pi^3 \, \delta^{(2)} \, (\Sigma \, k_{jT}) .$$

"Distribution amplitude" is the name given to ϕ, and the wave function ψ is normalized by

$$\int [dx] [dk_T] \, |\psi(x,k_T)|^2 = P_{3q} .$$

The "hard scattering amplitude", T_H, is calculated as a form factor for three parallel quarks going into three parallel quarks. A fuller expression for G_M might include some dependence on initial and final quark transverse momenta, but for large Q, the effect of these transverse momenta on T_H is small so they can be integrated as done implicitly above. The Q-dependence of $\phi(x,Q)$ is expected to be weak (i.e., logarithmic) and the $1/Q^4$ behavior comes from T_H.

There are 42 diagrams that can be drawn for T_H, but only 14 are non-zero, and only the 4 drawn in Fig. 9b need be calculated, the others being obtained by symmetries. For e_j being an operator that gives the charge of quark j, we get

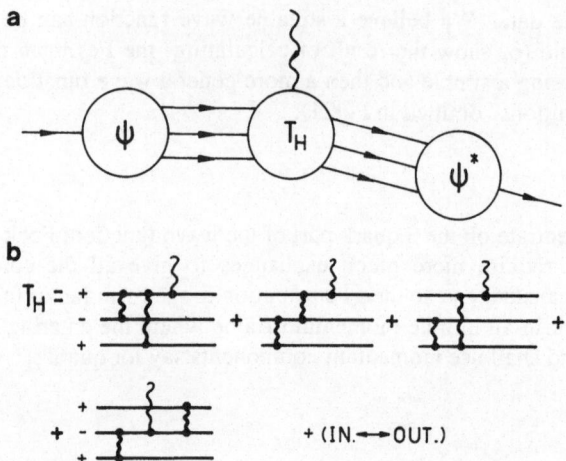

Fig. 9   (a) The proton form factor calculation. (b) Perturbation theory diagrams for
the hard scattering amplitude.

$$T_H = \left[\frac{16\pi\alpha_s(Q^2)}{3Q^2} \right]^2 \sum_{j=1}^{3} \{e_j T_j + (x \Leftrightarrow y)\}$$

with

$$T_1 = \frac{1}{x_3(1-x_1)^2} \frac{1}{y_3(1-y_1)^2} + \frac{1}{x_2(1-x_1)^2} \frac{1}{y_2(1-y_1)^2}$$

$$- \frac{1}{x_2 x_3(1-x_3)} \frac{1}{y_2 y_3(1-y_1)} = T_3(1 \Leftrightarrow 3)$$

and

$$T_2 = \frac{1}{x_1 x_3(1-x_1)} \frac{1}{y_1 y_3(1-y_3)} .$$

We note the $1/Q^4$, the additional $\log(Q^2/\Lambda^2)$ dependence within the strong coupling
parameters α_s, and the singularities near the kinematic boundaries of x_i and
y_i $[0 \le x_i, y_i \le 1]$. They could be controlled by giving the quarks and gluons a small mass,
but in any case should not be allowed to play a strong role or else perturbation theory
cannot apply.

Simple Choice of Wave Function

A simple and factorizable form of the wave function is

$$\psi(x,k_T) = N(x_1 x_2 x_3)^\eta \, e^{-\sum k_{iT}^2/2\alpha^2}$$

whereupon

$$\phi(x) = N'(x_1 x_2 x_3)^\eta .$$

This is really a one-parameter family of wave functions, the parameter being the power
η, and we will use it to begin learning how the form factor depends on the chosen wave
function. The parameter α that controls the transverse momentum should be some

reasonable value; one non-relativistic model[16] suggested $\alpha = 0.32$ GeV.

The integrals to get G_M can be done analytically. The power must satisfy $\eta > \dfrac{1}{2}$ to make those integrals converge. One way to begin looking at the results is to examine the ratio G_{Mp}/G_{Mn}, plotted in Fig. 10. The proton form factor has a zero at $\eta = 1$ and the neutron form factor has a zero at $\eta = \dfrac{1}{2} + \dfrac{1}{\sqrt{12}} \approx 0.8$. Of course, the neutron and proton form factors have opposite signs, and this constrains the value of η we may choose. Further, since it is the proton form factor that is positive, we are constrained to

$$\frac{1}{2} < \eta \le 0.8 .$$

(Incidentally, there was for a while a trivial error in the literature in the overall sign of T_H and hence of the calculated G_{Mp}. This had the effect of requiring $\eta > 1$ with consequent large effect on the normalization.)

We can quote the result

$$Q^4 G_{Mp} (Q^2) = \frac{1}{\pi^2} \alpha_s^2 \alpha^4 \frac{64}{27} \frac{1-\eta}{\eta} N_\eta^2 X^2$$

with

$$N_\eta^2 = \frac{(6\eta+2)!}{[(2\eta)!]^3}$$

and

$$X = \frac{(\eta!)^2 (\eta-1)!(2\eta-2)!}{(2\eta)!(3\eta-1)!} .$$

For $\eta = 0.6$, $\alpha_s = 0.25$, and $\alpha = 0.32$ GeV, this gives[17]

$$Q^4 G_{Mp} \approx 0.8 \text{ GeV}^4$$

which is near the data for $Q^2 \ge 5$ GeV^2.

More General Choice of Wave Function

The proceeding simple wave function is perhaps too simple. It demonstrates that the observed size of the form factor can be obtained, at least at high Q^2, from PQCD. On the other hand, the experimental value[18] of G_{Mp}/G_{Mn} is about -2 at the highest Q^2 where

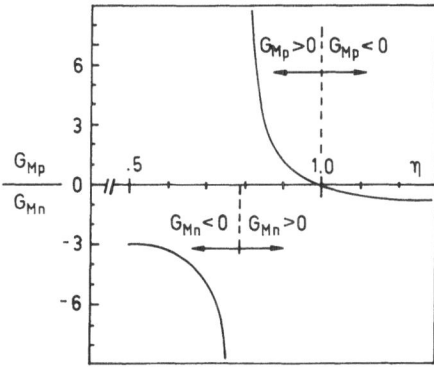

Fig. 10   Calculated G_{Mp}/G_{Mn} for the simple symmetric and factorizable wave function with parameter η.

measurements are available and this value cannot be obtained with the previous wave function as one may see from Fig. 10 (as $\eta \to \infty$ the ratio asymptotically approaches -1). We will in fact discover that G_{Mn}/G_{Mp} seems to require that the nucleon wave function is unsymmetric among the three quarks.

We shall continue to use a factorizable wave function

$$\psi(x,k_T) = N \, \phi(x) \, e^{-\sum k_{iT}^2/2\alpha^2}$$

and shall expand $\phi(x)$ in polynomials (times a weight factor $x_1 x_2 x_3$ to decrease sensitivity to the end point singularities).

For the present purposes, the particular choice of polynomials is not crucial, but we might as well use polynomials which are standard[13,14] to the subject and which would be useful if we were to ask further questions. The wave equation for $\phi(x,k_T)$ has a kernel for which perturbation theory is invalid in general, but is dominated by one gluon exchange for high k_T. This observation can be turned into an "evolution equation" that governs how the distribution amplitude $\phi(x,Q^2)$ changes with Q^2 for high Q^2. The evolution equation is solvable by the separation of variables technique to yield

$$\phi(x,Q^2) = x_1 x_2 x_3 \sum N_i \, \bar{\phi}_i(x)$$

where

$$N_i = N_i(Q^2) = n_i \, e^{-\gamma_i \ln \ln(Q^2/\Lambda^2)} \, .$$

The n_i are non-calculable constants but the γ_i are calculable[13,14] and are positive and monotonically increasing with i. The first six "Appel polynomials" are

$$\bar{\phi}_0 = 1$$

$$\bar{\phi}_1 = x_1 - x_3$$

$$\bar{\phi}_2 = 2 - 3(x_1 + x_3)$$

$$\bar{\phi}_3 = 2 - 7(x_1 + x_3) + 8(x_1^2 + x_3^2) + 4x_1 x_3$$

$$\bar{\phi}_4 = x_1 - x_3 - \frac{4}{3}(x_1^2 - x_3^2)$$

$$\bar{\phi}_5 = 2 - 7(x_1 + x_3) + \frac{14}{3}(x_1^2 + x_3^2) + 14 \, x_1 x_3 \, .$$

(Quarks 1 and 3 are the ones with parallel spin and some of the above are symmetric and some antisymmetric under $1 \leftrightarrow 3$).

It is now straight forward although time consuming to calculate the form factors and normalization condition. We will quote the results[19,20] even though the formulas are a bit long. We have

$$Q^4 G_{Mp}(Q^2) = 64\pi^2 \, C_B^2 \, \alpha_s^2 \, \frac{1}{1296} \, \{20N_1^2 - 42\sqrt{3} \, N_0 N_1$$

$$+ 36 \, N_2^2 + 28\sqrt{3} \, N_1 N_2 - 54 \, N_0 N_2$$

$$+ 188 \, N_3^2 - 144 \, N_2 N_3 - \frac{220}{\sqrt{3}} \, N_1 N_3 + 198 \, N_0 N_3$$

$$+ \frac{26}{27} \, N_4^2 + \frac{46}{\sqrt{3}} \, N_3 N_4 - \frac{22}{\sqrt{3}} \, N_2 N_4 - \frac{26}{3} \, N_1 N_4 + 10\sqrt{3} \, N_0 N_4$$

$$+ \frac{77}{9} N_5^2 - \frac{41}{9\sqrt{3}} N_4 N_5 - 59 N_3 N_5 + 11 N_2 N_5 + \frac{35}{\sqrt{3}} N_1 N_5 - 42 N_0 N_5 \}$$

$$Q^4 G_{Mn}(Q^2) = 64\pi^2 C_B^2 \alpha_s^2 \frac{1}{1296} \{ 54 N_0^2$$

$$+ 22 N_1^2 + 42\sqrt{3} N_0 N_1$$

$$- 6 N_2^2 - 28\sqrt{3} N_1 N_2 + 54 N_0 N_2$$

$$- \frac{170}{3} N_3^2 + 60 N_2 N_3 + \frac{220}{\sqrt{3}} N_1 N_3 - 30 N_0 N_3$$

$$+ \frac{26}{27} N_4^2 - \frac{46}{\sqrt{3}} N_3 N_4 - \frac{26}{3} N_1 N_4 - 10\sqrt{3} N_0 N_4$$

$$- \frac{145}{54} N_5^2 + \frac{41}{9\sqrt{3}} N_4 N_5 + \frac{65}{3} N_3 N_5 - \frac{5}{3} N_2 N_5 - \frac{35}{\sqrt{3}} N_1 N_5 + 20 N_0 N_5 \}$$

where $C_B = \frac{2}{3}$. Of course, we cannot choose the N_i as large as we wish but must satisfy the normalization condition

$$\frac{1}{165} [165 N_0^2 + 11 N_1^2 + 33 N_2^2 + 17 N_3^2 + \frac{1}{3} N_4^2$$

$$+ \frac{55}{9} N_5^2 - 44 N_0 N_3 + 2 N_2 N_3 + \frac{2}{3} N_1 N_4 - \frac{22}{3} N_0 N_5$$

$$- \frac{14}{3} N_2 N_5 + \frac{2}{3} N_3 N_5] = \frac{7!}{48\pi^4} \alpha^4 P_{3q} .$$

The formulas being long, let us break them down with the following observations:

(i) There is no N_0^2 term for the proton. This is the same as the zero at $\eta = 1$ in the previous wave function.

(ii) If only one N_i is non-zero, we get the most form factor for the normalization condition for i=3. Indeed, with $P_{3q} = 1$, $\alpha_s = 0.3$, and $\alpha = 0.39$ Gev, the proton form factor has the value indicated for the data at $Q^2 \geq 5$ GeV^2.

(iii) A single $\tilde{\phi}_i$ does not give the right ratio G_{Mp}/G_{Mn}. For symmetric $\tilde{\phi}$ the ratio is either zero or more negative than -3; for anti-symmetric $\tilde{\phi}_1$ the ratio is about +1. Pairs of symmetric or pairs of anti-symmetric $\tilde{\phi}_i$ do not work either.

(iv) A distribution amplitude that works quite well at giving both G_{Mp} and G_{Mn} is

$$\phi_M(x) = (0.38 \text{ GeV}^2) \, x_1 x_2 x_3 \, (\tilde{\phi}_3 - \tilde{\phi}_1) .$$

This amplitude is perhaps surprising in that it is asymmetric, and below we shall speculate a bit as to how this can come about. For now we should emphasize that the apparent asymptotic G_{Mn} has a size as well as a Q^2 falloff that is easy to match in PQCD with reasonable values of the QCD coupling constant and quark transverse momentum.

Soft Contributions

The PQCD expression for the form factor can be derived as an approximation to the impulse approximation. At high Q^2 only the "tail" or high k_T part of the wave function is important, and this is the piece that can be calculated in PQCD and substituted in the usual impulse approximation to obtain the PQCD result for G_M. We should mention that the impulse approximation is the dominant contribution[14] to the form factor at high Q^2; the same is not true at low Q^2.

There remain low k_T parts of the wave function that can make contributions to G_M. These are the "soft contributions"[15] and are not included in PQCD. It is (or will be) clear enough that they fall faster with Q^2 than the PQCD or "hard" contributions. The question remains: How big are they at Q^2's where experiments are done? For the wave function we shall use, the answer will be that they are big - unfortunately or fortunately - but not dominant[15].

We will work with the factorized form of the wave function

$$\psi(x,k_T) = \phi(x)\, g(k_T)\,,$$

where

$$\int^{Q^2} [d^2k_T]\, g(k_T) = 1$$

and write g as a sum of two parts,

$$g = g_{soft} + g_{hard}\,.$$

The "hard" part is the tail of the wave function and can be calculated using perturbative QCD, and using only it leads to the hard scattering formula for G_M given above. Let us ignore g_{hard}, setting it to zero, continue to approximate g_{soft} by a gaussian,

$$g_{soft} = \frac{192\pi^4}{\alpha^4}\, \exp(-\textstyle\sum k_{iT}^2/2\alpha^2)\,,$$

and see how the soft contributions behave.

The soft contributions are

$$Q^4\, G_{soft} = Q^4 \sum e_i \int [dx][d^2k_T]\Psi(x,k_T)\,\Psi(x,1_T^{(i)})$$

$$= 48\pi^4 Q^4/\alpha^4 \int [dx]\, \phi^2(x)\, \exp[-(x_1^2 + x_2^2 + x_2 x_2)\, Q^2/2\alpha^2]$$

$$= f(\zeta)$$

where Ψ is ψ times the spin-isospin function, $1_T^{(i)}$ represents the quark transverse momenta of the final nucleon in the case where the ith quark is struck and e_i is the charge of the struck quark. For simplicity of presentation, the second formula above is given only for the case of symmetric ϕ (although we give numerical results for a mixed symmetry case), and $\zeta \equiv Q^2/2\alpha^2$. At high Q^2, $f(\zeta)$ falls like $1/Q^2$.

For a given ϕ the soft contributions to $Q^4\, G_{Mp}$ depend only on ζ. However, for the distribution function ϕ_M suggested above, the peak of $f(\zeta)$ is about 4 GeV^4 or about four times the data, and for $\alpha=0.39$ GeV the peak in $f(\zeta)$ occurs at Q^2 of 20 GeV^2. This is not good. Let us proceed.

Adding the tail has a significant effect on the linear constraint on $g(k_T)$. As a model, let

$$g = g_{soft} + g_{hard}$$

$$= 768\pi^4\{A/4\alpha^4 \exp -\textstyle\sum k_{iT}^2/2\alpha^2$$

$$+ B\,\theta\,(p_{\rho T} > \lambda)\theta(p_{\lambda T} > \lambda)/(p_{\rho T}^2 + p_{\lambda T}^2)(p_{\lambda T}^2 + \lambda^2)\}$$

where $p_\rho = (k_1-k_2)/\sqrt{2}$, $p_\lambda = (k_1+k_2-2k_3)/\sqrt{6}$, and we have the tail only above momentum λ. The constraints read

$$\int [d^2k_T]g = 1 = A + (1/2)B \; \text{ln}^2(Q^2/2\lambda^2)$$

$$\int [dxd^2k_T] \; |\psi|^2 = p_{3q} = 48\pi^4\{A^2/\alpha^4 + 2\,B^2/\lambda^2\} \int [dx]\phi^2 .$$

Using this model wave function and seeing that g_{soft} peaks more sharply than g_{hard} gives

$$Q^4 G_{Mp} = A^2 \, f(\zeta) + 2{<}g_{soft}{>} \int [dx] \, \phi^2(x)$$

$$\times Q^4 \; g_{hard} \; [p_{\rho T} = (x_1-x_2)Q/\sqrt{2}; \; p_{\lambda T} = (3/2)^{1/2}(x_1+x_2)Q]$$

$$= A^2f(\zeta) + 512\pi^4 \; AB \int [dx] \; \phi^2/(x_1^2 + x_2^2 + x_1x_2)(x_1 + x_2)^2 .$$

The second term above is independent of Q^2 and should be the hard scattering result. Taking ϕ to be our standard ϕ_M and comparing to the hard scattering result given earlier leads to $AB = 2.6 \; \alpha_s^2/\pi^2$. This in turn leads to $A = 1/2$ and $\alpha \sim 0.38$ GeV/$\sqrt{2} = 0.27$ GeV.

Thus, among the effects of considering the tail of the transverse wave function has been a reduction of the "soft contributions" to G_{Mp} by a factor of four so that at their peak they are equal to the data, and the peak comes at about 10 GeV^2.

Discussion

We shall make some remarks.

Factorized Wave Function We have been working for simplicity with a factorized form for the wave function. This is not necessary. The hard and soft regions of transverse momenta could easily have different x-dependences, and this can give us significant extra freedom to manipulate the hard and soft contributions.

Asymptotic Ratio G_{Mp}/G_{Mn} It has been noted[13] that at very, very high Q^2 (i.e. ln ln $Q^2{\gg}1$) only the zeroth Appel polynomial survives and the proton form factor goes to zero relative to the neutron form factor. It should also be noted that in this limit the neutron form factor is positive, so that the neutron form factor may have a zero[17] at some large but finite Q^2. At present Q^2 one may then expect that the neutron form factor is falling relative to the proton, rather than vice versa, and it is interesting to note that the present data[18] on the neutron form factor shows just this effect. The ration G_{Mn}/G_{Mp} falls by about 30% as Q^2 goes from zero to 10 GeV, the limit of the data.

Chernyak and Zhitnitsky Distribution Amplitude[21] Chernyak and Zhitnitsky have proposed a distribution amplitude for the proton. Their distribution amplitude is got by supposing an expansion in terms of the six lowest Appel polynomials and fitting to six moments that are calculated using QCD sum rules, and is

$$\phi(x) = x_1x_2x_3 \; (0.111\bar\phi_0 - 0.274\bar\phi_1 - 0.212\bar\phi_2$$

$$+ 0.248\bar\phi_3 + 0.221\bar\phi_4 + 0.002\bar\phi_5) \; (\text{GeV})^2 .$$

An examination of the hard scattering expression for G_{Mp} shows that every single term there is positive if N_0, N_s and N_4 have one sign and N_1, N_2, and N_5 have the opposite sign. This is just the sign pattern in the Chernyak-Zhitnitsky distribution amplitude, excepting the last term whose coefficient is too small to be significant. The QCD sum rules have thus led to a distribution amplitude which satisfies one clear criterion for maximizing G_{Mp}.

Asymmetric Wave Function The sorts of wave functions that fit both the neutron and proton form factors are quite asymmetric in the three quarks. This is perhaps a surprise, and it may be worth speculating how it may come about. First, we should note that the distribution amplitude, which is a transverse momentum integrated wave function, is dominated by the high k_T part of the wave function (if the wave function falls in k_T as expected from PQCD). At the same time, the normalization (which unlike the distribution amplitude is gotten by squaring the wave function before integrating) is dominated by low k_T. Our expectation of near symmetry among the quark wave functions come from calculations of things like the charge radius or magnetic moment that are like the normalization in being dominated by the low k_T part of the wave function, and the x-dependence associated with this could be quite symmetric.

Why then might one expect an asymmetry at high k_T? Think of quark-quark scattering, or equally well, electron-electron scattering at very high energies. There is a large and angle dependent spin dependence. At 90° in the c.m., the amplitude for scattering two same helicity electrons is twice the magnitude of the amplitude for opposite helicity electrons. When we have high k_T quarks, they are got by a hard scattering from low k_T quarks, and now we know this amplitude is spin dependent. The pair of quarks with same helicity are more likely to scatter each other out to high k_T than other parts of quarks, and this same scattering will likely also scatter the quarks forward and backward so that one of the same helicity quarks will have a large share of the longitudinal momentum. This is just what is seen.

ELECTROMAGNETIC N–Δ TRANSITIONS AT HIGH–Q^2

The process here is $e + N \rightarrow e + \Delta$. At low Q^2, recent interest in this reaction[22,23] has been motivated by the possibility of seeing "deformations" in the quark wave function of the nucleon and delta. "Deformations" here means admixtures of spatial states other than the lowest S-states in the N and Δ.

The idea at low Q^2 is as follows: Three multipole amplitudes, the M1, E2 and C2, can contribute to a $J^P = \frac{1}{2}^+ \frac{3}{2}^+$ transition such as the N–Δ. If we neglect recoil and if there is no deformation, then

$$F_{E2} = F_{C2} = 0$$

since both of these involve l=2 spherical harmonics. The M1 requires just a spin flip, and so is non-zero and dominant. Now if there is some tensor interaction coming from a pion cloud or gluon exchange or anywhere, there will be some admixtures of D-state in the N and Δ. The size of this admixture can be measured by measuring the size of the E2 and C2 amplitudes.

The perturbative QCD result, valid at high Q^2, sits in contrast to the small E2 and C2 amplitudes seen at low Q^2. The results, which follow just from hadronic helicity conservation, are that[24]

$$F_{E2} = \sqrt{3} F_{M1}$$

and that the contributions to the cross section from F_{C2} are small.

A second result is also obtainable. We will learn the asymptotic Q^2 dependence of the helicity amplitudes, or equivalently of the M1, E2, and C2 amplitudes. Without the underlying theory, one cannot predict the Q^2 dependence or determine the relevance of any comparison between some experimentally determined N–Δ form factor and the nucleon form factors. There is, for example, a temptation to look at the M1 amplitude, divide out the threshold behavior to obtain a form factor that is non-zero at $Q^2=0$, just like the nucleon G_M, and then compare this N–Δ form factor to the dipole form. It falls faster than the dipole form. With the help of perturbative QCD, we will be able to show that the N–Δ form factors in question should fall like $1/Q^6$ asymptotically, and so the observed falloff is not a surprise.

We now proceed to obtain the results we have introduced in the following subsections on kinematics, simple QCD results, comments on data, and normalized QCD results.

Kinematics

We shall work in the Breit frame, wherein the nucleon, photon, and delta are collinear, with the incoming nucleon and outgoing delta having the opposite direction but same magnitude 3-momentum. The three independent helicity amplitudes are (Fig. 11)

$$G_m = <\Delta,\lambda' = m - \tfrac{1}{2}\,|\,\epsilon_m^\mu \cdot J_\mu^\gamma\,|\,N,\lambda = \tfrac{1}{2}>/2m_N$$

for $m = +, 0$ and $-$ with

$$\epsilon_{+-} = (0, +-1, -i, 0)/\sqrt{2}\ .$$

ϵ_0 is normalized and satisfies $\epsilon_0 \cdot q = \epsilon_0 \cdot \epsilon_{+-} = 0$ and J^γ is the electromagnetic current. Angular momentum conservation guarantees that $m = \lambda + \lambda'$ and the factor $1/2m_N$ makes G_m dimensionless.

The differential cross section including the possibility of polarized electrons and nucleons is

$$\frac{d\sigma}{d\Omega} = \frac{d\Sigma}{d\Omega} + hP\,\frac{d\Lambda}{d\Omega}$$

where h and P are the electron longitudinal polarization and nucleon polarization, respectively, and

$$\frac{d\Sigma}{d\Lambda} = \frac{\sigma_0 f_\Delta^{-1}}{1 + \tau^*}\ \{G_0^2 + (1/2)\,[1+2(1+\tau^{*)}\,\tan^2\theta/2]\,(G_+^2 + G_-^2)\}$$

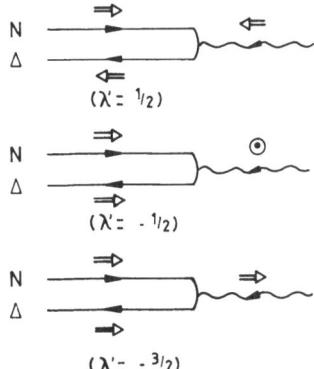

Fig. 11   The three independent helicity amplitudes for electromagnetic N–Δ transitions.

$$\frac{d\Delta}{d\omega} = \frac{\sigma_0 f_\Delta^{-1}}{1+\tau^*} \{\sqrt{2}(1+\tau^*)^{1/2} \tan\theta/2 \sin\beta \cos\alpha \, G_+G_0$$

$$+ [(1+\tau^*)(1+\tau^* \sin^2\theta/2)]^{1/2} \tan\theta/2 \sec\theta/2 \cos\beta (G_-^2 - G_+^2)\}$$

$$\sigma_0 \equiv \frac{4\alpha^2}{Q^4} \frac{E'^3}{E} \cos^2\theta/2$$

$$f_\Delta \equiv 1 - \frac{m_\Delta^2 - m_N^2}{2m_N E}$$

Here E and E' are the initial and final electron energies

$$\tau^* = v^2/Q^2 = (Q^2 + m_\Delta^2 - m_N^2)^2 / 4m_N^2 Q^2,$$

and α and β are the azimuth and polar angles of the nucleon polarization direction in the nucleon rest frame (with the z-axis defined by the incoming photon momentum and the x-axis lying in the scattering plane)[25].

The same cross section written in terms of C2, M1 and E2 multipole amplitudes may be got with the substitutions[26,27]

$$G_0 = \sqrt{(4\pi)} \frac{Q}{|\vec{q}|} F_{C2} = \frac{\sqrt{(4\pi)}}{\sqrt{(1+\tau^*)}} F_{C2},$$

$$G_+ = \sqrt{(4\pi)} (F_{M1} + \sqrt{3}F_{E2})/2$$

$$G_- = \sqrt{(4\pi)} (-\sqrt{3}F_{M1} + F_{E2})/2$$

where $\vec{q}$ is the photon 3-momentum in the nucleon rest frame (lab frame).

First QCD Predictions

The most elementary prediction of what happens at high Q^2 follows from the quark, and, therefore, the only helicity amplitude with the same helicity for the outgoing delta as for the incoming nucleon, it is the one that is large and G_0 and G_- are asymptotically zero relative to it. From the latter immediately follows the result[24]

$$\lim_{Q^2 \to \infty} F_{E2} = \sqrt{3}F_{M1} .$$

This is an interesting contrast to the non-relativistic result where, to the extent that the nucleon and delta have spherically symmetric spatial wave functions and recoil can be neglected, the E2 and C2 amplitudes are both zero and the M1, requiring just a spin flip, dominates.

We can also establish more specifically that at high Q^2

$$G_0 = O(m/Q) \times G_+$$

$$G_- = O(m^2/Q^2) \times G_+$$

where m is some mass scale. These follow from the same rules that were used in treating the deuteron form factors. Noticing that $|\vec{q}| \sim Q^2/m$, we see that the C2, M1 and E2 amplitudes fall at the same rate asymptotically. However, because of the kinematic factors involved, the C2 amplitudes will not contribute significantly to the cross sections at high Q^2.

Comment on Q^2 - Dependence and the Data

There is data on the N–Δ transition for Q^2 up to 2.4 GeV^2 (and upper limits beyond that). A form factor, which we will call F_{expt} below, can be extracted from this data in a plausible way[27], and it has been noted that this form factor falls with Q^2 at a rate faster than the dipole form factor for the nucleon. Now with the guidance of perturbative QCD we can examine the data and this particular extraction of form factor once again. We will discover that the form factor F_{expt} should fall like $1/Q^6$ asymptotically, and so we should feel no surprise that it falls faster than a dipole.

The procedure used to get F_{expt} is as follows[27]: Since the C2 and E2 amplitudes are (or were!) expected to be small, omit them from the expression for $d\sigma/d\Omega$. Then use $d\sigma/d\Omega$ to get "F_{M1}" from the data. Next remove a threshold factor $|\vec{q}^*|$ from the amplitude "F_{M1}" and also a normalization factor $[(E^* + m_N)/2m_N]^{1/2}$; $\vec{q}^*$ and E^* are the photon 3-momentum and nucleon energy in the Δ rest frame. Then,

$$C_3(q^2) = \left[\frac{2m_N}{E^* + m_N} \right]^{\frac{1}{2}} \frac{"F_{M1}"}{2|\vec{q}^*|}$$

and we end by just normalizing to unity at $q^2 = 0$,

$$F_{expt}(q^2) = C_3(q^2)/C_3(0) .$$

The data analyzed this way are compared to the dipole form $(1 + Q^2/0.71 \text{ GeV}^2)^2$ in Fig. 12.

The asymptotic Q^2 dependence is given first by the asymptotic Q^2 dependence of F_{M1}, or equivalently of G_+, which may be analyzed using the rules we learned when studying the deuteron form factors in section 1. We have

$$\lim_{Q^2 \to \infty} F_{M1} \sim 1/Q^3 .$$

Then work out that

$$E^* = \frac{m_\Delta^2 + m_N^2 + Q^2}{2m_\Delta}$$

and that $|\vec{q}^*| = (E^{*2} - m_N^2)^{1/2}$ to see that E^* and $|\vec{q}^*|$ both increase like Q^2 at high Q^2.

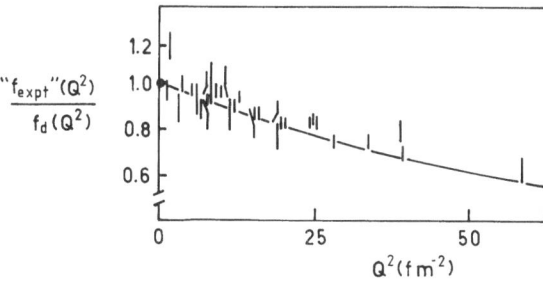

Fig. 12   Data for the magnetic dipole N–Δ transition (obtained as described in the text) divided by the nucleon dipole form factor and plotted vs. Q^2. This figure comes by courtesy of T. W. Donnelly.

Thus

$$\lim_{Q^2 \to \infty} F_{expt}(Q^2) \sim 1/Q^6 \,,$$

and it is no surprise to see F_{expt} falling faster with increasing Q^2 than the dipole form.

(We should mention that from the same era as F_{expt} there was a dispersion relation calculation of the N–Δ form factors that predicted the data extremely well[28].)

More Detailed QCD Predictions[24]

The helicity amplitude G_+ is the one which can be most easily calculated at high Q^2 in perturbative QCD. All masses can be set to zero in obtaining the leading term, greatly reducing one's labor. For the sake of justifying some notation, we should note that in the elastic scattering case, the relation between the analog of, G_+ and the magnetic form factor G_M is

$$G_M = \frac{m_N \sqrt{2}}{Q} G_+ \,.$$

It is this G_M that falls like $1/Q^4$ at high Q^2. So for the N$\to\Delta$ transition we may talk about the analog of G_M as being

$$G_M^{N \to \Delta} = \frac{m_N \sqrt{2}}{Q} G_+$$

and this $G_M^{N \to \Delta}$ will also fall like $1/Q^4$ at high Q^2.

One may do the Lorentz transformations to the infinite momentum frame to see that

$$2p^+ G_M^{N \to \Delta} = <\Delta, \lambda' = 1/2 | J^+ | N, \lambda = 1/2>$$

where p is the momentum of the incoming nucleon, which is moving very fast along the z-direction, with $p^+ = p^0 + p^3$ and $J^+ = J^0 + J^3$. The bulk of the work needed for this calculation has already been done. The hard scattering, high Q^2, result may be given in terms of the distribution amplitudes for the N and Δ and the hard scattering amplitude T_H,

$$G_M^{N \to \Delta} = \int [dx][dy] \, \phi_\Delta(x) \, T_H(x,y,Q) \, \phi_N(y)$$

where

$$\phi_\Delta(x) = \phi_\Delta(x) \, (1/\sqrt{3}) \, |uud + udu + duu> \uparrow\downarrow\uparrow$$

with ϕ_Δ being necessarily symmetric under the interchange $x_1 \leftrightarrow x_3$ and the analogous expression for the proton is a mixture of terms symmetric and antisymmetric under the same interchange.

If we expand the distribution amplitudes in Appel polynomials, as

$$\phi_\Delta = x_1 x_2 x_3 \sum N_i^* \, \phi_i(x) \,,$$

then we can obtain

$$Q^4 G_M^{N \to \Delta} = 64\pi^2 \, C_B^2 \, \alpha_s^2 \, \frac{\sqrt{2}}{3} \, \frac{1}{1296} \sum E_{ij} N_i^* N_j$$

where the coefficients E_{ij} are given in Table 1 for i and j ≤ 5.

Table 1

j	0	1	2	3	4	5
i						
0	-162	$63\sqrt{3}$	-81	45	$-15\sqrt{3}$	-30
2	-81	$-42\sqrt{3}$	18	-90	$11\sqrt{3}$	5/2
3	45	$110\sqrt{3}$	-90	170	$-23\sqrt{3}$	-65/2
5	-30	$-35\sqrt{3}/2$	5/2	-65/2	$41/(6\sqrt{3})$	145/18

Let us also calculate the soft contributions in this case, again for definiteness using a Gaussian for the transverse momentum wave function

$$\psi_\Delta(x,k_T) = \phi_\Delta(x)(192\pi^4/\alpha^4) \exp(-\sum k_{iT}^2/2\alpha^2) .$$

With $\zeta \equiv Q^2/2\alpha^2$ and $x_{ij} = x_i^2 + x_j^2$ we have for $p \to \Delta^+$

$$G_{M,soft}^{p\to\Delta} = (2/3) \int [dx] \{\phi_\Delta(\phi_S + \sqrt{3}\phi_A)e^{-x_{23}\zeta} - \phi_\Delta\phi_S e^{-x_{13}\zeta}\} .$$

The above expression is zero at zero Q^2 for any wave function, in agreement with the expected threshold behavior for this form factor. Also note that if the proton and delta distribution amplitudes are completely symmetric in x_1, x_2 and x_3 then the above soft contributions are zero for *any* Q^2. At last we have a clear example where the soft contributions are much much less than the hard scattering contributions!

More interestingly, the result that $G_{M,soft}^{N\to\Delta}$ is zero for wave functions that are completely symmetric, coupled with the non-zero result for the hard scattering $G_M^{N\to\Delta}$ in the same case means that the wave function's tail is not symmetric in x_1, x_2 and x_3. This is quite possible since the hard gluon exchanges that give the tail are sensitive to the relative helicity of the quarks they connect and thus distinguish the antiparallel quark from the other two. This also shows that a factorizable form of the wave function, $\psi = \phi \cdot g$, which has the same $\phi(x)$ and the same symmetry among the x_i at all k_T, cannot be right for a completely symmetric $\phi(x)$ and is at least suspect in any other case.

Summary

We have the prediction that at high Q^2, $F_{E2} = \sqrt{3} F_{M1}$. If born out, this will be quite dramatic. The present data on F_{E2}, all at Q^2 below 1 GeV^2, do show F_{E2}/F_{M1} rising with Q^2 but well below unity. The expected increase in F_{E2} is in a different language a recoil effect. (Even if the delta were spherically symmetric, in the nucleon rest frame it would appear squashed by a Lorentz transformation.) The large size of the effect should encourage thinking about the effects of recoil at lower Q^2 along with the effects of deformation.

We have also seen the asymptotic Q^2 dependence of the form factors, for example $G_M^{N\to\Delta} \sim 1/Q^4$ and seen that the data is not out of line with expectation.

REFERENCES

[1]   J. D. Walecka, this volume.
[2]   S. J. Brodsky and G. P. Lepage, Phys. Rev. **D24,** 2848 (1981).
[3]   S. J. Brodsky and G. R. Farrar, Phys. Rev. Lett. **31,** 1153 (1973), and Phys. Rev.
      D11, 1309 (1975; V. A. Matveev, R. M. Muradyan, and A. V. Tavkheldize, Lett.
      Nuovo Cim. **7,** 719 (1973).
[4]   A. I. Vainshtein and V. I. Zakharov, Phys. Lett. **72B,** 368 (1978).
[5]   C. E. Carlson and F. Gross, Phys. Rev. Lett. **53,** 127 (1984).
[6]   See for example R. G. Arnold, C. E. Carlson and F.Gross, Phys. Rev. **C21,** 1426
      (1980).
[7]   J. S. Levinger, Acta Phys. **33,** 135 (1973); T. Brady, E. Tomusiak and J. S. Lev-
      inger, Can. J. Phys. **52,** 1322 (1974); M. J. Moravcsik and P. Ghosh, Phys. Rev.
      Lett. **32,** 321 (1974).
[8]   G. R. Farrar and F. Neri, Phys. Lett. **130B,** 109 (1983).
[9]   S. J. Brodsky and B. T. Chertok, Phys. Rev. **D14,** 3003 (1976); S. J. Brodsky, C.
      -R. Ji and G. P. Lepage, Phys. Rev. Lett. **51,** 83 (1983).
[10] R. Woloshyn, Phys. Rev. Lett. **36,** 220 (1976); M. Gari and J. Hyuga, Nucl. Phys.
      A264, 409 (1976).
[11] R. G. Arnold *et al.,* Phys. Rev. Lett. **35,** 776 (1975); the last data point comes
      from using an exclusive-inclusive connection (R. G. Arnold, private communica-
      tion).
[12] See references 3 and 14, but also P. V. Landshoff, Phys. Rev. **D10,** 1024 (1974);
      and P. Cvitanovic, **ibid,** 338 (1974).
[13] G. P. Lepage and S. J. Brodsky, Phys. Rev. Lett. **43,** 545 and 1625 (E), (1979);
      V. A. Avdeendo, V. L. Chernyak and S. A. Koronblit, Yad. Fiz. **33,** 481 (1981).
[14] G. P. Lepage and S. J. Brodsky, Phys. Rev. **D22,** 2157 (1980).
[15] N. Isgur, C. Llewellyn-Smith, Phys. Rev. Lett. **52,** 1080 (1984).
[16] N. Isgur and G. Karl, as reported in N. Isgur, in New Aspects of Subnuclear Phy-
      sics, ed. A. Zichichi (Plenum, New York, 1980).
[17] C. E. Carlson, in Proceedings of the BUTG Workshop, M.I.T. July 1984, G.
      Rawitscher, ed.
[18] S. Rock *et al.,* Phys. Rev. Lett. **49,** 1139 (1982).
[19] A. Andrikopoulou, J. Phys. G: Nucl. Phys. **11,** 21 (1985).
[20] C. E. Carlson and F. Gross, Lund preprint LU TP 85-12 (1985) and CEBAF PR-
      85-005 (1985).
[21] V. L. Chernyak and I. R. Zhitnitsky, Nucl. Phys. **B246,** 52 (1984).
[22] V. Vento, G. Baym and A. D. Jackson, Phys. Lett. **102G,** 97 (1981); S. S. Gersh-
      tein and D. V. Dzhikiya, Sov. Jour. Nucl. Phys, **34,** 870 (1981); N. Isgur, G. Karl
      and R. Koniuk, Phys. Rev. **D25,** 2394 (1982).
[23] J. Dey and M. Dey, Phys. Lett. **138B,** 200 (1984); M. Weyrauch and H. J. Weber,
      contribution to 1985 CEBAF Summer Study and to be published.
[24] C. E. Carlson, Lund preprint LU TP 85-17 (1985).
[25] T. W. Donnelly, this volume.
[26] L. Durand, P. C. DeCelles and R. B. Marr, Phys. Rev. **126,** 1882 (1962).
[27] A. J. Dufner and Y. S. Tsai, Phys. Rev. **168,** 1801 (1968).
[28] J. D. Walecka and P. A. Zucker, Phys. Rev. **167,** 1479 (1968); P. L. Pritchett, J.
      D. Walecka, and P. A. Zucker, Phys. Rev. **184,** 1825 (1969); J. D. Walecka, Acta
      Phys. Pol. **B3,** 117 (1972).

SPECTROSCOPY OF N–Δ EXCITED STATES AND THE QUARK-MODEL

R. K. Bhaduri

Physics Department
McMaster University
Hamilton, Ontario, Canada L8S 4M1

DATA ON NUCLEON RESONANCES

My lectures here will expound on the simple quark model of baryons to understand the variety of baryon resonances that are encountered experimentally. We would mainly concentrate on the excited states of the nucleon (N) and the delta (Δ), that are "seen" in πN scattering, and in the pion photoproduction and electroproduction experiments. Resonances are associated with peaks in cross sections, and are manifest, for example, in a plot of the total πN cross section, σ_T, as a function of the pion laboratory momentum or the CM energy squared, s. In Fig. 1, $\sigma_T(\pi^+p)$, which is entirely in the isospin I = 3/2 channel, is plotted. It has a huge bump corresponding to the Δ(1232), and some minor peaks where many other overlapping Δ- resonances contribute. In Fig. 2, $\sigma_T(\pi^-p)$ is plotted, and here both the I = 1/2 and I = 3/2 channels contribute in the ratio 2:1. This plot has more structure, and some of the N-resonances can be identified. It is clear from these plots, however, that other experiments and more careful analysis of the data are necessary to isolate the characteristics of these excited states. In case of elastic πN scattering,

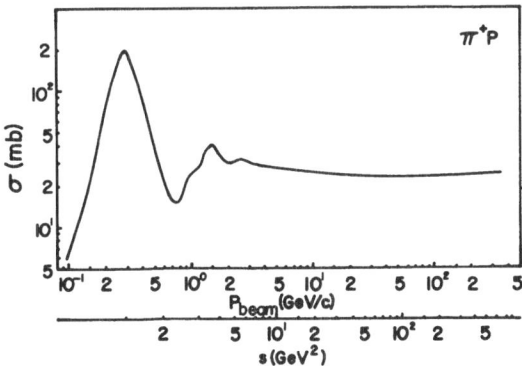

Fig. 1   Schematic plot of total π^+p cross section against the pion beam momentum in the laboratory frame. The corresponding centre-of-mass energy squared s is also shown.

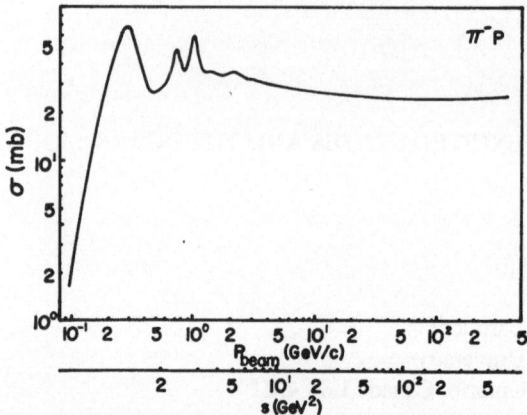

Fig. 2   The π^-p total cross section. The legend is the same as in Fig. 1.

since the pion has no spin, the scattering amplitude f (which must be a scalar), may simply be written as

$$f = g(k^2,\theta) + ih(k^2,\theta)\vec{\sigma}\cdot\vec{n}, \tag{1}$$

where $\vec{n} = (\vec{k}\times\vec{k}')/|\vec{k}\times\vec{k}'|$. Here $\vec{\sigma}$ is the Pauli spin-operator for the nucleon, and $\vec{k}$, $\vec{k}'$, are the initial and final pion three-momenta in the pion-nucleon CM-system, and θ is the angle between them. The functions g and h are complex, so at each energy and angle, there are four independent real parameters. One of these may be absorbed in an overall phase factor, leaving three real parameters to be determined by three independent sets of experiments. These experiments may involve unpolarized targets, with measurement of differential cross section and the recoil nucleon polarization, or experiments with polarized targets. For analysis of the data, the usual procedure is to make a partial wave expansion of the scattering amplitude, and express the partial wave amplitude in terms of complex phase shifts to take account of inelasticity. The complex partial wave amplitude is drawn on an Argand diagram as a function of energy, and the widths and positions of the resonances are deduced from the behaviour of these diagrams. For a clear explanation of these diagrams, see the old article by Donnachie[1] . It should be mentioned that in the inelastic region, a unique set of partial waves cannot be obtained from the data alone, and it is necessary to apply theoretical constraints of analyticity from dispersion relations, etc.. There are extensive analyses by the CMU–LBL[2] and the Karlsruhe-Helsinski[3] groups, and you may look up the Particle Data Tables[4] for further details. The πN - partial waves are denoted by the symbol $L_{2I,2J}$, I is the isospin and J the angular momentum of the system. For example, F_{15} would denote a partial wave with L = 3, I = 1/2 and J = 5/2. There is a nucleon resonance N(1680) that contributes to this partial wave (there is another at 2000 MeV), with spin 5/2 but *even* parity. Remember that the intrinsic parity of the pion is negative with respect to the nucleon. In Figs. 1 and 2, many such partial wave resonances overlap in a bump in the cross section. But when the real part of the F_{15} partial wave amplitude is plotted against the imaginary part in an Argand diagram, as in the schematic diagram 3, an individual resonance in the channel can be isolated. The amplitude moves rapidly around the circle in an anticlockwise direction with increasing energy. The peak of the resonance is at energy $E^{(R)}$, corresponding to the top of the circle, and its width is given by (E_2-E_1) as marked in the figure. A purely elastic amplitude (almost pure is P_{33} in the vicinity of $\Delta(1236)$), on the other hand, would follow the unitary circle with centre i and unit radius, as shown by the dotted circle.

It is not my intention to spend the major part of the talk on reviewing the methods of analysis of the data. The idea is to give you enough explanation so that you may learn to understand and use the Particle Data book[4]. Consider real photons, which may be absorbed by the nucleon to excite resonances (of N and Δ), or, conversely, may be emitted in the photodecay of the resonances. Real photons have no longitudinal polarizations, so if we take the propagation direction along the z-axis, then the transverse polarization vectors may be taken along the x- and y-directions. Constructing the circular polarization vectors as

$$\varepsilon^{(+)} \propto (\varepsilon^{(x)} + i\varepsilon^{(y)}), \quad \varepsilon^{(-)} \propto (\varepsilon^{(x)} - i\varepsilon^{(y)}),$$

and writing these in spherical polar coordinates, we see that $\varepsilon^{(+)} \propto \sin\theta e^{i\phi}$, $\varepsilon^{(-)} \propto \sin\theta e^{-i\phi}$. Thus the polarization vectors of the real photon make its wave function go like $Y_1^{m=1}$ and $Y_1^{m=-1}$, but the m = 0 part is absent. Although a high energy photon may carry away many units of angular momentum, the helicity (or the projection of the spin along the direction of propagation, which is taken as the z-direction here) can only be ±1. Thus, for example, if a resonance like $N(1680)F_{15}$ decays to the ground state $N(940)$ which has spin J = 1/2, the angular momentum J changes by 2 units of $\hbar$, but the helicity (i.e. the

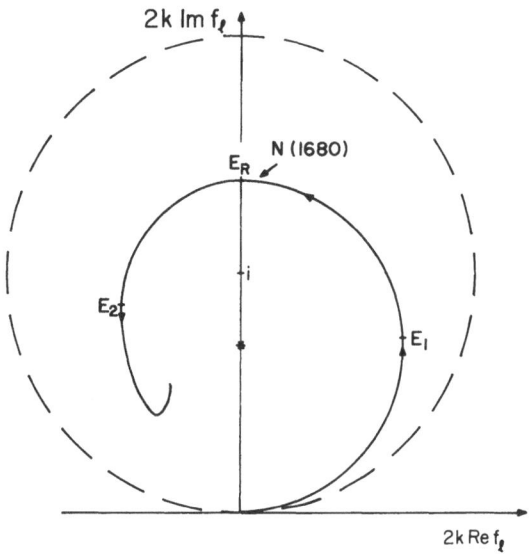

πN F15 Amplitude

Fig. 3   An Argand diagram for the π- N scattering amplitude in a given partial wave. For elastic scattering, the scattering amplitude is $f_l(k) = (e^{2i\delta_l}-1)/2ik$, where $\delta_l(k)$ is the real phase shift. In this case the plot is the dashed circle, with the highest point corresponding to $\delta_l = \pi/2$. For inelastic scattering, either a complex phase shift, or an additional real inelasticity parameter η_l ($0\leq\eta_l\leq1$) is introduced, such that $f_l = (\eta_l e^{2i\delta_l}-1)/2ik$. The full curve is a schematic drawing for the πN scattering amplitude F_{15}. There is a hint of another resonance contributing at the tip of the turning tail.

m-quantum number) may only change by ±1 unit. Therefore, the m = ±1/2 states of N(940) may only be reached by m = 3/2 or m = 1/2 states of N(1680)F_{15}, and not by the m = 5/2 state. Of course, the m = -3/2 state may make the transition to the m = -1/2 state of N(940), but this amplitude is related to the m = 3/2 → m = 1/2 amplitude by time reversal. Thus for proton target coupling, there are two such amplitudes $A_{3/2}^{(p)}$ and $A_{1/2}^{(p)}$ for resonances with J≥3/2, and similarly two more for neutron targets. These amplitudes are real, being matrix-elements of the hermitian electromagnetic coupling $j_\mu A_\mu$ with respect to the nucleon and resonance wave functions. The γN decay helicity amplitudes are listed on page S216 of Ref. 4, and serve as a sensitive test of the quark model wave functions. A good discussion may be found in the textbook by Close[5]. In pion electroproduction experiments, since the photon is virtual and longitudinal components are present, the analysis is more involved. We shall not discuss this, but refer you to the review article[6] in case you are interested.

A large number of resonances of the nucleon and the delta have been detected by the above techniques, especially up to resonance energies of about 2600 MeV, with the quantum numbers I, J and the resonance channel identified. Even in this energy range, there is reason to believe that many resonances remain undetected, because they are hard to excite by conventional methods. The experimentally well established N, Δ– resonances and their nominal masses are listed in Table 1, where we have grouped the states of the same parity together. We do not list the widths, typically about 200 MeV, and the decay modes, which are dominantly $N\pi$, $N\pi\pi$ and $N\rho$.

HOW MANY DEGREES OF FREEDOM?

Before proceeding further, we may ask: how many quarks are involved[7] in the generation of the observed intrinsic spectrum of Table 1? Are gluonic degrees of freedom active in these states that have been detected? To answer these questions, we examine the high temperature behaviour of the specific heat of the system[8] that is derived from these states. A knowledge of the specific heat often gives direct information about the active number of degrees of freedom. This is manifest, for example in the high temperature behaviour of the molar specific heat of a solid (Law of Dulong and Petit), and in the analysis of molecular specific heats in terms of rotational and vibrational degrees of freedom.

To obtain the specific heat of the nucleon (including the Δ–channel), we construct a partition function Z(β) for the system:

$$Z = \sum_J \sum_I (2J+1)(2I+1)\exp(-\beta E_{IJ}^{(R)}), \tag{2}$$

where $E_{IJ}^{(R)}$ are the experimental resonance energies of the truncated set in Table 1, and $\beta = 1/\tau$ in MeV^{-1}. Then

$$<E> = -\frac{1}{Z}\left(\frac{\partial Z}{\partial \beta}\right),$$

and the specific heat is

$$C_{N\Delta} = \frac{\partial}{\partial \tau}<E> = \beta^2 \frac{\partial^2}{\partial \beta^2} \ln Z. \tag{3}$$

This is easy to calculate using Eq. 2 as a function of β, and is plotted in Fig. 4. Unlike the classic Dulong-Petit law, the specific heat does not saturate to a limiting value here, but has a maximum at $\tau \approx 150$ MeV. To interpret this result, we adopt a model in which we assume that the specific heat may be described by the independent motion of n

Table 1   The excited states of N(940) and Δ(1232). The nominal masses in MeV are given in brackets. Weak (*) states of the particle data tables, Ref. 4, are not included.

N	3/2⁻(1520)	1/2⁻(1535)	1/2⁻(1650)	5/2⁻(1675)	3/2⁻(1700)
Δ	1/2⁻(1620)	3/2⁻(1700)			
N	1/2⁺(1440)	5/2⁺(1680)	1/2⁺(1710)	3/2⁺(1720)	7/2⁺(1990)
	5/2⁺(2000)				
Δ	3/2⁺(1600)	5/2⁺(1905)	1/2⁺(1910)	3/2⁺(1920)	7/2⁺(1950)
	9/2⁺(2300)	11/2⁺(2420)			
N	3/2⁻(2080)	7/2⁻(2190)	5/2⁻(2200)	9/2⁻(2250)	11/2⁻(2600)
Δ	1/2⁻(1900)	5/2⁻(1930)	9/2⁻(2400)	11/2⁻(2750)	

"particles" confined in a mean field. This mean field is specified only by a truncated single-particle spectrum ε_i , with which we may calculate a single-particle partition function $z = \sum_i \exp(-\beta\varepsilon_i)$. For example, for the motion of a zero mass Dirac particle in a spherical bag of fixed radius R, we take the $1s_{1/2}(2.04)$, $1p_{3/2}(3.20)$, $1p_{1/2}(3.71)$, $1d_{5/2}(4.33)$, $1d_{3/2}(5.12)$, $2s_{1/2}(5.40)$, and the $1f_{7/2}(5.43)$ single-particle states, where the energies are given in units of $\hbar c/R$. The truncation of the single-particle states is a matter of some judgement, but it is assumed that the odd- and even-parity states listed in Table 1 may be generated through these single-particle states. Then (with $x = \beta\hbar c/R$)

$$z = [2e^{-2.04x} + 4e^{-3.20x} + 2e^{-3.81x} + 6e^{-4.33x} + 4e^{-5.12x} + 2e^{-5.40x} + 8e^{-5.43x}]. \tag{4}$$

The single-particle specific heat, $C_1 = \beta^2 (\partial^2 \ln z/\partial\beta^2)$ is then easily calculated. At high temperatures, using classical statistics, the n-particle partition function is

$$Z_n = (z)^n/n!. \tag{5}$$

The specific heat of the n-particle system is

$$C_n = \beta^2 \frac{\partial^2}{\partial\beta^2} \ln Z_n = n\beta^2 \frac{\partial^2}{\partial\beta^2} \ln z = nC_1. \tag{6}$$

In other words, in the independent particle model, the n-particle specific heat is n-times the single-particle specific heat. We may now match this C_n with the "experimental" $C_{N\Delta}$ to deduce the number of particles n. In Fig. 4, the dark circles correspond to the calculated values of C_n with $n = 2$. Note that $C_{N\Delta}$ peaks at $\beta \sim 0.0065$ MeV^{-1} , whereas C_1, calculated from Eq. 4, peaks at $x = \beta\hbar c/R = 1.603$. Matching the position of the peak at $\beta = 0.0065$ MeV^{-1} yields R = 0.0065 × 197.3/1.603 = 0.8 fm. If we had taken n = 3, or n=1, the fit to $C_{N\Delta}$ would be totally destroyed. There is, of course, nothing special about the bag model. We could have generalized the set of single-particle states through an oscillator as well. It could also be applied to other systems (like Λ–Σ) or mesons. For more details, look up Ref. 8.

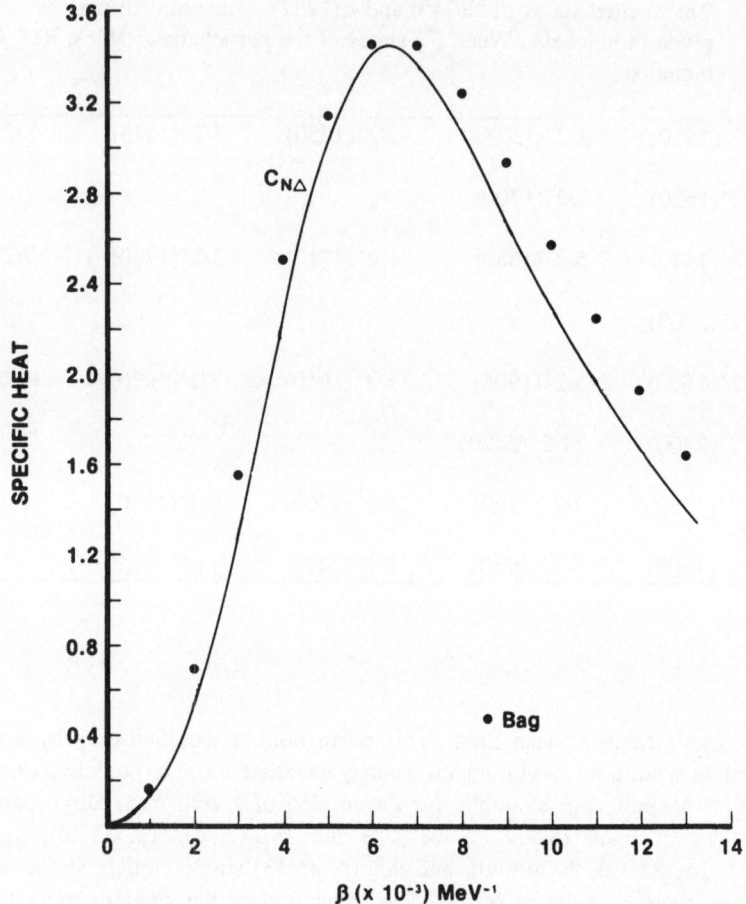

Fig. 4   Variation of the specific heat $C_{N\Delta}$ of the N–Δ system with inverse temperature
β. The specific heat is dimensionless since β is expressed in MeV^{-1}. The continuous curve is calculated from states in Table 1. The dark circles are the
points calculated from the bag-model, and correspond to <u>two</u> times the single-
particle specific heat C_I.

The analysis shows that the specific heat, derived from the experimentally detected
states of Table 1, may be understood from the independent particle motion of only two
identical particles! When one takes three valence quarks in the harmonic oscillator
model[9], the centre-of-mass motion of the quarks may be eliminated cleanly, and the orbital motion of the three quarks is described effectively by two "particles", each moving in
a harmonic field. We seem to be confirming this in a more model-independent manner.
It also appears that the gluonic degrees of freedom, if present, are *not* contributing to the
specific heat of the system in this energy range. This means that in the states listed in
Table 1, it is unlikely that the gluons have independent degrees of freedom. It does not
rule out the possibility that glueballs, or hybrid states of quarks and gluons may be
present in this energy range, but have so far escaped detection.

Table 2   The quantum numbers associated with the light quarks

Flavour	u	d
Charge	2/3	-1/3
Isospin	1/2	1/2
I_3	1/2	-1/2
Strangeness	0	0
Baryon number	1/3	1/3

SYMMETRY CONSIDERATIONS IN THE QUARK MODEL

We now proceed to build a model, in which there are three confined valence quarks that carry all the quantum numbers, and no explicit gluonic or "sea" quark degrees of freedom are present. The quantum numbers carried by the three flavours of quarks are listed in Table 2. It is clear from the considerations of the earlier section that it is important to eliminate the centre-of-mass motion in spectroscopy. This is more easily done in the nonrelativistic oscillator model. Before going into details, we would first like to figure out, purely from symmetry considerations, how many low-lying odd and even parity excitations of a baryon may be expected. This can be done through well known properties of group representations, but we will use more elementary methods. In nuclear physics, the neutron and proton may be considered as identical particles when the isospin quantum numbers are introduced. Similarly, the light quarks u, d and s may be treated on an equal footing but having different "flavours". This is convenient for classifying the states, as we wish to do now. In practice, flavour symmetry is broken appreciably, but this would only mean that there will be mixing between flavour symmetry states, but the number of states would not alter.

Going back to nuclear physics, consider the three-nucleon systems ^{3}H and ^{3}He . The total wave function consists of space, spin and isospin parts, and has to be antisymmetric with respect to the exchange of any pair of nucleons. In the three-quark system, again the total wave function has to have this antisymmetry property. But now, in addition to space, spin and flavour, it is assumed that each flavour or quark comes in three different colours. In a baryon, the colour part of the wave function separates out from the rest and is antisymmetric under the exchange of any pair. Actually any hadron is supposed to be colour-singlet, but the wave function separates out only in $q\bar{q}$ and qqq systems. This implies that for the baryon, the space × spin × flavour wave function is *symmetric* under the exchange of any pair of (identical) quarks. Let us start from the simplest, the spin part.

Spin

Three spin -1/2 quarks may give rise to s = 3/2 (symmetric, denote by S, 4 states) and s = 1/2 (two types of mixed symmetric states, denote by M_ρ and M_λ, 2x2 states). Some explanation is necessary at this stage. You may easily convince yourself that the s = 3/2 state is symmetric, and of course, it has (2s+1) = 4 states. For example, denoting the spin

```
        1      2      3
        x      x      x
        u      u      u
        d      d      d
        s      s      s
```

Fig. 5   Construction of the flavour wave function. The three quarks are denoted by
          crosses and numbered. Each quark may have flavour u, d or s, as shown in a
          column under it.

wave function by χ, the s = 3/2 state with projection 3/2 is obviously

$$\chi^s_{3/2} = \uparrow(1)\uparrow(2)\uparrow(3). \tag{7}$$

To obtain s = 1/2, you may proceed in two ways. Combine the spins of 1 and 2 to form
$s_{12} = 0$, and add spin to 3 to get s = 1/2. In this case the wave function is antisymmetric
with respect to the exchange of 1 and 2, but there is no overall symmetry with respect to
exchange of 1 and 3, etc.. In other words, you do not get back the same wave function
(within a sign) by exchanging 1 and 3, or 2 and 3. This is called a mixed symmetric
wave function, and we denote this symmetry by M_ρ, the subscript ρ indicating antisymmetry with respect to the interchange of 1 and 2. Obviously, there are (2s+1) = 2 such
wave functions with M_ρ–symmetry. The state with projection 1/2 is given by

$$\chi^\rho_{1/2} = \frac{1}{\sqrt{2}}(\uparrow\downarrow-\downarrow\uparrow)\uparrow. \tag{8}$$

Alternatively, we could couple $s_{12} = 1$, and combine the spin of the third particle to yield
s = 1/2. This is again a mixed symmetric wave function, but it is symmetric with respect
to the interchange   of 1 and 2. We denote this by M_λ, and again there are two such
states. A spin-state with λ–type mixed symmetry, and s = 1/2, $s_z = 1/2$ is

$$\chi^\lambda_{1/2} = \frac{1}{\sqrt{6}}(\uparrow\downarrow\uparrow+\downarrow\uparrow\uparrow-2\uparrow\uparrow\downarrow). \tag{9}$$

Note that no totally antisymmetric spin is possible. In short,

$$2^3 = 8 = 4 + 2 + 2$$

$$\text{S}\quad\text{M}_\rho\quad\text{M}_\lambda. \tag{10}$$

Obviously, since each quark can be spin up or down, there are 2^3 states in a 3-quark system. Eq. 10 expresses the fact that 4 of these states are symmetric, and the rest mixed
symmetric. Note that since we are treating the three quarks on an equal footing, we
could have done the same analysis by coupling spins of particles 2 and 3 to s_{23}, etc.. The
basis generated thereby is an alternate, but not an independent one from the earlier construct.

Flavour

Consider again three quarks, each of which may have one of the three flavours, u, d
or s. There are, in this case, $3^3 = 27$ states . To classify them, let us look at Fig. 5.
Denote the three quarks by three crosses, numbering them, as before, by 1, 2, 3. Quark 1
may have any of the three flavours u, d and s, which are arranged under 1 in a column.

Similarly for the quarks 2 and 3. How many symmetric states can we make? We may have states like u(1)u(2)u(3), and two more like this from the second and third row of Fig. 5. We may also construct symmetric states in which two of the quarks have the same flavour, and one of a different flavour. There are six such states, with flavour combinations udd, uss, duu, dss, suu, and sdd. There is also one symmetric combination of the type uds, where all three flavours are different. Thus there are, in all, 10 states of the symmetric type. Furthermore, since we are considering three quarks and three flavours, there is only one determinantal combination which is totally antisymmetric under the exchange of any pair. The rest of the 16 states are of mixed symmetric type, 8 of them of ρ–type (antisymmetric in 1 and 2), and 8 of λ–type (symmetric in 1 and 2). We may thus write

$$3^3 = 27 = 10 + 8 + 8 + 1$$

$$\text{S} \quad \text{M}_\rho \quad \text{M}_\lambda \quad \text{A.} \tag{11}$$

In the nonstrange sector, isospin takes the role of flavour, and the wave functions are exactly similar to the spin-case of the earlier section. Denoting the flavour wave-function by ϕ, we have (for the maximum projection states)

$$\phi_\Delta^s = uuu, \ \phi_p^\rho = \frac{1}{\sqrt{2}} \ (ud{-}du)u, \ \text{and} \ \phi_p^\lambda = -\frac{1}{\sqrt{6}} \ (udu{+}duu{-}2uud). \tag{12}$$

Spin $\times$ Flavour Wave Function

Knowing the symmetry properties of the spin-and flavour states, it is easy to classify the symmetry of the $3^3 \times 2^3 = 216$ states . In group theoretical language, we are considering the $[SU(3)_{\text{flavour}} \times SU(2)_{\text{spin}}]$ states. Examining Eqs. 10 and 11, we see that we may combine the four symmetric spin states with the 10 symmetric flavour states to obtain 40 states which are symmetric in the Hilbert space of spin and flavour. Denote these by $^4 10$, and note that these have spin s = 3/2. Similarly, the mixed symmetric spin states may be combined with the mixed-symmetric flavour states to yield other symmetric combinations. The combination, for the nucleon, that is symmetric under exchange of any pair is $(\chi^\rho \phi^\rho + \chi^\lambda \phi^\lambda)$. Such combinations yield 2x8 symmetric states. These are denoted by $^2 8$, and have spin s = 1/2. We may now write

$$^4 10 + {}^2 8 = 56 \ \text{(S)}, \tag{13}$$

indicating the 56 symmetric states. These correspond, of course, to the ground state spin 1/2 octet and the spin -3/2 decuplet. Similar consideration will show that

$$^4 1 + {}^2 8 = 20 \ \text{(A)}, \tag{14}$$

where the antisymmetric combination $^2 8$ is of the type $(\chi^\rho \phi^\lambda - \chi^\lambda \phi^\rho)$. The rest of the 140 states are mixed symmetric, split equally between ρ and λ–types :

$$^2 10 + {}^4 8 + {}^2 8 + {}^2 1 = 70 \ \text{(M}_\rho), $$

and

$$^2 10 + {}^4 8 + {}^2 8 + {}^2 1 = 70 \ \text{(M}_\lambda). \tag{15}$$

For the nucleon, the spin-isospin states with M_ρ–type of symmetry are of the type $(\chi^\rho \phi^\lambda + \chi^\lambda \phi^\rho)$, and the ones with M_λ–symmetry are $(\chi^\rho \phi^\rho - \chi^\lambda \phi^\lambda)$. The next step is to combine these states with the spatial part to form symmetric wave functions.

The Spatial Wave Function in the Oscillator Model

The oscillator model for the orbital motion of the quarks, developed some twenty years back[7,9] has been used for extensive spectroscopic analysis by Isgur and Karl[10,11]. It has the advantage that the centre-of-mass motion is eliminated trivially, and the wave functions are expressed analytically. Therefore, the symmetries of the wave functions are apparent[12] and the calculation of matrix-elements easily done. The Hamiltonian, assuming a two-body confinement force, and ignoring all other interquark forces, is

$$H_0 = \sum_i p_i^2/2m_i + \frac{1}{2} K \sum_{i<j} |\vec{r}_i - \vec{r}_j|^2. \tag{16}$$

Define the intrinsic coordinates $\vec{\rho}, \vec{\lambda}$ and the centre-of-mass coordinate $\vec{R}$:

$$\vec{\rho} = \frac{1}{\sqrt{2}} (\vec{r}_1 - \vec{r}_2), \ \vec{\lambda} = \frac{1}{\sqrt{6}} (\vec{r}_1 + \vec{r}_2 - 2\vec{r}_3),$$

and

$$\vec{R} = \frac{m(\vec{r}_1 + \vec{r}_2) + m'\vec{r}_3}{(2m+m')} . \tag{17}$$

Here, to include baryons other than N and Δ, it is assumed that particles 1 and 2 have mass m, whereas the third quark has mass m'. For N or Δ, m = m'. Dropping the centre-of-mass kinetic term,

$$H_0 = p_\rho^2/2m_\rho + p_\lambda^2/2m_\lambda + \frac{3}{2} K \ (\rho^2 + \lambda^2), \tag{18}$$

where

$$m_\rho = m, \quad m_\lambda = 3mm'/(2m+m'),$$

$$\vec{P}_\rho = m_\rho \frac{d}{dt} \vec{\rho}, \quad \vec{P}_\lambda = m_\lambda \frac{d}{dt} \vec{\lambda} \ .$$

Thus the intrinsic orbital motion of the quarks is described by two independent oscillators, in the $\vec{\rho}$–coordinate (antisymmetric in 1 and 2), and in the $\vec{\lambda}$–coordinate (symmetric in 1 and 2). It is clear that in systems like Λ, Σ or Ξ , the oscillator frequencies $\omega_\rho = \sqrt{3K/m_\rho}$ and $\omega_\lambda = \sqrt{3K/m_\lambda}$ are different, whereas in N, Δ and Ω^- (since we assume $m_u = m_d$) there is only one oscillator frequency $\omega_0 = \sqrt{3K/m}$. We shall concentrate on the N–Δ sector here. It is not our intent to write out all the wave functions of the low-lying states; you can look up these in the papers of Isgur and Karl[10,11]. We want to write out, however, a few representative wave functions and their symmetries. An orbital state is a product of the ρ–oscillator and the λ–oscillator wave functions. Its energy is specified by the total number of excitation quanta N:

$$N = N_\rho + N_\lambda , \tag{19}$$

with N_ρ the excitation quanta of the ρ–oscillator, and likewise for the λ–oscillator . The ground state is one for which both N_ρ and N_λ are zero. The lowest odd-parity states have N = 1, so either the $\rho-$ or the λ–quantum has been excited, with an excitation energy of $\hbar\omega$. For the N = 2 even-parity states, we may choose to excite two quanta of the same type and none of the other, or one of each type. Following Refs. 10 and 11, the orbital wave function is denoted by $\psi_{LM}^{()} (\rho, \lambda)$, where the superscript stands for the symmetry-type of the wave function. We display a few examples.

EXCITATION	ORBITAL WAVE FUNCTION	SYMMETRY

$N=0$

$$\psi_{00}^S = \frac{\alpha_0^3}{\pi^{3/2}} \exp\{-\frac{\alpha_0^2}{2}(\lambda^2+\rho^2)\}$$

S

$N=1$

$$\psi_{11}^\lambda = -\frac{\alpha_0^4}{\pi^{3/2}}(\lambda_x+i\lambda_y)\exp\{\cdots\}$$

M_λ

$$\psi_{11}^\rho = -\frac{\alpha_0^4}{\pi^{3/2}}(\rho_x+i\rho_y)\exp\{\cdots\}$$

M_ρ

$N=2$

$$\psi_{00}^\rho = \frac{2}{\sqrt{3}}\frac{\alpha_0^5}{\pi^{3/2}}(\vec{\rho}\cdot\vec{\lambda})\exp\{\cdots\}$$

M_ρ

$$\psi_{00}^\lambda = \frac{1}{\sqrt{3}}\frac{\alpha_0^5}{\pi^{3/2}}(\rho^2-\lambda^2)\exp\{\cdots\}$$

M_λ

$$\psi_{00}'^S = \frac{1}{\sqrt{3}}\frac{\alpha_0^5}{\pi^{3/2}}(\rho^2+\lambda^2-3\alpha_0^{-2})\exp\{\cdots\}$$

S

$$\psi_{11}^A = \frac{\alpha_0^5}{\pi^{3/2}}(\rho_+\lambda_z-\rho_z-\rho_z\lambda_+)\exp\{\cdots\}$$

A .   (20)

Here $\alpha_0 = \sqrt{m\omega_0}$, and $\rho_+ = (\rho_x+i\rho_y)$, etc.. We have not listed the N=2, L=2 states which may also have S, M_ρ or M_λ type symmetries. A few observations are relevant at this point. Note first that there is no symmetric N=1 orbital state, since this is necessarily of the type $(\vec{r}_1+\vec{r}_2+\vec{r}_3)\exp\{\cdots\}$, which is spurious. Next, check that ψ_{11}^A and ψ_{00}^ρ are states in which $N_\rho = 1$, $N_\lambda = 1$, so that the $\rho-$ and $\lambda-$ coordinates have been excited simultaneously. This implies that at least two of the three particles are excited, which cannot be done easily by operators which couple to single quarks. This would mean that the even-parity resonances having these symmetries would be difficult to reach experimentally through photoexcitation or pion scattering. It is now straightforward to construct totally symmetric 3-quark wave functions combining the spin-flavour states of Eqs. 13-15 with the spatial functions[16]. In essence, we have the symmetric 56 (^{2}8,^{4}10), mixed symmetric 70 (^{2}8,^{4}8,^{2}10,^{2}1) and antisymmetric 20 (^{2}8,^{4}1) spin-flavour states. These have to be combined with the N=0, L=0$^+$ (symmetric), N = 1, L = 1$^-$ (mixed symmetric), and N = 2, L = 0$^+$, 2$^+$ (symmetric and mixed symmetric), and 1$^+$ (antisymmetric) orbital states in the oscillator model. For simplicity, we have restricted the states up to N=2. More general forms are to be found in Ref. 12. In Table 3, the predicted N and Δ resonances of the model are displayed according to their symmetries, together with the π–N partial wave amplitudes in which these are expected to appear. In reality, there would be considerable mixing between these states due to interaction between the quarks. For example, the N=2, 56 P$_{11}$ and 70 P$_{11}$ states may get mixed by a spin-dependent q-q interaction. Other anharmonic components of the force would also split the degeneracies of the harmonic spectrum. Nevertheless, we do expect 5 low-lying odd-parity N states and 2 odd-parity Δ states, all of which are found. According to our classification, we expect 13 nucleon and 8 delta even-parity resonances belonging to the N=2 excitation.

Table 3   Predicted N–Δ spectrum from symmetry considerations using the oscillator model. The notation is fully explained in the text.

N=0, L=0⁺	56	$^2 8(P_{11})$	$^4 10(P_{33})$
N=1, L=1⁻	70	$^2 8(D_{13},S_{11})$	
		$^4 8(D_{15},D_{13},S_{11})$	$^2 10(D_{33},S_{31})$
N=2, L=0⁺	56	$^2 8(P_{11})$	$^4 10(P_{33})$
	70	$^4 8(P_{13})$	
		$^2 8(P_{11})$	$^2 10(P_{31})$
L=2⁺	56	$^2 8(F_{15},P_{13})$	$^4 10(F_{37},F_{35},P_{33},P_{31})$
	70	$^4 8(F_{17},F_{15},P_{13},P_{11})$	
		$^2 8(F_{15},P_{13})$	$^2 10(F_{35},P_{33})$
L=1⁺	20	$^2 8(P_{13},P_{11})$	

COMPARISON WITH EXPERIMENTAL DATA

Our task now is to confront the experimental data, which were listed already in Table 1. In Fig. 6, we show the low-lying odd-parity states of the N and Δ, together with the ground-state energies. The widths in this range of energies are typically of the order of 150 MeV, and are not shown. The following points should be noted from Fig. 6:

(a) The ground states of N and Δ are not degenerate, but split by about 300 MeV. This indicates the presence of strong spin-dependent q-q force, which pushes the S = 3/2 state up with respect to the S = 1/2 state.

(b) The same spin-dependent force should be responsible for the observed gap between the s = 3/2 states (S_{11},D_{15},D_{13}) belonging to $^4 8$ and the s = 1/2 states (S_{11},D_{13}) of the nucleon. But this gap is only about 150 MeV, half the ground state splitting. This implies that at least a sizable part of the spin-dependence force is of very short range, being less effective in the relative $l = 1$ state.

(c) The spin-orbit splitting in the odd-states of the nucleon is very small, and in the Δ is somewhat larger. This will loom as an important problem when we examine the force arising from the one-gluon exchange between the quarks.

(d) From the spectrum of Fig. 6, we deduce that the oscillator frequency $\hbar\omega \approx 500$ MeV. In fact it is gratifying that all five 70 , L = 1⁻ states of Table 2 are experimentally seen. However, we face a very serious problem when we find two more well-established odd-parity Δ–states, Δ(1900)S_{31} and Δ(1930)D_{35} which may only arise from the N = 3 shell in the oscillator model, and therefore should be at a much higher energy. In fact, there is also a less established odd-parity state in the nucleon, N(2080)D_{13} , which poses the

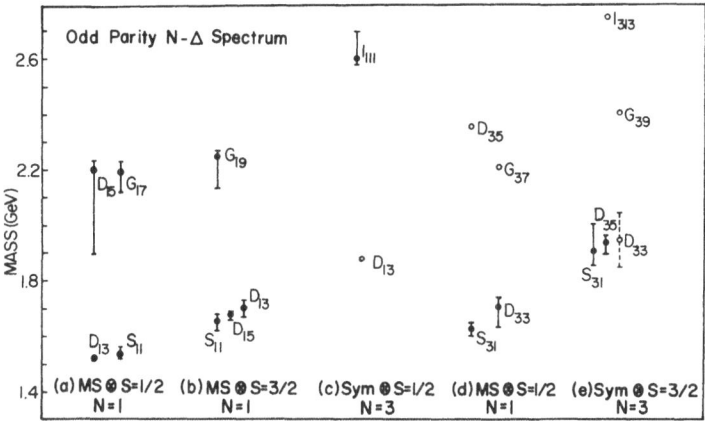

Fig. 6   The low-lying odd-parity resonances of the nucleon and the delta. The uncertainties in the nominal masses are shown by vertical lines. The widths are not shown. The open circles correspond to weak and less established states. In the data tables (Ref. 4) these correspond to one- or two-star resonances.

same question.

Finally, in Figs. 7 and 8, we plot the relative positions of the low-lying even-parity N and Δ states that have been seen. We have also included the few high spin states at higher energies that are seen clearly. The following features stand out:

(a) Why are almost all the even-parity excitations at about the same energy as the odd-parity states?

(b) Why, in particular, is the "Roper" resonance $N(1440)P_{11}$ so low? It is almost 100 MeV lower than the lowest odd-parity resonance $N(1520)D_{13}$. In the oscillator picture, the $L = 0$ and $L = 2$ states should have been degenerate, but a gap of more than 200 MeV is observed between $N(1440)P_{11}$ and the $L = 2$ multiplets, $N(1680)F_{15}$ and $N(1720)P_{13}$.

(c) What is the origin of the large splitting between $N(1440)P_{11}$ and $N(1710)P_{11}$?

(d) Practically no spin-orbit splitting is seen, both in the N and the Δ even-parity resonances.

(e) Most importantly, about half the 21 predicted N = 2, N–Δ resonances are clearly observed experimentally. Except for the $N(1710)P_{11}$, most of the even-parity mixed symmetry 70 states are either weak or missing, and so also the antisymmetric 20 states in the nucleon. There is a very weak resonance $N(2100)P_{11}$, which may belong to the 20 , $^2 8$ multiplet.

THE ONE-GLUON-EXCHANGE POTENTIAL AND THE OSCILLATOR MODEL

We have partially answered the last question already. To try to answer some of the other questions, it is necessary to examine the nature of the quark-quark interaction. It is

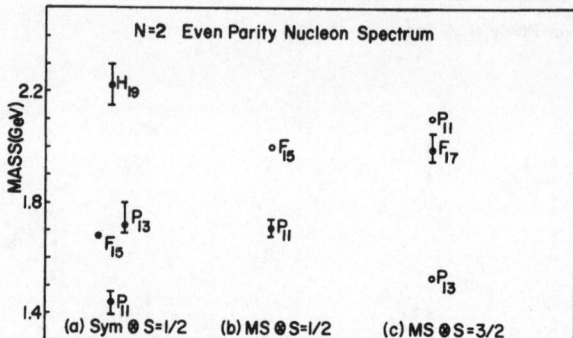

Fig. 7   The low-lying even-parity resonances of the nucleon (I = 1/2-channel). Legend
          same as in Fig. 6.

known that the relevant coupling constant is asymptotically free[13], i.e., it gets weaker as
the two quarks move closer. For low-energy spectroscopy, we may assume that the
baryonic volume is small, and use some r-independent value of the coupling constant,
and denote it by α_s . Only the one-gluon exchange potential (OGEP) will be considered,
and momentum-dependent terms will be neglected for simplicity. The one gluon-
exchange potential is very successful in reproducing the ground-state masses of the
baryons[14] . Between two equal-mass quarks, it is given by[15] (we set, from now on,
$\hbar = c = 1$):

$$V_{OGE}(\vec{r}) = -\frac{2}{3}\alpha_s[\frac{1}{r} - \frac{\pi}{m^2}\delta^3(\vec{r}) - \frac{1}{4m^2}\frac{1}{r^3}S_{12} - \frac{2\pi}{3m^2}(\vec{\sigma}_1 \cdot \vec{\sigma}_2)\delta^3(\vec{r})$$

$$-\frac{1}{4m^2r^3}\{(\vec{\sigma}_1 + 2\vec{\sigma}_2)\cdot(\vec{r}\times\vec{p}_1) - (\vec{\sigma}_2 + 2\vec{\sigma}_1)\cdot(\vec{r}\times\vec{p}_2)\}]. \tag{21}$$

The factor of -2/3 is special to baryons, arising from the colour part of the force. For
mesons, this factor is -4/3. The tensor operator S_{12} is $[3(\vec{\sigma}_1 \cdot \vec{e})(\vec{\sigma}_2 \cdot \vec{e}) - (\vec{\sigma}_1 \cdot \vec{\sigma}_2)]$, where
$\vec{\sigma}$'s are the Pauli spin-matrices and $\vec{e}$ the unit vector $\vec{r}/r$. Note that the OGEP is flavour-

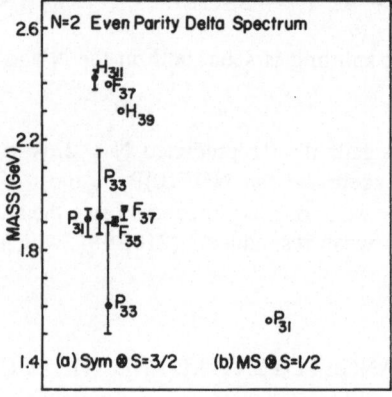

Fig. 8   The low-lying even-parity resonances of the delta (I = 3/2 channel).

independent. From Eq. 21, we see that it consists of the Coulomb potential, a zero range Darwin term, a long range tensor force, a spin-spin "hyperfine" interaction, and a spin-orbit potential. It is the spin-dependent hyperfine interaction that would give rise to the N−Δ splitting in the ground state. Its short range also explains why the splitting between the s = 3/2 and s = 1/2 odd-parity resonances is smaller. Its matrix-element with respect to oscillator states may easily be shown to be proportional to $\alpha_s \alpha_0^3/m^2$, where $\alpha_0 = \sqrt{m\omega_0}$. Note from Eq. 21 that matrix-elements of the Darwin, tensor and spin-orbit potentials are also proportional to the same factor $\alpha_s \alpha_0^3/m^2$, and only the Coulomb term goes like $\alpha_s \alpha_0$. Therefore, fitting the N−Δ splitting in the ground state also determines the strength of the tensor and spin-orbit parts of the force. This OGEP spin-orbit potential is far too strong, and would result in a splitting of a few hundred MeVs between the J = 5/2 and J = 1/2 states, obtained by coupling L = 1$^-$ to s = 3/2. From Fig. 5, we see, on the other hand, that in the nucleon, the three states S_{11}, D_{15} and D_{13} all come within 50 MeV. Interestingly enough, the OGEP spin-orbit strength is about right to yield the gap between Δ(1700)D_{33} and Δ(1620)S_{31}.

There is another way of generating a spin-orbit force. Consider, for simplicity, the motion of a relativistic quark, confined by a one-body scalar potential V_s. Through the Dirac equation, we know that there would be spin-orbit splitting. In a nonrelativistic reduction[16], the so-called Thomas term is obtained:

$$V_{LS}(r_i) = -\frac{1}{4m^2} \frac{1}{r_i} \frac{dV_s(r_i)}{dr_i} (\vec{l}_i \cdot \vec{\sigma}_i). \qquad (22)$$

Here, the -ve sign results from the scalar nature of the potential (V_s transforms like mass m under Lorentz transformation). This could cancel a large part of the OGEP spin-orbit force in the nucleon odd-parity states. However, the effect goes the other way for the Δ -states[11]. Therefore the spin-orbit splittings remain a problem. A few solutions to this problem have been suggested in recent years[17–19]. We shall come back to these in the last lecture.

One successful and simple prescription (followed by Isgur and Karl[11,12]) is to ignore all spin-orbit forces. These authors actually take only the hyperfine and the tensor components of the OGEP (see Eq. 21), and diagonalise these in the basis states of the harmonic Hamiltonian, Eq. 18. For the odd-parity states, only the N = 1 states are taken, while for the even-parity states, the diagonalisation is done in the basis of the N = 0 and the N = 2 states. The central components of V_{OGEP} are taken into account only for the diagonal matrix elements (i.e., in first order perturbation). This has the effect of bringing down the largely space symmetric N = 2, L = 0$^+$ state, much more than the N = 1, L = 0$^+$ state. This could explain the occurrence of the low-lying N(1440)P_{11} in relation to the odd-parity states, and also the gap between N(1440)P_{11} and N(1710)P_{11}, the latter being mostly mixed symmetric in this model. In this simple form, the Isgur-Karl model has been very successful, and of much use to the experimentalists. The model has been extended by taking more phenomenological pieces in the qq force, and extending the basis to the N = 3 states by Forsyth and Cutkosky[20].

At a somewhat deeper level, though, many of the questions raised earlier in the lecture remain unanswered. If the off-diagonal matrix-element of the central OGEP connecting the N = 0 and N = 2 symmetric states is not ignored, the "Roper" resonance is pushed up by more than a hundred MeV. Additionally, discrepancies persist[21] in the γN decay amplitudes $A^p_{1/2}$ and $A^n_{1/2}$, for the resonances N(1440)P_{11} and N(1710)P_{11}. Unless the qq-force is modified in an ad hoc manner for the N = 3 states, one cannot obtain the low-lying Δ states, S_{31} and D_{35}, at 1900 MeV.

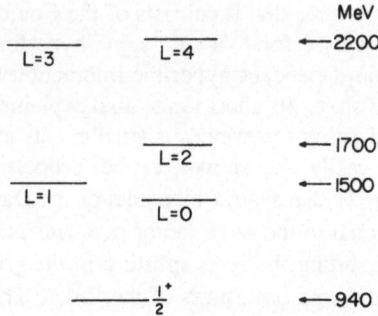

Fig. 9   A schematic diagram of the orbital excited states of the nucleon, which, combined with the total spin s = 3/2 or 1/2, would give rise to the observed resonances. The mixed symmetry even-parity N = 2 band is not shown.

We shall therefore put some more physical ingredients in this oscillator quark-model
in an attempt to overcome some of these difficulties.

DEFORMATION IN EXCITED STATES

Let us concentrate on some essential features of the baryon spectrum from the maze
of the experimental data. Stripping off the complications of spin, we draw a schematic
diagram (Fig. 9) for the orbital excited states of the nucleon. These states, when combined with spin (L-S coupling), may give rise to the observed resonances. We have, for
simplicity, omitted here the mixed symmetry even-parity states. The important features
shown here are the following:
   a) The $L=0^+$ excitation is lower than the lowest $L=1^-$ excited state;
   b) There is a gap of about 200 MeV between the $L=0^+$ and 2^+ excited states; and,
   c) The sequence of states (excluding the ground state) is similar to a rotational
       spectrum.

In fact, these features are not uncommon in the spectra of nuclei, and we give two examples that are found in the low-lying states of ^{4}He and ^{16}O . Both these are closed-shell
nuclei: ^{4}He filling the lowest $1s_{1/2}$ shell and the nucleons in ^{16}O occupying the
$1s_{1/2}$, $1p_{3/2}$, and $1p_{1/2}$ shells. In both nuclei, from the oscillator picture, one expects the
lowest excited state to be of odd-parity, whereas experimentally the even-parity 0^+ excitation comes lower in energy. This is very surprising in view of the shell gap - in ^{16}O,
the single-particle gap between the $1p_{3/2}$ and $2s_{1/2}$ levels is more than 10 MeV. So an
even-parity excitation due to promotion of two particles across the gap, should have been
about 20 MeV above the ground state. This was a hot problem in the sixties, and it took
a few years to realize that the spectra of ^{16}O could be explained naturally by assuming
that in the excited state the nucleus was getting appreciably deformed, although the
ground state is spherical. The members of the rotational band in ^{16}O could be excited by
the scattering of α–particles on ^{12}C . In ^{4}He, the situation seems to be similar, although
the rotational spectrum has not been experimentally established.

In the earlier sections, a "shell-model" for the quarks with L-S coupling scheme was
developed, in which the valence quarks moved in a spherical oscillator potential, and the
OGEP interaction between them was treated perturbatively in a truncated oscillator

space. We know, however, that a rotationlike spectrum may be generated by doing shell-model in a deformed potential[22]. Since orbital angular momentum is no longer a good quantum number in a deformed potential, one has to project out states of good L from the many-particle product states[23]. For large deformations, the energies of these states follows the rotational sequence $\hbar^2 L(L+1)/2I$, where I is the moment of inertia. In the quark model that we now develop, we adopt the above viewpoint. The details of this model may be found in Refs. 24 and 25.

The main modification from the standard oscillator model comes now from the assumption that the flavour-independent confinement potential is deformable. In the oscillator model, a spherically symmetric harmonic confinement was assumed, as in Eq. 16. We may take a more general attitude by taking a triaxial shape, with spring constants K_x, K_y and K_z. This would be very unsatisfactory if these were introduced as free parameters. We would, in fact, determine the ratios between the various spring constants by minimising the energy of the "intrinsic" state (i.e., a state where the number of excitation quanta N of Eq. 19 is specified), subject to the condition that the volume of the baryon in an intrinsic state does not increase due to deformation. Such a model is well-known in the study of deformed nuclei[26], and has also been used[27] to derive simple expressions for inertial parameters like the moment of inertia.

The Hamiltonian H_0 of Eq. 18 may now be written as

$$H_0 = \frac{p_\rho^2}{2m_\rho} + \frac{p_\lambda^2}{2m_\lambda} + \frac{3}{2} [K_x(\rho_x^2+\lambda_x^2) + K_y(\rho_y^2+\lambda_y^2) + K_z(\rho_z^2+\lambda_z^2)]. \tag{23}$$

For the N−Δ case of our interest, we put $m_\rho = m_\lambda = m$. Then Eq. 23 reduces to

$$H_0 = \frac{1}{2m} (p_\rho^2+p_\lambda^2) + \frac{1}{2}m \sum_{i=x,y,z} \omega_i^2(\rho_i^2+\lambda_i^2), \tag{24}$$

with $\omega_i = \sqrt{3K_i/m}$.

An intrinsic state is defined by specifying the quanta of excitation:

$$N_x = (n_{\rho_x} + n_{\lambda_x}), \ N_y = (n_{\rho_y} + n_{\lambda_y}), \ N_z = (n_{\rho_z} + n_{\lambda_z}), \tag{25}$$

with $N = N_x + N_y + N_z$.

The intrinsic energy of the system, excluding the centre-of-mass motion, is clearly,

$$E_i = \hbar\omega_x(N_x+1) + \hbar\omega_y(N_y+1) + \hbar\omega_z(N_z+1). \tag{26}$$

For a given set (N_x,N_y,N_z), this energy is minimised by varying ω_x, ω_y and ω_z, but subject to the condition that[25,26]

$$\omega_x\omega_y\omega_z = \omega_0^3. \tag{27}$$

This leads to the equilibrium condition:

$$\omega_x(N_x+1) = \omega_y(N_y+1) = \omega_z(N_z+1). \tag{28}$$

For the ground state, $N_x = N_y = N_z = 0$, and it follows from the above equations that $\omega_x = \omega_y = \omega_z = \omega_0$. The ground state remains spherical, as in the conventional model. In Table 4, we display the equilibrium intrinsic shapes, the corresponding energies (calculated from Eq. 26), the moments of inertia and the expectation values of the operator

Table 4   The intrinsic states in terms of the oscillator excitation quanta $N=N_x+N_y+N_z$. A given set of N_x,N_y,N_z defines an intrinsic state, the first few of which are listed in the first column. The equilibrium shape is shown in the third column. The corresponding intrinsic energies (Eq. 26) obeying the equilibrium conditions (27) and (28) are listed in the fourth column. The moment of inertia I is evaluated from Eq. 29, and the factor $\hbar^2/2I$ is shown in the fifth column. The last column lists $<L^2>$, obtained through Eq. 30.

Number of excitations, N	Multiplet structure (parity)	Equilibrium configuration	Intrinsic energy	$\dfrac{\hbar^2}{2I}$	$<L^2>$
$N=0$ $N_x=N_y=N_z=0$	56^+	$\omega_x=\omega_y=\omega_z$ (spherical)	$3\hbar\omega_0$		0
$N=1$ $N_x=N_y=0,N_z=1$	70^-	$\omega_x=\omega_y=2\omega_z$ (prolate)	$3.780\hbar\omega_0$	$0.126\hbar\omega_0$	3
$N=2$ $N_x=N_y=0,N_z=2$	$56^+,70^+$	$\omega_x=\omega_y=3\omega_z$ (prolate)	$4.327\hbar\omega_0$	$0.072\hbar\omega_0$	8
$N_x=N_y=1,N_z=0$	20^+	$\omega_x=\omega_y=\dfrac{\omega_z}{2}$ (oblate)	$4.762\hbar\omega_0$	$0.159\hbar\omega_0$	3
$N=3$ $N_x=N_y=0,N_z=3$	$56^-,70^-$	$\omega_x=\omega_y=4\omega_z$	$4.762\hbar\omega_0$	$0.047\hbar\omega_0$	15

L^2 (where $\vec{L}=\vec{L}_\rho+\vec{L}_\lambda$) with respect to the intrinsic states of interest. Note the following points:

(a) The shape of the system changes with excitation. The $N=1$ state is prolate and axially symmetric, as are the symmetric and mixed symmetric $N=2$ states. On the other hand, the antisymmetric $N=2$ state is oblate and comes higher in energy. Only the lowest $N=3$ state (prolate) is listed. Two more solutions, triaxial and spherical, come at higher energies.

(b) The deformation of the baryon increases as more and more quanta are excited, and the corresponding energies come down more relative to the spherical model.

(c) The moment of inertia and $<L^2>$ can be calculated quantum mechanically for each intrinsic state[25-27]. The moment-of-inertia is about an axis perpendicular to the symmetry axis, and at equilibrium deformation, is given by

$$I = \frac{(N_z+1)}{\omega_z} + \frac{(N_y+1)}{\omega_y}, \tag{29}$$

where the shape is symmetric about the z-axis. This is the same as the rigid-body moment of inertia for that shape. For axial symmetry,

$$<L^2> = (\frac{\omega_>}{\omega_<})^2 - 1 \, , \tag{30}$$

where $\omega_>(\omega_<)$ is the larger (smaller) of ω_y and ω_z .

(d) As in the spherical model, there is only one free oscillator parameter ω_0 in the model. All other quantities are determined by the number of excitation quanta, and may be calculated explicitly in terms of these and ω_0.

(e) Although the effective Hamiltonian H_0 ceases to be rotationally invariant in the excited states, the fundamental Lagrangian of the system must be rotationally invariant. This means that physically meaningful states are obtained by projecting out states of good angular momenta from the intrinsic states. To a good approximation, the energies of these projected states follow a rotational pattern. The energy of a state with angular momentum L is related to the intrinsic energy by

$$E_L = E_i - \frac{<L^2>}{2I} + \frac{L(L+1)}{2I} \, . \tag{31}$$

Corresponding to each intrinsic state, there is a rotational band. Here L may take values 0,2,4, etc. in the even-parity prolate states, and L = 1,3,5,... in the odd-parity prolate or even-parity oblate states.

The way the N = 2 and N = 3 states come down in energy is shown schematically in Fig. 10. For illustration, we take $\hbar\omega_0 = 550$ MeV. In the spherical oscillator model, the shells are each $\hbar\omega_0$ apart, as shown in the first column of Fig. 10. In the next column, the intrinsic energies, E_i , for the prolate configurations (see Table 4) are shown. Note already the appreciable lowering in energy of the N = 2 and N = 3 excitations. A further lowering in energy of the band-head (the lowest-L member of the band) is obtained through projection, and this is shown in the last column. The band-head energies are easily estimated by using Eq. 31. Although we do not show the next members of the bands, you may check that our choice of $\hbar\omega_0 = 550$ MeV yields the right spacings. For example, 1/2I for the N = 1 prolate state is found to be (through Eq. 29) about 70 MeV

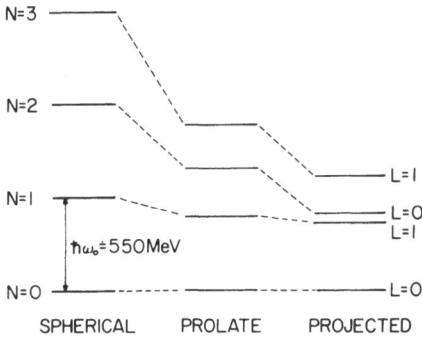

Fig. 10   The lowering in energy of the excited states due to deformation and projection. In column 1, the energies of the spherical oscillator states are shown. In column 2, the intrinsic energies of the lowest prolate configurations are displayed, for assumed $\hbar\omega_0 = 550$ MeV (see Table 4). The N = 0 state remains spherical. The corresponding projected band-heads are shown in the last column. The higher-L members of the band are not shown. There is no band built on the spherical ground state.

for $\hbar\omega_0 = 550$ MeV. The gap between $L = 3$ and $L = 1$ states in a rotational spectrum is $10/2I \sim 700$ MeV. This is what is found experimentally, see Fig. 9. Similarly, for N=2, $1/2I \sim 40$ MeV. This yields 6 times $1/2I \sim 240$ MeV between $L = 2$ and $L = 0$ states, and 20 times $1/2I \sim 800$ MeV between $L = 4$ and $L = 0$ states in the band. These again are about right, as seen in Fig. 9.

THE SPIN-ORBIT PROBLEM. ARE PIONS NEEDED?

In the deformed baryon model, the various intrinsic states may mix by the q-q interaction. Although we indicated in the last section how projection may help in getting agreement with experimental data, such mixings must be taken into account before projection. We will not give the details of the calculation here, since these may be found in Refs. 18, 21, and 25. The results of such calculations[21], ignoring the non-central forces, are shown in Figs. 11 and 12.

I would, at this point, like to go back to the spin-orbit problem described earlier. A detailed description of this problem in the context of the oscillator model will be found in the articles by Reinders[28] and Gromes[17]. Gromes had suggested that spin-orbit forces may arise due to the nonlocality of the confining potential, and this may help resolve the problem. We shall discuss two other possibilities here.

Going back to the OGE potential given by Eq. 21, note that the hyperfine interaction is of zero-range. This form of interaction is not suitable for any exact calculation. If you solve the Schrödinger equation with a zero range attractive potential, then the system will collapse, with no lower bound in energy. For this reason, in any nonperturbative calculation, one should replace $\delta^3(\vec{r}) \rightarrow f(r)$, where $f(r)$ is some short-range function[29,30]. Thus the interaction (Eq. 21) should be regarded as an "effective potential" with which to perform shell-model calculations in a *truncated* Hilbert space. If one took $f(r)$ to be of very short range (less than half a Fermi), and performed matrix-diagonalisation in a basis of states, then the results would be very sensitive to the size of the basis. Let us suppose

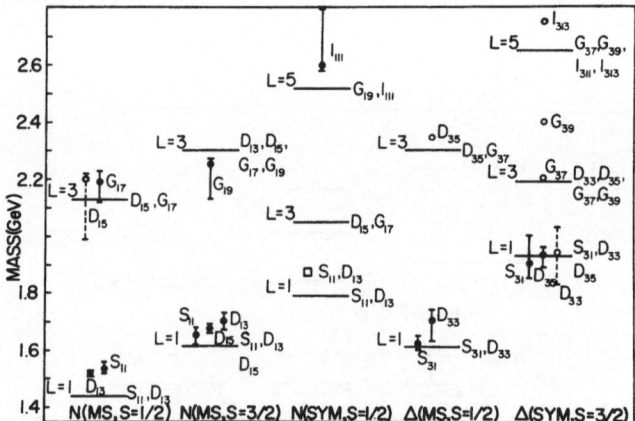

Fig. 11   Odd-parity $N = 1$ (MS) and $N = 3$ (SYM) nucleon and delta states. The horizontal lines are the results of the calculation with the deformed model described earlier, but ignoring the tensor and spin-orbit forces (see Ref. 21 for details). The permutation symmetries (SYM and MS) refer only to the main component in the wave function. The experimentally determined states are denoted the same way as in Fig. 6.

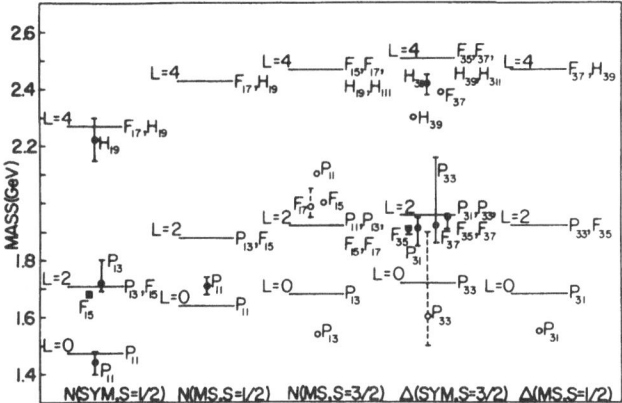

Fig. 12   The even-parity nucleon and delta spectra in the deformed model. Legend same as in Fig. 11.

you want to adjust α_s so as to get the Δ–N splitting. If you just did a perturbation estimate in the N = 0 shell, you would require some value of α_s, let us say $\alpha_s \approx 1$ for $\omega_0 \approx 500$ MeV. If now you did a matrix-diagonalisation of this very short range hyperfine interaction in a much bigger space, e.g., including N = 0, 2, 4, 6 shells, the same splitting between Δ and N may be obtained for a much smaller value of α_s. One should check, however, that f(r) is not of such short range that the results are unstable. This should be done, for example, by doubling the basis space and checking that you get the same result. The inadequacy of the perturbative method for very short-range attractive interactions was noted in Ref. 29. Now we see a possible solution of the spin-orbit problem. If f(r) is of very short range, then an exact treatment would entail a much smaller α_s than that required by perturbation to yield the Δ–N mass splitting. Note, however, that the spin-orbit and tensor forces are of long range. A reduced value of α_s suppresses these interactions, and a perturbative treatment is adequate for these long range forces. Such a position has been taken recently by Capstick and Isgur[19]. There must be some truth in this. The objection to the approach, however, is the following: By fine tuning the range of f(r), one can make the spin-orbit and tensor parts very weak indeed! And the results become uncomfortably sensitive to the range of f(r).

There is another factor that may require taking a reduced value for the quark-gluon coupling constant α_s, and thereby solve the spin-orbit problem. The pion may be looked upon as a fundamental field that couples to the u- and d-quarks directly to conserve the axial vector current[31]. If such a view is taken, then the pion should be treated on the same footing as the gluon, and there would be the one-pion-exchange potential (OPEP) between the nonstrange quarks. In the static approximation, and with pseudoscalar coupling, this potential is[22]

$$V_{OPE} = -\frac{\alpha_\pi}{12m^2}\,(\vec{\tau}_1 \cdot \vec{\tau}_2)[(\vec{\sigma}_1 \cdot \vec{\sigma}_2)\{4\pi\delta^3(\vec{r}) - m_\pi^2\frac{e^{-m_\pi r}}{r}\}$$

$$-S_{12}(1 + \frac{3}{m_\pi r} + \frac{3}{m_\pi^2 r^2})m_\pi^2\frac{e^{-m_\pi r}}{r}]\,. \qquad (32)$$

Here $\alpha_\pi = \dfrac{g_{qq\pi}^2}{4\pi}$ is the strength of the pseudoscalar quark-pion interaction, and m_π is the

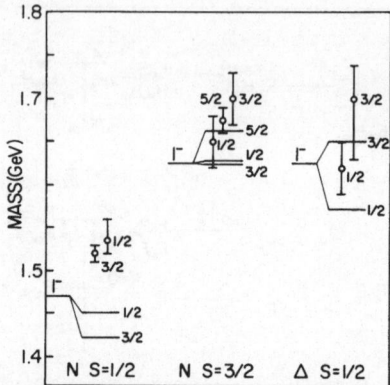

Fig. 13   Odd-parity N–Δ spectrum. Comparison of the deformed model calculations (horizontal lines) with experimental data. Only the N = 1 states are shown. The tensor and spin-orbit components of V_{OGE} (Eq. 21) and the tensor part of V_{OPE} (Eq. 32) are now included. See Ref. 18 for details.

pion mass. One may, in this approach, take this "pion" to be massless ($m_\pi \to 0$) and structureless if one likes, or alternatively, to have the characteristics of the physical pion. The main point in Eq. 32 is that even though V_{OPE} has a strong spin-dependent force, it has no spin-orbit component! It is possible, in this scheme of calculation, that about half or more of the splitting between Δ–N may come from the pion coupling. The OGEP is then required to be responsible for the rest of the Δ–N splitting, so that a much smaller value of α_s is needed[18,21]. This, in turn, reduces the spin-orbit force in OGEP.

Detailed calculations for the N = 1 and N = 2 N–Δ states have been performed in the deformed baryon model with this approach[18]. Such results for the low lying odd and even states are shown in Figs. 13, 14 and 15. To our knowledge, these are the first calculations of spin-orbit splittings for the N = 2 states.

We should mention that in order to extend this approach to baryons with strangeness, it is not enough to just consider pions (and gluons) between u, d-quarks. Even the ground

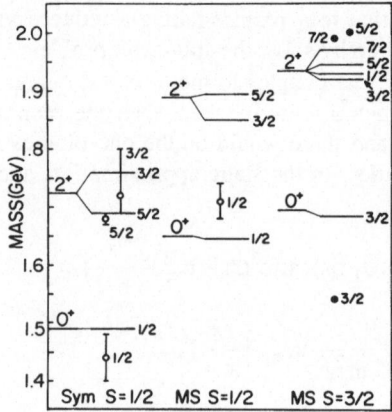

Fig. 14   Even-parity nucleon spectrum. Legend same as in Fig. 13.

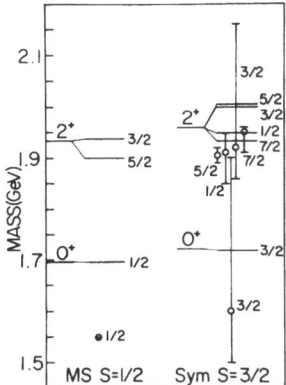

Fig. 15   Even-parity delta spectrum. Legend same as in Fig. 13.

state masses of the baryons cannot be fitted in such a model. One may, however, take the whole pseudoscalar octet of mesons as Goldstone bosons, and couple these to quarks[32]. The strengths of the couplings of π, K and η fields to the quarks are not independent parameters if one uses $SU(3)_{flavour}-$ symmetry. Such calculations are in progress, and yield good results[33].

CONCLUSIONS

Recently, an exact calculation of the baryon spectra in the nonrelativistic quark model (with linear confinement, and OGEP without tensor spin-orbit forces) has been performed[34]. It was found that the position of the low-lying $N(1440)1/2^+$ and $\Delta(1900)1/2^-$ and $5/2^-$ states cannot be explained in such a model even if the problem is solved exactly. We think that deformation of the baryon in the excited states is responsible for such effects. This deformation in shape, however, is not the result of a simple q-q interaction like the OGEP, but may be due to many-body effects[24]. Probably enhanced electromagnetic decay rates between members of a band (like the E2-transition between $N(1680)5/2^+ \rightarrow N(1440)1/2^+$) , if detected experimentally, would constitute the validity of the deformed model.

We are not so sure about the role of the pions in the solution of the spin-orbit problem. It does offer a solution, but the prospect of all the Goldstone bosons interacting with the quarks does not make me too happy. There is some tendency in baryon spectroscopy to make claims to the effect that everything has been solved, including the spin-orbit problem. Such claims are misleading. Only more refined experimental data will shed more light on the problem.

ACKNOWLEDGEMENTS

The author would like to thank Dr. M. V. N. Murthy for many discussions. Much of the work described in the last lecture was a collaborative effort with him and other colleagues whose names appear in the referred papers. This work was supported by the Natural Sciences and Engineering Council of Canada.

REFERENCES

[1]   A. Donnachie, in "Hadronic Interaction of Electrons and Photons", Proceedings of the eleventh Scottish Universities Summer School in Physics, 1970, edited by J. Cumming and H. Osborn (Academic Press, New York) p. 109.

[2]   R. E. Cutkosky et al., Phys. Rev. **D20,** 2804 and 2839 (1979).

[3]   E. Pietarinen, Nucl. Phys. **B107,** 21 (1976); R. Koch and E. Pietarinen, Nucl. Phys. **A336,** 331 (1980).

[4]   Reviews of Particle Properties, in Rev. Mod Phys. **56,** S206 (1984).

[5]   F. E. Close, "An Introduction to Quarks and Partons", (Academic Press, New York, 1979).

[6]   F. Foster and G. Hughes, Rep. Prog. Phys. **46,** 1445 (1983).

[7]   R. H. Dalitz, in Les Houches Lectures, 1965 (Gordon and Breach, New York, 1965).

[8]   R. K. Bhaduri and Mira Dey, Phys. Lett. **125B,** 513 (1983).

[9]   D. Faiman and A. W. Hendrey, Phys. Rev. **173,** 1720 (1968).

[10] N. Isgur and G. Karl, Phys. Rev. **D18,** 4187 (1978).

[11] N. Isgur and G. Karl, Phys, Rev. **D19,** 2653 (1979).

[12] G. Karl and E. Obryk, Nucl. Phys. **B8,** 609 (1968).

[13] N. K. Nielsen, Am. J. Phys. **49,** 1171 (1981).

[14] A. De Rujula, H. Georgi and S. L. Glashow, Phys. Rev. **D12,** 147 (1975).

[15] For derivation, see, for example, V. Berestetskii, E. M. Lifshitz and L. P. Pitaeveskii, "Relativistic Quantum Theory, Park I" (Pergamon Press, Oxford 1979) p. 280.

[16] L. I. Schiff, "Quantum Mechanics" (McGraw-Hill, New York, 1968), p. 482. The sign of the Thomas term depends on whether the potential transforms like a scalar or a vector under Lorentz transformation.

[17] D. Gromes, Z. Phys. **C18,** 249 (1983).

[18] M. V. N. Murthy, M. Brack, R. K. Bhaduri and B. K. Jennings, Z. Phys. C (1985) to be published.

[19] S. Capstick and N. Isgur, University of Toronto preprint (1985).

[20] C. P. Forsyth and R. E. Cutkosky, Z. Phys. **C18,** 219 (1983).

[21] M. V. N. Murthy and R. K. Bhaduri, Phys. Rev. Letter. **54,** 745 (1985).

[22] For a review, and references to original papers, see M. A. Preston and R. K. Bhaduri, "Structure of the Nucleus", (Addison-Wesley, Reading, Mass. 1975) p. 463.

[23] R. E. Peierls and J. Yoccoz, Proc. Phys. Soc. (London) **70,** 381 (1957).

[24] R. K. Bhaduri, B. K. Jennings and J. C. Waddington, Phys. Rev. **D29,** 2051 (1984).

[25] M. V. N. Murthy, M. Dey, J. Dey and R. K. Bhaduri, Phys. Rev. **D30,** 152 (1984).

[26] A. Bohr and B. R. Mottelson, "Nuclear Structure", Benjamin, Reading, Mass. 1975) Vol. II, p. 77.

[27] B. R. Mottelson, in "The Many Body Problem", Les Houches lectures, 1958 (John Wiley, New York, 1959) p. 313.

[28] L. J. Reinders, in "Baryon 1980", edited by N. Isgur (University of Toronto, 1980).

[29] R. K. Bhaduri, L. E. Cohler and Y. Nogami, Phys. Rev. Lett. **44,** 1369 (1980).

[30] R. K. Bhaduri, L. E. Cohler and Y. Nogami, Nuovo Cimento **65,** 376 (1981).

[31] G. E. Brown and M. Rho, Phys. Lett. **82B,** 177 (1979); S. Théberge, A. W. Thomas and G. A. Miller, Phys. Rev. **D22,** 2838 (1980); M. Weise, in Fifth Tropical School on Quarks, Mesons and Isobars in Nuclei, Motril, Granada, Spain, 1982.

[32]   Such an approach was suggested by A. Manohar and H. Georgi, Nucl. Phys. **B234,** 189 (1984).

[33]   M. V. N. Murthy, R. K. Bhaduri and E. Tabarah, to be published.

[34]   B. Silvestre-Brac and C. Gignoux, Phys. Rev. **D32,** 743 (1985).

POLARIZATION DEGREES OF FREEDOM IN ELECTRON SCATTERING FROM NUCLEI*

T. W. Donnelly

Center for Theoretical Physics
Laboratory for Nuclear Science and Department of Physics
Massachusetts Institute of Technology
Cambridge, Massachusetts 02139   U.S.A.

INTRODUCTION AND GENERAL OVERVIEW

Various aspects of polarization degrees of freedom in electron scattering are discussed in these lectures, including what happens when incident or scattered electrons are polarized and what happens when the target nucleus is polarized or when some part of the final-state polarization is measured (for instance, the polarization of the recoilling nucleus or of some emitted particle). In addition, even when the polarizations of the external particles are not themselves determined directly, there are important polarization degrees of freedom to consider such as those for the exchanged virtual photon or those implied by angular distributions in coincidence reactions. All of these aspects are touched upon here.

The material presented in these lectures has been organized in the following way: In the present section an overview of electron scattering including polarizations is given. The object is to see the general structure of this problem first and then to return in the succeeding sections to understand in more detail how this structure comes about. In particular, in Section II the leptonic tensors involved in electron scattering are discussed, with some attention paid to what happens when the electrons are polarized and why simplifications occur when they are ultra-relativistic particles. In Section III the hadronic (nuclear) side of the problem is considered and the "Super-Rosenbluth Formula" derived. Section IV forms a major part of the discussion and focusses in more detail on the subject of inclusive electron scattering from polarized targets. Finally, in Section V two special topics are further discussed, namely the fifth response function in $(\vec{e}, e'x)$ reactions and the complementary of $(e, e'\gamma)$ studies to studies of (e, e') with polarized targets.

At the end is given a list of references (with comments to guide the readers) to relevant conference proceedings and articles on the general subject presented here.[1−12]

---

*   This work is supported in part through funds provided by the U. S. Department of Energy (D.O.E.) under contract #DE-AC02-76ER03069.

Typeset in $T_{\!E}\!X$ by Roger L. Gilson

Let us begin by summarizing the general structure of electron scattering in order of increasing exclusivity:

$$(e, e') \quad \leftrightarrow \quad \text{inclusive} \quad \leftrightarrow \quad \text{single-arm scattering}$$
$$(e, e'x_1) \quad \leftrightarrow \quad \text{exclusive-1} \quad \leftrightarrow \quad \text{two-arm coincidence}$$
$$(e, e'x_1x_2) \quad \leftrightarrow \quad \text{exclusive-2} \quad \leftrightarrow \quad \text{triple coincidence}$$
$$\vdots \qquad\qquad\qquad \vdots \qquad\qquad\qquad \vdots$$

We shall also refer to inclusive scattering as an exclusive-0 reaction so that, in general, exclusive-n corresponds to having n particles detected in coincidence with the scattered electrons. Furthermore, any of the particles involved may be polarized $(\vec{e}, \vec{e}', \vec{x}_1, \vec{x}_2, \ldots)$.

First consider the simplest of electron scattering reactions, namely inclusive (or single-arm) experiments of the type $A(e, e')$ where only the scattered electron is detected and where nothing is polarized. To date, the vast majority of experimental work falls into this class. The kinematics in this case are illustrated in Fig. 1. Here an electron with 3-momentum $\vec{k}$ and energy ϵ is scattered through in angle θ_e to 3-momentum $\vec{k}'$ and energy ϵ'. In the process a virtual photon corresponding to 3-momentum transfer $\vec{q} = \vec{k} - \vec{k}'$, and energy transfer $\omega = \epsilon - \epsilon'$, is exchanged with the target nucleus A. The 4-momentum transfer* is then specified, $Q^2 = Q_\mu Q^\mu = \omega^2 - q^2$, where $q \equiv |\vec{q}|$, and from the kinematics of the scattering process must satisfy $Q^2 \leq 0$. Following standard procedures as outlined in Sections II, III and IV involving the construction of leptonic and hadronic tensors, $\eta_{\mu\nu}$ and $W^{\mu\nu}$, respectively, in this one-photon-exchange approximation, the cross section can be cast in the familiar Rosenbluth form:

$$\frac{d^2\sigma}{d\Omega_e d\epsilon'} = \frac{1}{M_i} \sigma_{\text{Mott}} f_{\text{rec}}^{-1} \left\{ v_L W^L + v_T W^T \right\} . \tag{1}$$

Here M_i is the target mass, σ_{Mott} is the elementary Mott cross section,

$$\sigma_{\text{Mott}} = \left\{ \alpha \, \cos \frac{\theta_e}{2} \Big/ 2\epsilon \, \sin^2 \frac{\theta_e}{2} \right\}^2 \tag{2a}$$

and a kinematic recoil factor may be included,

$$f_{\text{rec}} = 1 + 2\epsilon \, \sin^2 \frac{\theta_e}{2} \Big/ M_i . \tag{2b}$$

The electron kinematical factors v_L and v_T become in the extreme relativistic limit $(\epsilon, \epsilon' \gg m_e)$, which is assumed in all that follows,

$$v_L = \left(Q^2/q^2 \right)^2 \tag{3a}$$

$$v_T = -\frac{1}{2}\left(Q^2/q^2 \right) + \tan^2 \frac{\theta_e}{2} . \tag{3b}$$

---

* We use the conventions of Bjorken and Drell[13] plus the additional convention that 4-vectors are denoted by capital letters, $A^\mu = (A^0, A^1, A^2, A^3)$. We have $A_\mu B^\mu = A^0 B^0 - \vec{a} \cdot \vec{b}$.

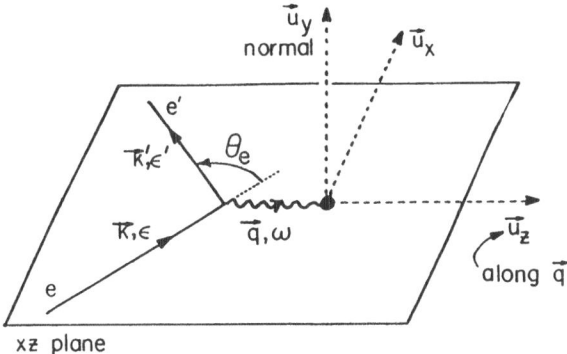

Fig. 1:   Inclusive electron scattering, $A(e, e')$.

Here, "L" stands for projections along $\vec{u}_z$ in Fig. 1 (that is, along $\vec{q}$), whereas "T" stands for projections in the $\vec{u}_x$, $\vec{u}_y$ plane; these are longitudinal and transverse projections, respectively. The hadronic (nuclear) information is then summarized by two structure functions W^L and W^T each depending on the variables (q, ω), but not on θ_e. Thus, a Rosenbluth separation can be effected when q and ω are fixed but θ_e varied so that v_L remains fixed and v_T varies, allowing W^L and W^T to be determined separately. Alternative rewritings of these basic expressions involve decompositions into W_1 and W_2 or A and B or F_L^2 and F_T^2 (see below), depending on the context within which they are used. In whatever form, however, the essential point is that two and only two structures are experimentally accessible in such inclusive unpolarized experiments.

Specifically, when discrete states are involved $(J_i^{\pi_i} \to J_f^{\pi_f})$ we have

$$\frac{d\sigma}{d\Omega_e} = 4\pi \sigma_{\text{Mott}} f_{\text{rec}}^{-1} \left\{ v_L F_L^2 + v_T F_T^2 \right\} \quad , \tag{4}$$

where

$$F_L^2(q) = \sum_{J \geq 0} F_{CJ}^2(q) \tag{5a}$$

$$F_T^2(q) = \sum_{J \geq 1} \left\{ F_{EJ}^2(q) + F_{MJ}^2(q) \right\} \tag{5b}$$

and where the Coulomb, electric and magnetic form factors are given in terms of reduced matrix elements of the respective electromagnetic multipole operators:*

$$F_{CJ}(q) = \frac{1}{[J_i]} \langle J_f \| \hat{M}_J(q) \| J_i \rangle \tag{6a}$$

$$F_{EJ}(q) = \frac{1}{[J_i]} \langle J_f \| \hat{T}_J^{\text{el}}(q) \| J_i \rangle \tag{6b}$$

$$F_{MJ}(q) = \frac{1}{[J_i]} \langle J_f \| i\hat{T}_J^{\text{mag}}(q) \| J_i \rangle \quad , \tag{6c}$$

---

* For a treatment of the "standard results" of unpolarized electron scattering in the Bjorken and Drell metric, see Ref. 13; see also Refs. 14-16 where another metric is used.

where $[x] \equiv \sqrt{2x+1}$. With our phase conventions these form factors are all real. Note in particular that only two quantities are accessible in unpolarized inclusive electron scattering, $viz.$ F_L^2 and F_T^2, each of which involves a sum of squares of the individual form factors. Thus, for example, we cannot separate the various Coulomb form factors from one another. This is to be contrasted to what happens when the target is polarized as we shall see in due course.

Next, let us discuss what occurs when polarizations are specified, all the while staying within the context of inclusive scattering. First, suppose that only the incident electron is polarized; namely, consider the reaction $A(\vec{e}, e')$ illustrated in Fig. 2. We retain only ultra-relativistic electrons as mentioned above and so the relevant type of electron polarization is longitudinal in the sense that it is the electron's helicity $h = \pm 1$ (projection of spin along the momentum direction $\vec{k}$) which is presumed to be specified now (see the discussions in Section II of what happens in the regime where $\epsilon \not\gg m_e$). If only the parity-conserving electromagnetic interaction is present, then the cross section here does not depend on the helicity h. Only when parity-violation effects due to interferences involving the weak interaction are allowed for does the situation become more interesting, for then we have[17, 18]

$$\sigma^h = \sigma^{\text{unpol.}} + h \left\{ v_L W_{AV}^L + v_T W_{AV}^T + v_{T'} W_{VA}^{T'} \right\} , \qquad (7)$$

where $\sigma^{\text{unpol.}}$ is the unpolarized cross section discussed above and the new terms which are proportional to h all come from vector (V)/axial-vector (A) electroweak interferences (the first subscript on the W's corresponds to the leptonic coupling, while the second corresponds to the hadronic coupling). In the course of obtaining such general expressions a new electron kinematic factor has been introduced (Cf. Eqs. (3) and see below):

$$v_{T'} = \tan \frac{\theta_e}{2} \sqrt{-(Q^2/q^2) + \tan^2 \frac{\theta_e}{2}} . \qquad (8)$$

Such interesting parity-violation effects can then be isolated by forming the asymmetry

$$a = \frac{\sigma^{+1} - \sigma^{-1}}{\sigma^{+1} + \sigma^{-1}} . \qquad (9)$$

Ideally in the general situation, having isolated the bracketed expression in Eq. (7), the electron scattering angle could be varied keeping q and ω (the arguments of the W's as above) constant and so varying the kinematical factors v_T and $v_{T'}$. A plot versus $\tan^2 \theta_e/2$ will no longer be linear as it was in the purely electromagnetic Rosenbluth decomposition discussed above and so, in principle, the three structure functions in Eq. (7) can be separated. In practice, of course, these parity-violating cross sections are very small, and such complete decompositions would require heroic experimental efforts.

Once again focussing on discrete-state transitions we may write

$$a = \Delta/\Sigma , \qquad (10)$$

where

$$\Sigma = \left(\frac{d\sigma}{d\Omega_e} \right)^{\text{unpol.}} \qquad \text{[Eq. (4)]} \qquad (11a)$$

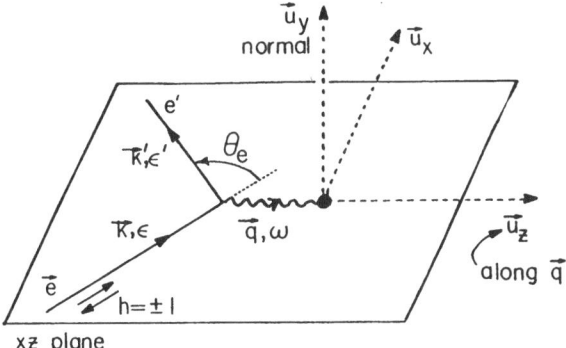

Fig. 2:   Inclusive electron scattering with polarized electrons $A(\vec{e}, e')$.

and

$$\Delta = 4\pi\sigma_{\text{Mott}} f_{\text{rec}}^{-1} \left(\frac{G|Q^2|}{2\pi\alpha\sqrt{2}} \right)$$

$$\times \left\{ a_A \left[v_L \sum_{J\geq 0} F_{CJ} F_{CJ}^{(0)} + v_T \sum_{J\geq 1} \left(F_{EJ} F_{EJ}^{(0)} + F_{MJ} F_{MJ}^{(0)} \right) \right] \right. \tag{11b}$$

$$\left. - a_V \left[v_{T'} \sum_{J\geq 1} \left(F_{EJ} F_{MJ_5}^{(0)} - F_{MJ} F_{EJ_5}^{(0)} \right) \right] \right\} \quad .$$

Here, each electromagnetic operator has an isoscalar and an isovector piece:

$$\hat{M}_{JM_J} = 1 \cdot \hat{M}_{JM_J;\, T=0,\, M_T=0} + 1 \cdot \hat{M}_{JM_J;\, T=1,\, M_T=0} , \quad \text{etc.} \tag{12a}$$

For the weak neutral current form factors (denoted "(0)") we have analogous expressions for the operators, but now with hadronic weak neutral current couplings $\beta_V^{(T)}$, $T = 0, 1$:

$$\hat{M}_{JM_J}^{(0)} = \beta_V^{(0)} \cdot \hat{M}_{JM_J;\, T=0,\, M_T=0} + \beta_V^{(1)} \cdot \hat{M}_{JM_J;\, T=1,\, M_T=0} , \quad \text{etc.} \tag{12b}$$

In addition, we have (real) axial-vector form factors (Cf. Eqs. (6))

$$F_{EJ_5}(q) = \frac{1}{[J_i]} \langle J_f \| - i\hat{T}_J^{el_5}(q) \| J_i \rangle \tag{13a}$$

$$F_{MJ_5}(q) = \frac{1}{[J_i]} \langle J_f \| \hat{T}_J^{mag_5}(q) \| J_i \rangle \tag{13b}$$

and their neutral current partners which have axial-vector couplings $\beta_A^{(T)}$, $T = 0, 1$. In the standard model, the leptonic and hadronic couplings are, respectively[18]

$$a_V = -(1 - 4\sin^2\theta_W), \quad a_A = -1$$
$$\beta_V^{(0)} = -2\sin^2\theta_W, \quad \beta_V^{(1)} = 1 - 2\sin^2\theta_W, \quad \beta_A^{(0)} = 0, \quad \beta_A^{(1)} = 1 \quad .$$

Next, let us turn to the situation illustrated in Fig. 3 where we again have inclusive scattering of unpolarized electrons, but now from polarized targets, $\vec{A}(e, e')$.

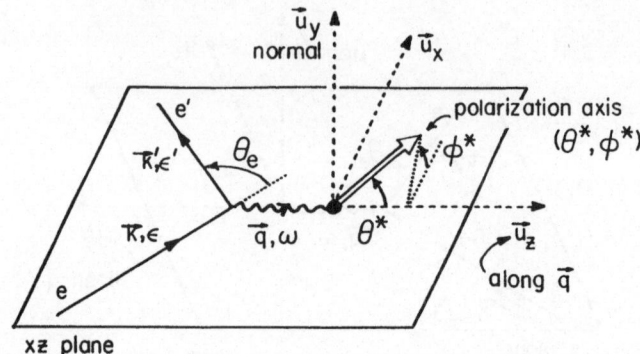

Fig. 3:   Inclusive electron scattering from polarized targets, $\vec{A}(e, e')$.

As shown in the figure, we presume the target to be polarized in some way with respect to a specific axis whose polar and azimuthal angles in the chosen coordinate system are θ^* and ϕ^*, respectively. Note that these angles are referred to the direction of momentum transfer $\vec{q}$, and when the electron kinematics change so does the direction of $\vec{q}$, and hence to keep θ^* and ϕ^* fixed the target polarization direction in the laboratory system will also have to change. We will return to the nature of the polarization shortly.

The procedures followed in factoring the leptonic (electron) and hadronic (nuclear) response functions proceed as before[11] and lead to the following general form (see Section IV for details) for the cross section which we denote by Σ_{fi}, where fi indicates that the initial nuclear state is polarized whereas the final nuclear state's polarization is not determined:

$$\Sigma_{fi} \sim v_L W_{fi}^L + v_T W_{fi}^T + v_{TL} \cos \phi^* W_{fi}^{TL} \\ + v_{TT} \cos 2\phi^* W_{fi}^{TT} \ . \tag{14}$$

Here we have ceased writing the Mott cross section, etc., for clarity. In addition to the electron kinematical factors v_L and v_T introduced above in Eqs. (3), we now have two new ones

$$v_{TL} = \frac{1}{\sqrt{2}}(Q^2/q^2)\sqrt{-(Q^2/q^2) + \tan^2 \frac{\theta_e}{2}} \tag{15a}$$

$$v_{TT} = \frac{1}{2}(Q^2/q^2) \tag{15b}$$

corresponding to transverse-longitudinal and transverse-transverse interferences, respectively. These electron kinematical factors each multiply nuclear response functions labelled the same way, where the latter quantities depend on q and ω as before and now on θ^* as well. Note, however, that they do not contain any ϕ^*-dependence; this is explicitly contained in the $\cos \phi^*$ and $\cos 2\phi^*$ factors in Eq. (14). Thus, an extended form of the Rosenbluth formula is obtained where q, ω and θ^* may be kept fixed and θ_e varied (to change the v's) and ϕ^* varied (to change the cosine factors) to achieve a separation of the *four* basic response functions which can occur in general.

Proceeding one step further into the problem, we may display the θ^* dependences as expansions in Legendre or associated Legendre polynomials and Fano tensors. The latter quantities are simply obtained by summing over the probabilities that the target is in states labelled J_i, M_{J_i} with Clebsch-Gordan coefficients as weighting factors (see Section IV). The nature of the target polarization determines these quantities and so we presume that they are known and controllable. The tensor expansions are the following:

$$W^L_{fi} = f^{(i)}_0 (W^L_0)_{fi} + \sum_{\substack{J \geq 2 \\ \text{even}}} f^{(i)}_J P_J(\cos \theta^*)(W^L_J)_{fi} \tag{16a}$$

$$W^T_{fi} = f^{(i)}_0 (W^T_0)_{fi} + \sum_{\substack{J \geq 2 \\ \text{even}}} f^{(i)}_J P_J(\cos \theta^*)(W^T_J)_{fi} \tag{16b}$$

$$W^{TL}_{fi} = \sum_{\substack{J \geq 2 \\ \text{even}}} f^{(i)}_J P^1_J(\cos \theta^*)(W^{TL}_J)_{fi} \tag{16c}$$

$$W^{TT}_{fi} = \sum_{\substack{J \geq 2 \\ \text{even}}} f^{(i)}_J P^2_J(\cos \theta^*)(W^{TT}_J)_{fi} \tag{16d}$$

where the quantities labelled $(W^K_J)_{fi}$ are referred to as *reduced response functions* and depend only on q and ω, but not on θ^* or ϕ^*. The summations run over all *even* values of J satisfying $2 \leq J \leq 2J_i$ and we have separated out the rank-zero terms in L and T, since these are especially simple:

$$f^{(i)}_0 (W^L_0)_{fi} = F^2_L \tag{17a}$$

$$f^{(i)}_0 (W^T_0)_{fi} = F^2_T \ , \tag{17b}$$

that is, from the $J = 0$ terms we simply recover the usual result for electron scattering from *unpolarized* targets. By varying the polar angle θ^* by rotating the polarization direction, we may, in principle, make Legendre decompositions of each of the four responses (and these may be separated from one another as mentioned above) and so obtain the basic model-independent quantities in such experiments, namely the reduced response functions. These latter quantities are then directly related to bilinear combinations of the electromagnetic form factors. We shall return in Section IV to consider these responses in more detail: We shall derive the longitudinal expression (Eq. (16a)) and obtain explicit forms for the reduced response functions.

Before proceeding with the overview, let us first consider a trivial, but important case, namely consider spin-1/2 targets (such as the proton itself!). We can form all tensors in Eqs. (16) satisfying $0 \leq J \leq 2J_i = 1$ and so J can be 0 or 1. However, only *even-rank* tensors can occur in Eqs. (16) and so only $J = 0$ is allowed; this just gives us back the unpolarized result again (see Eqs. (17)) and so nothing new is learned by polarizing a spin-1/2 target (recall: this statement refers only to the reaction $\vec{A}(e, e')$; for polarized electrons, see below). So the first interesting case for $\vec{A}(e, e')$ studies proceeding from low spin to high spin is encountered with $J_i = 1$, for then J can take on even values of 0 and 2 (see examples in Section IV).

For $\vec{A}(\vec{e}, e')$ reactions, illustrated in Fig. 4, the general analysis proceeds just as before. The cross section can be written in the form[11]

$$\sigma^h_{fi} = \Sigma_{fi} + h\Delta_{fi} \ , \tag{18}$$

where Σ_{fi} is what was discussed above for $\vec{A}(e, e')$ reactions and where the new contribution Δ_{fi} only occurs if both target and electrons are polarized. Using the

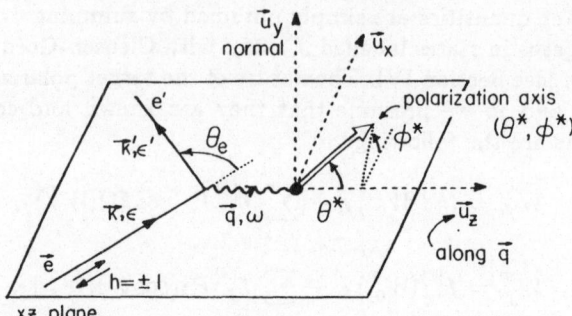

Fig. 4: Inclusive electron scattering from polarized targets with polarized electrons, $\vec{A}(\vec{e}, e')$.

explicit electron helicity dependence these two contributions can be isolated:

$$\Sigma_{fi} = \frac{1}{2}\left(\sigma_{fi}^{+1} + \sigma_{fi}^{-1}\right) \tag{19a}$$

$$\Delta_{fi} = \frac{1}{2}\left(\sigma_{fi}^{+1} - \sigma_{fi}^{-1}\right) \ . \tag{19b}$$

The new piece can be developed as before (Cf. Eq. (14)) yielding[11]

$$\Delta_{fi} \sim v_{T'} W_{fi}^{T'} + v_{TL'} \cos\phi^* W_{fi}^{TL'} \ , \tag{20}$$

where the primes indicate that polarized electrons are involved and where new electron kinematical factors are required (in fact, $v_{T'}$ already occurred in discussing parity violations, see Eqs. (7) and (8)). For convenience, we collect together the complete set of "Super-Rosenbluth" factors:

$$\begin{aligned}
v_L &= \left(Q^2/q^2\right)^2 \\[4pt]
v_T &= -\frac{1}{2}(Q^2/q^2) + \tan^2\frac{\theta_e}{2} \\[4pt]
v_{TL} &= \frac{1}{\sqrt{2}}(Q^2/q^2)\sqrt{-(Q^2/q^2) + \tan^2\frac{\theta_e}{2}} \\[4pt]
v_{TT} &= \frac{1}{2}(Q^2/q^2) \\[4pt]
v_{T'} &= \tan\frac{\theta_e}{2}\sqrt{-(Q^2/q^2) + \tan^2\frac{\theta_e}{2}} \\[4pt]
v_{TL'} &= \frac{1}{\sqrt{2}}(Q^2/q^2)\tan\frac{\theta_e}{2} \ .
\end{aligned} \tag{21}$$

The entire ϕ^*-dependence in Δ_{fi} is contained in the single factor $\cos\phi^*$ and so, having isolated Δ_{fi} using the electron helicity dependence, a further decomposition into $W_{fi}^{T'}$ and $W_{fi}^{TL'}$ can be accomplished. These in turn can be expanded as in Eqs. (16) into Legendre series:

$$W_{fi}^{T'} = \sum_{\substack{J \geq 1 \\ \text{odd}}} f_J^{(i)} P_J(\cos\theta^*)(W_J^{T'})_{fi} \tag{22a}$$

$$W_{fi}^{TL'} = \sum_{\substack{J \geq 1 \\ \text{odd}}} f_J^{(i)} P_J^1(\cos\theta^*)(W_J^{TL'})_{fi} \ , \tag{22b}$$

where again $J \leq 2J_i$ although now only *odd-rank tensors* occur. The complete decomposition of the cross section can then be done in a model-independent way down to the level of six different classes of reduced response functions $(L, T, TL, TT, T',$ and $TL')$, the first two with all even ranks satisfying $0 \leq J \leq 2J_i$, the second two with all of these except $J = 0$ and the last two with all odd ranks satisfying $1 \leq J \leq 2J_i$. In Section IV we shall return to discuss several illustrative examples of specific nuclear transitions to help in underlining the potential importance of such polarization studies. Included there is an expanded treatment of spin-1/2 targets where, as we have just seen, we need polarized electrons as well as polarized targets for new information to be forthcoming.

Now let us turn to electron scattering reactions where at least one particle is presumed to be detected in coincidence with the scattered electron, named exclusive-n scattering with $n \geq 1$. We begin with unpolarized exclusive-1 scattering as illustrated in Fig. 5. The electron scattering occurs as before (Cf. Fig. 1) and defines the xyz-coordinate system shown in these figures. Now in addition, particle x_1 is detected at angles θ_{x_1}, ϕ_{x_1} in this coordinate system and has momentum $p_{x_1} = |\vec{p}_{x_1}|$ and energy $E_{x_1} = (p_{x_1}^2 + M_{x_1}^2)^{1/2}$. A straightforward analysis (see Section III), similar to the one followed for the inclusive scattering processes discussed above, leads to the following general form for the cross section for such $(e, e'x_1)$ reactions:

$$\sigma \equiv {}^{(1)}\Sigma \sim v_L \, {}^{(1)}W^L + v_T \, {}^{(1)}W^T + v_{TL} \cos \phi_{x_1} \, {}^{(1)}W^{TL} \\ + v_{TT} \cos 2\phi_{x_1} \, {}^{(1)}W^{TT}, \tag{23}$$

where again the Mott cross section, etc., have been dropped for clarity and where a "(1)" has been added to indicate that we are discussing exclusive-1 electron scattering. Furthermore, we use the notation (as in the previous discussions of polarization) where Σ indicates that the electrons are not polarized; Δ will be considered as well below. To incorporate inclusive unpolarized scattering into this notation we could use a "0" $\leftrightarrow$ exclusive-0 and write in place of Eq. (1)

$$^{(0)}\Sigma \sim v_L \, {}^{(0)}W^L + v_T \, {}^{(0)}W^T \quad . \tag{24}$$

Returning to exclusive-1 scattering, in Eq. (23) we have a structure which is very similar to that found in discussing $\vec{A}(e, e')$ scattering (Cf. Eq. (14)). The same electron kinematical factors occur (Eqs. (21) summarize the complete set), although now instead of the polarization azimuthal angle ϕ^*, we have the azimuthal angle of the particle detected in coincidence with the electron, ϕ_{x_1}. This is the entire dependence on that angle and so, with appropriate out-of-plane coincidence measurements the four response functions in Eqs. (23) can be separated. Each one of them depends on the remaining variables of which there are four independent ones, say $\{q, \omega, E_{x_1}, \theta_{x_1}\}$.

When we add polarized electrons to the problem and consider $(\vec{e}, e'x_1)$ exclusive-1 scattering as illustrated in Fig. 6, we obtain a structure which parallels that found in discussing inclusive scattering (Cf. Eqs. (7) and (18)):

$$^{(1)}\sigma^h = {}^{(1)}\Sigma + h \, {}^{(1)}\Delta \tag{25}$$

As in Eqs. (19), the Σ and Δ pieces can be separated using the control provided by flipping the electron's helicity:

$$^{(1)}\Sigma = \frac{1}{2}({}^{(1)}\sigma^{+1} + {}^{(1)}\sigma^{-1}) \tag{26a}$$

$$^{(1)}\Delta = \frac{1}{2}({}^{(1)}\sigma^{+1} - {}^{(1)}\sigma^{-1}) \quad . \tag{26b}$$

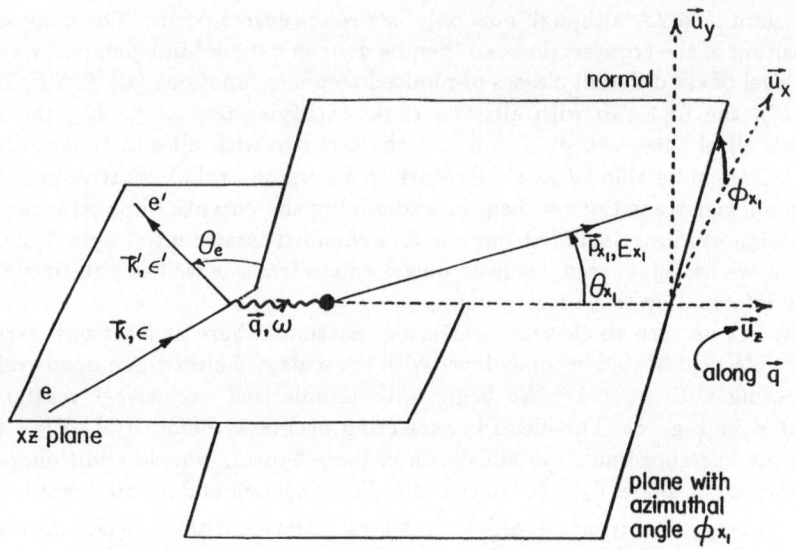

Fig. 5:   Exclusive-1 electron scattering, $A(e, e'x_1)$.

For parity-conserving exclusive-1 scattering, we obtain one extra "fifth response function":

$$^{(1)}\Delta \sim v_{TL'} \, \sin\phi_{x_1} \, ^{(1)}W^{TL'} \ . \tag{27}$$

Note that, in contrast to the polarized target inclusive scattering result in Eq. (20), which this resembles, a factor $\sin\phi_{x_1}$ occurs; to measure the fifth response function it is necessary to go out of plane. In general, the Δ response is non-zero in coincidence reactions (and of course inclusive polarized target studies, as we saw above) even when the process is purely parity conserving, and so only the unpolarized inclusive situation (Eq. (7)) is likely to be practical for weak interaction studies using electron scattering. The fifth response function is interesting for different reasons (see also Section V). Whereas $^{(1)}W^{TL}$ involves the real part of a TL interference, $^{(1)}W^{TL'}$ involves the imaginary part of the same thing. Two extremes are special: if only a single doorway dominates the final state (for example some isolated resonance) or when final-state interactions can be ignored, then $^{(1)}W^{TL}$ can be non-zero, but $^{(1)}W^{TL'}$ vanishes. Therefore, the latter (in general non-zero) quantity focuses on interesting interference physics. These ideas are expanded upon in Sec. V and Ref. 7; we shall not pursue them here, but rather go on with the general story.

Let us briefly go on to multiple-coincidence scattering, the prototype for which is the exclusive-2 (triple coincidence) reaction illustrated in Fig. 7. Here we have two particles in coincidence with the scattered electron, detected at angles $(\theta_{x_1}, \phi_{x_1})$ and $(\theta_{x_2}, \phi_{x_2})$ with energies E_{x_1} and E_{x_2}, respectively. The resulting response functions, $^{(2)}W^K$, depend on seven variables, say the set $\{q, \omega, \theta_{x_1}, E_{x_1}, \theta_{x_2}, E_{x_2}, \Delta\phi_{12} = \phi_{x_1} - \phi_{x_2}\}$. Now, the analog of the azimuthal angle dependence in Eqs. (23) and (27), which could be made explicit, is the *average azimuthal angle* $\Phi \equiv \frac{1}{2}(\phi_{x_1} + \phi_{x_2})$. The dependence on the difference $\Delta\phi_{12}$ is buried in the response functions themselves. In the general case we obtain a "Super-Rosenbluth" formula:[4]

$$^{(n)}\sigma^h = \, ^{(n)}\Sigma + h \, ^{(n)}\Delta \tag{28}$$

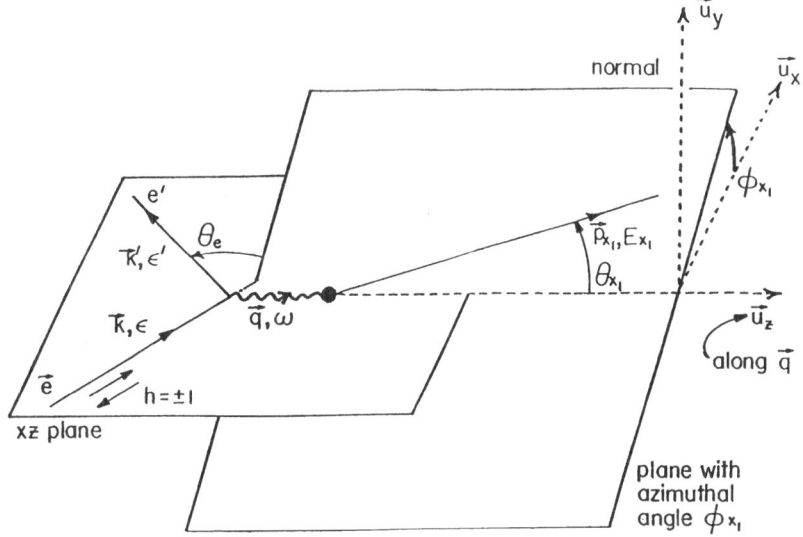

Fig. 6: Exclusive-1 electron scattering with polarized electrons, $A(\vec{e}, e'x_1)$.

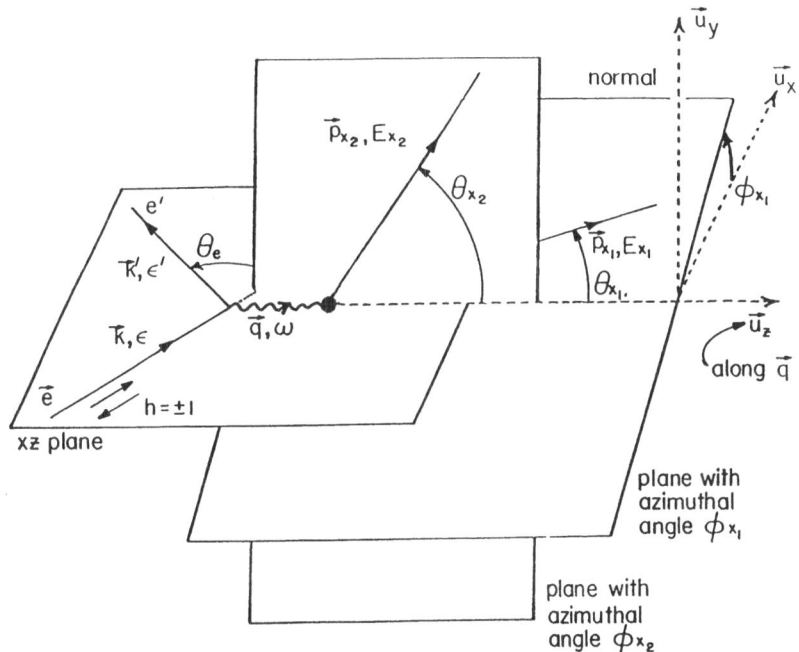

Fig. 7: Exclusive-2 electron scattering, $A(e, e'x_1x_2)$ or $A(\vec{e}, e'x_1x_2)$.

$$\begin{aligned}
{}^{(n)}\Sigma \sim\ & v_L\ {}^{(n)}W^L + v_T\ {}^{(n)}W^T \\
& + v_{TL}(\cos\Phi\ {}^{(n)}W^{TL} + \sin\Phi\ {}^{(n)}\tilde{W}^{TL}) \\
& + v_{TT}(\cos 2\Phi\ {}^{(n)}W^{TT} + \sin 2\Phi\ {}^{(n)}\tilde{W}^{TT})
\end{aligned} \tag{29a}$$

$$^{(n)}\Delta \sim v_{T'} {}^{(n)}\tilde{W}^{T'} + v_{TL'}(\sin\Phi \ ^{(n)}W^{TL'} + \cos\Phi \ ^{(n)}\tilde{W}^{TL'}) \ , \qquad (29b)$$

where

$$\Phi = \frac{1}{n}(\phi_{x_1} + \phi_{x_2} + \ldots + \phi_{x_n}) \qquad (30)$$

for $n \geq 1$ and where the response functions have as arguments the set of variables $\{q, \omega, \theta_{x_1}, E_{x_1}, \ldots E_{x_n}, \Delta\phi_{12}, \ldots, \Delta\phi_{1n}, \Delta\phi_{23}, \ldots, \Delta\phi_{2n}, \ldots, \Delta\phi_{n-1 \ n}\}$ for $n \geq$ 2. For $n = 1$ no $\Delta\phi_{ij}$ can occur in this set, and furthermore we see that Eqs. (29) simplify by having no response functions having tildas, that is, $^{(1)}\tilde{W}^K \equiv 0$. Finally, for (parity-conserving) inclusive unpolarized scattering only the L and T terms are non-zero. Equations (29) are the most complicated in form we can obtain, no matter how large n becomes; however, it should be recognized that each additional level of exclusivity adds significantly to the set of variables upon which the corresponding nine basic response functions depend.

For completeness, let us rewrite these results in two alternative forms. To condense things somewhat, let us absorb the explicit Φ-dependence and define response functions $^{(n)}R^K$:

$$^{(n)}R^L \equiv {}^{(n)}W^L, \quad {}^{(n)}R^T \equiv {}^{(n)}W^T, \quad {}^{(n)}R^{T'} \equiv {}^{(n)}\tilde{W}^{T'}$$

$$^{(n)}R^{TL} \equiv {}^{(n)}W^{TL}\cos\Phi + {}^{(n)}\tilde{W}^{TL}\sin\Phi$$

$$^{(n)}R^{TL'} \equiv {}^{(n)}W^{TL'}\sin\Phi + {}^{(n)}\tilde{W}^{TL'}\cos\Phi \qquad (31)$$

$$^{(n)}R^{TT} \equiv {}^{(n)}W^{TT}\cos 2\Phi + {}^{(n)}\tilde{W}^{TT}\sin 2\Phi \ .$$

Then, Eqs. (29) become

$$^{(n)}\Sigma \sim v_L \ ^{(n)}R^L + v_T \ ^{(n)}R^T + v_{TL} \ ^{(n)}R^{TL} + v_{TT} \ ^{(n)}R^{TT}$$

$$^{(n)}\Delta \sim v_{T'} \ ^{(n)}R^{T'} + v_{TL'} \ ^{(n)}R^{TL'} \ . \qquad (32)$$

Now first let us remove a factor v_T from every term and combine it with the multiplying Mott cross section (which we have not written in Eqs. (32)). Then, we obtain

$$\frac{d\sigma}{d\omega d\Omega_e d\Omega_1 \ldots} = \frac{1}{M_i}\sigma_{\text{Mott}} v_T \left\{ \left[\mathcal{E}(2\lambda \ ^{(n)}R^L) + ({}^{(n)}R^T) \right. \right.$$

$$\left. - \mathcal{E}({}^{(n)}R^{TT}) - \frac{1}{\sqrt{2}}\sqrt{\mathcal{E}(1+\mathcal{E})}(\sqrt{2\lambda} \ ^{(n)}R^{TL}) \right] \qquad (33)$$

$$\left. + h\left[\sqrt{1-\mathcal{E}^2}({}^{(n)}R^{T'}) - \frac{1}{\sqrt{2}}\sqrt{\mathcal{E}(1-\mathcal{E})}(\sqrt{2\lambda} \ ^{(n)}R^{TL'}) \right] \right\} \ ,$$

where

$$\lambda \equiv -\frac{Q^2}{q^2} = 1 - \left(\frac{\omega}{q}\right)^2 \to 0 \leq \lambda \leq 1 \qquad (34a)$$

represents the "virtualness" of the exchanged photon, and

$$\mathcal{E} \equiv \left[1 + \frac{2}{\lambda}\tan^2\frac{\theta_e}{2} \right]^{-1} \qquad (34b)$$

is usually referred to as the photon's degree of longitudinal polarization. If all kinematical variables except the electron scattering angle θ_e are fixed then a "Super-Rosenbluth" plot versus $\mathcal{E}$ will permit a decomposition of the cross section in Eq.

(33) into individual responses. Note, however, that the L and TT responses are both multiplied by $\mathcal{E}$ and so cannot be separated in this manner; in fact, the Φ-dependence in the TT responses (see Eqs. (31)) must be used in this case. Note also that this convenient rewriting of the general formula is accomplished by adjoining a factor $\sqrt{2\lambda}$ to each response in which there is an occurrence of the longitudinal projection of the current (and so a factor of 2λ for the L term where this projection occurs bilinearly). In some work these factors are absorbed in new definitions (for example, $^{(n)}S^L \equiv 2\lambda\, ^{(n)}R^L$) and so some caution must be exercised in interpreting just which type of response is meant.

In going to near-real-photon kinematics another rewriting of Eqs. (32) is convenient. This time let us extract a factor λ (Eq. (34a)) from every term to obtain

$$\frac{d\sigma}{d\omega\, d\Omega_e d\Omega_1 \dots} = \frac{1}{M_i}\sigma_{\text{Mott}}\lambda\left\{\left[\lambda\,^{(n)}R^L + \left(\frac{1}{2}+\xi\right)\,^{(n)}R^T\right.\right.$$
$$\left. - \frac{1}{2}\,^{(n)}R^{TT} - \frac{1}{\sqrt{2}}\sqrt{\lambda}\sqrt{1+\xi}\,^{(n)}R^{TL}\right] \tag{35}$$
$$\left. + h\left[\sqrt{\xi(1+\xi)}\,^{(n)}R^{T'} - \frac{1}{\sqrt{2}}\sqrt{\lambda}\sqrt{\xi}\,^{(n)}R^{TL'}\right]\right\} \ .$$

Here we have defined another kinematical combination

$$\xi \equiv \left\{\omega^2/4\epsilon\epsilon'\cos^2\frac{\theta_e}{2}\right\}\bigg/(1-\lambda) \ . \tag{36}$$

Now let us fix the incident electron energy ϵ, fix ω (and hence $\epsilon' = \epsilon - \omega$) to study some given nuclear excitation, but vary λ so that we approach near-real-photon kinematics: $\lambda \to 0$, corresponding to $q \to \omega$ (see Eq. (34a)). Since we have

$$\sin^2\frac{\theta_e}{2} = \left(\frac{\lambda}{1-\lambda}\right)\left(\frac{\omega^2}{4\epsilon\epsilon'}\right) \ , \tag{37}$$

we are going to kinematics where ϵ, ϵ', q and ω are of one scale and where $\sqrt{-Q^2}$ is much smaller, implying rather forward angle scattering (θ_e small). Note, however, that to use the extreme relativistic limit results as we do here that $\theta_e \not\ll \gamma^{-1}$, where $\gamma = \epsilon/m_e$, for in that angular regime the electron mass cannot be set to zero (see Ref. 11 for the kinematical factors in this more general case.) Under these circumstances the L, TL, and TL' responses in Eq. (35) are suppressed by factors λ or $\sqrt{\lambda}$ and $\xi \approx \omega^2/4\epsilon\epsilon'$, so that we obtain for the cross section

$$\left(\frac{d\sigma}{d\omega\, d\Omega_e \Omega_1 \dots}\right)^{\text{near-real-photon}} = \frac{1}{M_i}\sigma_{\text{Mott}}\lambda\left(\frac{1}{2}+\xi\right)\left\{\left[^{(n)}R^T - \cos\Gamma\,^{(n)}R^{TT}\right]\right.$$
$$\left. + h\sin\Gamma\,^{(n)}R^{T'}\right\} \ , \tag{38}$$

where only the transverse responses survive, as expected. Here we have defined an angle

$$\Gamma \equiv \sin^{-1}\left\{\frac{\epsilon^2 - (\epsilon')^2}{\epsilon^2 + (\epsilon')^2}\right\} \to 0 \leq \Gamma \leq \frac{\pi}{2} \ . \tag{39}$$

It should be emphasized that this is still electron scattering where we can define a scattering plane and hence the azimuthal angle Φ. If θ_e were strictly zero, then Φ

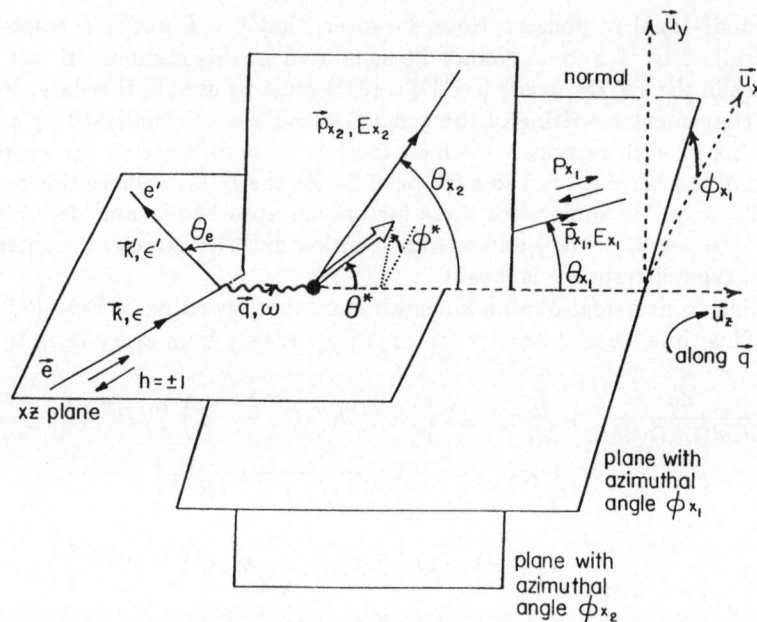

Fig. 8:   Exclusive-2 electron scattering from polarized targets and/or
with polarization of particle 1 measured $\vec{A}(\vec{e}, e'x_1x_2)$, $A(e, e'\vec{x}_1x_2)$,
$A(\vec{e}, e'\vec{x}_1x_2)$, $\vec{A}(e, e'x_1x_2)$, or $\vec{A}(\vec{e}, e'\vec{x}_1x_2)$.

would be undefined and the TT term, which contains $\cos 2\Phi$ and $\sin 2\Phi$ dependences,
would also be undefined. Also note that the T' term only enters when we have
exclusive-n scattering with $n \geq 2$ (see Eqs. (31): $^{(n)}R^{T'} = {}^{(n)}\tilde{W}^{T'}$). Thus, for
example, for $(e, e'x_1)$ reactions the T' response is absent. Finally, let us consider
$(e, e'x_1)$ reactions where θ_{x_1} is also small: this will frequently be the case for high
electron energies where relatively large cross sections tend to occur in the forward
direction. The leading dependences on θ_{x_1} are:

$$\sim \text{ constant for } L, T, T'$$
$$\sim \sin\theta_{x_1} \text{ for } TL, TL'$$
$$\sim \sin^2\theta_{x_1} \text{ for } TT .$$

Thus, we can drop the TT responses as well to obtain a handy relationship

$$\left(\frac{d^3\sigma}{d\Omega_e d\Omega_{x_1} dE_{x_1}}\right)^{\lambda, \theta_{x_1} \text{ small}}_{(e,e'x_1)}$$
$$= \frac{\alpha}{8\pi^2}\left\{1 + \left(\frac{\epsilon'}{\epsilon}\right)^2\right\} \frac{1}{E_\gamma \sin^2\frac{\theta_e}{2}}\left(\frac{d\sigma}{d\Omega_{x_1}}\right)_{(\gamma, x_1)} , \quad (40)$$

where $E_\gamma \simeq q \simeq \omega = \epsilon - \epsilon'$. This may be used, for instance, to relate $(e, e'K^+)$ and
(γ, K^+) studies of hypernuclei.[8]

Before turning to discussions of some of the details of how these results are
obtained, let us remark that the even more complicated situations shown in Fig.
8, possibly having polarized targets and/or measurements of final state particles'
polarizations (shown here for x_1), are very likely to prove to be quite interesting;
work is in progress on such reactions.

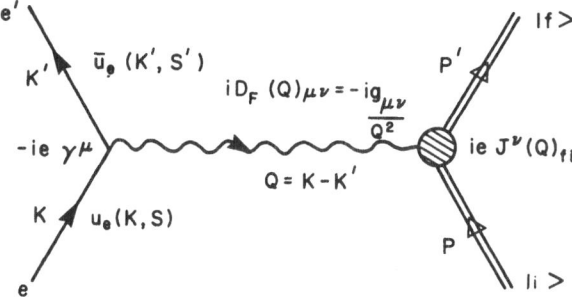

Fig. 9: Feynman diagram and rules for the electron-nucleus scattering process in the one-photon-exchange or first-order Born approximation.

LEPTONIC TENSORS – POLARIZED ELECTRON SCATTERING

Let us now go back to the beginning and see how some of the general structure summarized in Section I is obtained. The basic process of electron scattering in the one-photon-exchange or first Born approximation is shown diagrammatically in Fig. 9. In effect, we are doing first-order perturbation theory and (by Fermi's Golden Rule) the cross section goes as the square of the matrix element of the perturbation:

$$d\sigma \sim |H'_{fi}|^2 \ . \tag{41a}$$

When we apply the Feynman rules to the diagram in Fig. 9 (the associated factors are shown in the figure) and extract the photon propagator (this goes into making up the Mott cross section) we have an effective current-current interaction:

$$d\sigma \sim |j^e_\mu J^\mu_{fi}|^2 \ , \tag{41b}$$

when j^e_μ is the electron current and J^μ_{fi} is the fi matrix element of the nuclear electromagnetic current operator. Factoring this into electron (leptonic) and nuclear (hadronic) contributions, we have

$$d\sigma \sim \left(j^{e*}_\mu j^e_\nu\right)\left(J^{\mu*}_{fi} J^\nu_{fi}\right) \ . \tag{41c}$$

Now we must perform the appropriate average-over-initial and sum-over-final for the leptons and separately for the hadrons. "Appropriate" here will become clear when we look at polarization degrees of freedom. The leptonic sums yield a leptonic tensor,

$$\overline{\sum_{if}} j^{e*}_\mu j^e_\nu \rightarrow \eta_{\mu\nu} \tag{42a}$$

and the hadronic sums a corresponding hadronic tensor,

$$\overline{\sum_{if}} J^{\mu*}_{fi} J^\nu_{fi} \rightarrow W^{\mu\nu} \ . \tag{42b}$$

The cross section is then proportional to the contraction of these two second-rank Lorentz tensors:

$$d\sigma \sim \eta_{\mu\nu} W^{\mu\nu} \ . \tag{43}$$

In this section, we will focus on the leptonic tensor and then continue in Section III to discuss the nuclear part of the problem. We have a simple form for the electron's electromagnetic current:

$$j_\mu^e(K', S'; K, S) \sim \bar{u}_e(K', S')\gamma_\mu u_e(K, S) \ , \tag{44}$$

where the spinors are labelled by the 4-momenta K and K' and the spin projections S and S'. We note that this is a polar vector and that it satisfies the momentum space current conservation equation

$$Q^\mu j_\mu^e = 0, \quad \text{with} \quad Q^\mu = K^\mu - K'^\mu \ . \tag{45}$$

Substituting into Eq. (42a) we have for the leptonic tensor

$$\eta_{\mu\nu} = \sum_{if} \{\bar{u}_e(K', S')\gamma_\mu u_e(K, S)\}^* \{\bar{u}_e(K', S')\gamma_\nu u_e(K, S)\} \tag{46a}$$

$$= \sum_{if} \bar{u}_e(K, S)\gamma_\mu \underset{1}{\uparrow} u_e(K', S')\bar{u}_e(K', S')\gamma_\nu \underset{2}{\uparrow} u_e(K, S) \ , \tag{46b}$$

where properties of the γ-algebra[13] have been used to obtain Eq. (46b) from (46a). Now we can guarantee that only electrons occur (not positrons) by inserting projection operators in the positions marked 1 and 2 in Eq. (46b):

$$1: \quad \frac{1}{2m_e} (\not{K}' + m_e)$$

$$2: \quad \frac{1}{2m_e} (\not{K} + m_e) \ .$$

We obtain

$$\eta_{\mu\nu} = \frac{1}{4m_e^2} \sum_{if} \bar{u}_e(K, S)\gamma_\mu (\not{K}' + m_e)u_e(K', S')$$

$$\times u_e(K', S')\gamma_\nu (\not{K} + m_e)u_e(K, S) \ . \tag{47}$$

Unpolarized Electrons

If the electrons are completely unpolarized, then $\overline{\sum}_{if}$ implies doing the sum over all of the spin projections. Furthermore, since we have guaranteed that we cannot have any positrons in Eq. (47), we can extend these sums to all four spinor projections and then the result is a trace which can be easily evaluated[13]:

$$\eta_{\mu\nu}^{\text{unpol.}} = \frac{1}{2} \times \frac{1}{4m_e^2} Tr \{\gamma_\mu(\not{K}' + m_e)\gamma_\nu(\not{K} + m_e)\} \tag{48a}$$

$$= \frac{1}{2m_e^2} (K_\mu K'_\nu + K'_\mu K_\nu - g_{\mu\nu}(K \cdot K' - m_e^2)) \ . \tag{48b}$$

Two things are noted for future reference: (1) $\eta_{\mu\nu}^{\text{unpol.}}$ is symmetric under the interchange $\mu \leftrightarrow \nu$; (2) current conservation (Eq. (45)) implies that $Q^\mu \eta_{\mu\nu}^{\text{unpol.}} = 0$. In the extreme relativistic limit (ERL) where m_e/ϵ and m_e/ϵ' are very small and $k = |\vec{k}| \simeq \epsilon$, $k' = |\vec{k}'| \approx \epsilon'$, we can neglect the last term $\propto m_e^2$ in Eq. (48b).

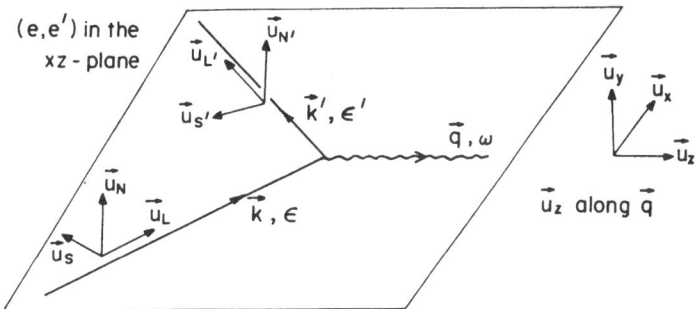

Fig. 10:   Coordinate systems used in electron scattering.

Polarized Electrons

Now let us see what happens when the electrons are polarized. In the most general discussions of the scattering of spin-1/2 particles, one usually begins with a density matrix in spin space and, for example, referring to Fig. 10, discusses the various polarization tensors which can occur: $L \to L'$, $L \to S'$, etc. However, as we shall see, there are at least two important aspects of electron scattering which simplify things considerably. Firstly, the *one-photon-exchange approximation* is quite good and the only communication between the electron and the nucleus is via the photon, an elementary spin-1 field. This, of course, is also true for the weak interaction mediated by $W^{\pm}$ or Z^0 exchanges. Secondly, the ERL is appropriate (we usually deal with electrons with energies ranging from 100's of MeV to 1000's of MeV compared to the electron mass $m_e = 0.511$ MeV). In this case, the scattering is *helicity conserving*, as we shall see explicitly below. (For a nice comparison, see the lectures by Carlson in this same Summer School where he discusses gluon exchanges between quarks and also invokes helicity conservation rules at relativistic energies.)

Returning to our basic expression for the leptonic tensor (Eq. (46b)), we now want to insert in the positions marked 1 and 2 not only the positive-energy projections as before, but now also want to guarantee that only the appropriate spin components are present[13]:

$$1: \quad \frac{1}{2}\left(1 + \gamma_5 \not{S}'\right)\frac{1}{2m_e}(\not{K}' + m_e)$$

$$2: \quad \frac{1}{2}\left(1 + \gamma_5 \not{S}\right)\frac{1}{2m_e}(\not{K} + m_e) \ .$$

The first factor in each case now projects the spin which is characterized by the 4-vector S^{μ} (S'^{μ}) satisfying[13]

$$S \cdot S = -1 = S' \cdot S' \tag{49a}$$

$$S \cdot K = 0 = S' \cdot K' \tag{49b}$$

We will return to these in a moment. With these projections in Eq. (46b) we can again extend the sums in $\sum_{if}$ to all four spinor projections (since only the "good" ones can occur anyway) and again obtain[11] an easily evaluated trace:

$$\eta_{\mu\nu}^{\text{pol.}} = \frac{1}{16m_e^2} Tr\left\{\gamma_{\mu}(1 + \gamma_5 \not{S}')(\not{K}' + m_e)\gamma_{\nu}(1 + \gamma_5 \not{S})(\not{K} + m_e)\right\} \tag{50a}$$

$$= \eta_{\mu\nu}^{s} + \eta_{\mu\nu}^{a} \ , \tag{50b}$$

where we now have both symmetric (s) and antisymmetric (a) tensors when considering the interchange $\mu \leftrightarrow \nu$:

$$
\begin{aligned}
\eta_{\mu\nu}^s = \frac{1}{8m_e^2} \{ &P_\mu P_\nu (1 - S \cdot S') \\
&+ g_{\mu\nu} Q^2 (1 + S \cdot S' - 2\Sigma \cdot \Sigma') \\
&+ Q^2 (\Sigma_\mu \Sigma_\nu' + \Sigma_\mu' \Sigma_\nu) + (P_\mu S_\nu + S_\mu P_\nu) \}
\end{aligned}
\tag{51a}
$$

$$
\eta_{\mu\nu}^a = \frac{i}{4m_e} \epsilon_{\mu\nu\alpha\beta} (S + S')^\alpha Q^\beta ,
\tag{51b}
$$

where $Q_\mu \equiv K_\mu - K_\mu'$, $P_\mu \equiv K_\mu + K_\mu'$, $\Sigma_\mu \equiv S_\mu - (Q \cdot S/Q^2) Q_\mu$ and $\Sigma_\mu' \equiv S_\mu' - (Q \cdot S'/Q^2) Q_\mu$. In stating the results here we have dropped contributions $\propto Q_\mu$ or $\propto Q_\nu$ since $\eta_{\mu\nu}$ must be contracted with $W^{\mu\nu}$ and current conservation requires that $Q_\mu W^{\mu\nu} = Q_\nu W^{\mu\nu} = 0$.

Let us simplify things a bit: let us consider only the incident electron to be polarized and then all terms in Eqs. (51) contining S' are absent. In fact, it is straightforward to retain the scattered electron's polarization as well, but the most practical situation is the one we consider here. In general, we can write the spin 3-vector in the following form:

$$
\vec{s} = hs(\cos \varsigma \vec{u}_L + \sin \varsigma \vec{u}_T) ,
\tag{52}
$$

where $s = |\vec{s}| \geq 0$, $h = \pm 1$ (to allow the spin to reverse easily in the formalism), and ς is an angle which specifies how transverse or longitudinal the polarization is (note that L and T as used in the present context mean *with respect to the incident electron's momentum* – see Fig. 10). Another angle can be introduced to specify where the transverse projections lie, that is, how much is along $\vec{u}_N$ and how much along $\vec{u}_S$ in Fig. 10:

$$
\vec{u}_T = \cos \eta \, \vec{u}_S + \sin \eta \, \vec{u}_N .
\tag{53}
$$

Now the conditions $S \cdot S = -1$ and $S \cdot K = 0$ tell us that

$$
s = \frac{1}{\sqrt{1 - \beta^2 \cos^2 \varsigma}} = \frac{\gamma}{\sqrt{\cos^2 \varsigma + \gamma^2 \sin^2 \varsigma}}
\tag{54a}
$$

$$
S^0 = h\beta s \cos \varsigma .
\tag{54b}
$$

Let us consider two extremes: (1) For longitudinally polarized electrons we have $\varsigma = 0$ and so

$$
(S^\mu)_{\text{long.}} = h\gamma(\beta, \vec{u}_L) .
\tag{55a}
$$

(2) for transversely polarized electrons we have $\varsigma = \pi/2$ and so

$$
(S^\mu)_{\text{trans.}} = h(0, \vec{u}_T) .
\tag{55b}
$$

In the ERL we have

$$
\beta = k/\epsilon \xrightarrow[\text{ERL}]{} 1
\tag{56a}
$$

$$
\frac{1}{\gamma} = \frac{m_e}{\epsilon} = \sqrt{1 - \beta^2} \xrightarrow[\text{ERL}]{} 0
\tag{56b}
$$

and so the longitudinal case is of order $\gamma \gg 1$ while the transverse case is of order 1. The implication is that the longitudinal projections provide the leading order, while the transverse corrections are down by a factor $1/\gamma$. Indeed, the factor $h = \pm 1$ now can be identified with the electron's helicity and we see that ERL electron scattering is helicity conserving with corrections of order $1/\gamma$.

Imposing these conclusions on Eqs. (51) leads to

$$\eta^s_{\mu\nu} \sim K_\mu K'_\nu + K'_\mu K_\nu - g_{\mu\nu} K \cdot K' \tag{57a}$$

just as above for unpolarized electrons (see Eq. (48b)), and

$$\eta^a_{\mu\nu} \sim ih\,\epsilon_{\mu\nu\alpha\beta} K^\alpha K'^\beta \ . \tag{57b}$$

Note that $Q^\mu \eta^a_{\mu\nu} = 0$ is automatically satisfied. For comparison, it is also instructive to note that exactly the same expressions are obtained in discussing the leptonic parts of weak interaction processes,[4] where, for example, in considering processes involving standard neutrinos, the helicity h in Eq. (57b) is fixed by the handedness of the neutrino. (Ref. 4 contains a broader view and includes general discussions of semi-leptonic electro-weak interactions in the same language).

We are now done with the leptonic tensor and can proceed to consider the hadronic (nuclear) responses in a little more detail.

HADRONIC TENSORS – "SUPER-ROSENBLUTH FORMULA"

The hadronic electromagnetic tensor is constructed from the current matrix elements by again performing appropriate averages-over-initial and sums-over-final as indicated in (42b). We obtain a second-rank Lorentz tensor $W^{\mu\nu}$ which can be decomposed into pieces which are symmetric (s) or antisymmetric (a) under the interchange $\mu \leftrightarrow \nu$:

$$W^{\mu\nu} = W^{\mu\nu}_s + W^{\mu\nu}_a \ . \tag{58}$$

Clearly, in contracting with the leptonic tensor no cross terms are allowed:

$$\eta_{\mu\nu} W^{\mu\nu} = \eta^s_{\mu\nu} W^{\mu\nu}_s + \eta^a_{\mu\nu} W^{\mu\nu}_a \ . \tag{59}$$

Furthermore, the nuclear electromagnetic current is conserved and so

$$Q_\mu W^{\mu\nu}_s = Q_\mu W^{\mu\nu}_a = 0 \ . \tag{60}$$

Let us now proceed to build these hadronic tensors from the available 4-momenta in the problem. We begin with simple familiar cases and progress towards the general "Super-Rosenbluth" forms.

Inclusive Electron Scattering, No Nuclear Polarizations

In this situation we are now interested in the nuclear vertex shown in Fig. 11. Here the virtual photon from the electron scattering (Fig. 9) brings in 4-momentum Q and causes the nucleus to go from state $|i\rangle$ to state $|f\rangle$. These hadronic states have 4-momenta P_i and P_f, respectively, as shown in the figure. Thus, we must build the hadronic tensors from the 4-momenta Q, P_i and P_f. In fact, we can use momentum conservation to eliminate one, say $P_f = P_i + Q$ leaving two independent 4-momenta: $\{Q, P_i\}$. The possible Lorentz scalars in the problem are Q^2, P_i^2 and $Q \cdot P_i$. Since we presumably know what the target is and since $P_i^2 = M_i^2$, we are left with two independent scalars to vary: $\{Q^2, Q \cdot P_i\}$. Moreover, since $Q^2 = \omega^2 - q^2$

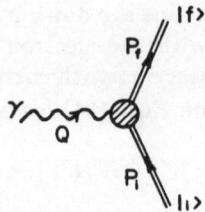

Fig. 11:   Hadronic (nuclear) vertex for inclusive electron scattering, (e, e').

and $Q \cdot P_i = \omega M_i$ in the laboratory system, we can regard our hadronic tensors to be functions of $\{Q^2, Q \cdot P_i\}$ or $\{q, \omega\}$.

Now we wish to write $W_s^{\mu\nu}$ and $W_a^{\mu\nu}$ in terms of the two independent 4-vectors Q^μ and P_i^μ. Alternatively, instead of P_i it turns out to be useful to employ the 4-vector

$$V_i^\mu \equiv \frac{1}{M_i} \left\{ P_i^\mu - \left(\frac{Q \cdot P_i}{Q^2} \right) Q^\mu \right\} , \qquad (61)$$

This is especially convenient because $Q \cdot V_i = 0$, by construction. Thus, we shall use Q^μ and V_i^μ to build the hadronic tensors.

To begin with, $W^{\mu\nu}$ must be a second-rank Lorentz tensor and so we can write the following general expansions:

$$W_s^{\mu\nu} = X_1 g^{\mu\nu} + X_2 Q^\mu Q^\nu + X_3 V_i^\mu V_i^\nu + X_4 (Q^\mu V_i^\nu + V_i^\mu Q^\nu) \qquad (62a)$$

$$W_a^{\mu\nu} = Y_1 (Q^\mu V_i^\nu - V_i^\mu Q^\nu) + i Y_2 \epsilon^{\mu\nu\alpha\beta} Q_\alpha V_{i\beta} , \qquad (62b)$$

where the scalar response functions (the X's and Y's) depend on the scalars discussed above:

$$X_i = X_i(Q^2, Q \cdot P_i) , \quad i = 1, \ldots, 4$$
$$Y_i = Y_i(Q^2, Q \cdot P_i) , \quad i = 1, 2 .$$

Next we note that, in the absence of parity violating effects from the weak interaction (see Section I), the hadronic electromagnetic current matrix elements are polar vectors and so the tensors here must have specific properties under spatial inversion. In particular, the ϵ-terms in Eq. (62b) have the wrong behavior and so Y_2 must vanish. Finally, we must make use of the current conservation conditions, Eqs. (60). This leads us to the following expressions (recall that $Q \cdot V_i = 0$)

$$Q_\mu W_s^{\mu\nu} = 0 = (X_1 + X_2 Q^2) Q^\nu + (X_4 Q^2) V_i^\nu \qquad (63a)$$

$$Q_\mu W_a^{\mu\nu} = 0 = (Y_1 Q^2) V_i^\nu \qquad (63b)$$

and so $X_1 + X_2 Q^2 = 0$, $X_4 = 0$ and $Y_1 = 0$, using the linear independence of Q^ν and V_i^ν. Defining $W_1 \equiv -X_1$ and $W_2 \equiv X_3$, to use more common nomenclature, we have then rederived the familiar results

$$W_s^{\mu\nu} = -W_1 \left(g^{\mu\nu} - \frac{Q^\mu Q^\nu}{Q^2} \right) + W_2 V_i^\mu V_i^\nu \qquad (64a)$$

$$W_a^{\mu\nu} = 0 . \qquad (64b)$$

Contracting the hadronic tensor with the previous results obtained for the electron (Eqs. (48b) or (57a)) leads to the well-known form for inclusive unpolarized electron scattering:[19]

$$\eta_{\mu\nu} W^{\mu\nu} = \eta_{\mu\nu}^s W_s^{\mu\nu} \sim W_2 + 2W_1 \tan^2 \frac{\theta_e}{2} , \qquad (65)$$

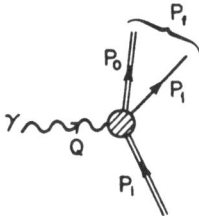

Fig. 12:   Hadronic (nuclear) vertex for exclusive-1 electron scattering, $(e, e'x_1)$.

with only two independent response functions, W_1 and W_2, which may be separated by fixing q and ω (and hence the scalar variables Q^2 and $Q \cdot P_i$) and varying the electron scattering angle θ_e to make a Rosenbluth decomposition. Alternatively, it is useful to speak of contributions which are transverse (T) or longitudinal (L) with respect to the direction $\vec{q}$ and to write the inclusive cross section in the form

$$\left. \frac{d\sigma}{d\Omega_e d\omega} \right|_{(e,e')}^{\text{lab, ERL, no nuclear polarizations}} = \frac{1}{M_i} \sigma_{\text{Mott}} f_{\text{rec}}^{-1} \left\{ v_L \, {}^{(0)}W^L + v_T \, {}^{(0)}W^T \right\} ,$$

$$(66)$$

where

$$^{(0)}W^L = (q^2/Q^2)W_1 + (q^2/Q^2)^2 W_2 \tag{67a}$$

$$^{(0)}W^T = 2W_1 , \tag{67b}$$

and where all other factors in Eqs. (66) were introduced in Section I.

Exclusive − 1 Electron Scattering, No Nuclear Polarization.

   Next, we consider $(e, e'x_1)$ reactions where a particle with 4-momentum P_1 is detected in coincidence with the scattered electron. Let us again assume that the total final nuclear state has 4-momentum P_f, so that all but the particle in coincidence (*i.e.*, the unobserved particles) have 4-momentum P_0 (see Fig. 12). Momentum conservation allows us to eliminate these two ($P_f = P_i + Q$ and $P_0 = P_f - P_1$), so that we are left with three independent 4-momenta from which to build the hadronic tensor: $\{Q, P_i, P_1\}$. The possible Lorentz scalars in the problem are Q^2, P_i^2, P_1^2, $Q \cdot P_i$, $P_1 \cdot P_i$ and $Q \cdot P_1$. Again, we know the target mass, M_i and the mass of the detected particle, M_1, and so, since $P_i^2 = M_i^2$ and $P_1^2 = M_1^2$ are fixed, we are left with four independent scalars to vary: $\{Q^2, \ Q \cdot P_i, \ P_1 \cdot P_i, \ Q \cdot P_1\}$. It is useful to express the kinematical dependence in the response functions in terms of laboratory system variables. We shall use the coordinate system in Fig. 5 to describe exclusive-1 electron scattering in which the detected particle has 3-momentum $p_1 = |\vec{p}_1|$ and is detected at angles (θ_1, ϕ_1) with respect to the chosen basis. The four scalar variables may be re-expressed in this laboratory system:

$$Q^2 = \omega^2 - q^2$$
$$Q \cdot P_i = \omega M_i$$
$$P_1 \cdot P_i = E_1 M_i$$
$$Q \cdot P_1 = \omega E_1 - q p_1 \cos \theta_1 ,$$

$$(68)$$

where $E_1 = \left(p_1^2 + M_1^2 \right)^{1/2}$ is the total energy of the detected particle.* Note that the azimuthal angle ϕ_1, does not occur here; that is, the internal functional dependence

* To simplfy the notation a bit, in this section we use $\theta_1 \equiv \theta_{x_1}$, $\phi_1 \equiv \phi_{x_1}$, etc.

involves θ_1, but not ϕ_1. Thus, the hadronic tensors for double coincidence reactions can be regarded to be functions of the scalar variables $\{Q^2, Q \cdot P_i, P_1 \cdot P_i, Q \cdot P_1\}$ or of the laboratory quantities $\{q, \omega, E_1, \theta_1\}$.

Now, as in the case of inclusive scattering, we wish to write $W_s^{\mu\nu}$ and $W_a^{\mu\nu}$ in terms of the independent 4-momenta in the problem. Instead of $\{Q^\mu, P_i^\mu, P_1^\mu\}$ we use the equivalent set $\{Q^\mu, V_i^\mu, V_1^\mu\}$, where

$$V_i^\mu \equiv \frac{1}{M_i}\left\{P_i^\mu - \left(\frac{Q \cdot P_i}{Q^2}\right)Q^\mu\right\} \tag{69a}$$

$$V_1^\mu \equiv P_1^\mu - \left(\frac{Q \cdot P_1}{Q^2}\right)Q^\mu , \tag{69b}$$

and these have the convenient properties, $Q \cdot V_i = Q \cdot V_1 = 0$.

Since $W^{\mu\nu}$ must be a second-rank Lorentz tensor, we have the following expansions:

$$\begin{aligned}W_s^{\mu\nu} &= X_1 g^{\mu\nu} + X_2 Q^\mu Q^\nu + X_3 V_i^\mu V_i^\nu + X_4 V_1^\mu V_1^\nu \\ &+ X_5(Q^\mu V_i^\nu + V_i^\mu Q^\nu) + X_6(Q^\mu V_1^\nu + V_1^\mu Q^\nu) \\ &+ X_7(V_i^\mu V_1^\nu + V_1^\mu V_i^\nu) \end{aligned} \tag{70a}$$

$$\begin{aligned}W_a^{\mu\nu} &= Y_1(Q^\mu V_i^\nu - V_i^\mu Q^\nu) + Y_2(Q^\mu V_1^\nu - V_1^\mu Q^\nu) \\ &+ Y_3(V_i^\mu V_1^\nu - V_1^\mu V_i^\nu) , \end{aligned} \tag{70b}$$

where we have not written any ϵ-terms (Cf. Eq. (62b)) using the parity properties of the electromagnetic tensor ($\sim VV$). The scalar response functions depend on the scalars discussed above:

$$X_i = X_i(Q^2, Q \cdot P_i, P_1 \cdot P_i, Q \cdot P_1) , \quad i = 1,\ldots,7$$
$$Y_i = Y_i(Q^2, Q \cdot P_i, P_1 \cdot P_i, Q \cdot P_1) , \quad i = 1,\ldots,3 .$$

Finally, we use the current conservation conditions (Eqs. (60)) to obtain the constraints

$$Q_\mu W_s^{\mu\nu} = 0 = (X_1 + X_2 Q^2)Q^\nu + (X_5 Q^2)V_i^\nu + (X_6 Q^2)V_1^\nu \tag{71a}$$
$$Q_\mu W_a^{\mu\nu} = 0 = (Q^2 Y_1)V_i^\nu + (Q^2 Y_2)V_1^\nu . \tag{71b}$$

and so $X_1 + X_2 Q^2 = 0$, $X_5 = 0$, $X_6 = 0$, and $Y_1 = 0$, $Y_2 = 0$ using the linear independence of Q^ν, P_i^ν and P_1^ν. For the symmetric hadronic tensor we obtain

$$\begin{aligned}W_s^{\mu\nu} &= X_1\left(g^{\mu\nu} - \frac{Q^\mu Q^\nu}{Q^2}\right) + X_3 V_i^\mu V_i^\nu \\ &+ X_4 V_1^\mu V_1^\nu + X_7(V_i^\mu V_1^\nu + V_1^\mu V_i^\nu) , \end{aligned} \tag{72a}$$

in which there are four independent response functions, to be compared with inclusive scattering (Eq. (64a)) in which only two appeared. For the antisymmetric tensor we obtain

$$W_a^{\mu\nu} = Y_3(V_i^\mu V_1^\nu - V_1^\mu V_i^\nu) , \tag{72b}$$

in which a fifth response function appears, in contrast to inclusive scattering (Eq. (64b)) where the antisymmetric tensor vanishes.

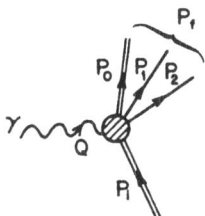

Fig. 13:   Hadronic (nuclear) vertex for exclusive-2 electron scattering, $(e, e'x_1x_2)$.

When these hadronic tensors are contracted with the previous results given in Eqs. (57) for the electron scattering leptonic tensor (we assume the ERL here for simpliticy), we obtain

$$\eta_{\mu\nu}W^{\mu\nu} = \eta^s_{\mu\nu}W^{\mu\nu}_s + \eta^a_{\mu\nu}W^{\mu\nu}_a , \qquad (73)$$

where for the symmetric contributions to exclusive-1 scattering ("(1)") we have

$$\eta^s_{\mu\nu}W^{\mu\nu}_s \sim v_L {}^{(1)}W^L + v_T {}^{(1)}W^T + v_{TT} {}^{(1)}W^{TT}\cos 2\phi_1 + v_{TL} {}^{(1)}W^{TL}\cos\phi_1 . \qquad (74a)$$

Now, in addition to longitudinal (L) and transverse (T) pieces (Cf. Eq. (66)), we also have a transverse-transverse interference (TT) and a transverse-longitudinal interference (TL). Of course, the four response functions here are linear combinations of X_1, X_3, X_4 and X_7, and depend on the four scalar quantities discussed above. The electron kinematical functions (the v's) were defined in Section I. For the antisymmetric contribution we have

$$\eta^a_{\mu\nu}W^{\mu\nu}_a \sim hv_{TL'} {}^{(1)}W^{TL'}\sin\phi_1 , \qquad (74b)$$

which is directly proportional to Y_3.

To obtain the inclusive cross section from these results we must integrate over the angle dependence (θ_1, ϕ_1) and sum over all open channels. Indeed, as is apparent from Eqs. (74), integrating over the explicit ϕ_1-dependence immediately causes the TT, TL, and TL' contributions to vanish, leaving only the L and T pieces as expected (see Eq. (66)).

Exclusive − 2 Electron Scattering, No Nuclear Polarizations

Now that the procedures are clear, let us proceed more rapidly to discussions which extend beyond the familiar (e, e') and $(e, e'x_1)$ analyses given above. Next we consider triple-coincidence, $(e, e'x_1x_2)$ reactions, where two particles having 4-momenta P_1 and P_2 are detected in coincidence with the scattered electron. As before P_0 will stand for the 4-momentum in the final state except for P_1 and P_2 (see Fig. 13)). Using momentum conservation we can eliminte the total final-state momentum, $P_f = P_i + Q$ and the unobserved particles' momentum $P_0 = P_f - (P_1 + P_2)$, This leaves four independent 4-momenta from which to build the hadronic tensor: $\{Q, P_i, P_1, P_2\}$. Assuming we know the target mass, M_i and the masses of particles 1 and 2, M_1 and M_2, respectively, we have three conditions, $P_i^2 = M_i^2$, $P_1^2 = M_1^2$ and $P_2^2 = M_2^2$, leaving seven independent scalar quantities upon which the response functions depend: $\{Q^2,\ Q\cdot P_i,\ P_1\cdot P_i,\ P_2\cdot P_i,\ Q\cdot P_1,\ Q\cdot P_2,\ P_1\cdot P_2\}$. Once again it is useful to employ laboratory variables as indicated in Fig. 7. Now there are two pairs of angles (θ_1, ϕ_1) and (θ_2, ϕ_2), for the two particles detected in coincidence with the scattered electron. These particles have 3-momenta $p_1 = |\vec{p}_1|$ and $p_2 = |\vec{p}_2|$ with energies $E_1 = \sqrt{p_1^2 + M_1^2}$ and $E_2 = \sqrt{p_2^2 + M_2^2}$, respectively. The seven scalar variables may then be re-expressed as

$$Q^2 = \omega^2 - q^2$$
$$Q \cdot P_i = \omega M_i$$
$$P_1 \cdot P_i = E_1 M_i$$
$$P_2 \cdot P_i = E_2 M_i \tag{75}$$
$$Q \cdot P_1 = \omega E_1 - q p_1 \cos \theta_1$$
$$Q \cdot P_2 = \omega E_2 - q p_2 \cos \theta_2$$
$$P_1 \cdot P_2 = E_1 E_2 - p_1 p_2 \{\cos \theta_1 \cos \theta_2 + \sin \theta_1 \sin \theta_2 \cos(\phi_1 - \phi_2)\} \ .$$

Note that now the azimuthal angles ϕ_1 and ϕ_2 occur as part of the scalar functional dependence of the response functions. In fact, it is useful to define

$$\text{Exclusive} - 2: \quad \Delta\phi_{12} \equiv \phi_1 - \phi_2 \tag{76a}$$

$$\Phi \equiv \frac{1}{2}(\phi_1 + \phi_2) \ , \tag{76b}$$

and then equivalently the response functions depend on the set $\{q, \omega, E_1, E_2, \theta_1, \theta_2, \Delta\phi_{12}\}$, but not on the average azimuthal angle Φ.

As before, when building the hadronic tensor, it is useful to replace the set of 4-vectors $\{Q^\mu, P_i^\mu, P_1^\mu, P_2^\mu\}$ with the projected set $\{Q^\mu, V_i^\mu, V_1^\mu, V_2^\mu\}$, where

$$V_i^\mu \equiv \frac{1}{M_i} \left\{ P_i - \left(\frac{Q \cdot P_i}{Q^2}\right) Q^\mu \right\} \tag{77a}$$

$$V_k^\mu \equiv P_k^\mu - \left(\frac{Q \cdot P_k}{Q^2}\right) Q^\mu, \quad k = 1, 2, \tag{77b}$$

and where $Q \cdot V_i = Q \cdot V_1 = Q \cdot V_2 = 0$. Using the previous arguments we may write

$$W_s^{\mu\nu} = X_1 g^{\mu\nu} + \ldots + X_{11} \left(V_1^\mu V_2^\nu + V_2^\mu V_1^\nu\right) \tag{78a}$$

$$W_a^{\mu\nu} = Y_1 \left(Q^\mu V_i^\nu - V_i^\mu Q^\nu\right) + \ldots + Y_6 \left(V_1^\mu V_2^\nu - V_2^\mu V_1^\nu\right) \ , \tag{78b}$$

where no ϵ-terms are allowed using the parity properties of the tensors and where all 17 response functions depend on the seven scalar quantities discussed above. Finally, imposing the current conservation conditions on these general forms and contracting the results with the electron scattering leptonic tensor given in Eqs. (57), we are led to the following results:

$$\eta_{\mu\nu}^s W_s^{\mu\nu} \sim v_L \, {}^{(2)}W^L + v_T \, {}^{(2)}W^T + v_{TT} \left({}^{(2)}W^{TT} \cos 2\Phi + {}^{(2)}\tilde{W}^{TT} \sin 2\Phi\right)$$
$$+ v_{TL} \left({}^{(2)}W^{TL} \cos \Phi + {}^{(2)}\tilde{W}^{TL} \sin \Phi\right) \tag{79a}$$

$$\eta_{\mu\nu}^a W_a^{\mu\nu} \sim v_{T'} \, {}^{(2)}\tilde{W}^{T'} + v_{TL'} \left({}^{(2)}W^{TL'} \sin \Phi + {}^{(2)}\tilde{W}^{TL'} \cos \Phi\right) \ , \tag{79b}$$

now with the full set of electron kinematical factors (see Eqs. (21)).

Exclusive $- 3, -4, \ldots$ Electron Scattering, No Nuclear Polarization.

If we try to generalize the above ideas to situations with three or more particles detected in coincidence with the scattered electron, we find that the problem does not continue to get more and more complicated in form. For example, consider exclusive-3 scattering as indicated in Fig. 14. Now we must work with the 4-vectors $\{Q^\mu, P_i^\mu, P_1^\mu, P_2^\mu, P_3^\mu\}$. From these we can form 10 independent dynamical scalars (i.e., using the known masses M_i, M_1, M_2 and M_3 to eliminate four possibilities as before). In the general case of exclusive-n electron scattering there are $3n + 1 +$

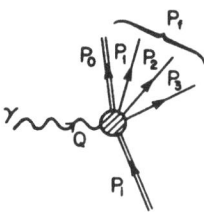

Fig. 14:   Hadronic (nuclear) vertex for exclusive-3 electron scattering, $(e, e' x_1 x_2 x_3)$.

δ_{n0} such independent scalar quantities; equivalently the following set of laboratory variables may be used:

$$\{q, \omega\} \ , \quad n \geq 0$$
$$\{E_k, \theta_k, \ k = 1, \ldots, n\} \ , \quad n \geq 1$$
$$\{\Delta\phi_{12}, \Delta\phi_{23}, \ldots, \Delta\phi_{n-1n}\} \ , \quad n \geq 2 \ ,$$

where $\Delta\phi_{kk'} \equiv \phi_k - \phi_{k'}$. Only the azimuthal angle differences occur as dependences contained within the response functions. This leaves the average azimuthal angle

$$\Phi \equiv \frac{1}{n}(\phi_1 + \ldots + \phi_n) \ , \quad n \geq 1 \qquad (80)$$

as the one kinematical aspect of the detected particles' momenta which does not occur as an argument in the response functions, but appears explicitly in the cross section. If we continue with exclusive-3 scattering and try to build the tensor $W^{\mu\nu}$ from the momenta $\{Q^\mu, P_i^\mu, P_1^\mu, P_2^\mu, P_3^\mu\}$ we immediately observe that a certain saturation has occurred. At the level of exclusive-2 scattering we had four independent 4-momenta; now for $n \geq 3$ we are trying to use five or more independent vectors in a 4-dimensional space. We cannot do so, since the space is spanned by only four. In the case of $n = 3$, for example, we can write

$$P_3^\mu = aQ^\mu + bP_i^\mu + cP_1^\mu + dP_2^\mu \ , \qquad (81)$$

where a, b, c and d are scalar quantities. Thus, we are back to having only the four momenta that were used for the $n = 2$ case. Indeed, this is true for all $n \geq 3$. The general result for the scattering of electrons (polarized or not) where no nuclear polarizations are specified is then similar in form to Eqs. (79). The "Super-Rosenbluth" formula is then the result given in Section I (Eqs. (28) and (29)) or above in Eqs. (79), but with "(2)" → "(n)" and Φ as generalized in Eq. (80). For inclusive scattering $(n = 0)$ only the L and T terms contribute (see Eqs. (66)). For exclusive-1 scattering, $\Phi = \phi_1$ and the response functions $^{(1)}\tilde{W}^{T'}$, $^{(1)}\tilde{W}^{TT}$, $^{(1)}\tilde{W}^{TL}$ and $^{(1)}\tilde{W}^{TL'}$, are all absent (see Eqs. (74)). As before, in considering exclusive-n scattering, if we integrate over the angle dependence (θ_n, ϕ_n) and sum over open channels insofar as particle n is conserved, then we shall recover the exclusive-$(n-1)$ results.

In the general case we may imagine using the helicity dependence to separate the T' and TL' from the L, T, TT, and TL terms. After this, the Φ-dependence may be used to separate $v_L \, {}^{(n)}W^L + v_T \, {}^{(n)}W^T$, $^{(n)}W^{TT}$, $^{(n)}\tilde{W}^{TT}$, $^{(n)}W^{TL}$ and $^{(n)}\tilde{W}^{TL}$ and to separate $^{(n)}\tilde{W}^{T'}$, $^{(n)}W^{TL'}$ and $^{(n)}\tilde{W}^{TL'}$. Finally, the θ_e-dependence in v_T

may be used in making a Rosenbluth decomposition of $^{(n)}W^L$ and $^{(n)}W^T$. Thus, all nine response functions are, in principle, experimentally accessible. Note that the original tensor $W^{\mu\nu}$ was constructed from bilinear products of the electromagnetic current matrix elements (Eq. (42b)). In turn, the currents are 4-vectors which satisfy the continuity equation, $Q_\mu J_{fi}^\mu = 0$; this implies* that $\omega J_{fi}^0 = q J_{fi}^3$, so that only three components of J_{fi}^μ are independent (say J_{fi}^1, J_{fi}^2 and J_{fi}^3, with $J_{fi}^0 = (q/\omega)J_{fi}^3$). Thus, there should be 3×3 independent terms in the cross section, and that agrees with the structure seen in Eq. (79). A similar analysis[4] of the weak interaction hadronic tensor leads to a general structure with $4 \times 4 = 16$ terms, since in that case we have axial-vector as well as vector currents and the former are not conserved.

Let us go into these L/T decompositions a little further. Before performing $\overline{\sum}_{if}$ we have for specific initial and final nuclear states three independent current matrix elements, $\rho_{fi}(\vec{q})$, $J_{fi}^z(\vec{q})$ and $J_{fi}^y(\vec{q})$, with $J_{fi}^z(\vec{q}) = (\omega/q)\rho_{fi}(\vec{q})$. Equivalently, we can choose to deal with the three independent quantities

$$J_{fi}^0(\vec{q}) \equiv J_{fi}^z(\vec{q}) \tag{82a}$$

$$J_{fi}^{\pm 1}(\vec{q}) \equiv \mp \frac{1}{\sqrt{2}} \left(J_{fi}^x(\vec{q}) \pm i J_{fi}^y(\vec{q}) \right) \ , \tag{82b}$$

which transform as a rank-1 spherical tensor under rotations. With our conventions for the electron kinematical factors in Eqs. (21) we have for the hadronic tensors, for specific states i and f, the following

$$W_{fi}^L = |\rho_{fi}(\vec{q})|^2 = (q/\omega)^2 |J_{fi}^0(\vec{q})|^2$$

$$W_{fi}^T = |J_{fi}^{+1}(\vec{q})|^2 + |J_{fi}^{-1}(\vec{q})|^2$$

$$W_{fi}^{TT} = 2\,\mathrm{Re}\left\{ J_{fi}^{+1}(\vec{q})^* J_{fi}^{-1}(\vec{q}) \right\}$$

$$W_{fi}^{TL} = -2\,\mathrm{Re}\left\{ \rho_{fi}(\vec{q})^* \left(J_{fi}^{+1}(\vec{q}) - J_{fi}^{-1}(\vec{q}) \right) \right\}$$

$$= -2(q/\omega)\,\mathrm{Re}\left\{ J_{fi}^0(\vec{q})^* \left(J_{fi}^{+1}(\vec{q}) - J_{fi}^{-1}(\vec{q}) \right) \right\} \tag{83}$$

$$W_{fi}^{T'} = |J_{fi}^{+1}(\vec{q})|^2 - |J_{fi}^{-1}(\vec{q})|^2$$

$$W_{fi}^{TL'} = -2\,\mathrm{Re}\left\{ \rho_{fi}(\vec{q})^* \left(J_{fi}^{+1}(\vec{q}) + J_{fi}^{-1}(\vec{q}) \right) \right\}$$

$$= -2(q/\omega)\,\mathrm{Re}\left\{ J_{fi}^0(\vec{q})^* \left(J_{fi}^{+1}(\vec{q}) + J_{fi}^{-1}(\vec{q}) \right) \right\}$$

We can also define three more contributions:

$$W_{fi}^{\overline{TT}} = 2\,\mathrm{Im}\left\{ J_{fi}^{+1}(\vec{q})^* J_{fi}^{-1}(\vec{q}) \right\}$$

$$W_{fi}^{\overline{TL}} = -2\,\mathrm{Im}\left\{ \rho_{fi}(\vec{q})^* \left(J_{fi}^{+1}(\vec{q}) + J_{fi}^{-1}(\vec{q}) \right) \right\}$$

$$= -2(q/\omega)\,\mathrm{Im}\left\{ J_{fi}^0(\vec{q})^* \left(J_{fi}^{+1}(\vec{q}) + J_{fi}^{-1}(\vec{q}) \right) \right\} \tag{84}$$

$$W_{fi}^{\overline{TL'}} = -2\,\mathrm{Im}\left\{ \rho_{fi}(\vec{q})^* \left(J_{fi}^{+1}(\vec{q}) - J_{fi}^{-1}(\vec{q}) \right) \right\}$$

$$= -2(q/\omega)\,\mathrm{Im}\left\{ J_{fi}^0(\vec{q})^* \left(J_{fi}^{+1}(\vec{q}) - J_{fi}^{-1}(\vec{q}) \right) \right\} \ .$$

---

* In fact, the nomenclature "longitudinal" used above is really not accurate: both longitudinal ($\mu = 3$) and "time" (charge, $\mu = 0$) components enter and are related by this current conservation identity.

By rewriting these expressions in cartesian components we can verify that

$$L, T, TT, TL; \underline{TT}, \underline{TL} \quad \leftrightarrow \quad \text{symmetric under } \mu \leftrightarrow \nu$$

$$T', TL'; \underline{TL'} \quad \leftrightarrow \quad \text{antisymmetric under } \mu \leftrightarrow \nu$$

In fact, in can be shown[11] that, in addition to the six terms we have been discussing which make up the ERL "Super-Rosenbluth" form,

$$\sum_K v_K W^K \quad , \quad K = L, T, TT, TL, T' \text{ and } TL' \ ,$$

we actually have three more terms with $K = \underline{TT}, \underline{TL}$ and $\underline{TL'}$, for a total of nine. However, the electron kinematical factors involved ($v_{\underline{TT}}$, $v_{\underline{TL}}$ and $v_{\underline{TL'}}$ are all higher order in γ^{-1} and so may be neglected in the ERL. The question arises: Is there additional physics in these three terms which are suppressed in the ERL? To answer this, we can explore the properties of Eqs. (83) and (84) under rotations about the z-direction (the direction of $\vec{q}$). Rotating the x- and y-axis (see Fig. 8, for instance) is equivalent to varying the average azimuthal angle Φ (Eq. (80)). It is straightforward to show that W_{fi}^L, W_{fi}^T and $W_{fi}^{T'}$ have no Φ-dependence, to show that the TL-interfernces have the following structures,

$$W_{fi}^{TL}(\Phi) = \text{Re } A_{fi}^{TL} \cdot \cos \Phi + \text{Re } \tilde{A}_{fi}^{TL} \cdot \sin \Phi$$

$$W_{fi}^{TL'}(\Phi) = -\text{Im } A_{fi}^{TL} \cdot \sin \Phi + \text{Im } \tilde{A}_{fi}^{TL} \cdot \cos \Phi \qquad (85a)$$

$$W_{fi}^{\underline{TL}}(\Phi) = \text{Re } A_{fi}^{TL} \cdot \sin \Phi - \text{Re } \tilde{A}_{fi}^{TL} \cdot \cos \Phi$$

$$W_{fi}^{\underline{TL'}}(\Phi) = \text{Im } A_{fi}^{TL} \cdot \cos \Phi + \text{Im } \tilde{A}_{fi}^{TL} \cdot \sin \Phi \qquad (85b)$$

and to show that the TT-interferences have the following structures,

$$W_{fi}^{TT}(\Phi) = \text{Re } A_{fi}^{TT} \cdot \cos 2\Phi + \text{Im } A_{fi}^{TT} \cdot \sin 2\Phi \qquad (86a)$$

$$W_{fi}^{\underline{TT}}(\Phi) = -\text{Re } A_{fi}^{TT} \cdot \sin 2\Phi + \text{Im } A_{fi}^{TT} \cdot \cos 2\Phi \ . \qquad (86b)$$

Thus, if the ERL responses in Eqs. (85a) and (86a) are measured and decomposed using the Φ-dependence, then the non-ERL responses in Eqs. (85b) and (86b) *are completely determined.*

One final observation from these rotational properties, which we shall return to in Section V, is that for exclusive-1 scattering where $\Phi \to \phi_{x_1}$ and where we have only $\cos \phi_{x_1}$ dependence for TL and $\sin \phi_{x_1}$ dependence for TL' we have

$$^{(1)}W_{fi}^{TL}(\phi_{x_1}) = \text{Re} \left(^{(1)}A_{fi}^{TL} \right) \cdot \cos \phi_{x_1} \ , \qquad (87a)$$

$$^{(1)}W_{fi}^{TL'}(\phi_{x_1}) = -\text{Im} \left(^{(1)}A_{fi}^{TL} \right) \cdot \sin \phi_{x_1} \ ; \qquad (87b)$$

that is, the $TL(TL')$ response is proportional to the Re (Im) part of the *same* amplitude, $^{(1)}A_{fi}^{TL}$.

INCLUSIVE ELECTRON SCATTERING FROM POLARIZED TARGETS

Now let us consider inclusive electron scattering from polarized targets in a little more detail. We begin by assuming that a specific nuclear transition, $J_i^{\pi i} \to J_f^{\pi f}$, is involved and only at the end in discussing spin-1/2 initial states will we presume that final states with various values of $J_f^{\pi f}$ are all present. The nuclear responses are built from the basic expressions given in the last sections (Eqs. (83)). To bridge the gap between these general expressions and the forms introduced in Section I (Eqs. (14) – (22)), one case will be studied here in some detail: specifically, let us consider the purely longitudinal or Coulomb responses

$$W_{fi}^L = |\rho_{fi}(\vec{q})|^2 \ .$$

We have for the Fourier transform

$$\rho_{fi}(\vec{q}) = \int d\vec{x} \ e^{i\vec{q}\cdot\vec{x}} \langle f|\hat{\rho}(\vec{x})|i\rangle \ . \tag{88}$$

Using the familiar expression[20] for multipole expanding the plane wave,

$$e^{i\vec{q}\cdot\vec{x}} = 4\pi \sum_{\substack{J\geq 0 \\ M}} i^J j_J(qx) Y_J^M(\Omega_x) Y_J^M(\Omega_q)^* \ , \tag{89}$$

we have

$$\rho(\vec{q})_{fi} = 4\pi \sum_{\substack{J\geq 0 \\ M}} i^J Y_J^M(\Omega_q)^* \langle f| \int d\vec{x} \ j_J(qx) Y_J^M(\Omega_x) \hat{\rho}(\vec{x})|i\rangle \ . \tag{90}$$

This leads to the introduction of the Coulomb operator[1, 14–16]

$$\hat{M}_{JM}(q) \equiv \int d\vec{x} \ j_J(qx) Y_J^M(\Omega_x) \hat{\rho}(\vec{x}) \ . \tag{91}$$

Putting all of this together we have

$$W_{fi}^L = |\rho_{fi}(\vec{q})|^2 = (4\pi)^2 \sum_{\substack{J'\geq 0 \\ M'}} \sum_{\substack{J\geq 0 \\ M}} (-i)^{J'} i^J$$

$$\times Y_{J'}^{M'}(\Omega_q) Y_J^M(\Omega_q)^* \langle f|\hat{M}_{J'M'}(q)|i\rangle^* \langle f|\hat{M}_{JM}(q)|i\rangle \ . \tag{92}$$

First, let us use this to recover the familiar results when the nucleus is *unpolarized*. We may use the Wigner-Eckart theorem[20] to write

$$\langle J_f M_f|\hat{M}_{JM}(q)|J_i M_i\rangle = (-)^{J_f-M_f} \begin{pmatrix} J_f & J & J_i \\ -M_f & M & M_i \end{pmatrix} \langle J_f\|\hat{M}_J(q)\|J_i\rangle \ , \tag{93}$$

making all of the M-dependence explicit in the $3-j$ symbol. Inserting this into Eq. (92) and performing the average-over-initial and sum-over-final, where these mean

$$\overline{\sum_{fi}} \leftrightarrow \left(\frac{1}{2J_i+1} \sum_{M_i}\right) \left(\sum_{M_f}\right) \ ,$$

making use of the orthogonality relationships for $3-j$ symbols[20]

$$\sum_{M_i M_f} \begin{pmatrix} J_f & J' & J_i \\ -M_f & M' & M_i \end{pmatrix} \begin{pmatrix} J_f & J & J_i \\ -M_f & M & M_i \end{pmatrix} = \delta_{J'J}\delta_{M'M}\Big/(2J+1) \ , \tag{94}$$

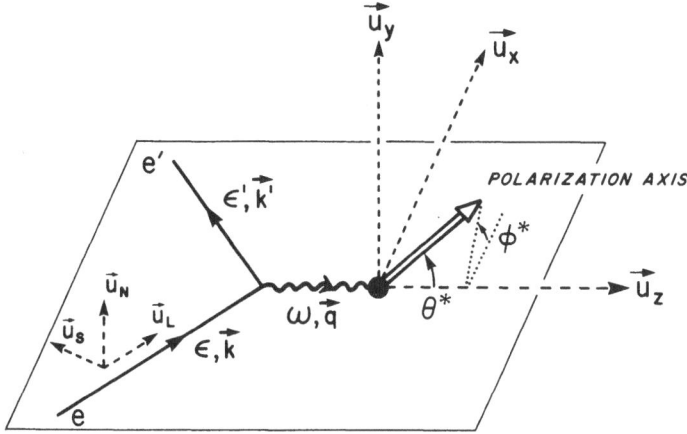

Fig. 15:   Kinematics and coordinate systems for the scattering of polarized electrons from polarized nuclear targets.

we obtain

$$\overline{\sum_{if}} W_{fi}^L = (W^L)^{\text{unpol.}}$$

$$= \frac{4\pi}{2J_i + 1} \sum_{J \geq 0} |\langle J_f \| \hat{M}_J(q) \| J_i \rangle|^2 \quad . \tag{95}$$

This is the familiar standard result. Only the incoherent sum of the squares of the multipole form factors occurs:

$$(W^L)^{\text{unpol.}} = 4\pi F_L^2(q) \tag{96a}$$

$$F_L^2(q) = \sum_{J \geq 0} F_{CJ}^2(q) \tag{96b}$$

$$F_{CJ}(q) = \frac{1}{[J_i]} \langle J_f \| \hat{M}_J(q) \| J_i \rangle \quad , \tag{96c}$$

see Eqs. $(4) - (6)$.

Now let us consider the situation when the target is *polarized*, but when no polarizations are measured in the final state. We can still do the sum-over-final, $\sum_{M_f}$, but we have to treat the initial state differently. Our general expression, (Eq. (92)) is still valid; however, we cannot directly go on to use the Wigner-Eckart theorem as in Eq. (93). In the matrix element

$$\langle J_f M_f | \hat{M}_{JM}(q) | J_i M_i \rangle \quad ,$$

M is referred to the usual z-axis (along $\vec{q}$ in our conventions) and M_f can be presumed to be referred to this same z-axis, since we are going to sum over it anyway. On the other hand, M_i is referred to some other axis of quantization specified by the angles θ^* and ϕ^* as illustrated in Fig. 15. Furthermore, M_i is not summed over. Before being able to use the Wigner-Eckart theorem we must relate the initial state to the standard z-axis projections and this calls for a rotation,[20]

$$|J_i M_i \rangle = \sum_{M_i'} D_{M_i' M_i}^{(J_i)}(\Omega^*) |J_i M_i' \rangle \quad . \tag{97}$$

Here, the states on the right-hand side of the equation involve projections labelled M'_i specified with respect to the usual z-axis. The typical matrix element is then

$$\langle J_f M_f | \hat{M}_{JM}(q) | J_i M_i \rangle = \sum_{M'_i} \mathcal{D}^{(J_i)}_{M'_i M_i}(\Omega^*) \langle J_f M_f | \hat{M}_{JM}(q) | J_i M'_i \rangle \qquad (98a)$$

$$= \sum_{M'_i} \mathcal{D}^{(J_i)}_{M'_i M_i}(\Omega^*)(-)^{J_f - M_f} \begin{pmatrix} J_f & J & J_i \\ -M_f & M & M'_i \end{pmatrix} \langle J_f \| \hat{M}_J(q) \| J_i \rangle , \qquad (98b)$$

where is is now possible to use the Wigner-Eckart theorem under the sum to go from Eq. (98a) to (98b), since all M-values are referred to the same z-axis. Our result so far is

$$|\rho_{fi}(\vec{q})|^2 = (4\pi)^2 \sum_{J' J \geq 0} \sum_{M' M} (-i)^{J'} i^J Y^{M'}_{J'}(\Omega_q) Y^M_J(\Omega_q)^*$$

$$\times \langle J_f \| \hat{M}_{J'}(q) \| J_i \rangle^* \langle J_f \| \hat{M}_J(q) \| J_i \rangle$$

$$\times \sum_{M'_i M''_i} \begin{pmatrix} J_f & J' & J_i \\ -M_f & M' & M''_i \end{pmatrix} \begin{pmatrix} J_f & J & J_i \\ -M_f & M & M'_i \end{pmatrix} \mathcal{D}^{(J_i)}_{M''_i M_i}(\Omega^*)^* \mathcal{D}^{(J_i)}_{M'_i M_i}(\Omega^*) .$$

$$\tag{99}$$

Furthermore, the rotation matrices can be combined[20]

$$\mathcal{D}^{(J_i)}_{M''_i M_i}(\Omega^*)^* \mathcal{D}^{(J_i)}_{M'_i M_i}(\Omega^*) = (-)^{M'_i - M_i}$$

$$\times \sum_{JM}(2J+1) \begin{pmatrix} J_i & J_i & J \\ M''_i & -M'_i & M \end{pmatrix} \begin{pmatrix} J_i & J_i & J \\ M_i & -M_i & 0 \end{pmatrix} \mathcal{D}^{(J_i)}_{M0}(\Omega^*) , \qquad (100)$$

where in fact the rotation matrix $\mathcal{D}^{(J_i)}_{M0}$ is just proportional to a spherical harmonic.[20] Finally, we must perform the sum-over-final, $\sum_{M_f}$, and for the initial state presume that the M_i substates are populated in some way. Let us suppose that the $\{M_i\}$ (referred to the axis of quantization having angles θ^* and ϕ^*, remember) are populated with probabilities $\{p_{(i)}(M_i)\}$, where

$$\sum_{M_i} p_{(i)}(M_i) = 1 . \qquad (101)$$

Then we also must perform the *weighted* sum, requiring altogether

$$W^L_{fi} = \sum_{M_i} p_{(i)}(M_i) \sum_{M_f} |\rho_{fi}(\vec{q})|^2 . \qquad (102)$$

In fact, it proves useful to introduce the Fano statistical tensors

$$f^{(i)}_I \equiv \sum_{M_i} p_{(i)}(M_i)(-)^{J_i - M_i} \langle J_i M_i J_i - M_i | (J_i J_i) I 0 \rangle \qquad (103a)$$

which has as its inverse

$$p_{(i)}(M_i) = \sum_I (-)^{J_i - M_i} [I] \begin{pmatrix} J_i & J_i & I \\ M_i & -M_i & 0 \end{pmatrix} f^{(i)}_I , \qquad (103b)$$

which we shall substitute for $p_{(i)}(M_i)$ in Eq. (102). This appears to have made things very complicated, with many sums over J's, M's, etc. In fact, using properties of the $3-j$'s, $6-j$'s and so on from Ref. 20 (and this is a good exercise for those who are not too familiar with such manipulations), we obtain

$$W^L_{fi} = 4\pi(-)^{J_f - J_i} \sum_J [J] f_J^{(i)} P_J(\cos\theta^*)$$

$$\times \sum_{J'J \geq 0} (-)^{\frac{1}{2}(J'-J)} [J'][J] \begin{pmatrix} J' & J & J \\ 0 & 0 & 0 \end{pmatrix} \begin{Bmatrix} J' & J & J \\ J_i & J_i & J_f \end{Bmatrix} t_{CJ'}(q) t_{CJ}(q) ,$$

$$(104)$$

where we use the notation

$$t_{CJ}(q) = \langle J_f \| \hat{M}_J(q) \| J_i \rangle = [J_i] F_{CJ}(q) \tag{105}$$

and recall that $[x] \equiv \sqrt{2x+1}$.

Several things should be noted at this point. Firstly, this response depends on θ^* through the Legendre polynomial $P_J(\cos\theta^*)$, but does not depend on ϕ^* – we remarked on this in Section I. Secondly, the $3-j$ symbol vanishes unless $J'+J+J =$ even and, since the parity properties of the Coulomb multipoles tell us that J' and J are either both even or both odd, we find that $J =$ even. Thirdly, the $6-j$ symbol summarizes the angular momentum triangle conditions and in particular tells us that $0 \leq J \leq 2J_i$.

Consider the $J = 0$ term in detail. We have $J' = J$ and

$$[0] = 1 , \quad P_0(\cos\theta^*) = 1$$

$$\begin{pmatrix} J & J & 0 \\ 0 & 0 & 0 \end{pmatrix} = (-)^J / [J], \quad \begin{Bmatrix} J & J & 0 \\ J_i & J_i & J_f \end{Bmatrix} = (-)^{J_f + J_i + J} / [J][J_i]$$

so that

$$(W^L_{fi})_{J=0} = f_0^{(i)} \frac{4\pi}{[J_i]} \sum_{J \geq 0} t^2_{CJ}(q) . \tag{106}$$

Now, from the normalization of the probabilities $p_{(i)}(M_i)$ to unity (Eq. (101)), it is easy to show that

$$f_0^{(i)} = 1/[J_i] , \tag{107}$$

regardless of the actual populations. So we have

$$(W^L_{fi})_{J=0} = 4\pi F^2_L(q) . \tag{108}$$

Furthermore, for *unpolarized* targets we have a uniform population,

$$\left(p_{(i)}(M_i)\right)^{\text{unpol.}} = \frac{1}{2J_i + 1} , \tag{109}$$

and it is easy to see that only the rank-0 Fano tensor is non-zero:

$$\left(f_I^{(i)}\right)^{\text{unpol.}} = \delta_{I0}/[J_i] . \tag{110}$$

This recovers for us the familiar result for the unpolarized response

$$W^L_{fi} = (W^L_{fi})_{J=0} = 4\pi F^2_L(q) . \tag{111}$$

The remaining nuclear responses which stem from the basic expressions (Eqs. (83)) are all dealt with in a similar manner. The resulting general forms were introduced in Section I in Eqs. (16), (17) and (22), where now we have explicit expressions for the *reduced response functions* (see Ref. 11 for details):

$$W_J^L(q)_{fi} = (-)^{J_i+J_f}[J] \sum_{J'J\geq 0} (-)^{\frac{1}{2}(J'-J)}[J'][J] \begin{pmatrix} J' & J & J \\ 0 & 0 & 0 \end{pmatrix} \begin{Bmatrix} J' & J & J \\ J_i & J_i & J_f \end{Bmatrix}$$

$$\times\, t_{CJ'}(q)t_{CJ}(q) \qquad\qquad J = \text{even}, \geq 0 \qquad\qquad\qquad (112a)$$

$$W_J^T(q)_{fi} = -(-)^{J_i+J_f}[J] \sum_{J'J\geq 1} [J'][J] \begin{pmatrix} J' & J & J \\ 1 & -1 & 0 \end{pmatrix} \begin{Bmatrix} J' & J & J \\ J_i & J_i & J_f \end{Bmatrix}$$

$$\times \left\{ (-)^{\frac{1}{2}(J'-J)} P_{J'+J}^{+} (t_{EJ'}(q)t_{EJ}(q) + t_{MJ'}(q)t_{MJ}(q)) \right.$$

$$\left. + (-)^{\frac{1}{2}(J'-J+1)} P_{J'+J}^{-} (t_{EJ'}(q)t_{MJ}(q) - t_{MJ'}(q)t_{EJ}(q)) \right\}$$

$$J = \text{even}, \geq 0 \qquad\qquad\qquad (112b)$$

$$W_J^{TT}(q)_{fi} = -\frac{(-)^{J_i+J_f}[J]}{\sqrt{(J-1)J(J+1)(J+2)}} \sum_{J'J\geq 1} [J'][J] \begin{pmatrix} J' & J & J \\ 1 & 1 & -2 \end{pmatrix} \begin{Bmatrix} J' & J & J \\ J_i & J_i & J_f \end{Bmatrix}$$

$$\times \left\{ (-)^{\frac{1}{2}(J'-J)} P_{J'+J}^{+} (t_{EJ'}(q)t_{EJ}(q) - t_{MJ'}(q)t_{MJ}(q)) \right.$$

$$\left. -(-)^{\frac{1}{2}(J'-J+1)} P_{J'+J}^{-} (t_{EJ'}(q)t_{MJ}(q) + t_{MJ'}(q)t_{EJ}(q)) \right\}$$

$$J = \text{even}, \geq 2 \qquad\qquad\qquad (112c)$$

$$W_J^{TL}(q)_{fi} = \frac{2\sqrt{2}(-)^{J_i+J_f}[J]}{\sqrt{J(J+1)}} \sum_{\substack{J'\geq 0 \\ J\geq 1}} [J'][J] \begin{pmatrix} J' & J & J \\ 0 & 1 & -1 \end{pmatrix} \begin{Bmatrix} J' & J & J \\ J_i & J_i & J_f \end{Bmatrix}$$

$$\times\, t_{CJ'}(q) \left\{ (-)^{\frac{1}{2}(J'-J)} P_{J'+J}^{+} t_{EJ}(q) - (-)^{\frac{1}{2}(J'-J+1)} P_{J'+J}^{-} t_{MJ}(q) \right\}$$

$$J = \text{even}, \geq 2 \qquad\qquad\qquad (112d)$$

$$W_J^{T'}(q)_{fi} = W_J^{T}(q)_{fi}, \quad \text{but with } J = \text{odd}, \geq 1 \qquad\qquad\qquad (112e)$$

$$W_J^{TL'}(q)_{fi} = -W_J^{TL}(q)_{fi}, \quad \text{but with } J = \text{odd}, \geq 1 \qquad\qquad\qquad (112f)$$

where the real transverse matrix elements are given by

$$t_{EJ}(q) = [J_i]F_{EJ}(q) \qquad\qquad\qquad\qquad\qquad\qquad (113a)$$

$$t_{MJ}(q) = [J_i]F_{MJ}(q) \;\;, \qquad\qquad\qquad\qquad\qquad (113b)$$

and where $P_K^{\pm} \equiv \frac{1}{2}(1 \pm (-)^K)$.

We have seen that having polarized targets (with or without having polarized electrons as well) leads to new information beyond the familiar unpolarized results (the $J = 0$ responses above). What is the nature of this new information? We shall answer this question by considering a few specific examples, for then the potential importance of such studies should become clear. Throughout, to be explicit, let us consider the targets to be 100% polarized in the (θ^*, ϕ^*) direction, in which case

$$f_I^{(i)} = \frac{(2J_i)![I]}{\sqrt{(2J_i+1+I)!(2J_i-I)!}} \qquad\qquad\qquad\qquad (114)$$

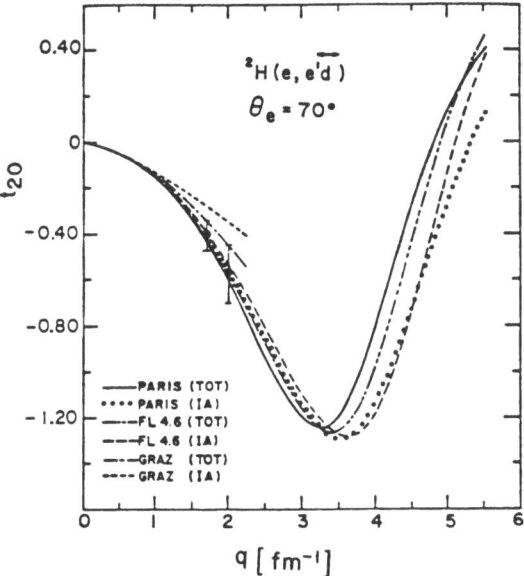

Fig. 16:   Recoil t_{20} tensor polarization for elastic scattering from the deuteron (from Ref. 21).

Let us begin with the case of unpolarized electrons, but polarized targets. As we noted in Section I and see again explicitly in Eqs. (112), the appropriate responses $(L, T, TT, \text{ and } TL)$ all have $J = $ even and $0 \le J \le 2J_i$. For spin-0 and spin-1/2 targets, only $J = 0$ responses are then present and these just give back the unpolarized cross sections. Thus, the first interesting case has $J_i = 1$ and a "three-star" example of this sort is elastic scattering from polarized ^2H. The multipoles involved are $C0$, $M1$ and $C2$ and for unpolarized scattering (or rank-zero responses) we have from Eqs. (17):

$$(W_0^L)_{fi} = \sqrt{3}\, F_L^2 = \sqrt{3}\, (F_{C0}^2 + F_{C2}^2) \tag{115a}$$

$$(W_0^T)_{fi} = \sqrt{3}\, F_T^2 = \sqrt{3}\, F_{M1}^2 \tag{115b}$$

Thus, lacking polarization we can hope to separate the L and T responses using the familiar Rosenbluth method, but cannot decompose the former into its $C0$ and $C2$ contributions. With tensor-polarized deuterium, however, we have in addition from Eqs. (112)

$$(W_2^L)_{fi} = -2\sqrt{3}F_{C2}\left(F_{C0} + \frac{1}{2\sqrt{2}}F_{C2}\right) \tag{116a}$$

$$(W_2^T)_{fi} = -\frac{1}{2}\sqrt{\frac{3}{2}}F_{M1}^2 \tag{116b}$$

$$(W_2^{TL})_{fi} = \frac{3}{\sqrt{2}}F_{M1}F_{C2} \tag{116c}$$

$$(W_2^{TT})_{fi} = \frac{1}{4}\sqrt{\frac{3}{2}}F_{M1}^2 \quad . \tag{116d}$$

Clearly, knowing a quantity such as $(W_2^L)_{fi}$ together with the unpolarized cross section will permit a complete separation of the problem into the fundamental form

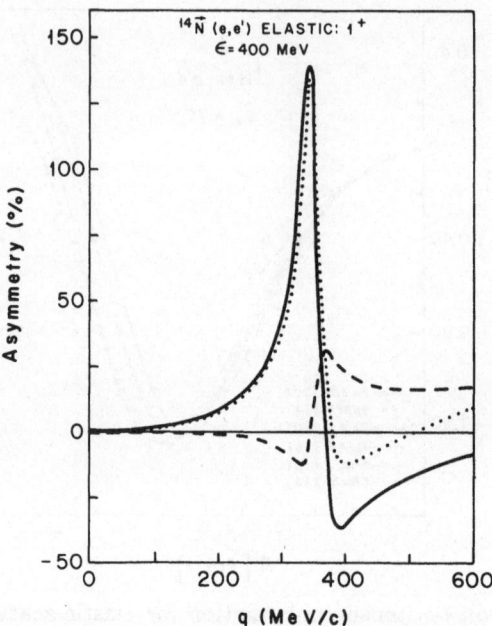

Fig. 17:   Elastic electron scattering from polarized $^{14}N(1^+)$. The asymmetries are defined to be $A_{ij} \equiv (\Sigma_i - \Sigma_j)/\Sigma_0$, where Σ_0 is the unpolarized cross section and we take $i, j = L, N, S$, the direction given with respect to the incident beam (see Figs. 10 and 15): A_{NS} (solid line), A_{LN} (dashed line), and A_{LS} (dotted line).

factors. In Fig. 16 are shown the results of a recent experiment[21] at Bates to measure t_{20} (with is simply related to the tensors used here.[10,11]). This was performed as a recoil polarization measurement rather than with a polarized target, although the two methods are directly related (see below). Clearly, pushing such measurements to higher momentum transfers where calculated responses show interesting model dependences is a high priority item for the future.

The spin-1 elastic scattering example mentioned here is characteristic of a wide class of polarized target experiments. In general, for non-trivial ground state spins we have a mixed multipole situation where unpolarized scattering alone can only give us two pieces of information W_0^L and W_0^T. With polarized targets, however, we get more information in the form of higher rank reduced response functions ($J = 2, 4, 6, \ldots,$) of all four kinds (L, T, TL and TT) and so can usually separate all of the multipoles from one another in a model-independent way. Thus, a prime use for polarization in the future in electron scattering is as a *multipole meter*.

Another example of elastic scattering from a spin-1 target, this time $^{14}\vec{N}$, is shown in Fig. 17. The three specific asymmetries are defined in the caption to the figure.

If we proceed to allow the electrons to be polarized as well as the spin-1 targets, then the odd-rank reduced response functions in Eqs. (112) are also available:

$$(W_1^{T'})_{fi} = -\frac{3}{2\sqrt{2}} F_{M1}^2 \qquad (117a)$$

$$(W_1^{TL'})_{fi} = -2\sqrt{3}F_{M1}\left(F_{C0} + \frac{1}{2\sqrt{2}}F_{C2}\right) \ . \tag{117b}$$

For instance, knowing $W_1^{TL'}$ in addition to the unpolarized cross section would provide a different way to separate the three elastic form factors in this case.

Before going on to discuss more complicated examples, let us at this juncture mention two things. One concerns the class of experiments of the sort $\vec{A}(e,\vec{e'})$: since ultra-relativistic electrons conserve their helicity in scattering, this contains exactly the same information as it contained in $\vec{A}(\vec{e},e')$ scattering. In the above discussions, it is only necessary to replace the initial electron helicity h by the scattered electron's helicity h'. The second point concerns *elastic* scattering where, instead of having a polarized target, one presumes that the nuclear recoil polarization is measured (in a second scattering using a polarimeter). A simple relationship exists[11] between the latter denoted $f \bar{\imath}$ and the former denoted $\bar{f} i$, with i and f interchanged:

$$(W_J^K)_{f\bar{\imath}} = \pm \left[\frac{2J_f+1}{2J_i+1}\right](W_J^K)_{\bar{f}f} \ , \tag{118}$$

with a plus sign for $K = L, T, TT$ and TL' and a minus sign for $K = TL$ and T'. The deuteron elastic scattering experiment[21] referred to above was in fact performed as a deuteron tensor recoil polarization measurement and is related to planned experiments with polarized targets using this "turn-around" relation.

Proceeding now to more complicated examples, let us next consider elastic scattering from polarized spin-3/2 targets where we have

$$F_L^2 = F_{C0}^2 + F_{C2}^2 \tag{119a}$$
$$F_T^2 = F_{M1}^2 + F_{M3}^2 \tag{119b}$$

and so both L and T form factors contain incoherent combinations of multipoles which cannot be separated without having polarization information. With this information, however, we also have from Eqs. (112) (see also Ref. 11):

$$(W_2^L)_{fi} = -4F_{C0}F_{C2}$$

$$(W_2^T)_{fi} = -\frac{4}{5}\left(F_{M1} + \sqrt{\frac{3}{2}}F_{M3}\right)^2$$

$$(W_2^{TL})_{fi} = \frac{4}{\sqrt{5}}F_{C2}\left(F_{M1} - \sqrt{\frac{2}{3}}F_{M3}\right)$$

$$(W_2^{TT})_{fi} = \frac{2}{5}\left(F_{M1}^2 - F_{M3}^2 + \frac{1}{\sqrt{6}}F_{M1}F_{M3}\right) \tag{120a}$$

for the even-rank tensors, and for the odd-rank tensors (requiring polarized electrons)

$$(W_1^{T'})_{fi} = -\frac{2}{\sqrt{5}}\left(F_{M1}^2 + F_{M3}^2\right)$$

$$(W_1^{TL'})_{fi} = -4\left\{F_{M1}\left(F_{C0} + \frac{2}{5}F_{C2}\right) + \frac{\sqrt{6}}{5}F_{C2}F_{M3}\right\}$$

$$(W_3^{T'})_{fi} = \frac{2}{\sqrt{5}}F_{M3}\left(F_{M3} + 2\sqrt{6}F_{M1}\right)$$

$$(W_3^{TL'})_{fi} = 2\sqrt{\frac{2}{3}}\left\{F_{M3}\left(F_{C0} + \frac{3}{5}F_{C2}\right) + \frac{\sqrt{6}}{5}F_{C2}F_{M1}\right\} \ . \tag{120b}$$

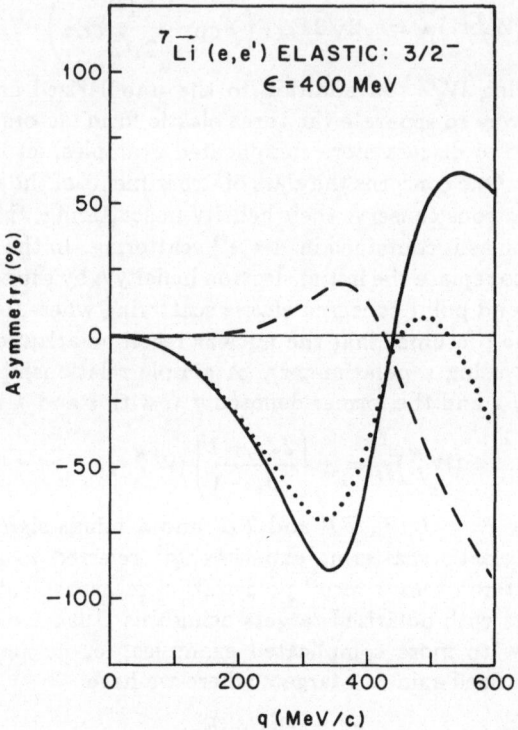

Fig. 18:   Elastic electron scattering from polarized ^7Li (3/2⁻). The three asymmetries defined in Fig. 17 are shown: A_{NS} (solid line), A_{LN} (dashed line), and A_{LS} (dotted line).

Clearly, the problem is overdetermined: having some of this polarization information in addition to the unpolarized results would permit a complete decomposition into the four basic form factors.

Examples are shown in Figs. 18 and 19 for polarized lithium, $\vec{^7Li}$. It is apparent at once that the asymmetries and polarizations are *large* in these typical situations.

We can proceed to study elastic scattering from higher spin targets (for example, results for $\vec{^{27}Al}$ with $J_i = \frac{5}{2}$ are displayed in Fig. 20); however, let us go on to a few examples of inelastic transitions. Using Eqs. (112) or the tables in Ref. 11 for the case $J_i^{\pi i} = \frac{3}{2}^{\pm} \rightarrow J_f^{\pi f} = \frac{1}{2}^{\pm}$, where we can have $M1$, $C2$, and $E2$ multipoles, yields

$$(W_2^L)_{fi} = -2F_{C2}^2$$

$$(W_2^T)_{fi} = -\left(F_{E2}^2 - F_{M1}^2 - 2\sqrt{3}F_{E2}F_{M1}\right)$$

$$(W_2^{TL})_{fi} = 2F_{C2}\left(F_{M1} + \frac{1}{\sqrt{3}}F_{E2}\right)$$

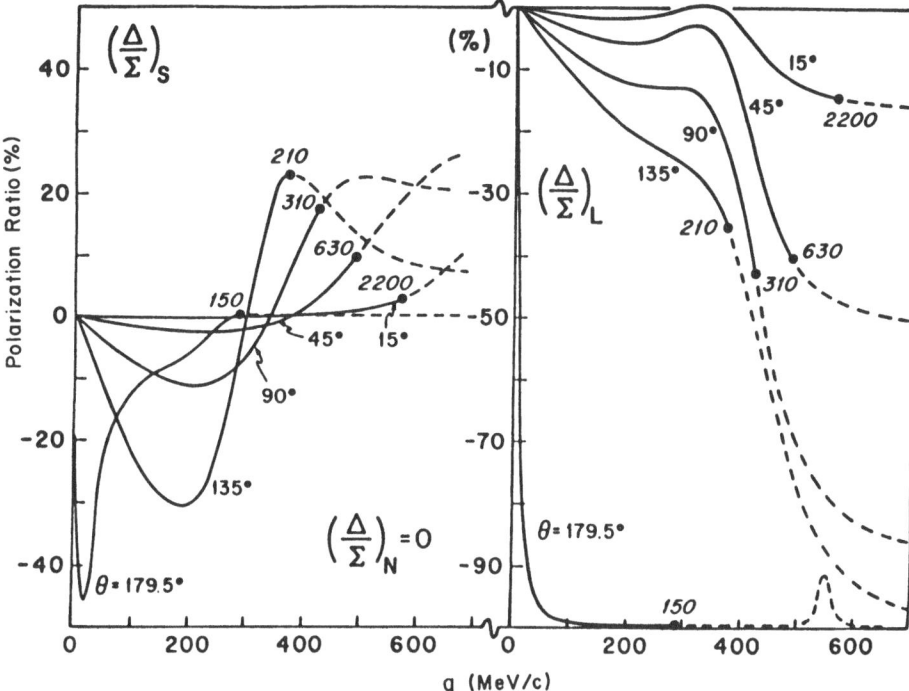

Fig. 19:   Elastic electron scattering from polarized ^7Li $(3/2^-)$. The polarization ratios are defined by $(\Sigma/\Delta)_i$, where $i = L, N, S$, the directions given with respect to the incident beam (see Fig. 10 and 15): $(\Delta/\Sigma)_L$ (solid line), and $(\Delta/\Sigma)_S$ (dashed line); $(\Delta/\Sigma)_N$ is identically zero.

$$(W_2^{TT})_{fi} = \frac{1}{2}\left(F_{E2}^2 - F_{M1}^2 + \frac{2}{\sqrt{3}}F_{E2}F_{M1}\right) \tag{121a}$$

$$(W_1^{T'})_{fi} = -\sqrt{5}\left(F_{M1}^2 + \frac{3}{5}F_{E2}^2 + \frac{2}{5}\sqrt{3}F_{E2}F_{M1}\right)$$

$$(W_1^{TL'})_{fi} = -\frac{2}{\sqrt{5}}F_{C2}\left(F_{M1} + 3\sqrt{3}F_{E2}\right)$$

$$(W_3^{T'})_{fi} = \frac{4}{\sqrt{5}}F_{E2}\left(F_{E2} - \sqrt{3}F_{M1}\right)$$

$$(W_3^{TL'})_{fi} = \frac{4}{\sqrt{15}}F_{C2}\left(F_{E2} - \sqrt{3}F_{M1}\right) \quad . \tag{121b}$$

Again, it is clear that the additional polarization information will permit a complete decomposition into the underlying multipole form factors to be achieved. Examples are shown in Figs. (21), (22), (23) and (24) for $\vec{Li}$ and in Fig. (25) for $^{39}\vec{K}$. Once again *very large asymmetries* are typically seen to occur.

Many more examples are considered in detail in Ref. 11 and the reader is directed there for more discussion (including a closer look at the examples mentioned

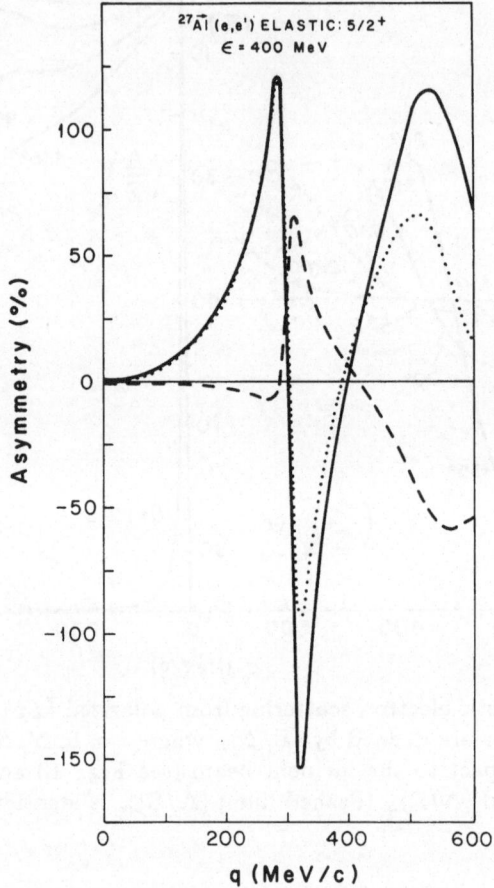

Fig. 20: Elastic electron scattering from polarized $^{27}Al(5/2^+)$. The three asymmetries defined in Fig. 17 are shown: A_{NS} (solid line), A_{LN} (dashed line), and A_{LS} (dotted line).

above). Here we will only select two points to make via illustrative examples. One concerns the question: Are the asymmetries the same for any given $J_i^{\pi i} \rightarrow J_f^{\pi f}$ transition, or is there interesting nuclear structure information reflected in different asymmetries? The answer is the latter as seen from the very different results obtained for two different $\frac{5}{2}^+ \rightarrow \frac{3}{2}^+$ transitions in $^{25}\vec{Mg}$ and shown in Figs. (26) – (29). The second concerns specific reasons for wanting to have a multipole meter. There are numerous reasons why it is of great interest to separate the underlying multipoles; let us consider just one idea here. Suppose we focus on the $\frac{3}{2}^- \rightarrow \frac{7}{2}^-$ (4.63 MeV) transition in ^7Li. In principle, we can have $C2$, $E2$, $M3$, $C4$, $E4$ and $M5$ multipoles. However, if the electromagnetic operators were purely one-body in nature and if the states in ^7Li contained active nucleons *only in the 1p-shell*, then the maximum multipolarity allowed would be 3 and there would be no $C4$, $E4$ or $M5$ multipoles. If, on the other hand, the electromagnetic operators contain *two-body* meson exchange current contributions (remember: our multipole analyses was quite general – it included the possibility of such currents), then the higher multipoles can occur as well. In fact, the leading MEC effects

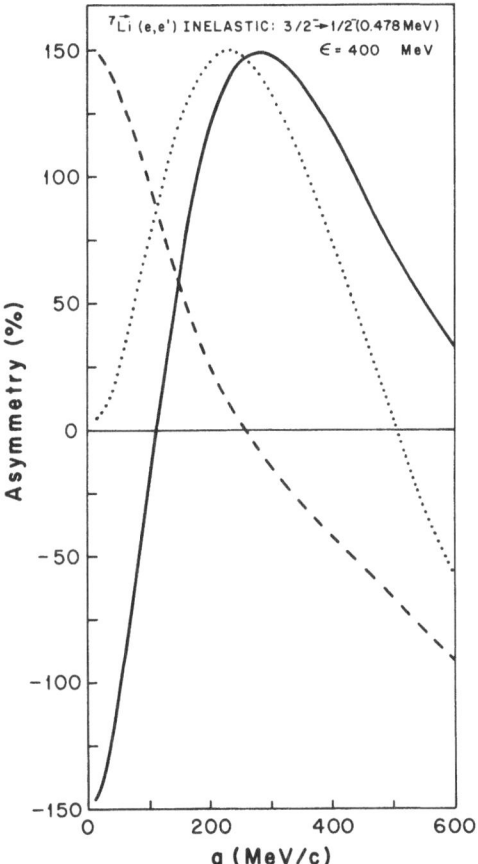

Fig. 21:   Inelastic electron scattering from polarized $^7Li(3/2^- \mapsto 1/2^-)$. The three asymmetries defined in Fig. 17 are shown: A_{NS} (solid line), A_{LN} (dashed line), and A_{LS} (dotted line).

are transverse[22] and so we expect $E4$ and $M5$ multipole strength, but not $C4$ strength. Finally, if *more than the 1p-shell participates* in the dynamics, then again the higher multipoles can occur, only now, in general, all three ($C4$, $E4$ and $M5$) will be present. Clearly, having separated the multipoles using polarized targets we could begin to understand the interplay of these different effects. In Fig. 30 we show the usual unpolarized transverse response, W_0^T along with the quantity W', where

$$W' \equiv W_0^T + \frac{22}{89}W_2^T - \frac{600}{89}W_2^{TT} - \frac{43}{89}W_1^{T'} \qquad (122)$$

and vanishes if no $E4$ or $M5$ multipoles are present. The results in Fig. 30 were obtained[11] using the MEC effects discussed in Ref. 22.

As a last example, let us return to the special important case of spin-1/2 targets. As we saw above, having polarized targets only yields no new information beyond the usual unpolarized cross section, since the only tensors permitted had to have ranks 0 or 1 and only the former is even. On the other hand, with polarized spin-1/2 targets *and* polarized electrons the rank-1 tensors are now accessible. Inserting

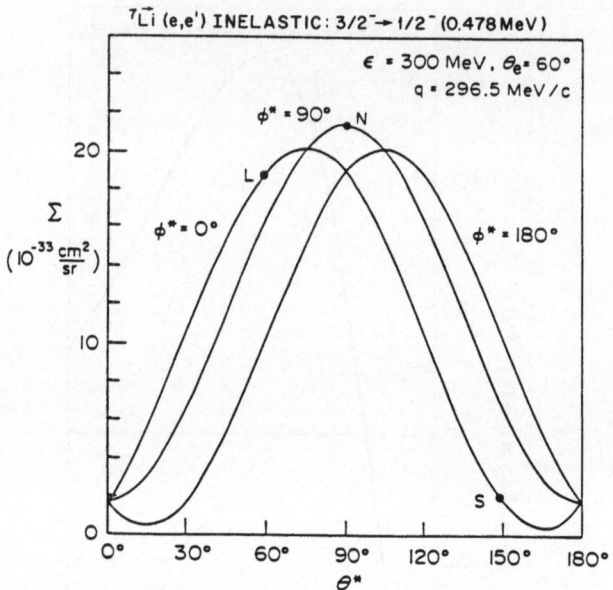

Fig. 22:   Inelastic electron scattering from polarized $^{7}Li(3/2^{-} \mapsto 1/2^{-})$.
The cross section Σ is displayed as a function of the polarization direction
of the nucleus (θ^{*}, ϕ^{*}) for given electron scattering kinematics.

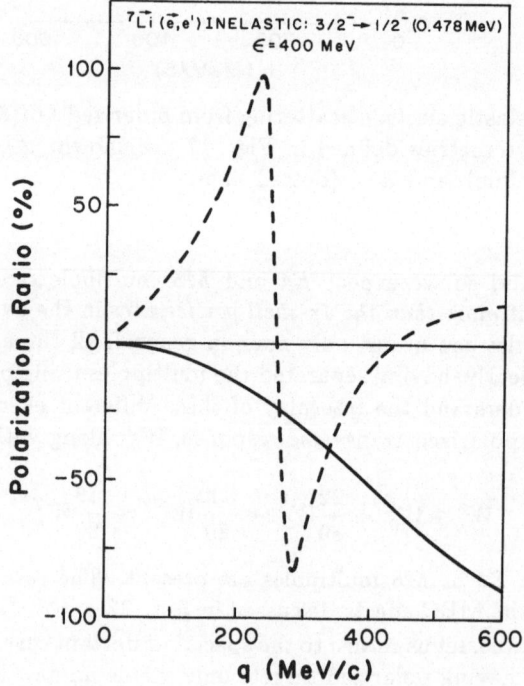

Fig. 23:   Inelastic electron scattering from polarized $^{7}Li(3/2^{-} \mapsto 1/2^{-})$.
The two polarization ratios defined in Fig. 19 are shown: $(\Delta/\Sigma)_{L}$ (solid
line), $(\Delta\Sigma)_{S}$ (dashed line); $(\Delta\Sigma)_{N}$ is identically zero.

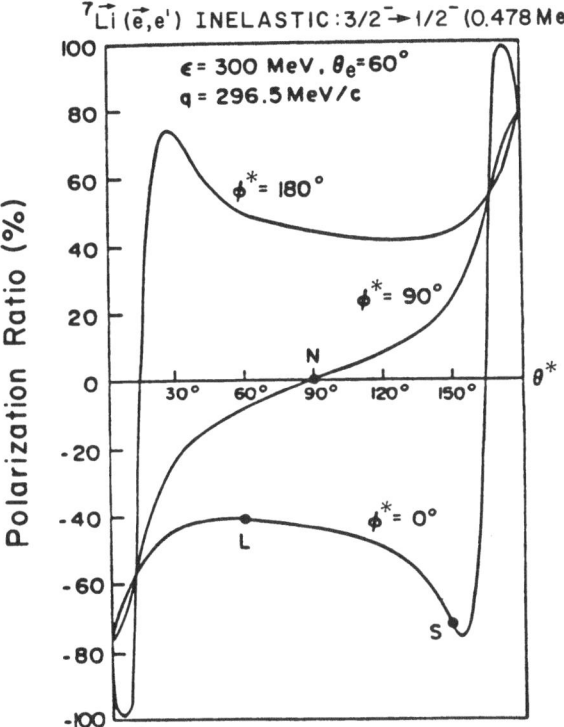

$^7\vec{Li}\,(\vec{e},e')$ INELASTIC : $3/2^- \to 1/2^-$ (0.478 MeV)

Fig. 24: Inelastic electron scattering from polarized $^7Li(3/2^- \mapsto 1/2^-)$. The polarization ratio Δ/Σ is displayed as a function of the polarization direction of the nucleus (θ^*, ϕ^*) for given electron scattering kinematics.

Eqs. (22) into Eq. (20) yields

$$
\begin{aligned}
\Delta_{fi} &\sim v_{T'}\, W^{T'}_{fi} + v_{TL'} \cos\phi^* W^{TL'}_{fi} \\
&= f_1^{(i)} \left\{ v_{T'} \cos\theta^* \left(W_1^{T'} \right)_{fi} + v_{TL'} \sin\theta^* \cos\phi^* \left(W_1^{TL'} \right)_{fi} \right\} \,,
\end{aligned}
\tag{123}
$$

where the Fano tensor is especially simple:

$$
f_1^{(i)} = \frac{1}{\sqrt{2}} \left\{ p_{(i)} \left(M_{J_i} = +\frac{1}{2} \right) - p_{(i)} \left(M_{J_i} = -\frac{1}{2} \right) \right\} \,,
\tag{124a}
$$

with

$$
\sqrt{2} f_0^{(i)} = p_{(i)} \left(M_{J_i} = +\frac{1}{2} \right) + p_{(i)} \left(M_{J_i} = -\frac{1}{2} \right) = 1 \,.
\tag{124b}
$$

Note that both $v_{T'}$ and $v_{TL'}$ and hence all of the right-hand side of Eq. (123), contain a factor $\tan\theta_e/2$ (see Eqs. (21)) and so become small as $\theta_e \to 0$.

Consider first elastic scattering from spin-1/2 targets where $C0$ and $M1$ form factors can occur. The unpolarized cross section involves the rank-0 reduced response functions (Eqs. (7)):

$$
(W_0^L)_{fi} = \sqrt{2} F_{C0}^2
\tag{125a}
$$

$$
(W_0^T)_{fi} = \sqrt{2} F_{M1}^2 \,.
\tag{125b}
$$

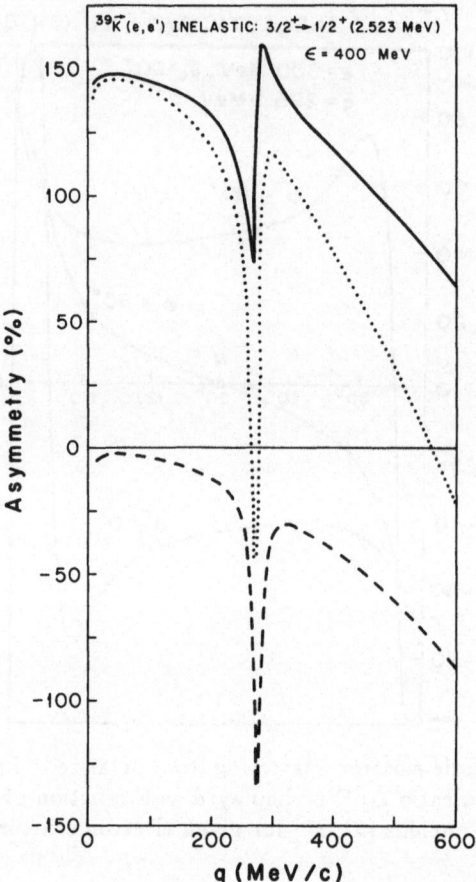

Fig. 25:   Inelastic electron scattering from polarized $^{39}K(3/2^+ \mapsto 1/2^+)$.
The three asymmetries defined in Fig. 17 are shown: A_{NS} (solid line),
A_{LN} (dashed line), and A_{LS} (dotted line.)

As long as we are not interested in the relative sign between F_{C0} and F_{M1}, then,
in principle, the usual Rosenbluth separation is all that is needed to separate the
Coulomb from the magnetic effects. On the other hand, in practice one form factor
may be very small in magnitude compared to the other, and it may be very difficult
to perform the separation. A fundamental example of this problem is the nucleon
itself, where for large momentum transfers the electron prefers to scatter magneti-
cally and the electric (i.e., Coulomb) response is relatively smaller. With polarized
targets and electrons we also have[11]

$$(W_1^{T'})_{fi} = -\sqrt{2}F_{M1}^2 \tag{125c}$$

$$(W_1^{TL'})_{fi} = -2\sqrt{2}F_{C0}F_{M1} \ , \tag{125d}$$

and it is especially this last TL' response which is of particular interest, as it involves
the interference between F_{C0} and F_{M1}. Going back to Eq. (123), we see that to
optimize the effect from this term we require $\phi^* = 0°$ or $180°$ (corresponding to
target polarization in the electron scattering plane) and $\theta^* = 90°$ (perpendicular to

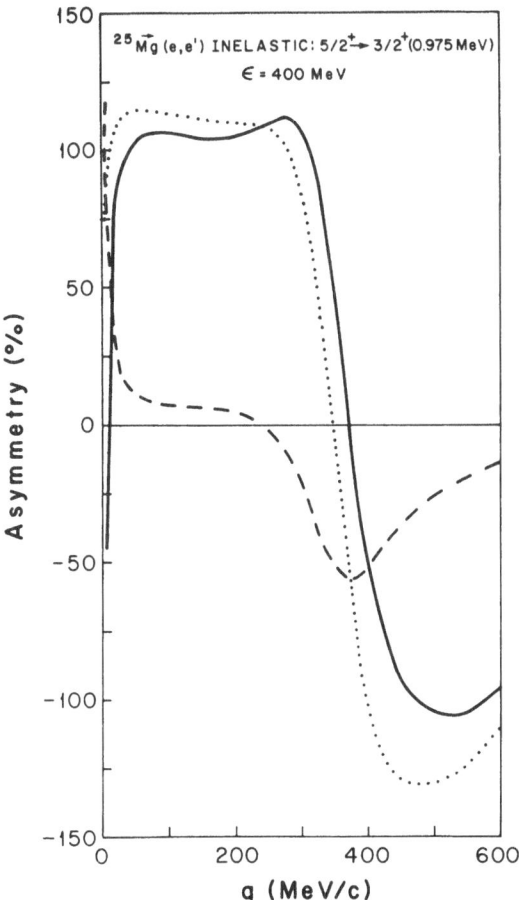

Fig. 26:   Inelastic electron scattering from polarized ^{25}Mg ($5/2^+ \mapsto 3/2^+$ #1). The three asymmetries defined in Fig. 17 are shown: A_{NS} (solid line), A_{LN} (dashed line), and A_{LS} (dotted line).

the momentum transfer direction $\vec{q}$). Thus, the TL' interference is ideally studied when the situation is one one shown in Fig. 31.

Consider as a second example the inelastic scattering transition $\frac{1}{2}^{\pm} \rightarrow \frac{3}{2}^{\pm}$ where $M1$, $C2$ and $E2$ multipoles all enter. The unpolarized scattering (rank-0) reduced response functions are

$$(W_0^L)_{fi} = \sqrt{2} F_{C2}^2 \tag{126a}$$

$$(W_0^T)_{fi} = \sqrt{2} \left(F_{M1}^2 + F_{E2}^2 \right) \quad . \tag{126b}$$

For the nucleon-to-delta transition which falls in this category, we have two problems: (1) in unpolarized scattering we cannot separate F_{M1}^2 and F_{E2}^2 in Eq. (126b); and (2) even if we were content only to separate W_0^L in a Rosenbluth decomposition, the F_{M1}^2 contribution dominates and, just as with elastic scattering, makes such separations hard to achieve in practice. On the other hand, with polarized targets and electrons we have for the $\frac{1}{2}^{\pm} \rightarrow \frac{3}{2}^{\pm}$ transition the additional responses[11]

$$(W_1^{T'})_{fi} = \frac{1}{\sqrt{2}} \left(F_{M1}^2 - F_{E2}^2 - 2\sqrt{3} F_{M1} F_{E2} \right) \tag{126c}$$

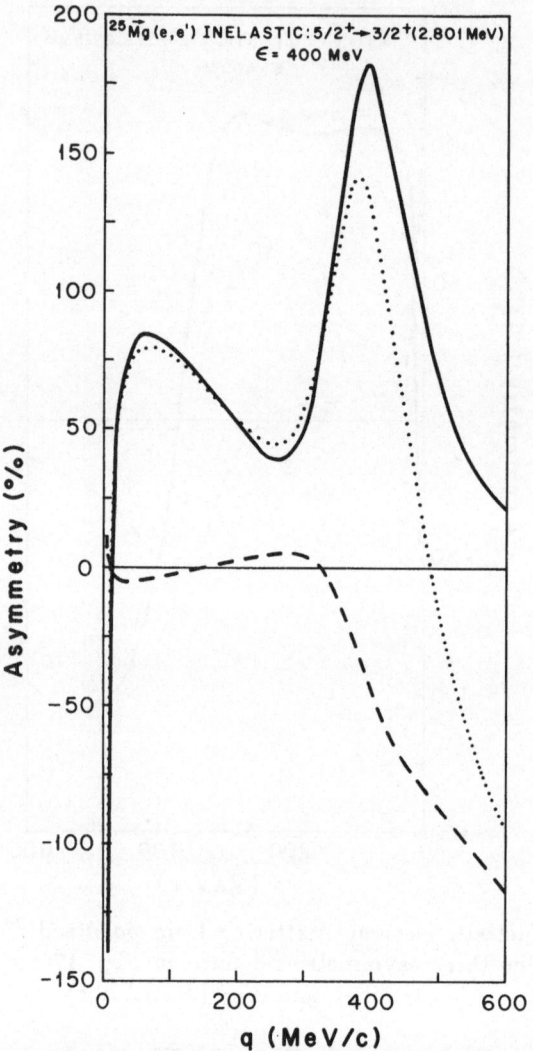

Fig. 27:   Inelastic electron scattering from polarized ^{25}Mg ($5/2^+ \mapsto$ $3/2^+$ #2). The three symmetries defined in Fig. 17 are shown: A_{NS} (solid line), A_{LN} (dashed line), and A_{LS} (dotted line).

$$(W_1^{TL'})_{fi} = -\sqrt{2}F_{C2}\left(F_{M1} + \sqrt{3}F_{E2}\right) \ . \tag{126d}$$

Once again we see the occurence of interference terms involving F_{M1} with F_{C2} or F_{E2} which should allow the latter small effects to be extracted. In passing, we note the differences between this situation $\left(\frac{1}{2}^\pm \to \frac{3}{2}^\pm\right)$ and the one discussed above $\left(\frac{3}{2}^\pm \to \frac{1}{2}^\pm$ in Eqs. (121)$\right)$: It makes a great deal of difference which state is polarized, $\frac{1}{2}$ or $\frac{3}{2}$, even though both examples only have $M1$, $C2$ and $E2$ multipoles.

Let us generalize these results to the situation where all partial waves can occur in the final state. We shall take the initial polarized state to have $J_i^{\pi_i} = \frac{1}{2}^+$, although the ideas are straightforwardly extended to $\frac{1}{2}^-$ as well. If we denote the

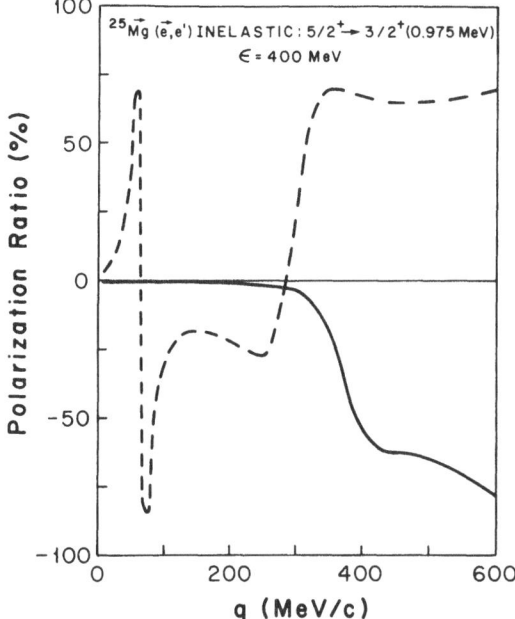

Fig. 28:   Inelastic electron scattering from polarized ^{25}Mg ($5/2^+ \mapsto 3/2^+$ #1). The two polarization ratios defined in Fig. 19 are shown: $(\Delta/\Sigma)_L$ (solid line), and $(\Delta/\Sigma)_S$ (dashed line); $(\Delta/\Sigma)_N$ is identically zero.

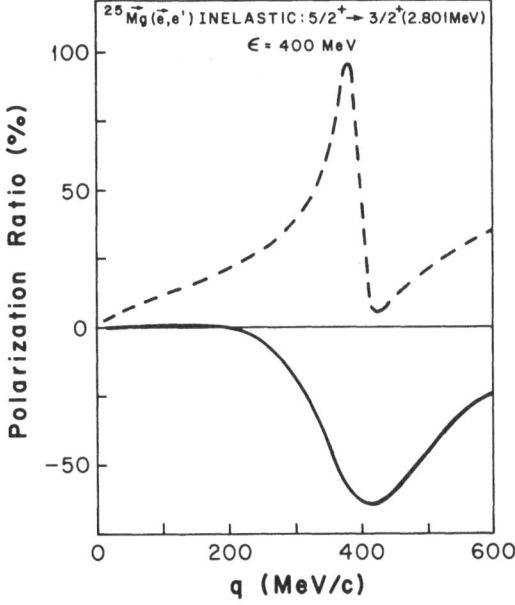

Fig. 29:   Inelastic electron scattering from polarized ^{25}Mg ($5/2^+ \mapsto 3/2^+$ #2). The two polarizations defined in Fig. 19 are shown: $(\Delta/\Sigma)_L$ (solid line), and $(\Delta/\Sigma)_S$ (dashed line); $(\Delta/\Sigma)_N$ is identically zero.

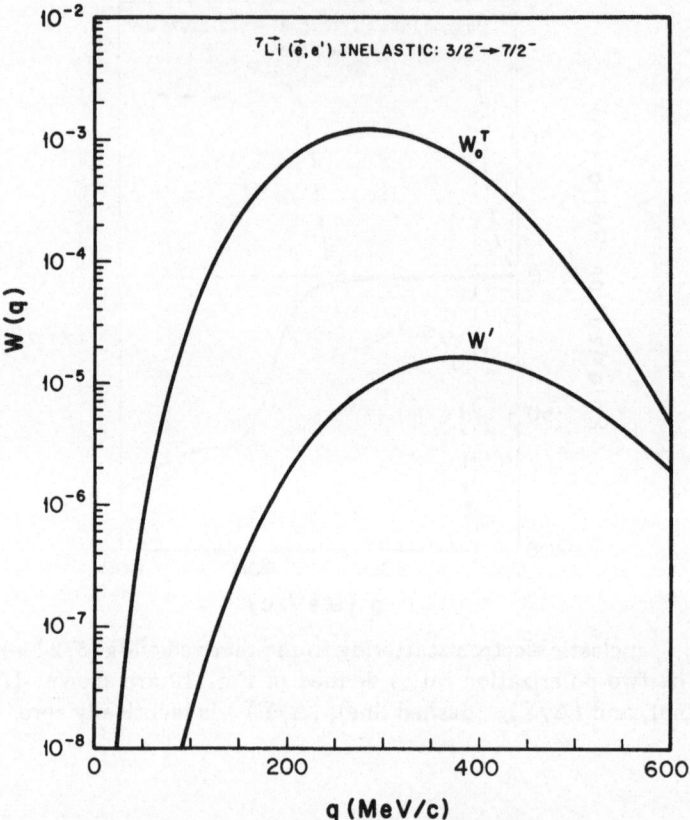

Fig. 30:   Inelastic electron scattering from polarized ^7Li $(3/2^- \mapsto 7/2^-)$.
The response function W_0^T is displayed for the shell model, while the function W' defined in Eq. (122) is displayed using results where the meson-exchange current effects are taken into account.

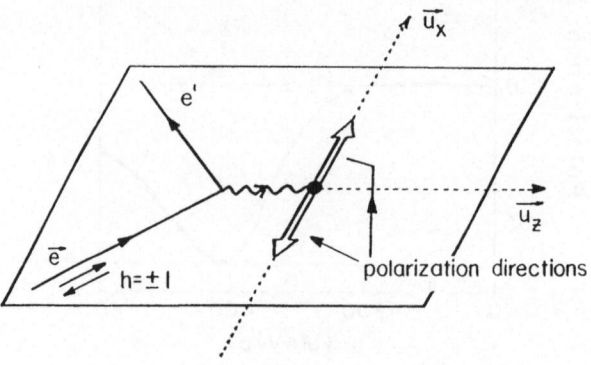

Fig. 31:   Polarized electron scattering from polarized spin-1/2 targets: special kinematics for optimizing the TL' response.

multipole matrix elements

$$t^{\pm}_{CJ}(q) \equiv \langle J_f \| \hat{M}_J(q) \| J_i \rangle \ , \tag{127}$$

with $J_f = J \pm \frac{1}{2}$ and with similar expressions for $t^{\pm}_{EJ}(q)$ and $t^{\pm}_{MJ}(q)$, then we can write (again using Eqs. (112)):

$$\sqrt{2}(W^L_0)_{fi} \equiv U^L = \sum_{J \geq 1} \{(t^+_{CJ-1})^2 + (t^-_{CJ})^2\} \tag{128a}$$

$$\sqrt{2}(W^T_0)_{fi} \equiv U^T = \sum_{J \geq 1} \{(t^+_{EJ})^2 + (t^-_{EJ})^2 + (t^+_{MJ})^2 + (t^-_{MJ})^2\} \tag{128b}$$

$$\sqrt{2}(W^{T'}_1)_{fi} \equiv U^{T'} = \sum_{J \geq 1} \left\{ \frac{1}{J+1}\left((t^+_{EJ})^2 + (t^+_{MJ})^2\right) - \frac{1}{J}\left((t^-_{EJ})^2 + (t^-_{MJ})^2\right) \right.$$
$$\left. + \frac{2\sqrt{J(J+2)}}{J+1}\left(t^+_{EJ}t^-_{MJ+1} - t^+_{MJ}t^-_{EJ+1}\right) \right\} \tag{128c}$$

$$\sqrt{2}(W^{TL'}_1)_{fi} \equiv U^{TL'} = \sqrt{2}\sum_{J \geq 1} \left\{ t^+_{CJ-1}\left(\sqrt{\frac{J-1}{J}}t^+_{EJ-1} - \sqrt{\frac{J+1}{J}}t^-_{MJ}\right) \right.$$
$$\left. - t^-_{CJ}\left(\sqrt{\frac{J+1}{J}}t^-_{EJ} + \sqrt{\frac{J-1}{J}}t^+_{MJ-1}\right) \right\} \ . \tag{128d}$$

Of course, results for specific final states $J^{\pi\prime}_f$ can be extracted from these sums. For example, we have for

$\underline{J^{\pi\prime}_f = \frac{1}{2}^+(C0, M1)}$ [Cf. Eqs. (125): elastic scattering from the nucleon; excitation $N \to N^*$ (1440), for example]

$$U^L = (t^+_{C0})^2$$
$$U^T = (t^-_{M1})^2$$
$$U^{T'} = -(t^-_{M1})^2$$
$$U^{TL'} = -2t^+_{C0}t^-_{M1}$$

$\underline{J^{\pi\prime}_f = \frac{1}{2}^-(C1, E1)}$ [$N \to N^*$ (1535), for example]

$$U^L = (t^-_{C1})^2$$
$$U^T = (t^-_{E1})^2$$
$$U^{T'} = -(t^-_{E1})^2$$
$$U^{TL'} = -2t^-_{C1}t^-_{E1}$$

$\underline{J^{\pi\prime}_f = \frac{3}{2}^+(M1, C2, E2)}$ [Cf. Eqs. (126): $N \to \Delta$ (1232), for example]

$$U^L = (t^-_{C2})^2$$
$$U^T = (t^+_{M1})^2 + (t^-_{E2})^2$$
$$U^{T'} = \frac{1}{2}\left\{(t^+_{M1})^2 - (t^-_{E2})^2 - 2\sqrt{3}t^+_{M1}t^-_{E2}\right\}$$
$$U^{TL'} = -t^-_{C2}(t^+_{M1} + \sqrt{3}t^-_{E2}) \ ,$$

etc. Alternatively, we can form four other linear combinations of the quantities in Eqs. (128):

$$U_1^\pm \equiv U^T \pm U^{T'} \tag{129a}$$

$$U_2^\pm \equiv \left\{ U^L + \frac{2}{3}\left(U^T - U^{T'}\right) \right\} \pm \frac{2}{\sqrt{3}} U^{TL'} . \tag{129b}$$

Either set $\{U^L, U^T, U^{T'}, U^{TL'}\}$ or $\{U_1^\pm, U_2^\pm\}$, is experimentally accessible in electron scattering when both the electrons and the spin-1/2 target are polarized. The set in Eqs. (129) has the advantage that

$$U_m^\pm \geq 0, \quad m = 1, 2 . \tag{130}$$

Several interesting properties may be obtained by examining Eqs. (128), (129):

(1) $U_1^+ = 0$ for $J_f^{\pi'} = \frac{1}{2}^+$ and $\frac{1}{2}^-$ contributions. Thus, in the rest of the sum which makes up U_1^+ we can focus on effects from $J_f \geq \frac{3}{2}$.

(2) Specific linear combinations can be formed where some dominant contribution is suppressed. For example, suppose we are studying the Δ-region where $(t_{M1}^+)^2$ is most important. Then in forming the combination $U^T - 2U^{T'} = \frac{1}{2}(3U_1^- - U_1^+)$ we have eliminated such terms and can focus on interferences $(t_{M1}^+) \times$ (something) and on contributions from other partial waves.

3. The *quadratic* combination

$$U_3 \equiv 2U^L(U^T - U^{T'}) - (U^{TL'})^2 \geq 0 \tag{131}$$

is of special interest. For any case where a single $J_f^{\pi'}$ partial wave is dominant, we have $U_3 = 0$. In fact, U_3 can only be non-zero by having cross terms between the responses from *different* partial waves. For example, suppose that only the $J_f^{\pi'} = \frac{1}{2}^+$ and $\frac{3}{2}^+$ partial waves are important (for instance, in the Δ-region with a $\frac{1}{2}^+$ background), then we find that

$$U_3 = \left\{ 2t_{M1}^- t_{C2}^- - t_{C0}^+(\sqrt{3}t_{E2}^- + t_{M1}^+) \right\}^2 ,$$

with interesting cross terms such as t_{M1}^+ (from the $\frac{3}{2}^+$) with t_{C0}^+ (from the $\frac{1}{2}^+$).

Before concluding this section, let us discuss some numbers of practical importance.

External Targets

Let us take a (large) external electron current

$$I_e^{ext} = 100\mu A = 6.2 \times 10^{14} \ e^-/s$$

and a "reasonable" target thickness (for decent resolution work)

$$t = 10 \text{ mg/cm}^2 = \frac{6.0}{A} \times 10^{21} \text{ nuclei/cm}^2$$

to obtain for the product the luminosity

$$\mathcal{L}^{\text{ext}} = \frac{38}{A} \times 10^{35} \text{ cm}^{-2} s^{-1} = \frac{33}{A} \times 10^{40} \text{ cm}^{-2} \text{ day}^{-1} \ .$$

With this we can set the scale for counting rates given the cross section in cm^2. Now if we wish to employ typical cryogenic polarized targets we have a problem: they can only tolerate $\sim$ 10's of nA! Thus, the luminosity is lower than the above number by four orders of magnitude.

Internal Targets

On the other hand with the new facilities being planned or built (Bates upgrade, CEBAF, EROS, etc.) which involve linac + stretcher ring combinations, we are discussing the situations where the electrons are effectively reused by having them circulate around the ring (typically making 1000 passes). So the internal electron current is

$$I_e^{\text{int}} = 1000 \times I_e^{\text{ext}} = 100 mA = 6.2 \times 10^{17} \ e^-/s \ .$$

If internal polarized targets are used (gas jets, containment bottles), then a canonical target thickness may be

$$t = 10^{15} \text{ nuclei/cm}^2 \ .$$

One would like higher densities, but the above number seems to be within the realm of possibility for the not-too-distant future. The luminosity in this case is

$$\mathcal{L}^{\text{int}} = 0.6 \times 10^{33} \text{ cm}^{-2} s^{-1} = 0.5 \times 10^{38} \text{ cm}^{-2} \text{ day}^{-1} \ ,$$

which, while smaller than $\mathcal{L}^{\text{ext}}$, is certainly in an interesting regime.

NeutrinoPhysics

For comparison, let us add the corresponding numbers for ν's. For a 5m cube of detector with mass 500 Mg we have a current of about $10^{10} - 10^{12}$ ν/s (the lower number for high-energy accelerators; the higher number for medium-energy high-intensity facilities such as LAMPF), we have a target thickness of 2.5 kg/cm^2 = 1.5×10^{27} nuclei/cm^2 and so a luminosity of $\mathcal{L}_\nu \sim 10^{37} - 10^{39}$ cm$^{-2} s^{-1}$ or $10^{42} - 10^{44}$ cm^{-2} day^{-1}.]

Many of the electromagnetic cross sections of interest are larger than 10^{-31} cm^2 sr^{-1} and range down to the lowest measured values of about 10^{-39} cm^2 sr^{-1}. (Neutrino cross sections for specific channels are typically about 10^{-40} cm^2, for comparison.) Thus, we see why polarized internal target electron scattering has become so topical: it appears that such studies will be feasible in the not-too-distant future.

SPECIAL TOPICS

We conclude these discussions by expanding a bit on two special aspects of exclusive-1 $(e, e'x)$ reactions.

The Reaction $(e, e'\gamma)$

Let us consider this purely electromagnetic coincidence reaction in a little more detail and show how, in fact, it has much in common with inclusive scattering from polarized nuclei. A more complete exposition of this subject may be found in Ref. 23.

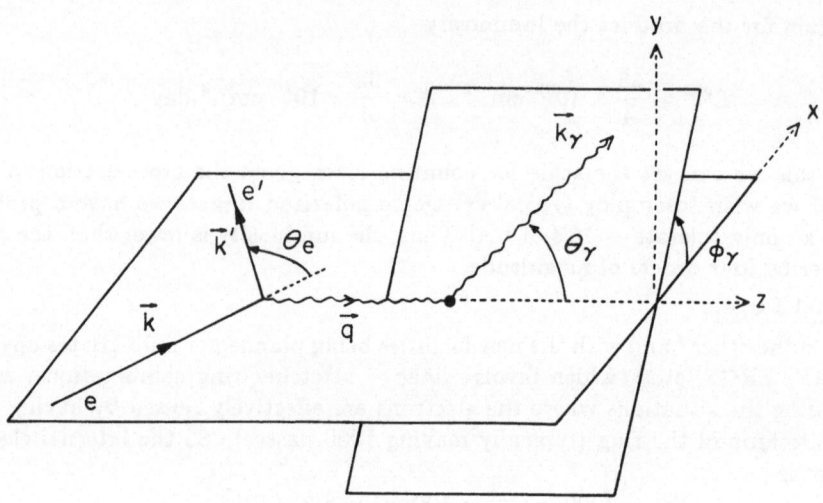

Fig. 32:   Kinematics for the $(e, e'\gamma)$ reaction.

The general kinematics for the process are shown in Fig. 32. To be specific, let us consider the sequence of steps involving discrete states, $i \xrightarrow{ee'} n \xrightarrow{\gamma} f$, illustrated in Fig. 33. Furthermore, we shall only consider the nuclear process and ignore effects of electron bremsstrahlung which may interfer with this process. Referring to the kinematical situation shown in Fig. 32, we have the real photon detected at angles $(\theta_\gamma, \phi_\gamma)$ with respect to the direction of the momentum transfer $\vec{q}$ and to the electron scattering plane specified by $\vec{k}$ and $\vec{k}'$ as indicated. In addition, the angular distribution function is labelled with the electron helicity and the real-photon circular polarization. We can write for the general cross section

$$\left(\frac{d^2\sigma}{d\Omega_e d\Omega_\gamma}\right)^{h,\sigma}_{i \to n \to f} \propto \frac{1}{2J_i + 1} \sum_{M_{J_i}, M_{J_f}}$$

$$\left| \sum_{M_{J_n}} \langle J_f M_{J_f} | H_\gamma(\Omega_\gamma, \sigma) | J_n M_{J_n} \rangle \langle J_n M_{J_n} | H_{ee'}(\vec{q}, h) | J_i M_{J_i} \rangle \right|^2 , \tag{132}$$

where we are taking the initial and final nuclear states to be unpolarized and so must average over initial and sum over final as indicated. Here, $H_\gamma(\Omega_\gamma, \sigma)$ is the interaction Hamiltonian for the emission of a (real) photon with circular polarization σ into the solid angle Ω_γ, while $H_{ee'}(\vec{q}, h)$ is the interaction Hamiltonian for the electro-excitation of the nucleus for a 3-momentum transfer $\vec{q}$ from the electron. The intermediate state need not have its magnetic substates uniformly populated by the excitation process, and so in general we are dealing with a polarized intermediate state; this fact is the underlying basis for the fundamental connection between the $(e, e'\gamma)$ process and the scattering of electrons from polarized nuclei. For this reason, we must label the nuclear states with their M_J values as well as their angular momentum and parity quantum numbers.

While we will not go into any detail concerning the derivation of the results of the rest of this section, we now briefly summarize the main aspects of the calculation,[23] which parallel the developments of Section IV and Ref. 11. The electro-excitation part of the process $i \to n$ involves $H_{ee'} \sim j_\mu J^\mu$, where j_μ and J^μ

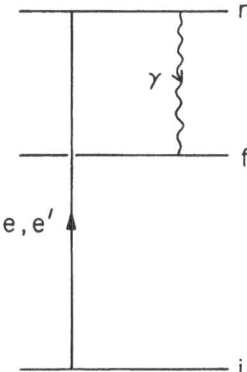

Fig. 33: Labelling of levels in discussing the $(e, e'\gamma)$ reaction.

are the electron and nuclear currents, respectively. Then

$$|H_{ee'}|^2 \sim \eta_{\mu\nu}\overline{W}^{\mu\nu}_{n'n} \quad , \tag{133}$$

where the electron tensor $\eta_{\mu\nu}$ was discussed in Section II for longitudinally polarized electrons in the extreme relativistic limit and the nuclear tensor is given by

$$\overline{W}^{\mu\nu}_{n'n} \sim \left(J^\mu_{n'i}(\vec{q})\right)^* \left(J^\nu_{ni}(\vec{q})\right) \tag{134}$$

where n' and n here refer to the $M_{J_{n'}}$ and M_{J_n} values required for Eqs.(132). Similarly, $H_\gamma \sim J_\mu A^\mu(\vec{k}_\gamma, \sigma)$, where J_μ is the nuclear current for the transition $n \to f$ and A^μ is the 4-potential of the real photon (in a frame where $A^0 = 0$). Then, $\vec{A} \sim \vec{e}(\vec{k}_\gamma; 1\sigma)e^{i\vec{k}_\gamma \cdot \vec{x}}$ in terms of the standard unit spherical vectors, and so

$$|H_\gamma|^2 \sim J_{fn'}(\vec{k}_\gamma; \sigma)^* J_{fn}(\vec{k}_\gamma; \sigma) \quad , \tag{135}$$

where the $J(\vec{k}_\gamma; \sigma)$ are the transverse spherical vector components of the nuclear current relative to $\vec{k}_\gamma$.

The remainder of the analysis of the excitation part of the process is similar to that of Section IV and Ref. 11. The nuclear current components are expanded in terms of matrix elements of the electromagnetic multipole operators, for which the M_J-dependence is extracted via the Wigner-Eckart theorem, leaving the usual reduced matrix elements. As far as the real-photon decay process is concerned, the corresponding nuclear states are specified relative to a coordinate system defined by the direction $\vec{q}$, and so the electromagnetic multipole operators, which are referred to $\vec{k}_\gamma$, must be rotated. The required rotation matrices involve the angles $(\theta_\gamma, \phi_\gamma)$, and the evaluation of all of the sums over the magnetic quantum numbers can be performed, leaving spherical harmonics and sums over J-values of bilinear combinations of the multipole matrix elements for the real-photon decay.

Upon carrying out the developments sketched above, we arrive at the double-differential cross sections for the exclusive (coincidence) reactions $(e, e'\gamma)$ and $(\vec{e}, e'\vec{\gamma})$, which may be written in the form

$$\left(\frac{d^2\sigma}{d\Omega_e d\Omega_\gamma}\right)^{h,\sigma}_{i \to n \to f} = \left\{ \frac{\Gamma^{n \to f}_\gamma}{\Gamma^{n \to anything}_{total}} \right\} \Sigma^{ni}_0 W^{h,\sigma}(\theta_\gamma, \phi_\gamma) \quad , \tag{136}$$

where $\Gamma_\gamma^{n\to f}$ is the γ-decay width for the transition $n \to f$ and $\Gamma_{\text{total}}^{n\to\text{anything}}$ is the total decay width for the state n. Of course, the ratio of these widths is frequently of order unity. The cross section is labelled with the helicity of the incident electron, $h = \pm 1$, and with the circular polarization of the detected (real) photon, $\sigma = \pm 1$. The other factors in Eq. (136) are the unpolarized (e, e') cross section,

$$\Sigma_0^{ni} = \left(\frac{d\sigma}{d\Omega_e}\right)_{(e,e')}^{\text{unpol.}} = 4\pi\sigma_M f_{\text{rec}}^{-1} F^2(q, \theta_e)^{ni} \ , \tag{137}$$

and an angular distribution function $W^{h,\sigma}(\theta_\gamma, \phi_\gamma)$ to be discussed below, where we have suppressed the dependence of $W^{h,\sigma}$ on the 4-momentum transfer Q for clarity.

Equation (136) is the basic factorization of the physics content of the problem into three contributions. Σ_0^{ni}, being the unpolarized (e, e') cross section, contains the overall probability for electro-exciting the state in question (transition $i \to n$). The ratio of the widths expresses the probability for having the particular γ-ray transition $n \to f$. The new physics (i.e., beyond just (e, e') or γ-decay separately) is contained in the angular distribution function $W^{h,\sigma}(\theta_\gamma, \phi_\gamma)$. As we shall see below, this function depends both on the excitation mechanism (through the q-dependence in the electron scattering form factors and the q- and θ_e-dependence involved in the electron kinematics) and on the decay process (through the mixing ratios for the γ-decay and the dependence on the angles θ_γ and ϕ_γ).

Proceeding now one step further with the analysis, the electron helicity and the real-photon circular polarization dependence of the angular distribution function can be made explicit (following the ground-work laid in Ref. 11 and sketched above). It turns out that, with the approximations which we have made, $W^{h,\sigma}(\theta_\gamma, \phi_\gamma)$ involves two terms, one which is independent of h and σ and a second one which is proportional to the product $h\sigma$. Thus, we are led to the fact that $W^{h,\sigma}(\theta_\gamma, \phi_\gamma)$ has an explicit (h, σ)-dependence of the form

$$W^{h,\sigma}(\theta_\gamma, \phi_\gamma) = \frac{1}{2}\{W_\Sigma(\theta_\gamma, \phi_\gamma) + h\sigma W_\Delta(\theta_\gamma, \phi_\gamma)\} \ , \tag{138}$$

where the meanings of the subscripts Σ and Δ will become clear in due course; note that the angular distribution functions W_Σ and W_Δ do not have any h- or σ-dependence.

Before considering W_Σ and W_Δ in any more detail, we proceed with a more general discussion of some of the experimental considerations which follow from Eq. (138). We first note that several types of experiments are possible:

Electrons and Photon Unpolarized

$$\left(\frac{d^2\sigma}{d\Omega_e d\Omega_\gamma}\right)_{i\to n\to f}^{\text{unpol.}} = \frac{1}{2} \sum_{h=\pm 1} \sum_{\sigma=\pm 1} \left(\frac{d^2\sigma}{d\Omega_e d\Omega_\gamma}\right)_{i\to n\to f}^{h,\sigma} \tag{139a}$$

$$= \left\{\frac{\Gamma_\gamma^{n\to f}}{\Gamma_{\text{total}}^{n\to\text{anything}}}\right\} \Sigma_0^{ni} W_\Sigma(\theta_\gamma, \phi_\gamma) \ , \tag{139b}$$

and so we are only able to measure the W_Σ angular distribution function in the absence of any electron and photon polarizations.

Electrons Polarized and Decay Photon Polarization Measured

For this case, we may define a polarization-difference cross section

$$
\left(\frac{d^2\sigma}{d\Omega_e d\Omega_\gamma}\right)^{\text{pol.}}_{i\to n\to f} \equiv \left(\frac{d^2\sigma}{d\Omega_e d\Omega_\gamma}\right)^{h=+1,\sigma=+1}_{i\to n\to f} - \left(\frac{d^2\sigma}{d\Omega_e d\Omega_\gamma}\right)^{h=-1,\sigma=+1}_{i\to n\to f} \tag{140a}
$$

$$
= \left(\frac{d^2\sigma}{d\Omega_e d\Omega_\gamma}\right)^{h=+1,\sigma=+1}_{i\to n\to f} - \left(\frac{d^2\sigma}{d\Omega_e d\Omega_\gamma}\right)^{h=+1,\sigma=-1}_{i\to n\to f} \tag{140b}
$$

with other obvious combinations of terms with $h = \pm 1$ and $\sigma = \pm 1$ possible, but not displayed here. From Eqs. (136) and (138), we have that

$$
\left(\frac{d^2\sigma}{d\Omega_e d\Omega_\gamma}\right)^{\text{pol.}}_{i\to n\to f} = \left(\frac{\Gamma^{n\to f}_\gamma}{\Gamma^{n\to \text{anything}}_{\text{total}}}\right) \Sigma^{ni}_0 W_\Delta(\theta_\gamma,\phi_\gamma) \ . \tag{140c}
$$

Thus, by having knowledge of **both** the electron's helicity and the decay photon's polarization, we can isolate the Δ contributions to the angular distribution function, as well as the contributions from the Σ pieces.

Electrons Polarized, but no Decay Photon Polarization Measured

This involves the sum

$$
\sum_{\sigma=\pm 1} W^{h,\sigma}(\theta_\gamma,\phi_\gamma) = W_\Sigma(\theta_\gamma,\phi_\gamma) \tag{141}
$$

and so leads to the same result as the completely unpolarized case (i); that is, no Δ contributions are present.

Electrons Unpolarized, but Decay Photon Polarization Measured

The sum in this case ("average-over-initial", in fact) is

$$
\frac{1}{2}\sum_{h=\pm 1} W^{h,\sigma}(\theta_\gamma,\phi_\gamma) = \frac{1}{2}W_\Sigma(\theta_\gamma,\phi_\gamma) \tag{142}
$$

and as in case (iii) above is proportional to the unpolarized angular distribution function.

Thus, there are really only two distinct classes of experiments possible here. One type has no polarizations specified and allows W_Σ to be measured, while the second type has **both** electrons and decay photon polarizations specified and allows a measurement of W_Δ in addition. In other words, we are interested in the reactions $(e, e'\gamma)$ and $(\vec{e}, e'\vec{\gamma})$, respectively. On the other hand, the reactions $(\vec{e}, e'\gamma)$ and $(e, e'\vec{\gamma})$ do not lead to new information beyond that obtained from the simpler unpolarized measurements. Two ingredients go into this result. The first involves the assumption that parity is conserved; it if is not conserved (as is true at the level of the weak interaction), then, for example, measurements with only polarized electrons will be of interest. Secondly, the real photon was taken to be a plane wave and so time-reversal invariance and hermiticity could be invoked. If final-state interactions are included for the outgoing photon, the above simplifications are not obtained. This discussion may be placed within the more general context of exclusive-1 electron scattering $(e, e'x)$ and $(\vec{e}, e'x)$, as discussed in the next subsection. When particle x experiences a final-state interaction, then the polarized electron coincidence cross section involves the "fifth response function" which does

not occur in the absence of electron polarization (only four response functions occur in this latter case). With x = hadron as in $(e, e'p)$ or $(\vec{e}, e'p)$ studies, final-state
interactions are generally important; of course, with x = photon the final-state interactions are usually rather weak and so we have neglected them throughout this
work.

Returning to a more detailed discussion of $W^{h,\sigma}(\theta_\gamma, \phi_\gamma)$, we recall that we have
found that the entire problem can be expressed in terms of the two angular dis-
tribution functions, $W_\Sigma(\theta_\gamma, \phi_\gamma)$ and $W_\Delta(\theta_\gamma, \phi_\gamma)$. Having factored the cross section
as given in Eq. (136), these two functions may be shown to satisfy the integral
conditions

$$\int d\Omega_\gamma W_\Sigma(\theta_\gamma, \phi_\gamma) = 1 \tag{143a}$$

and

$$\int d\Omega_\gamma W_\Delta(\theta_\gamma, \phi_\gamma) = 0. \tag{143b}$$

Following the developments outlined at the beginning of this section, we find that
the angular distribution functions may be expressed directly in terms of the response
functions which enter into discussions of electron scattering from polarized targets
(Ref. 11). In that previous study, and in Sec. IV, we found that the $(\vec{e}, e')$ cross
section could be written in the form

$$\left(\frac{d\sigma}{d\Omega_e}\right)^h_{ba} = \Sigma_{ba} + h\Delta_{ba} , \tag{144}$$

for inclusive electron scattering with incident electron helicity h and involving the
nuclear transition $a \to b$. Specifically, when the target was polarized but the final
nuclear polarization was not measured, we used the notation $\Sigma_{b/a}, \Delta_{b/a}$, etc., with a
slash to indicate which nuclear state (initial and/or final) did not have its polarization specified. The kinematical situation for this type of experiment is shown in
Fig. 15 where the angles (θ^*, ϕ^*) specify the orientation of the axis of polarization
relative to the momentum transfer $\vec{q}$ (we use exactly the same conventions in Fig.
32 above.) The cross section for this process contains the (θ^*, ϕ^*) dependence in
the following form

$$\Sigma_{b/a} = \Sigma_0^{ba}\left[1 + \sum_{\substack{J \geq 2 \\ \text{even}}} \left(P_J(\cos\theta^*)R_J^0(q, \theta_e)_{b/a} + P_J^1(\cos\theta^*)\cos\phi^* R_J^1(q, \theta_e)_{b/a}\right.\right.$$

$$\left.\left. + P_J^2(\cos\theta^*)\cos 2\phi^* R_J^2(q, \theta_e)_{b/a}\right)\right] \tag{145a}$$

and

$$\Delta_{b/a} = \Sigma_0^{ba}\left[\sum_{\substack{J \geq 1 \\ \text{odd}}} \left(P_J(\cos\theta^*)R_J^0(q, \theta_e)_{b/a} + P_J^1(\cos\theta^*)\cos\phi^* R_J^1(q, \theta_e)_{b/a}\right)\right] \tag{145b}$$

where Σ_0^{ba} is the usual unpolarized (e, e') cross section defined in Eq. (137).

The remaining polarization dependence (*i.e.*, beyond that due to the angles
(θ^*, ϕ^*)) in the cross section is contained in the polarization tensors $R_J^M(q, \theta_e)_{b/a}$.
These quantities can in turn be rewritten in terms of the reduced response functions
used in Section IV:

$$J = \text{even} \quad R_J^0(q, \theta_e)_{b/a} = f_J^{(a)}\left(v_L W_J^L(q)_{b/a} + v_T W_J^T(q)_{b/a}\right)\Big/ F^2(q, \theta_e)^{ba} \tag{146a}$$

$$R_J^1(q,\theta_e)_{\not b a} = f_J^{(a)} v_{TL} W_J^{TL}(q)_{\not b a}/F^2(q,\theta_e)^{ba} \quad, \tag{146b}$$

$$R_J^2(q,\theta_e)_{\not b a} = f_J^{(a)} v_{TT} W_J^{TT}(q)_{\not b a}/F^2(q,\theta_e)^{ba} \quad, \tag{146c}$$

$$J = \text{odd} \quad R_J^0(q,\theta_e)_{\not b a} = f_J^{(a)} v_T' W_J^{T'}(q)_{\not b a}/F^2(q,\theta_e)^{ba} \quad, \tag{146d}$$

$$R_J^1(q,\theta_e)_{\not b a} = f_J^{(a)} v_{TL}' W_J^{TL'}(q)_{\not b a}/F^2(q,\theta_e)^{ba} \quad, \tag{146e}$$

where $F^2(q,\theta_e)^{ba} = f_0^{(a)}\left(v_L W_0^L(q)_{\not b a} + v_T W_0^T(q)_{\not b a}\right)$ and so $R_0^0(q,\theta_e)_{\not b a} = 1$. Furthermore, as shown in Section IV, if the polarization of the final nuclear state is specified, rather than that of the target, then a useful "turn-around" relation can be used:

$$R_J^M(q,\theta_e)_{a\not b} = (-)^{J+M} R_J^M(q,\theta_e)_{\not b a} \quad; \tag{147}$$

of course, on the left-hand side of the equation we now presume that the initial state is b (unpolarized $\leftrightarrow$ $\not b$) and the final state is a (polarization specified $\leftrightarrow$ a).

In particular, let us choose $a = n$ and $b = i$ in Eq. (147); then $R_J^M(q,\theta_e)_{n\not i}$ may be related directly to $R_J^M(q,\theta_e)_{\not i n}$ using this turn-around relation. To stress the polarization dependence, let us write

$$(R_J^M)_{\not i n} = R_J^M\left(\theta^*,\phi^*;\left\{f_J^{(n)}\right\}\right)_{\not i n} \quad,$$

$$\Sigma_{\not i n} = \Sigma_{\not i n}\left(\theta^*,\phi^*;\left\{f_J^{(n)}\right\} \quad, J = \text{even}\right) \quad,$$

and

$$\Delta_{\not i n} = \Delta_{\not i n}\left(\theta^*,\phi^*;\left\{f_J^{(n)}\right\} \quad, J = \text{odd}\right) \quad,$$

with a completely analogous notation when $\not i n \to n \not i$, where we have suppressed the dependence on the electron kinematic variables for clarity. In all cases, $0 \leq J \leq 2J_n$.

We are now in a position to relate the angular distribution functions $W_\Sigma(\theta_\gamma,\phi_\gamma)$ and $W_\Delta(\theta_\gamma,\phi_\gamma)$ to the polarization quantities discussed above. The steps outlined at the beginning of this section lead to the following basic connection formulas:

$$W_\Sigma(\theta_\gamma,\phi_\gamma) = \frac{\Sigma_{n\not i}\left(\theta_\gamma,\phi_\gamma;\left\{f_J^{(n\to f+\gamma)}\right\} \quad, J = \text{even}\right)}{4\pi\Sigma_0^{ni}} \quad, \tag{148a}$$

$$W_\Delta(\theta_\gamma,\phi_\gamma) = \frac{\Delta_{n\not i}\left(\theta_\gamma,\phi_\gamma;\left\{f_J^{(n\to f+\gamma)}\right\} \quad, J = \text{odd}\right)}{4\pi\Sigma_0^{ni}} \quad, \tag{148b}$$

where now the reason for the notation W_Σ and W_Δ is apparent. In other words, the angular distribution functions are obtained using the inclusive electron scattering response functions, but with the polarization angles (θ^*,ϕ^*) replaced by the real-photon angles $(\theta_\gamma,\phi_\gamma)$ and the Fano tensors in the former case replaced by new quantities which are specified by the nature of the γ-decay, $n \to f + \gamma$ (see below). Note that the appropriate quantities are those which correspond to electron scattering from an **unpolarized** target $\not i$, but where the **final** nuclear polarization is measured (n). In effect, measuring the angular distribution of the decay photon amounts to analyzing the polarization of the intermediate state n.

All that remains is the specification of the effective Fano tensors which enter into Eqs. (146). These contain the rest of the γ-decay information (*i.e.*, beyond the angular dependence we have already considered). We find the effective Fano tensors are given by the following simple identities:

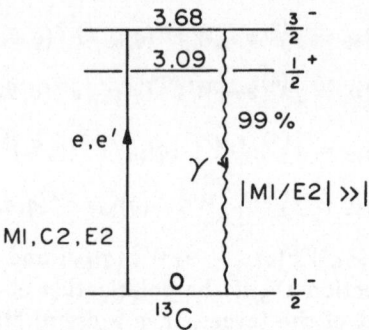

Fig. 34:   Specific example of the $(e, e'\gamma)$ reaction in ^{13}C.

$$f_J^{n \to f + \gamma} = \frac{1}{\sqrt{2J_n + 1}} \begin{cases} \left(\dfrac{W_J^T(q)_{fn}}{W_0^T(q)_{fn}} \right)_{q=\omega=E_\gamma} &, \quad \text{if } J \text{ is even} \\[4mm] \left(\dfrac{W_J^{T'}(q)_{fn}}{W_0^T(q)_{fn}} \right)_{q=\omega=E_\gamma} &, \quad \text{if } J \text{ is odd.} \end{cases} \tag{149}$$

In effect, it is as though we are again considering electron scattering from a polarized nucleus (state n), in this case to a lower-energy state (f), and evaluating the results on the real-photon line $q = \omega = E_\gamma$. Of course, once again all of the formalism of Ref. 11 may immediately be brought to bear on the problem, allowing the determination of the $W_J^K(q)_{fi}$. It should be noted that the results in Eqs. (149) involve evaluating the transition form factors for the transition $n \to f$ in the long-wavelength limit (*i.e.*, $q = \omega = E_\gamma$ is small compared to the typical nuclear momentum scale of about $1 \text{ fm}^{-1} \approx 200 \text{ MeV/c}$). Under such circumstances, usually only the lowest-multipolarity operators will be important, and hence the low values of J will usually correspond to larger values of $|f_J|$ than obtained when J is high. For example, if the γ-decay could proceed via the $E1$, $M2$, $E3$ and $M4$ multipoles (for instance when $J_n^{\pi_n} = \frac{5}{2}^-$ and $J_f^{\pi_f} = \frac{3}{2}^+$), then in principle J can take on a wide range of values (in this example, $0 \le J \le 2J_n = 5$). However, generally the $E1$ decay will be dominant and so only the terms with $J = 0$, 1 and 2 will be important; the f_J's with higher values of J will be suppressed. Thus, the dominance of the allowed real-photon multipoles will make it difficult to access the information in the complete $(e, e'\gamma)$ cross section contained in the high-J parts of the angular distribution functions. In contrast, inclusive studies with polarized targets do not have this low-q multipole ordering and so in general have important contributions from all values of J.

Let us consider just one example (more complete discussions are contained in Ref. 23) In Fig. 34 we give the level scheme for a particular $(e, e'\gamma)$ reaction in ^{13}C. For this $\frac{1}{2}^- \to \frac{3}{2}^- \to \frac{1}{2}^-$ sequence we need the reduced response functions for inclusive inelastic electron scattering where the $\frac{3}{2}^-$ state is polarized, *i.e.*, exactly the results given in Eqs. (121). Using the "turn-around" relations in Eqs. (147) and the rest of the results summarized in this subsection, we immediately have the angular distribution function for the $(e, e'\gamma)$ reaction. Some typical results are given in Fig. 35: again large (measurable) asymmetries are predicted for such implicit polarization measurements.

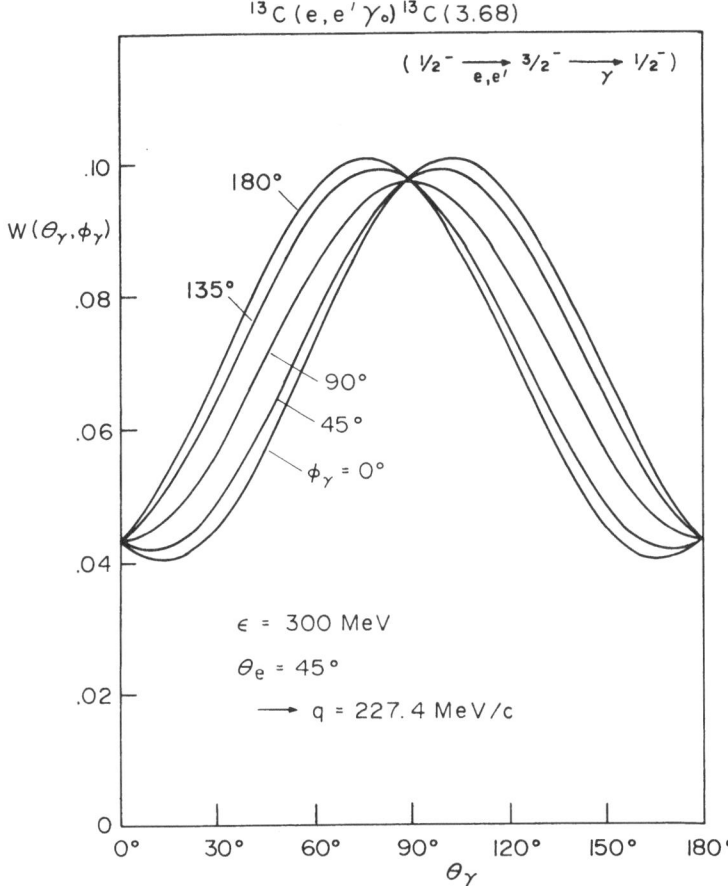

Fig. 35:   Angular distribution function $W(\Omega_\gamma)$.

The "Fifth" Response Function in $(\vec{e}, e'x)$Reactions

Finally, let us conclude with a brief discussion of this special response. If we do not have polarized electrons at our disposal we may study the four response functions given in Eq. (23) when considering exclusive-1 electron scattering. With electron polarization, a fifth response function (Eq. (24)) now becomes accessible. Note that the difference (electron-helicity-difference) cross section is in general not zero for exclusive (coincidence) electon scattering. This is to be contrasted with inclusive scattering where such terms vanish, as long as parity conservation is assumed. Consequently, measurements of such difference cross sections in the latter case are sensitive to parity-violating effects involving the weak interaction. The arguments which led to the usual inclusive electron scattering cross section assumed only the presence of the parity-conserving electromagnetic interaction; when both electromagnetic and weak interactions are taken into account, new terms enter which do not in general vanish, leading to antisymmetric response functions and non-zero helicity-difference cross sections. While such effects should also occur for exclusive scattering, they will generally be overwhelmed by the non-zero helicity dependence that occurs even when parity is conserved.

For the electron-helicity-difference cross section we have

$$^{(1)}\Delta \sim v_{TL'} \sin\phi_1\, {}^{(1)}W^{TL'} \quad , \tag{150}$$

requiring the detection of particle x in $(\vec{e}, e'x)$ out of the electron scattering plane. In contrast, the TL part of the electron-spin-averaged cross section may be isolated using its unique dependence on $\cos\phi_1$; we have

$$^{(1)}\Sigma \sim v_{TL} \cos\phi_1\, {}^{(1)}W^{TL} \quad . \tag{151}$$

As we saw in Eqs. (87) that the TL and TL' contributions come from calculating

$$Re(T^*L)\cos\phi_1 \quad , \quad \text{for } TL \tag{152a}$$

$$Im(T^*L)\sin\phi_1 \quad , \quad \text{for } TL' \tag{152b}$$

respectively, and so the former involves the real part of an interference between transverse and longitudinal (charge) components of the currents, whereas the latter involves a similar imaginary part; that is to say $^{(1)}W^{TL} \sim Re(T^*L)$, while $^{(1)}W^{TL'} \sim Im(T^*L)$. Now, if the $(e, e'x)$ reaction proceeds through a channel in which a single phase dominates for all projections of the current $(T \sim |T|e^{i\delta},\ L \sim |L|e^{i\delta}$, with the same $\delta)$, then T^*L is real and $^{(1)}W^{TL}$ may be non-zero while $^{(1)}W^{TL'}$ vanishes. Moreover, it happens that $^{(1)}W^{TL'}$ also vanishes in the absence of final-state interactions. Therefore, if $^{(1)}W^{TL'} \neq 0$, then interesting effects must be coming into play. For example, in the Δ-region, coincidence electron scattering will be driven to a large extent by the 33-amplitude (i.e., with one phase, δ_{33}) and, while $^{(1)}W^L$, $^{(1)}W^T$, $^{(1)}W^{TT}$ and $^{(1)}W^{TL}$ may all be non-zero, $^{(1)}W^{TL'}$ may be expected to vanish. To the extent that it does not vanish, one will be measuring the interferences of the 33-amplitude with amplitudes for some other channels and thus addressing interesting physics questions.

While calculations of coincidence $(\vec{e}, e'x)$ cross sections are in progress[24] and not yet so fully developed that it is possible to say where the most interesting addressable problems lie, it is nevertheless true that the preliminary results already suggest that the fifth response function is interestingly different from the other four. The special sensitivity to the nature of the final (unbound) nuclear state is clearly a desirable feature. Let us briefly consider the various regions of coincidence electron scattering (say $(\vec{e}, e'p)$ reactions) and what might be learned from the fifth response function. In Fig. 36, the general situation is sketched. At low excitation energies, roughly where the giant resonances occur, we have various competing reaction mechanisms involving different nuclear final states: for example, we have features such as the giant electric dipole (GDR), the giant electric dipole spin-flip (GDSR) or the giant electric quadrupole (GQR) resonances, at medium-to-high q the high-spin particle-hole states and underlying all of these a "direct" or "non-resonant" background. Interferences can occur between such structures and appear as important strength in the fifth response function. Referring to Eq. (150) we see that $^{(1)}\Delta$ is maximized by choosing $\phi_1 = 90°$. That is, the plane in which the proton is detected will be perpendicular to the electron scattering plane (see Fig. 6). Ideally, an array of proton detectors would be situated at various angles θ_1 in this plane, going from the direction $\vec{q}$ up through the normal to the electron scattering plane $(\vec{u}_y)$, and back to $-\vec{q}$; such a set-up might be referred to as the "Mohawk geometry". With an additional similar array of detectors also in the perpendicular

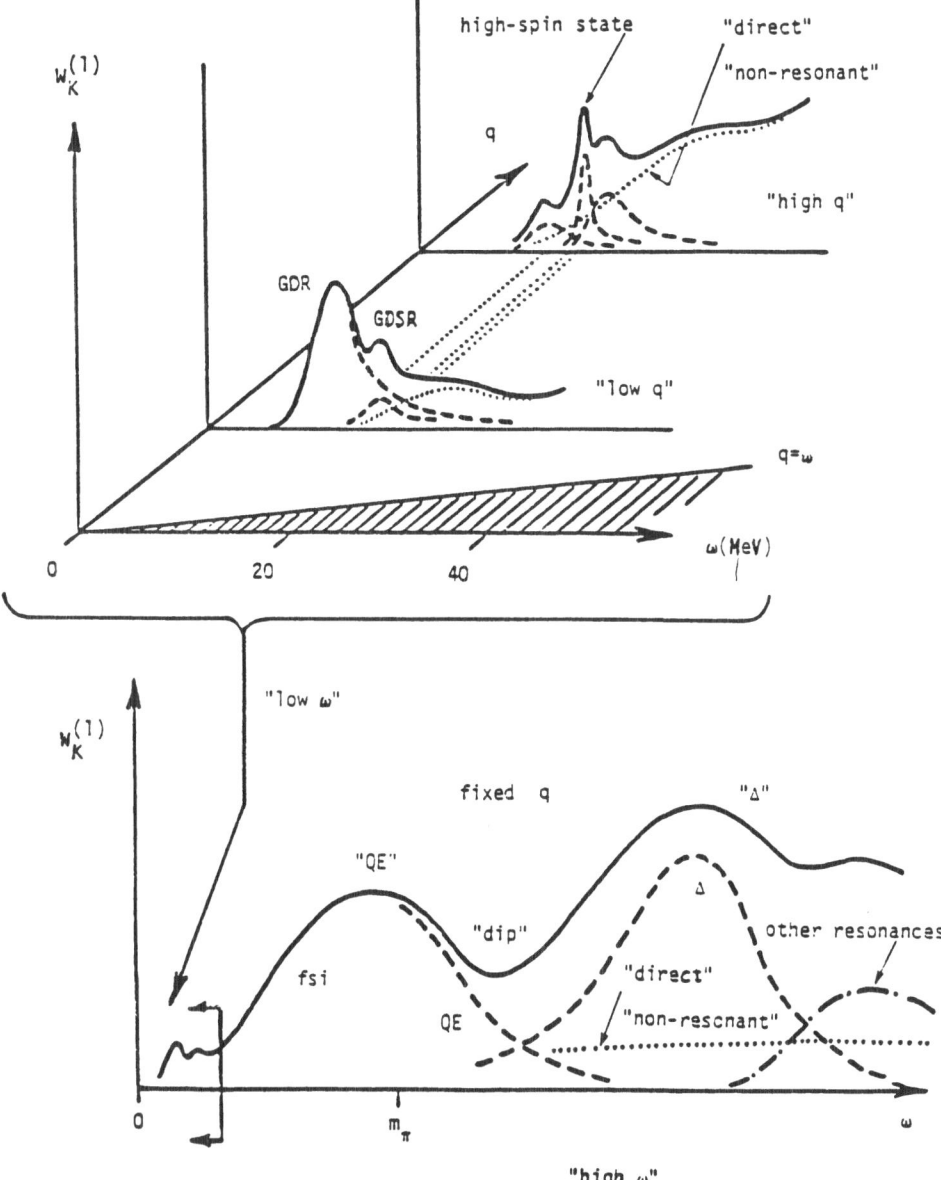

Fig. 36: Schematic representation of electromagnetic nuclear response functions for different regions in the (q, ω)-plane.

plane, but now with $\phi_1 = -90°$, the set-up could be called the "reflected Mohawk geometry". Without going into detail about what is interferring with what, it is still clear from initial studies that $^{(1)}\Delta$ contains different information from $^{(1)}\Sigma$ and that both are large enough in magnitude to be measured in practical experiments. In the latter, a resonance such as the GQR which is important in this region of q and ω can interfere with itself, where as in the former it cannot – only interferences of *something transverse with something else longitudinal* can occur in the fifth response function.

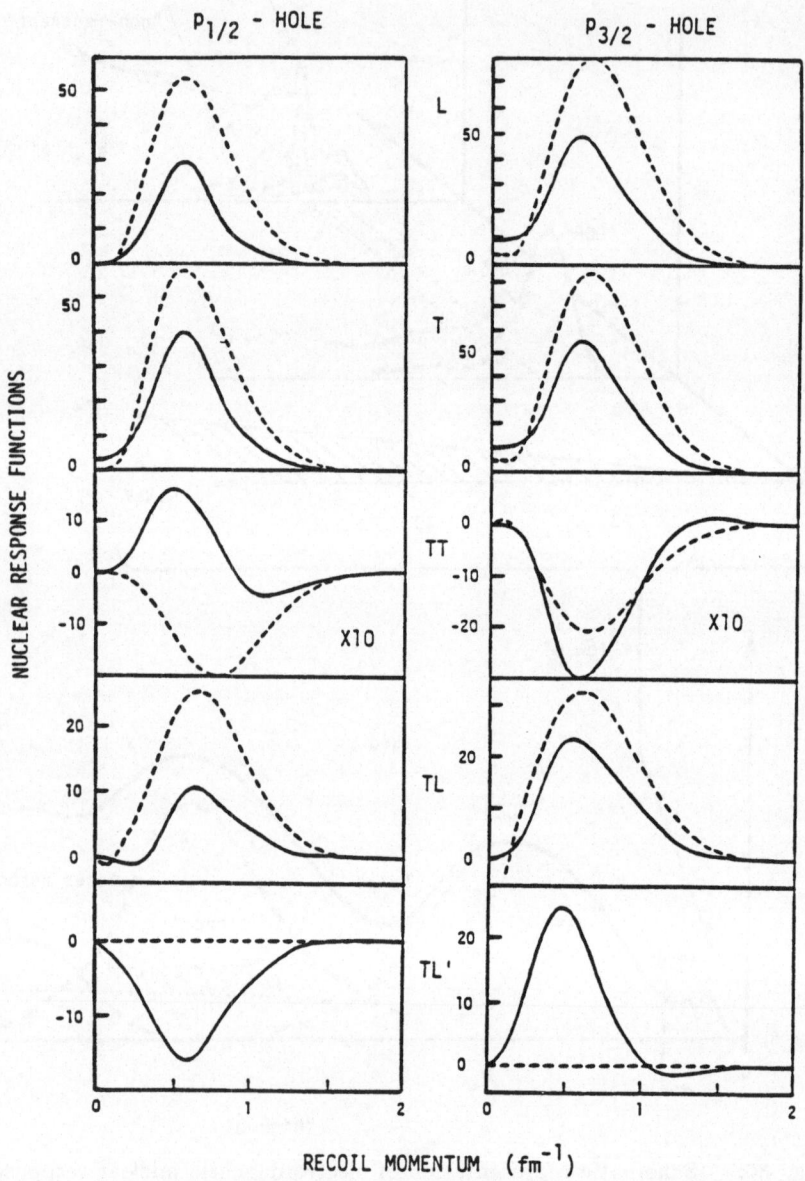

Fig. 37:   Nuclear response functions (fm^3) in the quasi-free scattering region for the reactions $^{16}0(e,e'p)^{15}N$ and $^{16}0(\vec{e},e'p)^{15}N$, redrawn from Ref. 25. The solid lines show results where final-state interactions are included; the dashed lines come from using plane-wave outgoing protons. Note that the major differences in shape between the $p_{1/2}$ and $p_{3/2}$ hole cases occur for the TL' (the "fifth response function") and TT interference terms; however, the latter are very small here (they have been multiplied by a factor of 10).

At higher excitation energies other reaction mechanisms come into play (see Fig. 36). In the quasi-elastic (QE) region, for instance, the primary focus may be on the direct final-state interactions felt by the outgoing intermediate-energy proton,[25] in contrast to the specific doorways through which the reaction proceeds in the giant resonance region. An example of what might be expected is shown in Fig. 37. Again, the fifth response function is seen to be relatively large even at the quasi-elastic peak.

At even higher excitation energies the threshold for pion production is exceeded and still other reaction mechanisms begin to play a role. While no calculations are yet available in the Δ-region, this might be one of the most interesting for studies with polarized electrons. In $^{(1)}\Sigma$ the delta is so dominant that other effects (the QE tail, tails of other higher-lying resonances, "direct, non-resonant" backgrounds, etc.) are masked. On the other hand, in $^{(1)}\Delta$ only interferences can occur and so these interesting effects may become accessible.

REFERENCES

References 1 – 7 are various conference proceedings where particular aspects of the general ideas presented here have been put forward previously. Some of this material was assembled for other Summer Schools and may be especially useful to students.

1. T. W. Donnelly, *Electromagmetic and Weak Currents in Nuclei* in "Symmetries in Nuclear Structure," (K. Abrahams, K. Allaart and A. E. L. Dieperink, eds.) p. 1, Plenum Press, New York, 1983. [Basic formalism done primarily for students; done in the Bjorken and Drell metric.]

2. T. W. Donnelly, *Considerations of Polarization in Electron Scattering*, in "Perspectives in Nuclear Physics at Intermediate Energies," (S. Boffi, C. Ciofi degli Atti and M. M. Giannini, eds.) p. 24, World Scientific Publishing Co., Trieste, Italy, 1984. [Basic formalism for inclusive scattering from polarized targets.]

3. T. W. Donnelly, *Physics with Polarized Electrons and Targets*, at the Workshop on Polarized Targets in Storage Rings, Argonne National Laboratory (ANL-84-50), (May, 1984), 17. [Extension of Ref. 2 with more detail on multipole counting, etc.]

4. T. W. Donnelly, *Electron Scattering and Neutrino Reactions in Nuclei*, in the International School for Nuclear Physics "Nuclear and Subnuclear Degrees of Freedom and Lepton Scattering," Erice, Sicily, (April, 1984); *Progress in Particle and Nuclear Physics* **13**, 183 (1985). ["Super-Rosenbluth" formula, etc.; includes discussion of ν reactions in nuclei.]

5. T. W. Donnelly, *Coincidence and Polarization Measurements with High-Energy Electrons*, at the CEBAF Workshop (June, 1984). [Electromagnetic part of Ref. 4, somewhat condensed; includes some discussion of "fifth" response function.]

6. T. W. Donnelly, *Super-Rosenbluth Physics*, at the BUTG Workshop "Nucleon and Nuclear Structure and Exclusive Electromagnetic Reaction Studies," Massachusetts Institute of Technology (July, 1984). ["Reader's Digest" version of Ref. 5.]

7. T. W. Donnelly, *Polarized Electron Scattering from Polarized Nuclei*, at the 6th Seminar "Electromagnetic Interactions of Nuclei at Low and Medium Energies," Moscow, USSR (December, 1984). [Specific focus on polarized electrons with polarized targets.]

8. T. W. Donnelly, *Photo- and Electro-Production of Kaons and the Study of Hypernuclei,* at the Workshop on Electron and Photon Interactions at Medium Energies, Bad Honnef, F. R. Germany (October, 1984) (to be published in "Lecture Notes in Physics," Springer-Verlag; MIT preprint CTP #1226. [Specific $(e, e'x)$ study, with $x = K^+$.]

9. T. W. Donnelly, *New Physics with Electrons,* at the CEBAF Workshop (June, 1985). [Polarization, coincidence and one person's view of the "three-star" experiments at CEBAF energies.]

Basic papers on some of the subjects discussed here:

10. T. W. Donnelly and I. Sick, *Rev. Mod. Phys.* **56**, 461 (1984). [Mainly elastic magnetic electron scattering with a chapter specifically on *elastic* scattering from polarized targets.]

11. T. W. Donnelly and A. S. Raskin, "Considerations of Polarization in Inclusive Electron Scattering from Nuclei," (to be published in *Annals of Physics*; MIT preprint CTP #1254). [Our basic paper on this subject.]

12. T. W. Donnelly, A. S. Raskin and J. F. Dubach, "Studies of the Reactions $(e, e'\gamma)$ and $(\vec{e}, e'\vec{\gamma})$," (to be published). [Relates these processes to the preceeding polarization studies (Ref. 11); specific examples discussed.]

ADDITIONAL REFERENCES

13. J. D. Bjorken and S. D. Drell, "Relativistic Quantum Mechanics," McGraw-Hill, New York (1964).

14. T. deForest, Jr. and J. D. Walecka, *Adv. in Phys.* **15**, 1 (1966).

15. T. W. Donnelly and J. D. Walecka, *Ann. Rev. Nucl. Sci.* **25**, 329 (1975).

16. J. D. Walecka, "Electron Scattering," Lectures given at Argonne National Laboratory (ANL-83-50), (January, 1984).

17. J. D. Walecka, *Nucl. Phys.* **A285**, 349 (1977).

18. T. W. Donnelly and R. D. Peccei, *Phys. Rep.* **50**, 1 (1979).

19. R. von Gehlen, *Phys. Rev.* **118**, 1455 (1960); M. Gourdin, *Nuov. Cim.* **21**, 1094 (1961); J. D. Bjorken (unpublished) (1960).

20. A. R. Edmonds, "Angular Momentum in Quantum Mechanics," (3rd edition), Princeton University Press, Princeton (1974).

21. M. E. Schultze, *et al.*, *Phys. Rev. Lett.* **52**, 597 (1984).

22. J. Dubach, J. H. Koch and T. W. Donnelly, *Nucl. Phys.* **A271**, 279 (1976).

23. T. W. Donnelly, A. S. Raskin and J. F. Dubach, (to be published).

24. G. Co', S. Krewald and T. W. Donnelly, (to be published).

25. J. W. Van Orden, in Workshop "Nucleon and Nuclear Structure and Exclusive Electromagnetic Reaction Studies," MIT (July, 1984), p. 210; A. Picklesimer, J. W. Van Orden and S. W. Wallace, University of Maryland preprint # 85-185.

TRINUCLEON BOUND STATES

J. L. Friar

Theoretical Division
Los Alamos National Laboratory
Los Alamos, NM   87545 USA

INTRODUCTION

The four bound few-nucleon systems (^{2}H, ^{3}H, ^{3}He, ^{4}He) have played a role in nuclear physics far out of proportion to their abundance on earth, and their study constitutes one of the oldest and most important subfields of our discipline. In one of the first review articles[1] treating nuclear physics, a separate section was reserved for the three-nucleon problem. Since that time many such articles have been written.

The special importance of these four nuclei stems from the great difficulty in solving the many-body problem. Special techniques exist for solving that problem when the number of particles becomes huge, a limit of no obvious relevance to nuclear physics. On the other hand we can also solve "exactly" (in the numerical sense) well-posed model problems with four or fewer nucleons. Our lack of ability to construct from first principles a tractable Hamiltonian for the interaction of a single pair of nucleons which describes all the phenomena associated with this system means that we routinely use semiphenomenological Hamiltonians, which incorporate physical constraints and some parameters which are fitted to two-nucleon experimental data. Thus, the three- and four-nucleon systems constitute a special testing ground for new ideas and concepts in nuclear physics, simply because we can solve for their wave functions and because their proper-ties have not been incorporated into our Hamiltonian models.

Of particular importance to us here is the electromagnetic interaction. Like the few-nucleon problem, electromagnetism is a relatively "clean" field, with constraints pro-duced by fundamental principles, and with a small coupling constant which makes com-plicated physical processes contribute only weakly. Thus, electromagnetic interaction results are "interpretable", particularly if wave functions are accurately known. This does not imply that our work is "cut and dried", with little room for innovation. Quite to the contrary, because so much is known, electromagnetic interactions in few-body sys-tems are the place to look for "exotic" phenomena. Because the technical aspects of the few-nucleon problem tend to obscure the many simple results, we will concentrate in the first lecture on understanding why three-body calculations are done the way they are, in what sense they are complicated, and in what sense they are not complicated. In the second lecture we will concentrate on electromagnetic interactions involving three nucleons, and other topics, including three-body forces.

Although much of the modern work in this field is formulated in momentum space, most of the older work and the work described in this lecture were formulated in configuration space (CS). Many techniques have been used to calculate CS wave functions, beginning with the august Rayleigh-Ritz variational principle[1]. Why do we and others work in configuration space? In our case the answer is simple: our physical intuition and insight are greatest there. There are, however, distinct advantages to momentum space for certain problems, such as relativistic treatments of few-nucleon systems. In what follows we will emphasize almost exclusively the bound few-nucleon systems in configuration space, and the approach of the Los Alamos-Iowa collaboration to solving the Schrödinger equation for these systems[2].

QUALITATIVE ASPECTS

No discussion of the three-nucleon problem is complete without a schematic discussion of the two-nucleon Hamiltonian. Many of the detailed quantitative features are irrelevant, while a few seemingly unimportant qualitative features determine most of the trinucleon properties.

The key underlying assumption is that few-nucleon dynamics is non-relativistic. This important simplification relies on the fact that typical values of *mean* internal nuclear momenta, $\bar{p}$, are 100-200 MeV/c, and thus $(v/c)^2 = (\bar{p}/Mc)^2$ for a nucleon of mass $Mc^2 = 939$ MeV is one-few percent. Since $(v/c)^2$ gives the scale of relativistic corrections, this estimate would indicate that a nucleus is largely nonrelativistic. The argument hides the fact that short-range potentials can be very strong and induce local momenta which are correspondingly large; the estimate above should only be interpreted as "in the mean". Moreover, our potential models "hide" the effects of relativity in the phenomenological parts, because parameters are fit to data.

There are three salient features of the two-nucleon potential which drastically, and unfavorably, effect our ability to solve the few-nucleon Schrödinger equation. These are:

1. Forces between like nucleons (e.g., pp or nn) are weaker than the forces between unlike nucleons (np);

2. The two-nucleon spin-triplet potential contains a strong tensor force which couples neighboring orbital waves;

3. The short-range force exhibits very strong repulsion, which makes the probability of nucleon-nucleon overlap at short distances very small.

Without these complications the few-nucleon Schrödinger equation is quite easy to solve. Feature 1 induces important spin and isospin correlations in the wave function. If the forces between all particles were identical, only a single (different) scalar function of the particle separations would describe each of the few-nucleon systems. With a tensor force present, the deuteron wave function has a tensor (d-wave) component, as do the triton and α-particle, which greatly complicates solving the Schrödinger equation. A strong short-range repulsion produces "holes" in the wave function. These holes must be accurately generated in any solution, which is thus rendered considerably more difficult.

In addition to these qualitative aspects of the nucleon-nucleon force, we note also that the odd-parity nucleon-nucleon partial waves (e.g., $^1P_1$, $^3P_{0,1,2}$), are relatively weak, and we will see later that they play a very small role in the triton.

A few basic principles motivate the procedures used to solve numerically various three-body problems. These are:

1. Nuclei (including the triton) are weakly bound, and average momenta are consequently small compared to the nucleon mass;

2. In the triton the average momentum is comparable to the inverse of the radius (R) and consequently the angular momentum barrier suppresses high partial waves of the nucleon-nucleon force;

3. Unlike the case of heavy nuclei, the Pauli principle doesn't play a particularly large role;

4. The details of the force are relatively unimportant in the overall binding, although they can severely complicate achieving a solution.

As we previously discussed, a nonrelativistic treatment of the triton should suffice, as indicated by 1. One estimate of the average momentum is $\bar{p} = \sqrt{ME_b} \equiv \hbar\kappa$ where $E_b = 8.5$ MeV is the binding energy, and consequently, $\bar{p} \cong 90$MeV/c. A typical trinucleon size is 2 fm, so that $\bar{p}R \sim 1$. Because Bessel functions of argument z and order l peak for $z > l$ it is clear that the angular momentum barrier will greatly suppress orbital angular momenta greater than 2 in the triton.

GEOMETRICAL ASPECTS

The geometry of the triton illustrates the greater difficulty in solving the Schrödinger equation for the triton compared to the deuteron. The deuteron is described by a single vector $\vec{r}$ separating the nucleons, and only its magnitude is relevant for a description of the two scalar functions, u(r) and w(r), which determine the s-wave and d-wave parts of the wave function. Fig. 1 shows the triton, where we have arbitrarily numbered the nucleons. Three points define a plane and thus only two vectors, $\vec{x}_1$ and $\vec{y}_1$, describe the system. Because the orientation of the plane is arbitrary, only three independent interparticle coordinates (x_1, y_1, θ_1) are required to specify the wave function. Our choice of vectors is arbitrary, however, since any set of the Jacobi coordinates formed from the nucleon coordinates $\vec{r}_i$ (i, j, k cyclic) is adequate:

$$\vec{x}_i = \vec{r}_j - \vec{r}_k \quad , \tag{1}$$

$$\vec{y}_i = \frac{1}{2}(\vec{r}_j + \vec{r}_k) - \vec{r}_i \quad . \tag{2}$$

Clearly the sums of the $\vec{x}_i$ or $\vec{y}_i$ vanish and they are linearly dependent. Traditionally, the set $(\vec{x}_1, \vec{y}_1)$ is relabeled as $(\vec{x}, \vec{y})$, where $\vec{x}$ and $\vec{y}$ are denoted the "interacting pair" and "spectator" coordinates, respectively.[3]

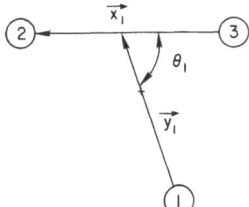

Fig. 1   Jacobi coordinates (x_1, y_1, θ_1) for the trinucleon problem.

Group theoretical methods[4] are used to classify in a well-defined way the wave function components which can occur for the positive parity, spin-1/2 trinucleons. Most of the important qualitative aspects of this scheme are rather obvious, however. Like the deuteron, the principal triton wave function component is s-wave in character. However, because there are several coordinates describing the problem, this can be further broken down into three distinct categories: (1) the S-state, completely symmetric under the interchange of spatial coordinates (i.e., the $\vec{r}_i$); (2) the S'-state, which has mixed spatial symmetry (neither symmetric nor antisymmetric); (3) the S"-state which has spatial antisymmetry. The last state has negligible size because the antisymmetry requires very large momentum components, which are lacking in the ground state, and because it is generated by the weak odd-parity nucleon-nucleon forces. The S'-state vanishes when the np, nn, and pp forces are identical, and for this reason it can be viewed as a space-isospin-(spin) correlation in the ground state. Its physical importance will be discussed later. The S-wave components are clearly spin doublet, since the trinucleons have spin-1/2; they are isodoublets if we ignore the Coulomb force in ^{3}He. There are also three independent spin-quartet D-wave components, analogous to the deuteron case. Unlike the deuteron case, it is possible to construct a positive parity vector ($\hat{x} \times \hat{y}$), and this leads to three quartet and one doublet P-state components, which are very small. Adding everything together, there are 10 S-, P-, and D-state components, specified by 16 scalar functions.

The Schrödinger equation for the deuteron involves 2 coupled equations in one variable (r). The Schrödinger equation for the triton is a set of 16 coupled partial differential equations in 3 independent variables. This large number of equations makes the problem roughly equivalent to a single 4-variable problem, which would require heroic efforts, even for modern supercomputers. The way to circumvent this seemingly intractable situation is to use our knowledge of the physics of the problem: the angular momentum barrier suppresses many of the problem's complexities.

Fig. 2 shows two of the energy scales of the triton. The upper graph illustrates the spin- and isospin-independent MT-V nucleon-nucleon potential model[5], plotted versus nucleon-nucleon separation, x, and for comparison, the centrifugal part of the kinetic energy (for $l = 2$): $\hbar^2 l(l+1)/Mx^2$. We see that the latter dwarfs the potential energy. Clearly, for higher values of l this mismatch is even greater. The implications for the binding of the triton are immediate: potential energy contributions for the higher nucleon-nucleon partial waves rapidly decrease as l increases. We can easily see by assuming a spin- and isospin-independent potential $V_{23}(x)$ between nucleons 2 and 3 and expanding this in a partial-wave series in both $\hat{x}$ and $\hat{y}$:

$$V_{23}(x) = \Sigma_\alpha \mid \alpha > V_{23}(x) < \alpha \mid \quad , \tag{3}$$

$$\mid \alpha > = [Y_l(\hat{x}) \otimes Y_l(\hat{y})]_0 \quad , \tag{4}$$

and the "channel"-label α is simply l in this case. This series is much simpler than the general case, because we have assumed the same potential in *every* partial wave. Taking the expectation value of the potential between all three pairs of nucleons gives

$$<V> = 3<V_{23}(x)> \equiv 3\Sigma_l \int_0^\infty dx\, x^2 C_l(x) V_{23}(x) = \Sigma_l <V_l> \quad , \tag{5}$$

where the partial-wave projected correlation function is

$$C_l(x) = \int \mid \, < \alpha \mid \Psi > \mid^2 y^2 dy \quad . \tag{6}$$

Only the completely space-symmetric S-state occurs in the wave function for this prob-

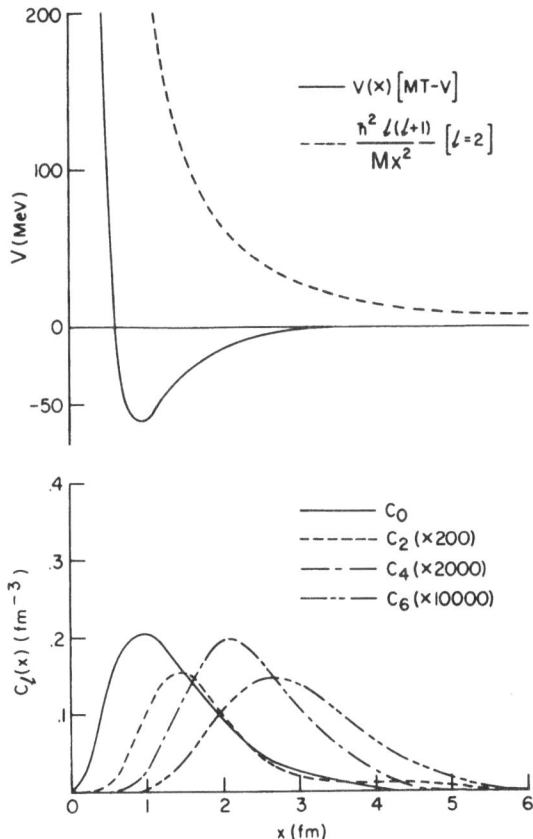

Fig. 2   Comparison of centrifugal kinetic energy with the MT-V potential (top) and partial-wave projected triton correlation functions for that potential (bottom).

lem, and only even values of l are nonvanishing because of this. The lower plot in Fig. 2 shows the first four C_l's, which rapidly decrease in size with increasing l. The dominant $C_0(x)$ is small at the origin because of the repulsion in $V(x)$, while the remaining $C_l(x)$'s behave as x^{2l} for small x. This means that only increasingly larger values of x contribute to the integrands in Eq. (5), which are suppressed by the finite range of the force. The values of $<V_l>$ (for $l = 0, 2, ..., 10$) for this simple potential model are given by [-36.5, -.163, -.019, -.002, -.0004, -.00008] MeV, dramatically illustrating the rapid convergence as l increases. Clearly it should be sufficient to restrict l to 4 or less. We will see later that this convergence rate also applies to more realistic potential models. We note that the sum of all the C_l's is the usual two-body correlation function, $C(x)$.

By expanding the potential in a series and then truncating the series after a reasonable number of terms, we have in effect reduced the problem to solving a set of coupled equations (for the partial waves) in *two* variables x and y, which makes the problem tractable. A good estimate of the time scale for numerically solving the deuteron problem, starting from scratch, is one or two months. The scale for the triton bound state is perhaps two years! The problem is still very difficult, and requires a substantial commitment of personal and computer time. For future reference we note that all calculations using the Faddeev approach (to be described next) decompose the nucleon-nucleon potential into partial waves and solve that (truncated) problem "exactly".

BOUNDARY CONDITIONS AND THE FADDEEV-NOYES EQUATION

We wish to solve a partial differential equation, the Schrödinger equation, for the triton bound state. It is sometimes forgotten by those who don't perform numerical calculations that such solutions require the imposition of well-defined boundary conditions. Simple bound-state problems only require the imposition of finiteness requirements for the wave function at the origin and at asymptotically large distances, where the wave function vanishes exponentially.

The scattering problem is more complex, and finiteness alone is not enough. Years ago, Foldy and Tobocman[6] showed that the three-body Lippmann-Schwinger (LS) equation (the Schrödinger equation rewritten as an integral equation) for scattering has no unique solutions, even when outgoing scattered waves are specified in the usual way. Even the two-body Lippmann-Schwinger equation has no unique solution, without further subsidiary conditions, if the problem is imposed in a particular way! The problem we pose is: what is the outgoing-wave solution for two nucleons with a total energy of 20 MeV? This is a "trick" question, because we have deliberately not specified the center-of-mass (CM) motion of the two nucleons. As stated, an arbitrary linear combination of wave functions for a deuteron with 22.2 MeV CM energy, two nucleons in a $^1S_0$ threshold state with 20 MeV CM energy, and two nucleons with an internal energy of 10 MeV and 10 MeV CM energy solves the problem. Trivially, we can avoid the problem by working in the CM frame, which fixes the relative two-nucleon energy. Unfortunately, even in the CM frame of the three-nucleon system this does not suffice, since the recoil of a third nucleon can compensate for the CM motion of the remaining pair in any state of internal motion commensurate with conservation of energy. Because of this, complicated phenomena are possible, which makes the *ad hoc* imposition of boundary conditions a dubious exercise. An incoming plane wave for a proton-deuteron system (pd) can scatter directly to a dp final state, or break up into a ppn final state, or the initial proton can pick up the neutron in the deuteron and that deuteron can escape. These many physical channels are not orthogonal and specifying outgoing waves is not enough. In the jargon of few-body physics, there are "disconnected diagrams", "dangerous δ-functions", "noncompact kernels", and "nonunique solutions". All these diseases are merely symptoms of the original problem.

Of particular importance is rearrangement, such as the neutron pickup example described above. We write the Schrödinger equation in the form

$$[E-(T+V_{12}+V_{13}+V_{23})]\Psi = 0 \ , \tag{7}$$

where T, E, and V_{ij} are the kinetic energy, total energy, and potential energy for the pair (ij), respectively. If both V_{23} and V_{13} can support a deuteron bound state, an initial plane-wave state of nucleon 1 and bound nucleons 2 and 3 [denoted (1;23)] can asymptotically become nucleon 2 plus a bound (13) pair [(2;13)]; the converse is also true and both wave functions contain both physical processes. The difficulty is that while the LS equation specifies that the (1;23) configuration has an incoming plane wave and outgoing spherical wave, it does not rule out incoming plane waves for (2;13). In order to achieve a unique solution the LS equation must be supplemented by additional *homogeneous* equation[7,8], which rule out unwanted incoming plane waves.

Faddeev provided the means to circumvent this dilemma[9]. Although Faddeev's procedure was developed in momentum space, Noyes[10] later cast that work into a physically equivalent configuration space form. We arbitrarily write

$$\Psi(\vec{x},\vec{y}) = \psi(\vec{x}_1,\vec{y}_1) + \psi(\vec{x}_2,\vec{y}_2) + \psi(\vec{x}_3,\vec{y}_3) \equiv \psi_1 + \psi_2 + \psi_3 \quad, \tag{8}$$

where the variables $(\vec{x}_i,\vec{y}_i)$ are the Jacobi coordinates defined earlier, and the function ψ in Eq. 8 is the *same* for all three terms. The original Schrödinger equation becomes three separate equations

$$(E-T-V_{23})\psi_1 = V_{23}(\psi_2+\psi_3) \quad, \tag{9}$$

$$(E-T-V_{13})\psi_2 = V_{13}(\psi_1+\psi_3) \quad, \tag{10}$$

$$(E-T-V_{12})\psi_3 = V_{12}(\psi_1+\psi_2) \quad. \tag{11}$$

Clearly, Eqs. 10 and 11 are simply permutations of Eq. 9, and we need solve only Eq. 9. Since that equation involves only V_{23} (and not V_{13}) the problem of the rearrangement reaction has disappeared for ψ_1. It is contained in ψ_2. By this clever mechanism, Faddeev showed that we only need to specify explicitly the much simpler boundary conditions for ψ_1, rather than for Ψ. Note that the sum of Eqs. 9, 10, and 11 reproduces Eq. 7.

This is seen most clearly in Fig. 3, where the regions of interest for the variables x and y are illustrated. The configuration (1;23) corresponds to an asymptotic state with $y\rightarrow\infty$, and $x<x_d$, the physical extent of the bound pair (23), and is denoted the "deuteron strip". Rearrangement corresponds to small $x_2 = |\vec{r}_1 - \vec{r}_3|$ (i.e., a bound state in 13) and this occurs when $\theta = 0$, and $y = x/2$ or $\theta' = 30°$ in terms of the polar coordinates

$$x = \rho\cos\theta' \quad, \tag{12a}$$

$$y = \frac{\sqrt{3}}{2}\rho\sin\theta' \quad. \tag{12b}$$

In complete analogy with the two-body problem, we can impose boundary conditions most easily for the reduced wavefunction

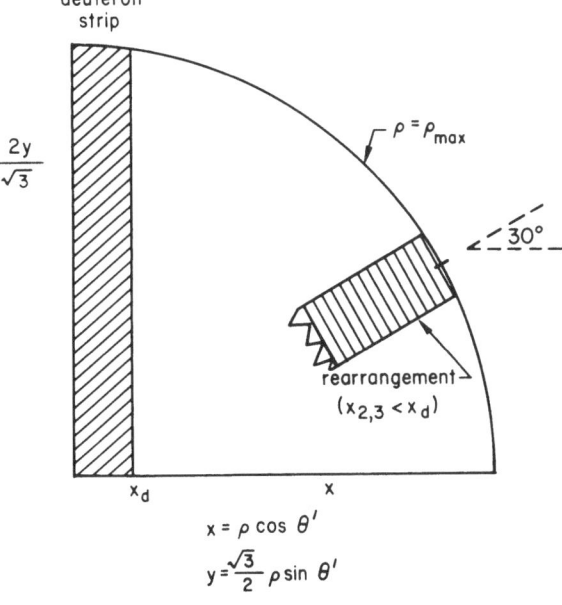

Fig. 3   Configuration space regions for Nd (nd or pd) scattering problem.

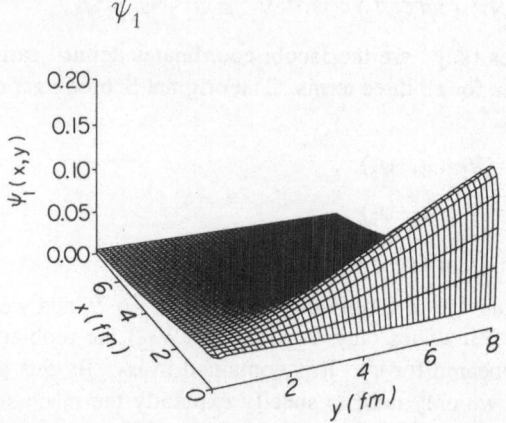

Fig. 4   Faddeev wavefunction for quartet nd scattering, Ψ_1, plotted versus x and y.

$$\phi_1 = xy\psi_1 \;\; , \tag{13}$$

by enforcing $\phi_1=0$ along x=0 and y=0, and outgoing wave boundary conditions[11] along $\rho=\rho_{max}$.

These physical considerations can be seen graphically in Fig. 4 and Fig. 5 for $\theta=0$, which depict wave functions for the scattering of zero energy neutrons and deuterons in the quartet spin state. The smooth function ψ_1 in Fig. 4 has structure only along the deuteron strip, while Fig. 5 depicts v_3, a component of the total wave function Ψ, which has structure along the deuteron strip and a ridge with "wings" along $\theta'=30°$, which is the outgoing wave in the rearrangement channel. It is clearly a simpler procedure to solve for ψ_1 than v_3, which has *much* more structure.

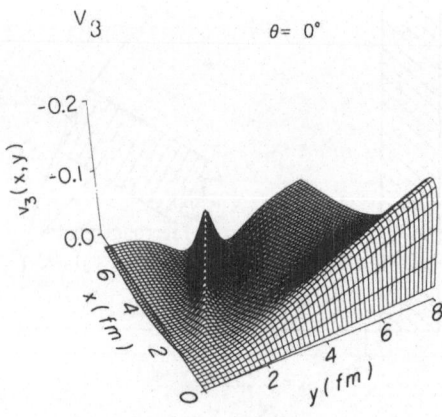

Fig. 5   Schrödinger wavefunction component, v_3, for $\theta=0°$ generated from Ψ_1 in Fig. 4, plotted versus x and y.

The bound-state problem[3] has much simpler boundary conditions: we need only make the wave function vanish for some large $\rho=\rho_{max}$. Nevertheless, the Faddeev motivations for the scattering problem work equally well for the bound state, and we anticipate that the Faddeev wavefunction ψ_1 will be smoother and easier to model numerically than Ψ.

Having made the decision to partial-wave project the nucleon-nucleon force, it is necessary to determine the consequence of this for the Faddeev-Noyes equation. For simplicity we assume a force which is independent of spin and isospin and acts only in the s-wave. In terms of our previous discussion, such a force looks like $|0>V(x)<0|$, where the projector $|0>$ refers to s-waves. This produces, with $E=\hbar^2K^2/M$,

$$\left[\frac{\partial^2}{\partial x^2} + \frac{3}{4}\frac{\partial^2}{\partial y^2} - U(x) + K^2\right]\phi(x,y) = U(x)\int_{-1}^{1}d\mu \left[\frac{xy}{x_2y_2}\right]\phi(x_2,y_2) \quad , \tag{14}$$

where $U(x) = MV(x)/\hbar^2$, $\mu=\cos\theta$, and $\phi(x,y) = \frac{1}{2}\int_{-1}^{1}d\mu\, \phi_1(x,y,\mu)$. Note that ϕ does not depend on μ; it is completely independent of θ. Moreover, for the s-wave force chosen, all higher partial waves of ϕ_1 must vanish, because V vanishes for those waves, and therefore $\Psi(x,y,\mu) = \phi(x,y) + \phi(x_2,y_2) + \phi(x_3,y_3)$. This is an extremely important result, since all of the angular (μ) dependence in Ψ comes from the permuted terms, $\phi(x_2,y_2)$ and $\phi(x_3,y_3)$, and the computation of a 3-variable function has been reduced to one of only two variables. When many partial waves are computed, one has coupled equations in the two variables x and y. Nevertheless, the angular momentum barrier makes the required number tractable, and the calculation possible.

NUMERICAL MODELLING

We still must make a choice of numerical methods in order to solve the equations. A technique which has proven exceptionally powerful in modern engineering applications is the finite element method, and its variant, the method of splines[12]. Fig. 6 depicts at the top a function which we wish to approximate for computational purposes, between the points x_0 and x_4, and for demonstration purposes we choose to do so by dividing the distance into 4 equally spaced regions or intervals. The finite element method consists of approximating the function in each interval by a (different) polynomial of order N and forcing the function and its first m derivatives to be continuous at the "breakpoints" between intervals. For definiteness we will use cubic polynomials (N=3) involving 4 parameters, and force the function and its derivative to be continuous. There are a total of 16 parameters, and 2 imposed conditions at each of 3 breakpoints, leaving 10 free parameters. The function is chosen to vanish at the end points, leaving 8 parameters which are chosen so that at two "collocation" points (indicated by x's) in each of the 4 intervals the function agrees exactly with the function we are modeling. If we are solving an equation for this function, we force the equation to be exactly satisfied at those points.

An alternative scheme is to use splines, which eliminates much of the labor. The finite elements in a given interval are grouped with those in a neighboring interval, which are then overlapped as shown in the middle of the figure. That is, at any point, x, the function is approximated as the sum of two overlapping functions, each defined in a double interval. These spline functions and their first m derivatives are required to vanish at

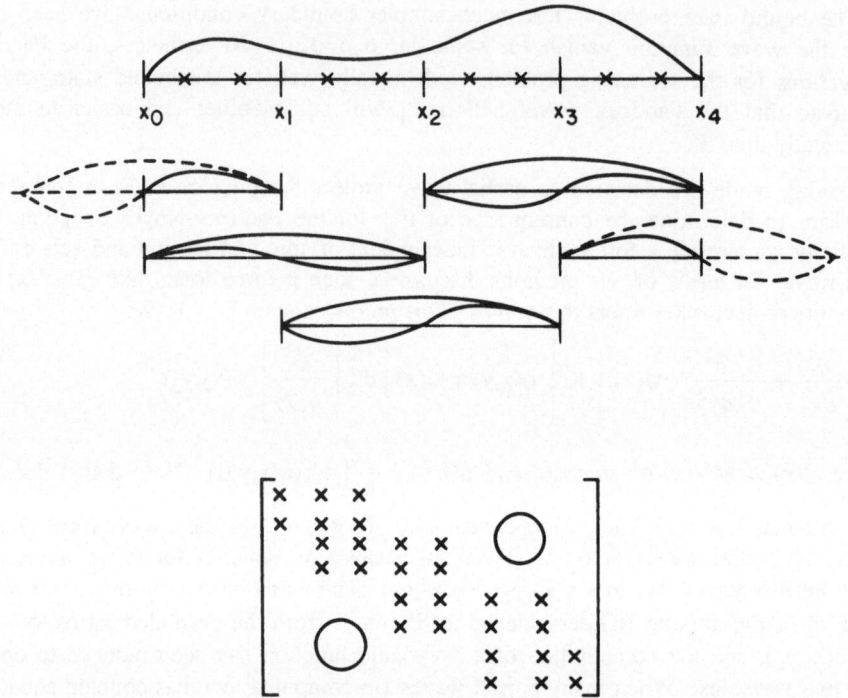

Fig. 6   The function at the top is approximated by the sum of 5 spline functions in the middle. The use of such splines with a second-order differential equation leads to the banded matrix shown at the bottom.

the right and left ends of the double interval and to be continuous at the middle boundary. For our case (N=3 and m=1) the 8 finite element parameters for any double interval are reduced to two by these six conditions. We have graphed these (Hermite) splines as even and odd functions in the double interval, and the remaining two parameters are simply the overall strengths of each of these functions. The beauty of this scheme is that the use of overlapping splines now guarantees that the function and its first derivative are everywhere continuous without any extra work! The boundary conditions are trivially satisfied by making the even function in the end intervals vanish, and the remaining 8 parameters in the 5 overlapping spline functions are determined at the collocation points, as before. The strength of this method is that the overall number of unknowns has been reduced to the minimum before we even set up matrix equations.

The orthogonal collocation method allows one to choose the collocation points so that the power of Gauss quadratures and splines can be combined[13]. If we were to perform an integral over the function in the figure, a natural way to do this would be to integrate between breakpoints and use a Gauss quadrature formula in each interval. Using those quadrature points as collocation points constitutes the method of orthogonal collocation, which substantially improves rates of convergence when solving equations using splines.

Because splines are local functions, separately defined in each double interval, the collocation conditions couple splines from neighboring intervals only. The complete set of such conditions for all parameters (8 in our example) constitutes a matrix equation, and this matrix has a very special form because of the locality; it is a "band" matrix, with most of the elements zero, as shown at the bottom of Fig. 6. Such matrices are much easier to invert than dense matrices, and should be preserved, if possible. In order to deal with the angular integral in Eq. 14, we transform from (x,y) coordinates to the polar coordinates (ρ, θ'). The integral destroys the double band structure in x and y; polar coordinates preserve this structure in the variable ρ.

There are a number of important advantages which accrue from using splines to model a function: (1) the spline approximant and a specified number of derivatives are automatically continuous; (2) the splines automatically provide an interpolating function at any point; (3) they lead to a band matrix; (4) they are "optimally" smooth; (5) it is easy to change from the equally spaced intervals of our example to any desired distribution; (6) the splines are easy to program on a computer; (7) boundary conditions are easy to impose; (8) the approximants exactly satisfy the constraint equations at the collocation points; (9) piecewise local functions such as splines do not propagate approximation errors, as global functions do; (10) the relative accuracy of the wave function and eigenvalue should be comparable. We also note that the use of overlapping double intervals corresponds closely to one derivation of the powerful Gregory's integration rule from Simpson's integration rule.

There is little difference in principle between solving Eq. 14 for a single nucleon-nucleon (NN) partial wave and using many partial waves. The size of the matrices becomes much larger, and the matrix bookkeeping becomes very tedious and intricate. In general for each nucleon-nucleon partial wave, there are two spectator partial waves associated with the two spin states of the latter, except for total angular momentum, J, equal to zero, which generates only one. The four NN partial waves (^SL_J) for each J (^1J_J, ^3J_J, $^3J-1_J$, $^3J+1_J$) thus generate 8 trinucleon channels, except for J=0, which has only two, associated with $^1S_0$ and $^3P_0$. As we indicated earlier, the $^1S_0$ and $^3S_1$ waves should be dominant, and we must also include the $^3D_1$ wave which is strongly coupled by the tensor force to the $^3S_1$ wave. This combination is the standard 5-channel calculation (all positive-parity NN waves with J≤1). The 9, 18, 26, and 34 channel cases correspond to positive parity waves with J≤2, all waves with J≤2, positive parity waves with 3≤J≤4 plus all waves with J≤2, and all waves with J≤4, respectively.

RESULTS FOR TRINUCLEON BOUND STATES

A brief summary of results[14] for the Reid Soft Core[15] (RSC), Argonne[16] V_{14} (AV14), Super-Soft-Core(C)[17] [SSC(C)], and Paris[18] potential models is given in Table 1 as a function of channel number. Several conclusions are obvious: (1) the 5-channel approximation gives most of the binding (within .2-.3 MeV); (2) the negative-parity NN waves don't have a large effect; (3) the binding is roughly 1 MeV below experiment; (4) the point-nucleon rms charge radii (i.e., the proton radii) for ^{3}He and ^{3}H are larger than experiment. Because the positive-parity waves dominate, this table doesn't demonstrate the rate of convergence of the partial-wave series. This is shown in Table 2 for the RSC 34-channel case, where <W> is broken down into contributions for fixed J and fixed parity. All but 1% of the total potential energy (indicated by Σ in the last column) is generated by the first 5 channels, and most of the rest from the remaining positive-parity waves. The small negative-parity NN forces give 200 keV more binding, which is not obviously reflected in Table 1 (compare 18 channels to 9 channels). The reason is that

Table 1   Binding energies, point charge rms radii in fm, and percentages of wave
           function components for various two-body force models[17].

	-E (MeV)					$\langle r^2\rangle^{1/2}_{He}$	$\langle r^2\rangle^{1/2}_{H}$	$P_{S'}$	P_D
Model	5	9	18	26	34	34	34	34	34
RSC	7.02	7.21	7.23	7.34	7.35	1.85	1.67	1.40	9.50
AV14	7.44	7.57	7.57	7.67	7.67	1.83	1.67	1.12	8.96
SSC(C)	7.46	7.52	7.49	7.54	7.53	1.85	1.68	1.24	7.98
Paris[19]	7.30		7.38						
Expt.		8.48				1.69^3	1.54^4	--	--

the negative-parity forces couple directly to the small components of the wave function
and this leads to nearly canceling contributions from first- and second-order perturbation
theory. First-order perturbation theory works well for all the other small force components.

The probabilities of the important S'- and D-state wave function components are
small. The D-state probabilities for the triton are very nearly 3/2 times the corresponding
D-state probabilities of the deuteron for each potential model. Clearly there is under-
binding, and the radii aren't correct either. The latter and other important observables
depend on the binding energy, and since that is wrong the observables can't be correct.
In order to investigate this problem which has plagued us for a decade, we anticipate
some of the results of the next section, and introduce a three-body force to increase binding. We don't need to know what it is at this stage. Our study of these observables will
allow us to gain a qualitative understanding of them at the same time.

Although a wide variety of bound-state calculations have been performed during the
previous two decades for a variety of potential models, many produced only binding
energies and no wave functions, and others required approximations whose reliability
was difficult to assess. The recent studies[20] of the Los Alamos-Iowa group have produced a large number of numerically accurate triton wavefunctions for four different
two-body potential models in combination with several different three-body force
models, that these model combinations accurately describe nature, the solutions at least

Table 2   Potential energies (in MeV) for the RSC 34-channel case broken down
           according to J (total nucleon-nucleon angular momentum) and parity, and
           the kinetic energy for comparison.

J	0	1	2	3	4	Σ
$\langle V_J\rangle$	-13.729	-43.647	-0.435	-0.115	-0.020	-57.946
$\langle V_J^+\rangle$	-13.553	-43.874	-0.188	-0.117	-0.014	-57.746
$\langle V_J^-\rangle$	-0.176	0.227	-0.247	0.002	-0.006	-0.200
$\langle T\rangle$						50.600
$\langle H\rangle$						-7.345

incorporate the correct quantum mechanical constraints. Moreover, the binding energies for the set of models extend from below to above the physical binding energy of the triton. This provides us for the first time with the opportunity to investigate how a variety of important ground-state observables depend on binding energy, and whether there is any "true" model dependence as well.

What are the important ground-state properties, besides the binding energy? A list of the most commonly calculated ones would include the (point) charge radii, $<r^2>_{He}^{1/2}$ and $<r^2>_{H}^{1/2}$, the probabilities of the various wave function components (which are not measurable[21]), the Coulomb energy of ^{3}He, E_c, the magnetic moments of ^{3}He and ^{3}H, their asymptotic norms (sizes of asymptotic wave function components), and the β-decay matrix element of ^{3}H. The magnetic moments depend on meson-exchange currents and on the S'- and D-state probabilities, $P_{S'}$ and P_D, as does the β-decay matrix element; we will discuss them later. The asymptotic norms depend on binding, but this has not been assessed in detail yet. The radii and Coulomb energy depend sensitively on the binding energy, and calculations of these observables which use models that underbind will produce inadequate predictions. We assess the status of these important physical quantities below, together with simple qualitative arguments that account for our conclusions.

For pedagogical purposes, the difference of the ^{3}He and ^{3}H charge radii can be understood in terms of the oversimplified picture in Fig. 7. The sketch at the top depicts a schematic ^{3}He when the nucleon-nucleon forces between all pairs are identical. This is represented by an equilateral triangle configuration, with shading depicting the protons. The charge or proton radius, R_p, measures the integrated probability of finding a proton at a distance r from the center-of-mass. In this simple example, the proton, neutron, and mass radii are all the same. When the forces between pairs are different, the appropriate pictures for ^{3}He and ^{3}H are those of Fig. 7b and Fig. 7c. The np forces are stronger than the nn or pp ones (only the np system has a two-body bound state) and this allows the protons in ^{3}He and the neutrons in ^{3}H to lie further from the center-of-mass than their counterparts (θ>60°). The resulting isosceles configuration is reflected in the appearance of an S'-state, which directly measures the isosceles-equilateral difference, and in the fact that R_p for ^{3}He increases, while that of ^{3}H decreases, and hence $<r^2>_{He}^{1/2}$ is greater than $<r^2>_{H}^{1/2}$, irrespective of any pp Coulomb force in ^{3}He.

These arguments can be made quantitative by decomposing the mean-square-radius in impulse approximation into isospin components[22]: the isoscalar part $<r^2>_s$ mirrors Fig. 7a and is determined by sums of squares of wave function components. The isovector component contains one part proportional to the isoscalar component and another part largely determined by the overlap of the S- and S'-states, which we denote $<r^2>_v$ (v does *not* mean isovector), and determines the difference between ^{3}He and ^{3}H. One finds for ^{3}He (Z=2) and ^{3}H (Z=1), with upper and lower signs, respectively,

$$Z<r^2> = Z<r^2>_s \pm <r^2>_v \ . \tag{15}$$

These quantities have very different behaviors. Radii in general are sensitive to the asymptotic parts of the wavefunction. If one assumes that the entire wavefunction is represented by the asymptotic form[3], $N \exp(-\kappa\rho)/\rho^{5/2}$, one finds that

$$<r^2>_s^{\frac{1}{2}} = \frac{1}{2\kappa} \sim E_B^{-\frac{1}{2}} \ . \tag{16}$$

Fig. 8 shows the results of calculating $<r^2>_s^{1/2}$, and $<r^2>_v^{1/2}$, together with the experimental data corrected for the nucleons' finite size[2]. The fit to the isoscalar points is

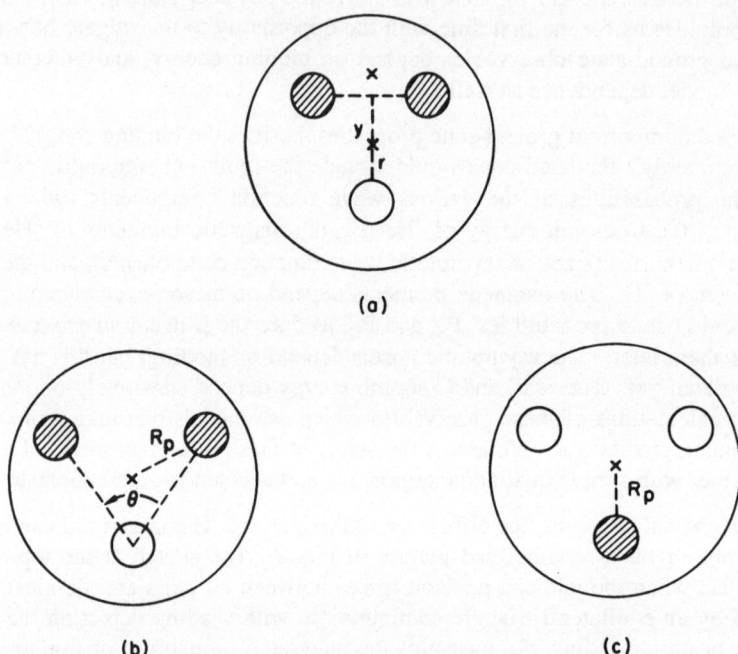

Fig. 7   Schematic trinucleons with identical forces between protons (shaded) and neu-
          trons in (a) and with different forces for ³He in (b) and ³H in (c).

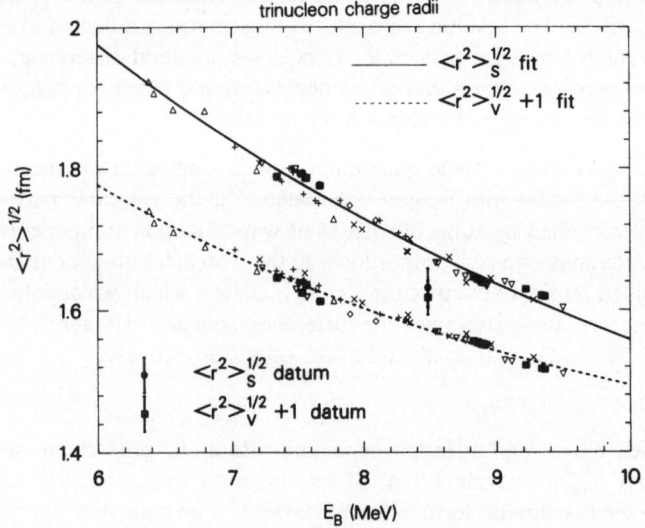

Fig. 8   Calculated trinucleon (point nucleon) rms charge radii decomposed into isos-
          calar (s) and difference (v) contributions in impulse approximation, together
          with data, plotted versus corresponding binding energy. The ³He calculations
          contained no Coulomb force.

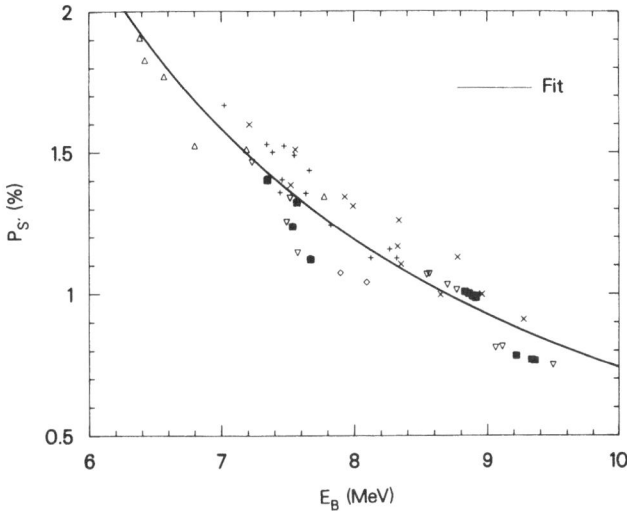

Fig. 9   Calculated trinucleon S'-state percentages plotted versus corresponding binding energy.

accurately represented by $.8E_B^{-.5}$, indicating that our simple argument was essentially correct. The difference radius is fit by $.14E_p^{-.9}$, and this different behaviour reflects different physics. Clearly, the amount of S'-state plays a significant role. The percentage of S'-state is plotted versus binding energy in Fig. 9, and the fit varies as $E_B^{-2.1}$. This decrease is expected, because as binding increases only the average force is important,

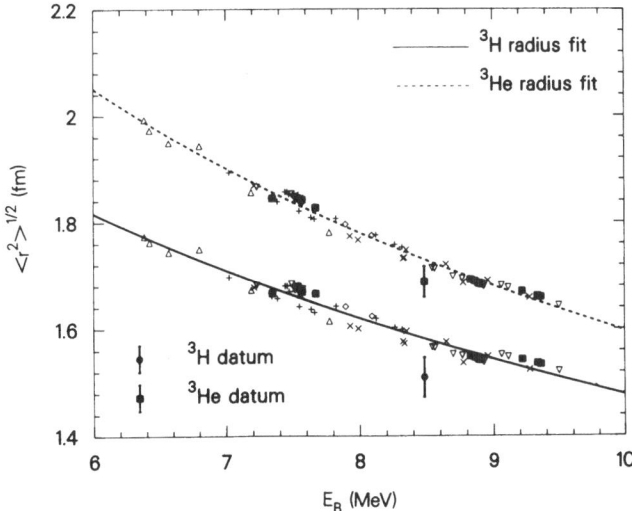

Fig. 10   Calculated trinucleon (point nucleon) rms charge radii in impulse approximation, plotted versus corresponding binding energy. The ^{3}He calculations contained no Coulomb force.

and the np-nn difference is less important. In a simple harmonic oscillator description, the S'-state is given in terms of excited state configurations, which decrease $\sim E_B^{-2}$ as the oscillator spacing increases with binding. Finally, the ^{3}He and ^{3}H results are shown in Fig. 10. If the small discrepancies between theory and experiment are real, they probably reflect a small breakdown of the impulse approximation.

The Coulomb force $V_c(x)$ between protons in ^{3}He is quite weak and can be accurately treated in perturbation theory. The second-order Coulomb effect[23] is estimated to be $\sim$ -4 keV, compared to a ^{3}He-^{3}H binding energy difference of 764 keV. Since $V_c \sim 1/R$, schematically, and since $R \sim E_B^{-1/2}$, we expect E_c to scale roughly as $E_B^{1/2}$. A better description is available, however, if we utilize Fig. 7a. In this schematic ^{3}He the distance x between protons is given by $\sqrt{3}r$, and thus $E_c = <V_c(x)> = \alpha<1/r>/\sqrt{3}$ where α is the fine structure constant. Consequently[24],

$$E_c \cong \frac{\alpha}{\sqrt{3}} \int \frac{d^3r}{r} [\rho_s(r) + \rho_v(r)] g(r) \equiv E_c^H \quad , \tag{17}$$

where we have added the effect of nucleon finite size[22], $g(r)$, and written the matrix element in terms of the scalar and difference *charge densities*. The accuracy of this hyperspherical approximation is demonstrated in Fig. 11. Although *a priori* a very implausible approximation, E_c^H overestimates E_c by only 1 percent. This is an important result, because the charge densities are experimentally measurable. Using these data[22] one finds $E_c = 638 \pm 10$ keV. This is significantly less than the binding energy difference and reflects the existence of nonnegligible charge-symmetry-breaking forces other than the Coulomb interaction.

CONCLUSIONS

Rapid and significant advances have been made in the few-nucleon problem recently. Many aspects of the bound states, including the Coulomb energy and charge radii, are now fairly well understood. Although we have concentrated on the trinucleon bound

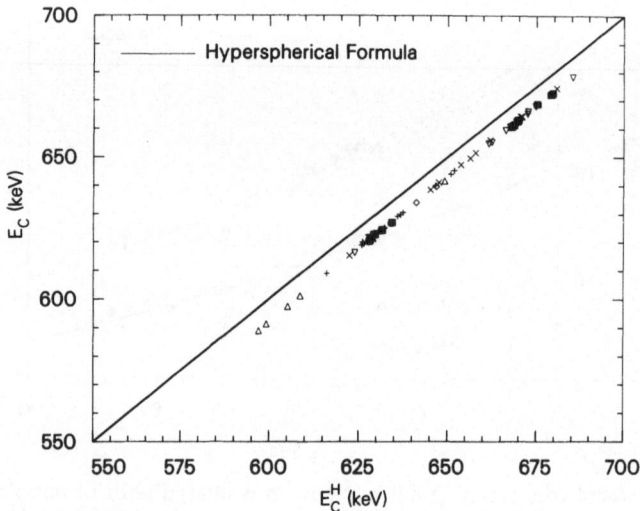

Fig. 11 ^{3}He Coulomb energy, E_c, plotted versus the corresponding hyperspherical approximation E_c^H.

states, the continuum is also important. Photonuclear reactions necessarily break up the triton and ^{3}He, and this is an important area of study. The continuum problem above breakup threshold is much more complicated than the bound-state problem, because the boundary conditions are difficult to implement in a tractable way. Nevertheless, the future of three-body physics lies in this regime.

NONTRADITIONAL NUCLEAR PHYSICS

For much of its 50 year existence nuclear physics has made tacit assumptions in its approach to problem solving. These assumptions, which comprise what I call traditional nuclear physics, are:

1. Nuclei are basically nonrelativistic and weakly bound, with average momenta being typically 100-200 MeV/c;

2. The binding of nuclei is produced primarily by two-body forces, which act only between pairs of nucleons at a time;

3. Only nucleon degrees of freedom are important, and nucleon sub-structure and meson or quark degrees of freedom can be ignored.

Although there were some early challenges to this approach to our field, it was only in the late 1960's that a serious, concerted effort was made to find exceptions to these "rules". The problem was that traditional nuclear physics was reasonably successful. Moreover, the curse of nuclear physics and related fields is our inability to accurately solve the many-body problem beyond the mean-field approximation, which meant that disagreements between theory and experiment were difficult to interpret. Were they due to poor wavefunctions, or to a poorly understood reaction mechanism?

The importance of the few-nucleon problem can be understood in this context. At the same time that modern intermediate energy (i.e., nontraditional) nuclear physics was being developed, great strides were being made in the few-nucleon problem. The early calculations by Tjon and collaborators[5] and by Kalos[25] used modern computational methods to solve for binding energies and wave functions; the latter were then available for computing electromagnetic matrix elements. This is still the strength of the field. We can solve the Schrödinger equation "exactly" for wave functions, and use the wave functions in calculations of electromagnetic processes, which are the most "interpretable" of all the types of reactions available to nuclear physicists. This is also our challenge for the next decade.

We have already estimated the size of relativistic corrections to be on the order of one to a few percent. The best evidence for relativistic corrections in low-energy nuclear reactions occurs for deuteron forward photodisintegration[26], where the proton is detected the dominant non-relativistic electric dipole (E1) reaction, so that a nominally one percent relativistic correction becomes a 20 percent effect! This points out one of the difficulties in challenging traditional nuclear physics: novel reactions or special regimes of known reactions must be sought in order to suppress the "ordinary" physics. The importance of relativistic effects will arise again in the context of three-nucleon forces.

One of the biggest success stories in all of nuclear physics during the decade of the 1970's was the convincing demonstration of meson degrees of freedom in electromagnetic reactions, and in particular, of the importance of the pion in exchange currents. Much of that story revolved around the threshold deuteron photo- and electrodisintegration and np radiative capture, all of which are magnetic dipole (M1) processes.

Riska and Brown[27] calculated the dominant pion-exchange processes and showed that the long-standing 10 percent discrepancy in np radiative capture could be largely understood from those processes alone. Moreover, many of the uncertainties in the pion's strong interaction had been eliminated as the consequences of chiral symmetry[28], which singles out the pion as a special particle, had unfolded during the decade of the 1960's. The possible importance of such meson-exchange currents had been known since the 1930's, when Siegert[29] demonstrated that the long-wavelength E1 current operator could be written in a form involving only the electric dipole operator (calculated from the charge density), which was shown to be accurately known in the nonrelativistic approximation. This combination is known as Siegert's theorem and is the backbone of photonuclear physics, because it allows a simple interpretation of reactions. Magnetic processes are very model dependent and sensitive to details of the current, while Seigert's current is not.

In what follows we will investigate in some detail in the context of the three-nucleon problem two of the nontraditional elements we listed earlier: three-nucleon forces and meson-exchange currents. We will see that both are linked to relativistic corrections and to each other. Regrettably, we must leave the interesting two-body problem to others.

THREE-BODY FORCES

Introduction

Before considering the evidence for three-body forces in nuclei, we first discuss whether such forces exist in other systems, and how they are defined. Most of the weaker fundamental forces, gravitational and electromagnetic, are basically two-body in nature. The considerations of Newton and Coulomb were based on that assumption. Is this assumption valid? We give two answers, which we will discuss in detail: (1) it is an excellent approximation; (2) it depends on your point of view.

We begin with a classical example, the earth-moon system with a small satellite orbiting the earth. We also assume, as Newton did, that each tiny particle of mass (atom) interacts with every other by two-body forces; that is, the interaction between two such particles is not affected by the presence of a third. This by itself is not enough to be able to solve for the coupled motion of our classical system, since there are enormous numbers of atoms in the problem we posed. It was Newton's genius that allowed him to see that the interaction of large bodies could be constructed from that of the individual tiny pieces, after he invented the necessary mathematics! We therefore reduce the problem to one of three *macroscopic* bodies interacting with each other. Does the position of the moon affect the force between satellite and the earth? If one neglects the tides, the answer is no, and the problem is simply one of 3 separate two-body forces between the composite objects. However, the tides caused by the moon affect the satellite motion in an observable way[30], and the position of the moon is clearly relevant, which means that the earth-moon satellite system exhibits a three-body force mediated by a deformation of the earth, namely the tides. The effect is very small, however.

A second example of three-body forces is the atomic Axilrod-Teller force[31]. Many-body calculations with groups of atoms are traditionally performed by assuming an effective interaction between *atoms,* rather than breaking the problem down into purely Coulombic two-body interactions between all the nuclei and electrons in all the atoms, which is much too complicated. Typical of two-body atom-atom interactions are the long-range van der Waals force, and the Lennard-Jones force. Having arranged the problem in this way, there will be forces between three atoms, between four atoms, ... , which arise from mutual distortion.[32] The long-range three-atom force is the Axilrod-Teller force, whose most salient feature is the strong dependence that it has on the relative angular orientation of the atoms. This is very typical of three-body forces, whatever their origin, and was a feature of our classical example. This property will be important to us later.

We see that many three-body forces are largely a matter of definition, rather than fundamental. In order to make calculations tractable, we deal with the interactions of composite systems, rather than their constituents. Although the *constituents* may interact via two-body forces, the *composite* objects can interact via effective many-body forces. For our purposes we define three-nucleon forces as those forces which depend in an irreducible way on the simultaneous coordinates of three nucleons, when only nucleon degrees of freedom are taken into account. One new element appears in our definition, the word "irreducible". In our classical example we separated the total force into two-body forces between pairs of objects (e.g., satellite-earth) and whatever was left over. It is important not to confuse the *sequential* interactions of two-body forces as a three-body force; that is the meaning of "irreducible". It is a simple concept but a complicated technical matter to put it into practice[33], and the latter requires much more time than we have here.

This brief introduction to a fascinating subject brings us to the topic of interest: three-nucleon forces. The types of processes which can contribute are illustrated in Fig. 12. We are primarily interested in pion-range forces, since the pion has the longest range ($\approx \hbar/\mu c$, where μ is the pion mass). We hope, on the basis of arguments to be presented later in connection with exchange currents, that the longest-range forces will dominate. Fig. 12a shows the generic two-pion-exchange three-nucleon $(2\pi-3N)$ force. A π^+ is

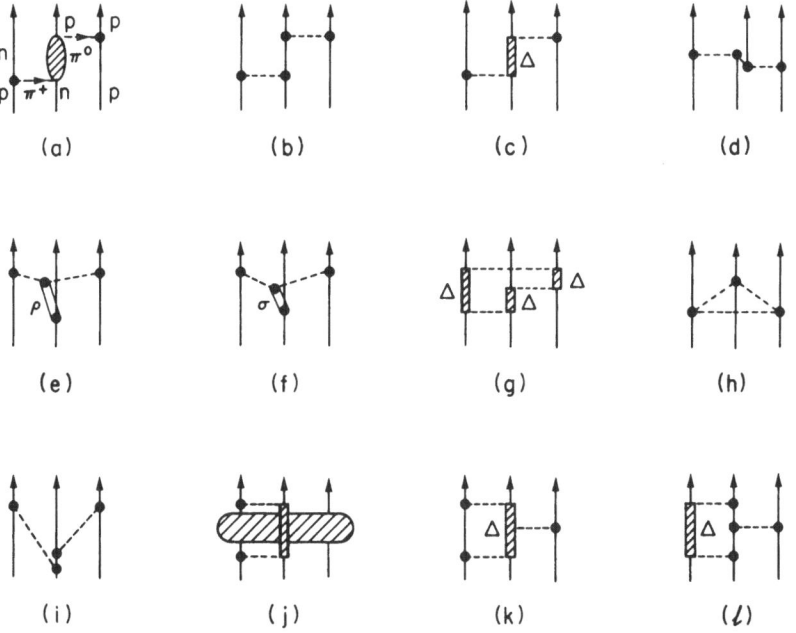

Fig. 12   Various physical processes contributing to three-nucleon forces. Solid, dashed, shaded and double lines depict nucleons, pions, isobars (nucleon excited states), and heavy mesons, respectively.

emitted by the proton on the left, propagates and scatters from the middle nucleon (the "blob" represents the scattering mechanism), then turns into a π° which is absorbed by the rightmost proton. The many possible combinations of pion charges means that this force has a complex isospin structure. Because we are dealing with pions, it also has a complicated spin structure. Moreover, the pion is not real, but virtual or "off-shell". Figs. 12b-f are possible components of Fig. 12a. The second process is "reducible": that is, it looks like two sequential exchanges of a pion, and hence is not fundamental. It must be discarded. Note that this reducible graph can be cut in two without breaking anything but *nucleon* lines. The next process is similar to the classical example; the second nucleon is "deformed" into a isobar by the pion exchange and leads to the conceptually important isobar-mediated three-body force. Fig. 12d has an intermediate nucleon-antinucleon pair, and leads to a force which is conceptually the same (but not structurally) as the atomic Primakoff-Holstein three-electron force[34]. In Figs. 12e and 12f the pion scatters from virtual mesons. The three-isobar force shown next is a 3π–3N force, which is conceptually similar to the Axilrod-Teller atomic force, because it is produced by the mutual distortion of three nucleons. The remaining processes can also contribute to 2π–3N and 3π–3N forces.[2]

Evidence

The results presented earlier strongly indicate that there is a defect in binding from conventional two-body forces. Moreover, the too large (calculated) radii are likely a symptom of this same problem, as we saw. There are several plausible explanations: (1) relativistic corrections have not been calculated; (2) three-body forces, which depend on the simultaneous coordinates of all 3 nucleons in the triton, have not been included; (3) our model Hamiltonians are simply inadequate, and the effects of nucleon structure or meson degrees of freedom should be taken into account. In fact, these categories are not distinct. Relativistic corrections can be broken down into one-body (kinetic-energy) terms, two-body (potential-energy) terms, and three-body (and higher) potential-energy terms. The size estimate we previously made of relativistic corrections (1-few percent), taken for the kinetic or potential energies (± 50 MeV), predicts a scale of 0.5-1 Mev. Those calculations that have been performed on the one- and two-body parts are consistent with this estimate, but find a tendency for cancellation between the attractive kinetic-energy correction and a repulsive potential-energy correction, leaving a small residue. It is also known that a substantial part of the two-pion-exchange three-body force is a relativistic correction[33] of order V_π^2/Mc^2, where V_π is the usual one-pion-exchange potential (OPEP). Moreover, the isobar part of the former force shown in Fig. 12c is due to nucleon substructure: a pion emitted by nucleon 1 (virtually) polarizes nucleon 2 into an isobar, which decays back to a nucleon plus a pion, which is absorbed by nucleon 3. Most of the currently popular three-nucleon forces[35-39] have been derived by considering meson degrees of freedom. These forces clearly exist in nature, but are they large enough to solve our binding problem?

Another long-standing problem has been a good theoretical understanding of the ^{3}He charge form factor, or the Fourier transform of the charge density. The form factor, (Fig. 22, to be shown later), has a typical diffraction shape, as a function of q, the momentum transfer, falling rapidly through zero, becoming negative in the secondary maximum, and then positive again. The difficulty has been that theoretical calculations have predicted too small a (negative) strength in the secondary maximum. The point-nucleon charge density $\rho_{ch}(r)$ constructed from the experimental form factor $F_{ch}(q^2)$ is consequently much lower than theoretical calculations near the origin[40], as shown in Fig. 13. This follows from the Fourier transform relationship

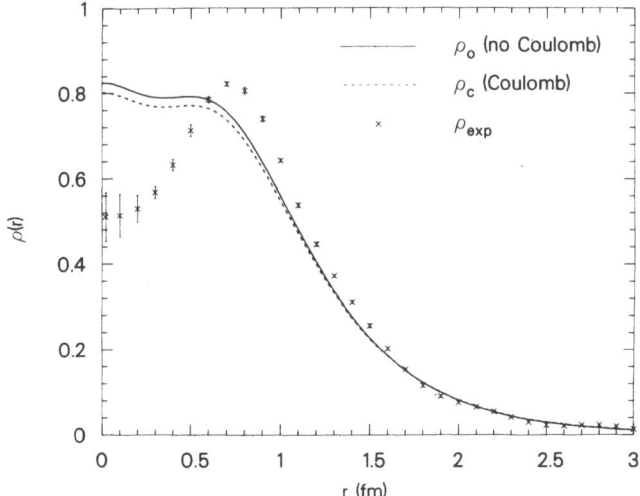

Fig. 13   Experimental (x's) and theoretical charge densities for ^{3}He. The theoretical curves correspond to including or not including a Coulomb force between the protons in ^{3}He.

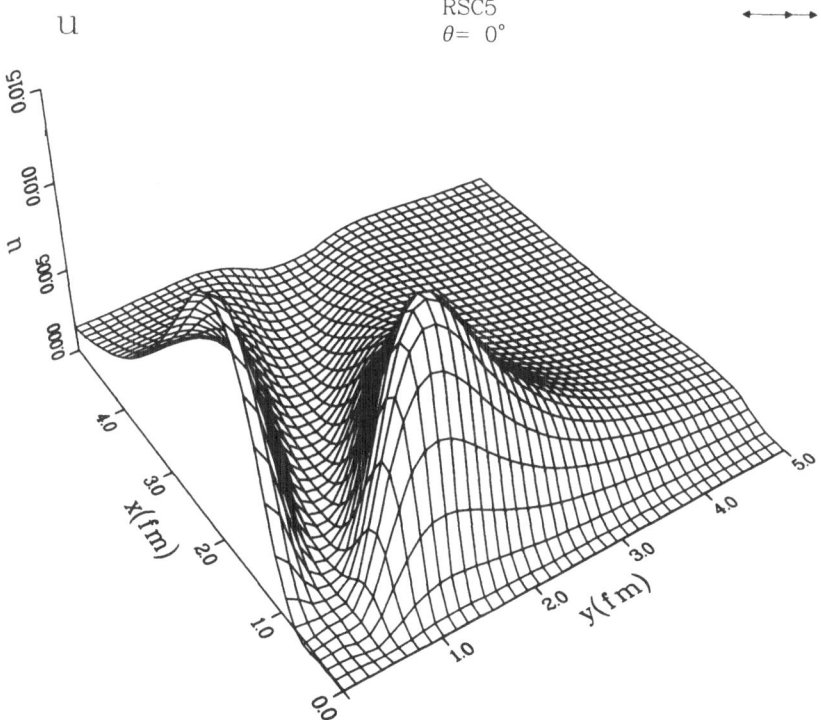

Fig. 14   The spatially symmetric (S-state) component, u, of the Schrödinger wave function from a 5-channel RSC calculation for $\theta = 0°$, plotted versus x and y.

$$\rho_{ch}(0) = \frac{1}{2\pi^2} \int_0^\infty F_{ch}(q^2)q^2 dq \quad . \tag{18}$$

Clearly a large negative contribution to F_{ch} lowers $\rho_{ch}(0)$. The argument that we have presented is somewhat controversial[2], because values of F_{ch} for very large q are needed in order to make the integral converge, and this requires considerable theoretical assumptions and extrapolation, some of which may be dubious. Nevertheless, there is a problem with the form factor, as we will see later.

In impulse approximation the charge density measures the probability of finding a proton at a distance r from the trinucleon center-of-mass, indicated by the x in Fig. 1. Taking nucleon 1 to be that proton, we have $r = 2y/3$, and forcing r to zero makes y zero. This is the condition of all three nucleons existing in a collinear configuration. Binding, on the other hand, prefers equilateral or isosceles configurations, so that each nucleon can be attracted by the short-range force of each of the other nucleons. Both of our problems with experiment could be solved if the three-nucleon force were attractive for equilateral configurations and repulsive for collinear ones. Schematic models of the force have this structure, and produce both effects, although other models may not. We note that r=0, or y=0, does not correspond to the "hole" in the wavefunction produced by the strong

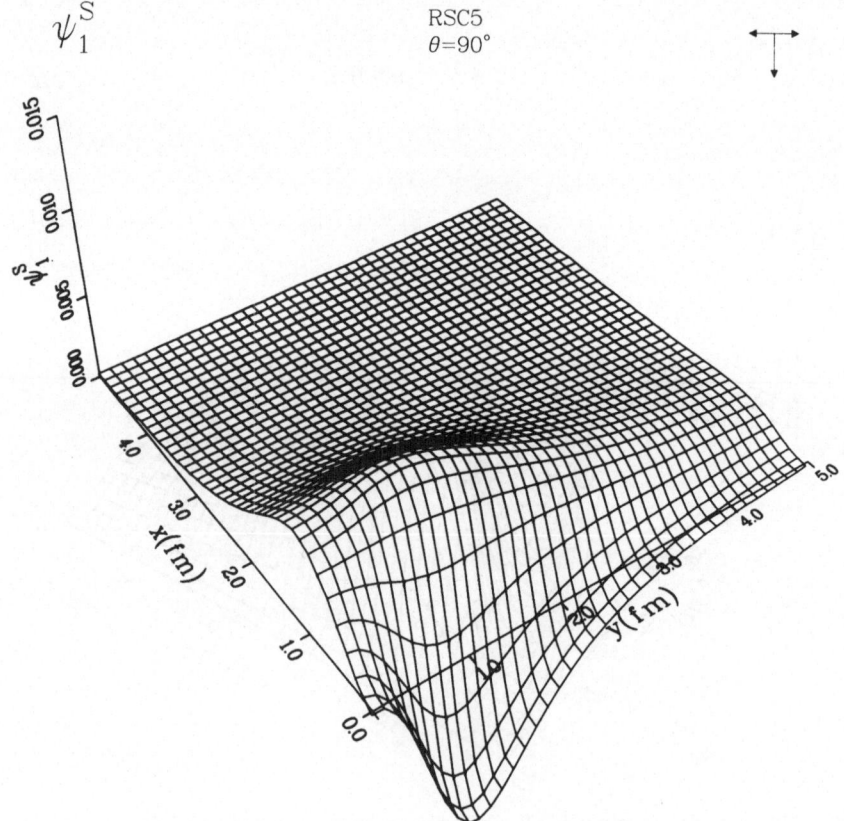

Fig. 15   The Faddeev wave function component, ψ_1^S, which generated u in Fig. 14, plotted versus x and y.

short-range repulsion. The S-state component of the wave-function for $\theta=0°$, corresponding to a 5-channel RSC potential calculation, is shown in Fig. 14. The deep valley at $\theta'=30°$ reflects that repulsion, while $\rho_{ch}(0)$ is given by an integral over x, along y=0. This Schrödinger wavefunction is generated by the much smoother Faddeev wavefunction component ψ_1^S shown in Fig. 15.

In addition to bound states, the trinucleons have a rich continuum structure. At very low (essentially zero) energy the scattering of a nucleon from the deuteron can be characterized by a single observable, the scattering length, a, which can be decomposed into spin-doublet (a_2) and spin-quartet (a_4) components. The latter is quite uninteresting, because it seems to depend only on the deuteron's binding energy due to the effect of the Pauli principle in the quartet state; consequently, all "realistic" force models produce nearly the same result. Calculated doublet scattering lengths, on the other hand, have been too large. Typical values are shown in Fig. 16, where a_2 has been calculated[41] for a variety of realistic and unrealistic two- and three-body force models. These results for pd and nd scattering separately fall on "Phillips lines"[42] when plotted versus the corresponding triton or ^{3}He binding energy. The fit to the nd results passes through the experimental datum; the pd result does not, which is a mystery at this time. The fact that all of the nd doublet results track the same Phillips line indicates that whatever physical mechanism corrects the binding defect will also produce a correct value for a_2, at least for the nd case.

Finally, analyses of the nn-scattering length, a_{nn}, from two separate experiments, n+n→(n+n)+p and π^-+d→(n+n)+γ, have produced three different values of a_{nn}. It has been argued[43] that three-nucleon forces, conspicuously missing in the latter reaction and not included in the analyses of the former reaction, might produce agreement among the values of a_{nn} from the different reactions. Only schematic calculations have been performed to date[44].

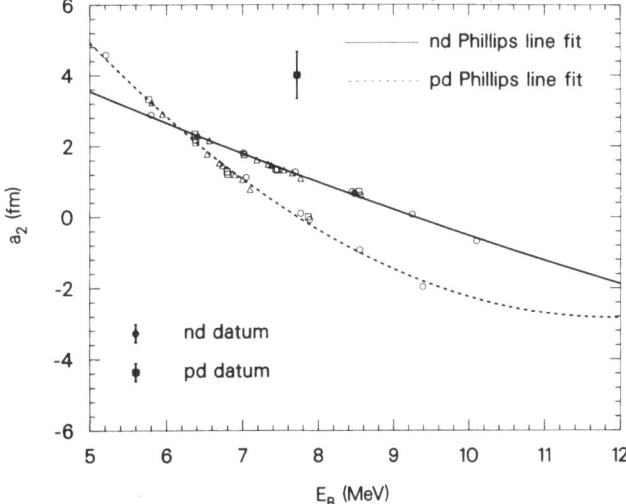

Fig. 16   Doublet Nd (nd and pd) scattering lengths plotted versus ^{3}H and ^{3}He binding energies, respectively. Individual points are from theoretical calculations (squares, triangles, and circles correspond to realistic two-body force models, the additional inclusion of three-body forces, and unrealistic two-body force models).

The evidence we have presented is tantalizing, but it is at best circumstantial. At present the best evidence exists in the properties of the bound state. Can current models of the three-nucleon force produce a substantial increase in binding? At least four such models have been used recently: (1) the Tucson-Melbourne (TM) two-pion-exchange force[35]; (2) the Brazilian (BR) two-pion-exchange force[36]; (3) the Urbana-Argonne (UA) schematic force[37]; (4) the Hajduk-Sauer isobar model[38]. Hajduk and Sauer do not explicitly include a separate three-body force in their model, but rather include isobar components in their wavefunctions. Three-body-force contributions, implicitly included in their model, must be deduced later. The TM and BR models incorporate Figs. 12c-12f into their forces.

Calculations

The early calculations used different force models and various approximations, which resulted in a chaotic situation, some calculations finding negligible additional binding and others finding more than one MeV. The situation has recently been clarified in part[45]. Most calculations had resorted to perturbation theory using 5-channel wave functions[46], which fails badly. Perturbation theory is inadequate for the TM model, giving results which are much too small. The 5-channel wave function approximation is also inadequate in general, as noted by Hajduk and Sauer[38], because the pion-exchange potentials tend to couple to small wave function components not adequately represented in the 5-channel approximation; 34 channels are required for complete convergence[45]. The latter calculations found approximately 1.5 MeV additional binding from both the TM and BR forces, in combination with two different two-body force models. Calculations of $\rho_{ch}(0)$ are not completed.

Although these results indicate a substantial three-body force effect, caution is required. Hajduk and Sauer find a small (-.3 MeV) three-body force effect. Their approach is very different from the TM and BR groups, and the physical reasons for the discrepancy are not known. Moreover, the "long-range" two-pion-exchange force is unfortunately quite sensitive to its short-range behavior, and it is possible to substantially lower the binding by making plausible modifications of this behavior. This field is in its infancy and much more work needs to be performed.

Finally, Fig. 17 shows a possible scheme[47] for determining the size of three-body forces by exploiting its angular dependence in the continuum. The initial pd configuration can be broken up into a p+p+n final state, which is measured in an equilateral configuration (b) and in a collinear one (c). This very difficult experiment might shed light on such forces, by looking for the expected additional attraction in the former configuration and repulsion in the latter.

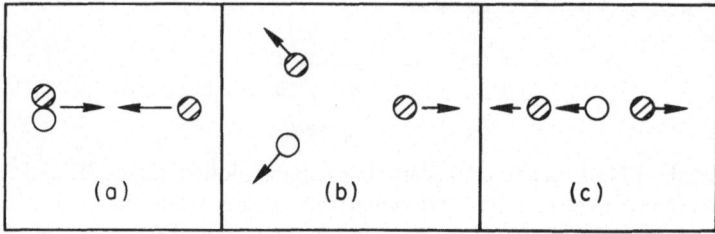

Fig. 17   Scenario for probing three-nucleon forces with pd initial state (a) becoming equilateral (b) and collinear (c) three-body breakup configurations.

ELECTROMAGNETIC INTERACTIONS

Relativistic Corrections

If relativistic effects are corrections (rather than dominant), an expansion of opera-tors[48–49] in powers of (v/c) could prove useful in explicating the physics of various processes. The nonrelativistic charge operator (ρ_0) has the form

$$\rho_0(\vec{x}) = \sum_{i=1}^{A} e_i \delta^3(\vec{x}-\vec{x}_i) \ , \tag{19}$$

which confines the charge at a point $\vec{x}$ to those nucleons located at $\vec{x}_i$ with $e_i = 1$ (pro-tons), rather than $e_i = 0$ (neutrons). The classical current operator has two components,

$$\vec{J}_0(\vec{x}) = \sum_{i=1}^{A} e_i \left\{ \frac{\vec{p}_i}{2Mc}, \delta^3(\vec{x}-\vec{x}_i) \right\} + \vec{\nabla} \times \sum_{i=1}^{A} \mu_i \frac{\vec{\sigma}(i)}{2Mc} \, \delta^3(\vec{x}-\vec{x}_i) \ , \tag{20}$$

where e_i, μ_i, $\vec{p}_i$, and $\vec{\sigma}(i)$ are the charge, magnetic moment (in nuclear magnetons), momentum, and (Pauli) spin operators for nucleon i, and M is the common nucleon mass. The first term is the convection current, produced by charges moving with velocity $\vec{p}/M$. The second is the magnetization current, produced by the elementary "bar magnets" that are the individual nucleons. Note the explicit factor of 1/Mc in each. The currents are smaller than the charge by factors of 1/c and 1/M, and M is a large number on the scale of nuclear momenta.

In order to count powers of 1/M as equivalent to powers of 1/c, we must use the fact that nuclei are weakly bound, and the potential and kinetic energies ($p^2/2M$) are nearly equal and opposite. Consequently, we expect momentum-(kinetic) and potential-dependent operators of the same order in (1/c) to be comparable. For both these reasons we reckon potential energy as order (1/M), and thus a series in 1/M is the same as a series in 1/c. These are our "rules of scale".

Corrections to the charge operator ρ_0 (order $(1/c)^0$) are of order $1/c^2$, or $1/M^2$, as ori-ginally argued by Siegert[29]. They include[26] the spin-orbit interaction charge density, the Darwin-Foldy term, and various meson-exchange contributions. The spin-orbit and Darwin-Foldy terms play an important role in the charge density differences of iso-topes[48], and the former produces a relativistic correction to the dipole operator, which is the dominant such correction to deuteron forward photodisintegration. Corrections to the current operator are of order $1/c^3$ (or $1/M^3$) and higher. They include meson-exchange contributions. The scheme we have listed here, with the components of the charge opera-tor being of order $(1/c)^0 + (1/c)^2 + (1/c)^4 + \cdots$, and the current being of order $(1/c)+(1/c)^3+...$, is not the only one possible, in general. The other possibility is for the leading-order current to be of order $(1/c)^0$ and the charge to be of leading order $(1/c)$. The former type of current, whose archetype is the electro-magnetic or vector current, is denoted[49] class I, while the latter is termed class II. Does any simple example of the latter class exist? The answer is yes and is exemplified by the axial vector current, which is important in β-decay. The Gamow-Teller β-decay operator is the axial current, $\vec{J}_A$, and is of nonrelativistic order $(1/c)^0$.

Current Conservation and Exchange Currents

The nonrelativistic currents we wrote in Eq. 20 are also the standard currents of atomic physics. They do not depend on the binding potential. Is the same true for nuclear physics? The answer is no, and points out an important qualitative difference between nuclear and atomic physics: binding in atomic physics is accomplished via exchange of *neutral* virtual quanta (photons), while in nuclear physics at least half of the binding arises from the exchange of charged quanta (e.g., mesons). The difference is qualitative, because the motion of any charged particle generates a current in both classical and quantum physics. In a weakly bound system of heavy particles, the binding quanta (mesons) move very rapidly compared to the nucleons, and hence the charge is largely confined to the heavy particles. The charge operator in Eq. 19 simply reflects this statement. Meson-exchange corrections to ρ_0 arise from nucleon recoil and the finite time of propagation of the mesons between nucleons, and are at least second order in $1/c^2$. On the other hand, the weak binding argument we produced earlier would indicate that the nonrelativistic nuclear current gets large (50% to 100%) contributions from potential-dependent currents. This estimate turns out to be too high because of an accident of nature. The big exchange-current effects in the two-body problem we discussed earlier were found in isospin-changing (isovector) magnetic dipole transitions, which primarily involve the magnetization current. The isovector part of that current is proportional to the isovector nucleon magnetic moment, $\mu_v^0 = \mu_p - \mu_n = 4.71$ n.m., whose large size suppresses the fractional exchange-current contribution. Were μ_v^0 of "normal" size like the isoscalar moment, $\mu_s^0 = \mu_p + \mu_n = 0.88$ n.m., exchange-current effects in nuclei would be typically 50%!

The single most important theoretical aspect of electromagnetism is gauge invariance, which follows from the masslessness of the photon. It must be possible to make a photon's wave function orthogonal to the Poynting vector (i.e., transverse) in any frame of reference, because the photon's helicity is an observable. A Lorentz transformation can alter the photon's wavefunction, however, and gauge invariance is the condition which restores transversality. For processes which involve only a single photon, real or virtual, gauge invariance is exactly equivalent to conservation of the electromagnetic current, whose components are ρ and $\vec{J}$:

$$\vec{\nabla} \cdot \vec{J}(\vec{x}) = -i[H, \rho(\vec{x})] \ , \tag{21}$$

where H is the strong interaction Hamiltonian.

If we write $H = T + V$, where T and V are the kinetic and potential energies, and use isospin notation for $e_i = (1 + \tau_z(i))/2$ in Eq. 19, we see that those parts of the potential between nucleons i and j which are isospin dependent $[(\vec{\tau}(i) \cdot \vec{\tau}(j))V_{ij}]$ will not commute with ρ, and hence there must be exchange or potential-dependent currents in $\vec{J}$ to make current conservation possible. The strong isospin dependence of the force guarantees large exchange currents, as we previously argued. We note that the magnetization current is divergenceless (solenoidal), and the convection current satisfies Eq. 21 in conjunction with the kinetic energy, T.

The existence of these currents does not mean that we can calculate them. Indeed, we are faced with the same problem that has confronted nuclear physics from its beginnings: without a tractable model of the strong interactions, we are able to calculate only in perturbation theory, which does not obviously converge. Fortunately, an accident of nature rescues us from this dilemma. The nucleon-nucleon interaction is strongly repulsive for small separations, and this makes the probability of finding nucleons in such configurations very unlikely. It also means that the matrix elements of any short-range current operators are greatly suppressed, and the longest-range operators should dominate. This is illustrated in Fig. 18, which shows the two-body trinucleon correlation function C(x), formed by integrating $|\Psi|^2$ over y and $\hat{x} \cdot \hat{y}$. Exchange currents would contribute to the ground state proportional to $\int C(x)\vec{J}(x)x^2 dx$. The maximum value of C(x) falls between one- and two-pion -range as indicated by the arrows, and ρ-meson

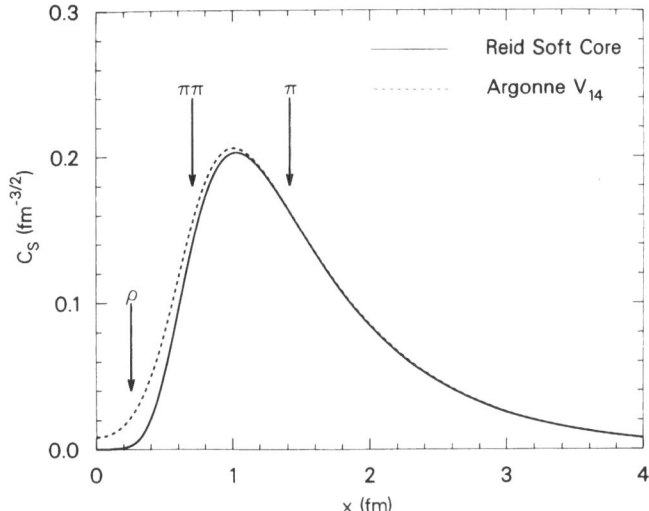

Fig. 18   Two-body triton (isoscalar) correlation functions for the RSC and AV14 potential models, together with the ranges of various meson exchanges.

range corresponds to a very small value of $C(x)$. The additional factor of x^2 in the
volume element further accentuates the long-range operators. We expect on the basis of
these arguments that the longest-range currents, the one-pion-exchange (OPE) currents,
should dominate, and explicit calculations bear this out.

The pion's mass is much smaller than that of any other meson, which appeared to be
accidental until the discovery of (approximate) chiral symmetry. Not only does that symmetry account for the small mass, it places constraints on the pion's interactions[50] with
other hadrons, which allows many calculations to be performed that would otherwise be
dubious. Because OPE currents play the dominant role in exchange currents, and
because of their past and continuing importance in our field, we derive them from "first
principles" in the following section.

<u>One-Pion-Exchange Currents</u>

Fig. 19 shows the four dominant processes involving the exchange of a single pion.
Fig. 19a depicts the OPE potential arising from π^+ exchange. In addition there are contributions from π^- and π^0 exchange. In few-nucleon systems the OPE potential is
extremely important, and dominates the binding; it is attractive and has a very strong tensor force. Figs. 19b-19d show how a pion influences the electromagnetic interaction of a
nucleus: the cross and wiggly line denote an external electromagnetic interaction, which
produces a pion (photopion production) on one nucleon that is later absorbed by a second
nucleon. Processes (b), (c), and (d) are the "seagull", "true-exchange", and "isobar"
portions of the pion-exchange current.

Because of their importance, we will derive the operators corresponding to Fig. 19.
We are only interested in the non-relativistic portions of these processes and consequently work in the static limit ($M\rightarrow\infty$). The basic building block we need is the pion-
nucleon vertex, $j_\pi^\alpha(\vec{x})$, two of which comprise OPEP. Because the pion has three charge
states (π^+,π^0,π^-), the vertex must be isovector, indicated by the (isovector) index α. This

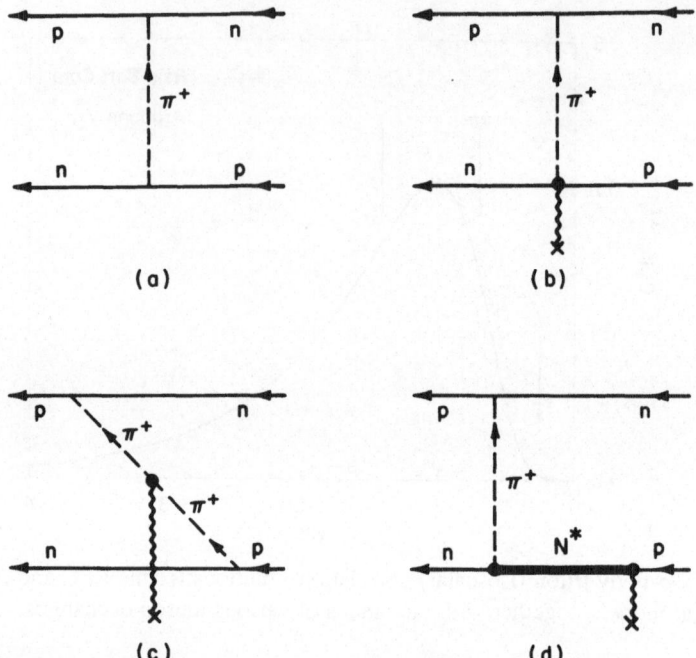

Fig. 19   One-pion-exchange processes contributing to OPEP in (a), and to the OPE
          currents in (b)-(d).

vertex determines the probability of a virtual pion, whose wave function is $\phi_\pi^\alpha(\vec{x})$, being
emitted from nucleon 1 at the point $\vec{x}$:

$$H_{\pi NN} = \int \Psi_f^\dagger(\vec{x}) \, j_\pi^\alpha(\vec{x}) \, \Psi_i(\vec{x}) \, \phi_\pi^\alpha(\vec{x}) \, d^3x \quad , \tag{22}$$

where

$$j_\pi^\alpha(\vec{x}) = \frac{-f}{\mu} \vec{\sigma}(1) \cdot \vec{\nabla}_1 \delta^3(\vec{x} - \vec{x}_1) \, \tau^\alpha(1) \quad . \tag{23}$$

The integrated probability of a pion of mass μ and charge state α being emitted by a
nucleon which moves from state i to state f is given by the Hamiltonian $H_{\pi NN}$, while the
various parts of the vertex operator j_π^α have simple physical interpretations. The δ-
function reflects the locality of the interaction; the pion can only be emitted from a
nucleon located at $\vec{x}_1$. The pion has spin 0 and negative parity. Because parity is conserved in the strong interactions, the vertex must include a compensating negative-parity
operator, and only $\vec{\nabla}$ (the pion's momentum operator) is available, because we have ruled
out the nucleon's momentum (we are working in the static limit). The vertex must be
Hermitian and a scalar and the only other vector available is $\vec{\sigma}(1)$, the nucleon's spin.
Alternatively, Eq. 23 reflects the pion's preferred p-wave interaction with a nucleon. The
nucleon isospin operator $\tau^\alpha(1)$ allows $H_{\pi NN}$ to be an isospin scalar of the form $\vec{\tau} \cdot \vec{\phi}_\pi$ or
$\tau^\alpha \phi_\pi^\alpha$. The dimensionless coupling constant f ($f_0^2 \equiv f^2/4\pi \cong 0.079$) determines the strength
of the interaction and the (-) sign is conventional. We see that the *static* j_π^α is uniquely
determined by invariance (spin, parity, isospin) arguments.

The other bit of physics we require is the equation of motion of the pion field, which allows us to propagate a pion from one point to another, or, equivalently, to "tie together" two vertices. This equation for the pion field is the usual wave equation for massive particles, and mirrors the analogous Maxwell equation for the electromagnetic vector potential:

$$(\vec{\nabla}^2 - \mu^2 - \frac{\partial^2}{\partial t^2}) \, \phi_\pi^\alpha(x) = j_\pi^\alpha(x) \quad , \tag{24}$$

where we have allowed for finite propagation time of the pion. This violates our static assumption and we should neglect all time dependence, resulting in

$$(\vec{\nabla}^2 - \mu^2) \phi_\pi^\alpha(\vec{x}) = j_\pi^\alpha(\vec{x}) \quad , \tag{25}$$

which can be solved for $\phi_\pi^\alpha(\vec{x})$:

$$\phi_\pi^\alpha(\vec{x}) = \frac{-1}{4\pi} \int h_0(|\vec{x} - \vec{y}|) J_\pi^\alpha(\vec{y}) \, d^3y = \frac{f}{4\pi\mu} \, \tau^\alpha(1)\vec{\sigma}(1) \cdot \vec{\nabla}_1 h_0(|\vec{x} - \vec{x}_1|) \quad , \tag{26}$$

when the static pion propagator is

$$h_0(z) = \frac{e^{-\mu z}}{z} \quad . \tag{27}$$

The integrated energy shift due to the exchange of a pion being emitted by nucleon 1 at $\vec{y}$ and absorbed by nucleon 2 at $\vec{x}$ is then simply given by perturbation theory:

$$\Delta E = \int j_\pi^\alpha(\vec{x}) \phi_\pi^\alpha(\vec{x}) d^3x = \frac{-1}{4\pi} \iint j_\pi^\alpha(\vec{x}) \, h_0(|\vec{x} - \vec{y}|) j_\pi^\alpha(\vec{y}) \, d^3x \, d^3y \quad . \tag{28}$$

In this exercise we have ignored the nucleon wave function (Ψ) in Eq. 22 because we are constructing an operator in the Hilbert space of the nucleons. Identifying ΔE as the OPE potential, $V_\pi(\vec{r})$, we obtain after performing the integrals,

$$V_\pi(\vec{r}) = \frac{f_0^2}{\mu^2} \, (\vec{\tau}(1) \cdot \vec{\tau}(2)) \, \vec{\sigma}(1) \cdot \vec{\nabla} \, \vec{\sigma}(2) \cdot \vec{\nabla} \, h_0(r) \quad , \tag{29}$$

where $\vec{r} = \vec{x}_1 - \vec{x}_2$. The isospin dependence is explicit and is responsible for exchange currents, as we indicated earlier. Moreover, the derivatives lead to the extremely important tensor force.

The model-independent pion electromagnetic interactions are also easily obtained. Typically, the electromagnetic interaction vertices can be obtained by means of the "minimal" substitution: all factors of momentum, $\vec{p}$, in the strong interaction Hamiltonian are replaced by $(\vec{p} - e\vec{A})$, where $\vec{A}$ is the electromagnetic vector potential and e is the fundamental charge. Indeed, this is the origin of the nucleon convection current via the kinetic energy, T. We also note that the magnetization current is special (it is solenoidal, or divergenceless), and does not follow from such arguments. In Eqs. 22 and 23, we resort to a trick and write

$$\vec{\nabla}_1 \phi_\pi(\vec{x}_1) \cdot \vec{\tau}(1) \text{ as } i[\vec{p}_1, \vec{\tau}(1) \cdot \phi_\pi(\vec{x}_1)], \text{ replace } \vec{p}_1 \text{ by } \vec{p}_1 - e_1 \vec{A}_1 \,,$$

and use the isospin form of e_1, which leads to

$$H_{\gamma\pi N} = -e \iint \Psi_f^\dagger(\vec{x}) \, \vec{J}_{SG}(\vec{x}) \Psi_i(\vec{x}) \cdot \vec{A}(\vec{x}) \, d^3x \quad , \tag{30}$$

where

$$\vec{J}_{SG}(\vec{x}) = \frac{-f}{\mu} \, \vec{\sigma}(1)\delta^3(\vec{x} - \vec{x}_1)(\vec{\tau}(1) \times \vec{\phi}_\pi)_z \quad , \tag{31}$$

and we have used $[\tau_z, \vec{\tau}\cdot\vec{\phi}_\pi] = -2i(\vec{\tau}\times\vec{\phi}_\pi)_z$. The electromagnetic current operator corresponding to Fig. 19b can now be easily calculated in perturbation theory using Eq. 26:

$$\Delta E = -e\int \vec{J}_{SG}(\vec{x})\cdot\vec{A}(\vec{x})\,d^3x \ , \tag{32}$$

where

$$\vec{J}_{SG}(\vec{x}) = \frac{f_0^2}{\mu^2}\vec{\sigma}(1)\delta^3(\vec{x}-\vec{x}_1)\vec{\sigma}(2)\cdot\vec{\nabla}_x h_0(\vec{x}-\vec{x}_2)(\vec{\tau}(1)\times\vec{\tau}(2))_z + (1\leftrightarrow 2) \ . \tag{33}$$

This result is quite obvious, given Eq. 31 and our previous derivation for OPEP; we simply replace ϕ_π^α in Eq. 31 by that in Eq. 26. There are *two* terms, because the process is not symmetrical in nucleons 1 and 2, unlike the OPEP case.

The remaining model-independent process is depicted in Fig. 19c, where the fundamental pion electromagnetic vertex is given by

$$H_{\pi\pi\gamma} = e\int[\phi_\pi^\dagger(\vec{x})\times\overleftrightarrow{\vec{\nabla}}_x\phi_{\vec{\pi}}(\vec{x})]_z\cdot\vec{A}(\vec{x})\,d^3x \ , \tag{34}$$

where $f\overleftrightarrow{\nabla}g \equiv f\vec{\nabla}g-g\vec{\nabla}f$. Up to a sign, the form is almost obvious, since the nucleon convection current is often written in the form: $(\Psi_f^\dagger\vec{p}\,\Psi_i + (\vec{p}\,\Psi_f)^\dagger\Psi_i)/2M$. The unusual aspect is the isospin dependence, which has been constructed[51] to give a (+) sign for the π^+ interaction, a (-) sign for the π^- interaction, and 0 for a π°. The true-exchange current is obtained by connecting *either* $\phi_{\vec{\pi}}$ ($\equiv\vec{\phi}_\pi$) to nucleon 1 and the other one to nucleon 2; because the interaction is symmetric in 1 and 2, we don't double count this way. We find

$$\vec{J}_{ex}(\vec{x}) = \frac{-f_0^2}{\mu^2}(\vec{\tau}(1)\times\vec{\tau}(2))_z\vec{\sigma}(1)\cdot\vec{\nabla}_x h_0(|\vec{x}-\vec{x}_1|)\overleftrightarrow{\vec{\nabla}}_x\vec{\sigma}(2)\cdot\vec{\nabla}_x h_0(|\vec{x}-\vec{x}_2|) \ , \tag{35}$$

which has the same isospin structure as $\vec{J}_{SG}(\vec{x})$. We also note that this process is semiclassical[51].

The remaining contribution is Fig. 19d. Because of its complexity, and the fact that it is model dependent, our derivation will be somewhat schematic. The Δ-isobar is a nucleon excited state with spin and isospin 3/2 and positive parity, which we treat as a static particle. The interaction depicted in Fig. 19d corresponds to the electromagnetic creation of the isobar and its subsequent decay by pion emission. We have not shown the additional process with the opposite time ordering, the electromagnetic interaction occurring last. The transition from $1/2^+$ to $3/2^+$ can be magnetic dipole or electric quadrupole, the latter being negligible. Current operators for magnetic dipole processes have the form

$$\vec{J}_\Delta(\vec{x}) = \vec{\nabla}_x\times\vec{M}_\Delta(\vec{x}), \quad \text{or} \quad H_{\Delta N\gamma} = -e\int\vec{M}_\Delta(\vec{x})\cdot\vec{B}(\vec{x})d^3x \ ,$$

where $\vec{B}$ is the magnetic field. Assuming that the electromagnetic interaction occurs on nucleon 1, the basic process is represented in second-order perturbation theory by

$$\vec{M}_\Delta(\vec{x}) = \frac{-f\mu_v}{2\mu M}\sum_{M_\Delta}\frac{<\frac{1}{2}M'|\vec{S}_\Delta^\alpha\cdot\vec{\nabla}_1\phi_\pi^\alpha(\vec{x}_1)|\frac{3}{2}\frac{3}{2}M_\Delta><\frac{3}{2}\frac{3}{2}M_\Delta|\hat{\vec{\mu}}_\Delta\delta^3(\vec{x}-\vec{x}_1)|\frac{1}{2}M>}{-\Delta M} .\tag{36}$$

where we have written the effective $(\pi N\Delta)$-vertex as $-(f/\mu)\vec{S}_\Delta^\alpha\cdot\vec{\nabla}_1\phi_\pi^\alpha(\vec{x}_1)$ in complete analogy with Eq. 22 ($\vec{S}_\Delta^\alpha$ is the $N\Delta$ transition-"spin" operator, which replaces $\vec{\sigma}_1$ in Eq. 23), the magnetic dipole *isovector* $(\gamma N\Delta)$-vertex as $\mu_v\hat{\vec{\mu}}_\Delta\delta^3(\vec{x}-\vec{x}_1)/2M$ in analogy with Eq. 20, and the energy denominator is simply the negative of the Δ-nucleon mass difference,

$-\Delta M = -300\,\text{MeV}$. The only complexity in Eq. 36 is the intermediate-state spin sum over the magnetic quantum numbers of both spin and isospin (M_Δ), which produces projection operators in the nucleon spin and isospin space. We give these without proof; the spin-projection operators for spin 1/2 or 3/2 intermediate states are:

$$\sum_{M_I} <\tfrac{1}{2}M' | \vec{A}_0\cdot\vec{V} | JM_I><JM_I | \vec{B}_0\cdot\vec{V}' | \tfrac{1}{2}M> \tag{37}$$

$$\equiv <\tfrac{1}{2}M' | \hat{P}_J | \tfrac{1}{2}M><\tfrac{1}{2}\,\tfrac{1}{2} | V_z | J\tfrac{1}{2}><J\tfrac{1}{2} | V_z' | \tfrac{1}{2}\,\tfrac{1}{2}> \;,$$

where $\vec{A}_0$ and $\vec{B}_0$ are constant vectors,

$$\hat{P}_{\frac{1}{2}} = \vec{A}_0\cdot\vec{B}_0 + i\vec{\sigma}\cdot\vec{A}_0\times\vec{B}_0 \;, \tag{38}$$

$$\hat{P}_{\frac{3}{2}} = \vec{A}_0\cdot\vec{B}_0 - \frac{i}{2}\vec{\sigma}\cdot\vec{A}_0\times\vec{B}_0 \;, \tag{39}$$

$\vec{V}$ and $\vec{V}'$ are any (vector) nuclear operators and $\vec{\sigma}$ is the nucleon spin operator. The isospin projection is analogous with $\vec{\sigma}\to\vec{\tau}$. The derivation is best performed using the Wigner-Eckart theorem in a brute-force manner.

We can now easily complete the derivation, using Eq. 39. The second time ordering, not shown in Fig. 19, is equivalent to the Hermitian conjugate of Eq. 36. Using this we find

$$\vec{J}_\Delta(\vec{x}) = \frac{\lambda\mu_v f_0^2}{4M\mu^2\Delta M}\vec{\nabla}\times\left\{ \delta^3(\vec{x}-\vec{x}_1)[(\vec{\tau}(1)\times\vec{\tau}(2))_z\vec{\sigma}(1)\times\vec{\nabla}_x - 4\tau_z(2)\vec{\nabla}_x]\vec{\sigma}(2)\cdot\vec{\nabla}_x h_0(\vec{x}-\vec{x}_2) \right.$$

$$\left. + (1\leftrightarrow 2) \right\} \;, \tag{40}$$

where

$$\lambda = <\tfrac{1}{2}\,\tfrac{1}{2} | \hat{S}_\Delta^{zz} | \tfrac{3}{2}\,\tfrac{1}{2}><\tfrac{3}{2}\,\tfrac{1}{2} | \hat{\mu}_\Delta^z | \tfrac{1}{2}\,\tfrac{1}{2}> \tag{41}$$

is calculated for a proton (isospin component $+1/2$) with spin component $+1/2$ using the z-component in both spin and isospin for the operators $\vec{S}^\alpha$ and $\vec{\mu}_\Delta$. Eq. 40 is model independent in the sense that only angular momentum arguments have been used in its construction. The operators $\vec{S}_\Delta^\alpha$ and $\vec{\mu}_\Delta$ were defined so that they are dimensionless, as is λ, which contains all of the model dependence. If we resort to the quark model for the nucleon and isobar, we find

$$\lambda^{QM} = \frac{16}{25} \tag{42}$$

while the Chew-Low model of the isobar[52] gives

$$\lambda^{CL} = \frac{4}{5} \;.$$

Deriving Eq. 42 is an excellent exercise (hint: $<\hat{S}_\Delta^{zz}> = 4\sqrt{2}/5$; $<\hat{\mu}_\Delta^z> = 2\sqrt{2}/5$).

This completes our derivation of the pion-exchange currents. The final result is

$$\vec{J}_\pi(\vec{x}) = \vec{J}_{SG}(\vec{x}) + \vec{J}_{ex}(\vec{x}) + \vec{J}_\Delta(\vec{x}) \;. \tag{44}$$

Another very good exercise is to verify that Eq. 21 holds for $\vec{J}_\pi$, with H replaced by OPEP.

Evidence for Exchange Currents

Most of the evidence for exchange currents centers on magnetic dipole processes, and in particular on static and transition magnetic moments. We have already mentioned[27] the isovector magnetic transition between the $^3S_1$ deuteron and the $^1S_0$ threshold state of the np system. What about the deuteron magnetic moment? Because the deuteron ground state is an isoscalar system, the (isovector) exchange currents we derived earlier do not contribute. The currents due to the exchange of positive and negative mesons *exactly* cancel; this is precisely the meaning of isoscalar. Only those exchange currents of (relativistic) order $(1/c^3)$ and higher contribute. Indeed, the deuteron magnetic moment is usually written[21] in the form

$$\mu_d = \mu_s^0 - \tfrac{3}{2} P(D)(\mu_s^0 - \tfrac{1}{2}) + \Delta\mu_d = .85774 \quad , \tag{45}$$

where the numerical value is experimental, $P(D)$ is the deuteron D-state probability and $\Delta\mu_d$ is the contribution from small relativistic corrections of various types. It is worth noting that relativistic corrections have an intrinsic ambiguity built into them; different methods for calculating them give different operators. This doesn't mean that observables, or matrix elements of these operators, are ambiguous. They are not, because the same ambiguities are contained in the nuclear potentials and the wavefunctions, and exactly cancel those in the operator. The ambiguity is, therefore, nothing more than a unitary transformation. It causes a complication in Eq. 45, however, since both $P(D)$ and $\Delta\mu_d$ are affected by it, although the ambiguity in both terms can be shown to cancel[21]. It does make it impossible to attribute any fundamental meaning to $P(D)$; that is, it is not measurable, and the division between the second and third terms in Eq. 45 is artificial. Nevertheless, the scale of the corrections $(\mu_d - \mu_s^0)/\mu_s^0$ is -.0251, and $P(D) = 3.9\%$ satisfies Eq. 45 if we arbitrarily set $\Delta\mu_d$ to zero.

The trinucleons have isospin 1/2, so that the magnetic moments can be broken down into an isoscalar component ($\mu_s = \mu(^3He) + \mu(^3H)$) similar to the deuteron case, and an isovector component ($\mu_v = \mu(^3He) - \mu(^3H)$). One finds that[53]

$$\mu_s = \mu_s^0 - 2P_D(\mu_s^0 - \tfrac{1}{2}) + \Delta\mu_s = 0.85131 \quad , \tag{46}$$

and

$$\mu_v = -\mu_v^0[1 - \tfrac{4}{3} P_s' - \tfrac{2}{3} P_D] + \Delta\mu_0 + \Delta\mu_v = -5.10641 \quad , \tag{47}$$

where $\Delta\mu_0$ is a very small contribution from orbital angular momentum, and $\Delta\mu_s$ and $\Delta\mu_v$ represent corrections to the impulse approximation. The isoscalar part is nearly identical to the deuteron case, and $P_D = 3.8\%$ produces equality between impulse approximation and experiment, when $\Delta\mu_s$ vanishes. The isovector case is rather different. Using reasonable values of $P_s' = 1\%$ and $P_D = 10\%$ produces $\Delta\mu_v/\mu_v^0 = -16.5\%$, compared to $\Delta\mu_s/\mu_s^0 = 5.4\%$. The scale of the two corrections is different, as is the sign.

The isoscalar discrepancy is consistent with a (large) correction of relativistic order, while the isovector case is much larger, and is indicative of nonrelativistic exchange currents. Note that the absolute size of $\Delta\mu_v$ would be comparable to the size of the impulse approximation, were μ_v^0 equal to 1, which is consistent with our previously discussed rules of scale for exchange currents. The size of the seagull part of the pion-exchange current is typically $\Delta\mu_v^{SG}/\mu_v^0 = -(14-15)\%$, or most of the discrepancy. The true-exchange and isobar part of the pionic current have opposite signs and typical values $\Delta\mu_v/\mu_v^0 \cong \pm 2\%$, with the upper and lower signs referring to the true-exchange and isobar contributions, respectively. The net theoretical result is slightly too small, but dramatically illustrates the importance of pion-exchange currents.

The analysis of the tritium β-decay matrix element is identical to that of the isovector magnetic moment in impulse approximation. The nuclear matrix elements are $(1+3g_A^2 M_A^2)$, where the superallowed Fermi part is 1, g_A is the axial vector coupling constant, and the Gamow-Teller matrix element has the form

$$|M_A| = 1 - \frac{4}{3}P_S' - \frac{2}{3}P_D + \Delta M_A = 1 - 0.042(8) \quad , \tag{48}$$

where the numerical value is experimental.[54] Using our previous estimate of probabilities we estimate[2] $\Delta M_A = 0.04(2)$. The size of this correction is consistent with a relativistic correction and our previous analysis that the impulse approximation axial vector current is class II, and has no nonrelativistic exchange currents.

Given the fact that the magnetic moments are very strongly affected by pion-exchange currents, we should also expect that the magnetic form factor, or Fourier transform of the magnetization density, is similarly affected. This is indicated for ^{3}He in Fig. 20 which shows three calculations by Hajduk, Sauer, and Strueve[39], together with the data. The shape of $|F_{mag}|$ as a function of the momentum transfer, q, is a typical diffraction structure. The impulse approximation (no exchange currents) based on the Paris potential has its diffraction minimum at much too small a value of q. Including the exchange currents (labeled isobar model) moves the diffraction minimum out toward the experimental results. Unfortunately, theoretical uncertainties in how to deal with the nucleon form factors (the large dots in Fig. 19) in the exchange currents lead to (at least) three different prescriptions, two of which are indicated in the figure (labelled F_1 and G_E).

The problem is seen most clearly in the charge form factor, which in impulse approximation has the structure

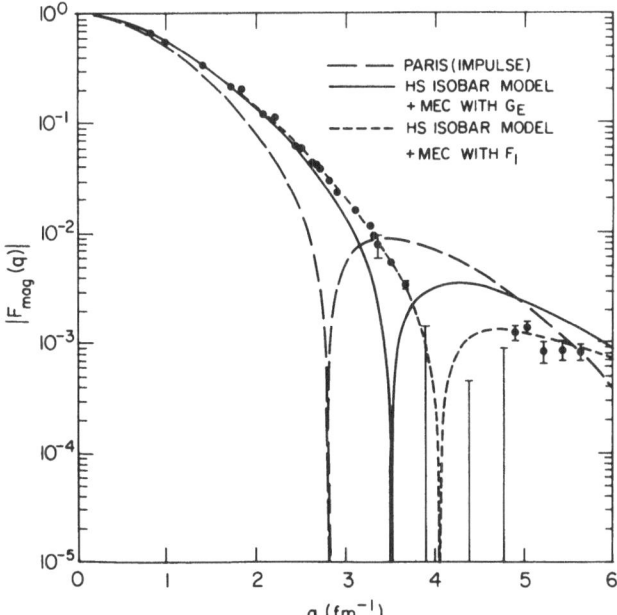

Fig. 20 ^{3}He magnetic form factor, together with 3 calculations by Hajduk, Sauer, and Strueve.

$$F_B(q^2) = \int d^3r \, e^{i\vec{q}\cdot\vec{r}} \rho_0(\vec{r}) \quad . \tag{49}$$

The subscript "B" indicates that this is the "body" form factor, the probability that, when one nucleon is struck and receives momentum q, the recoiling nucleus is capable of reconstituting itself in the ground state. That probability is rather small, because small momentum components, rather than large, are most probable in a nucleus, which is why form factors are always shown on semi-log plots! The most probable reaction is for the struck nucleon to be ejected. But in "grabbing" the nucleus the electromagnetic interaction must first grab a nucleon, and that does not have unit probability, but rather $G_N(q^2)(\leq 1)$, because the nucleon has its own structure. In accordance with accepted probability practice one takes the product of the probability distributions:

$$F_{ch}(q^2) = G_N(q^2)F_B(q^2) \quad . \tag{50}$$

In dealing with the pion-seagull exchange current, it is not known which type of nucleon form factor, G_N, to use, the electric (charge) form factors[48] F_1 or G_E, or the axial vector form factor, G_A. The experiment would appear to prefer F_1 or G_A (both are larger than G_E), but the calculations are not sufficiently unambiguous to allow that conclusion.

Similar results (see Fig. 21) are evident in the ^3H magnetic form factor together with new high momentum transfer data eagerly awaited for two decades[55]. The dashed curve

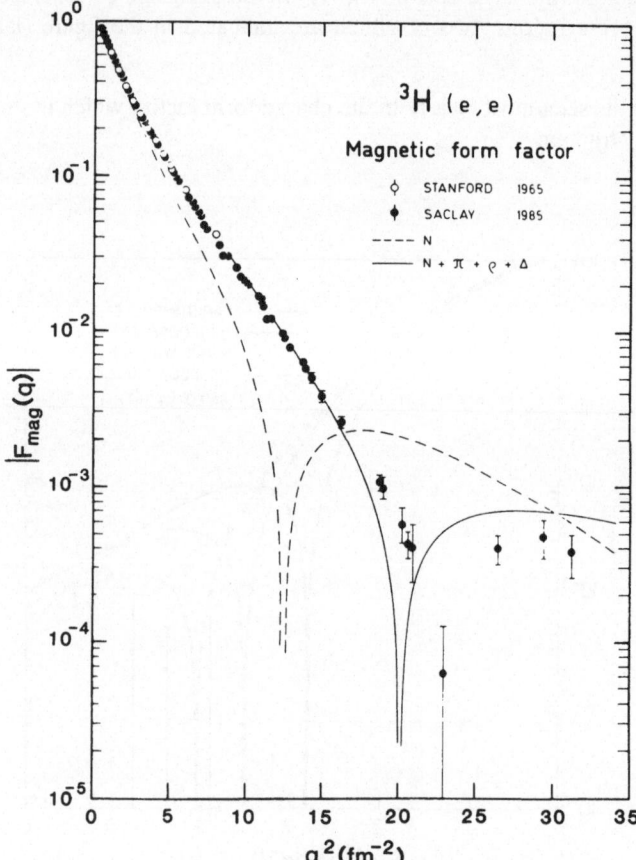

Fig. 21 ^3H magnetic form factor, together with 2 calculations by Hajduk, Sauer, and Strueve.

is the impulse approximation and the solid curve includes a variety of exchange currents. The agreement between theory and experiment is rather good.

Just as long-standing discrepancies between theory and experiment involving thermal np radiative capture pointed to exchange currents, so did problems with thermal nd radiative capture (n+d→³H+γ). In the latter case, the capture rate vanishes in impulse approximation if one assumes that all the forces between nucleons are identical. In that limit the (s-wave) ground state wavefunction is an eigenfunction of the magnetic moment operator, and the matrix element vanishes by orthogonality. This greatly suppresses the doublet part of the decay rate, which we see from Eqs. 46 and 47 if we drop the leading order μ_s^0- and μ_v^0-terms. The remaining probabilities are now the overlaps of the appropriate pieces of the two wavefunctions. In addition the decay can also proceed from the quartet part of the nd state. Recent measurements of the total rate summarized in Ref. 2 give a cross-section of .515(9) mb, 600 times smaller that the corresponding np case.

Extensive calculations by Torre and Goulard[56] have established that in impulse approximation the quartet rate is 20 percent larger than the doublet. The exchange currents lower the former by 20 percent while raising the latter by 500 percent, increasing the impulse approximation result of .2 mb to a total of .6 mb, in fairly good agreement with experiment. The seagull, isobar and true-exchange pionic currents contribute roughly in the ratio 3:2:-1.

We have concentrated almost entirely on the long-range pionic currents, motivated by the shape of the correlation function. Is this adequate? There are a wide variety of short-range contributions, most of which can be obtained from Fig. 19 by replacing a pion propagator by that of a heavy meson. Typically these contributions are 10-20 percent of the pionic ones, at least for small momentum transfers, and their calculation is much more model dependent.

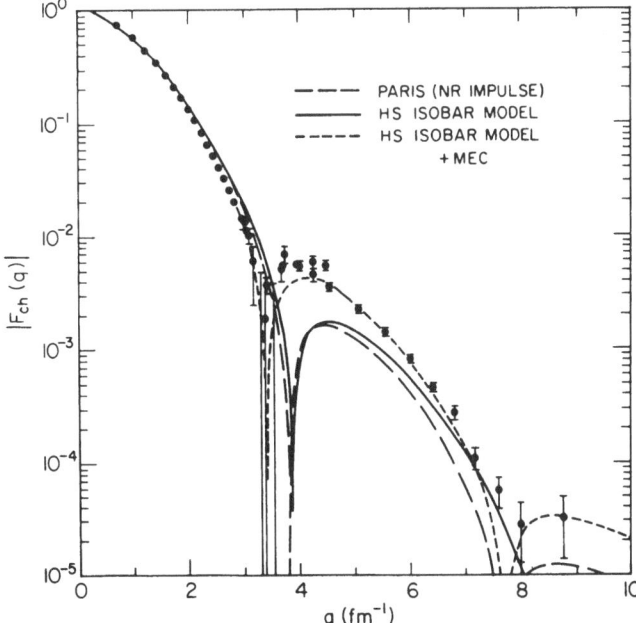

Fig. 22   ³He charge form factor, together with 3 calculations by Hajduk, Sauer, and Strueve.

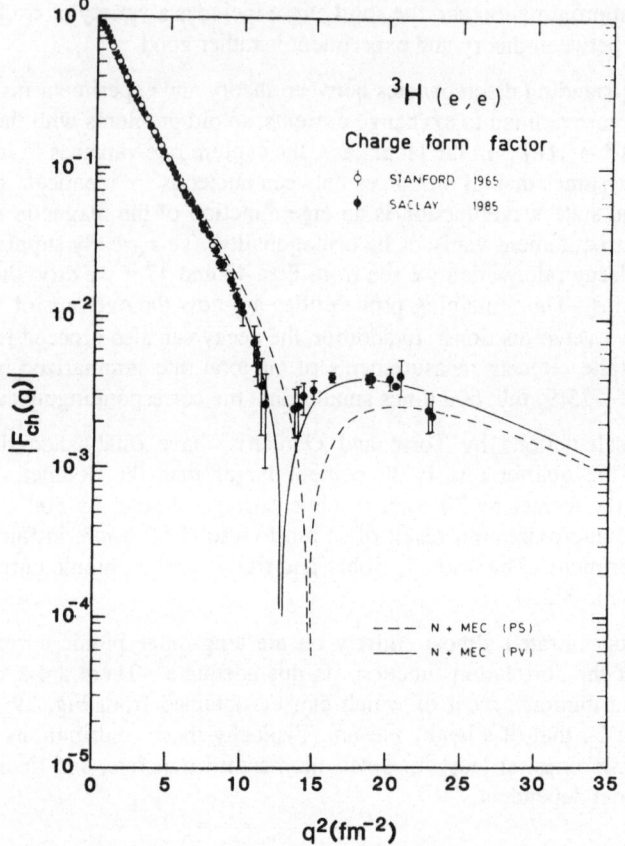

Fig. 23 ^{3}H charge form factor, together with 2 calculations by Hajduk, Sauer, and
          Strueve.

Finally, we show the charge form factors[55] of ^{3}He and ^{3}H in Figs. 22 and 23, together
with the calculations of Hajduk, Sauer, and Strueve[39]. The impulse approximation and
the HS isobar model for ^{3}He are deficient in the region of the secondary diffraction maximum. The addition of ambiguous exchange currents of relativistic order improves the
agreement. Unfortunately the "realistic" potential models which are used to calculate
wave functions don't have the correct form to accommodate relativistic corrections, and
the cancellations which must take place to eliminate the ambiguity from matrix elements
cannot take place. The results for ^{3}H are shown together with two calculations which
include exchange currents using different (nonstatic) models of the π-nucleon vertex (PS
and PV). This model dependence partially reflects the ambiguity discussed above. This
is a murky and technically complex subject, which the interested reader can find discussed elsewhere[26].

SUMMARY

We summarize by noting that in isovector magnetic dipole processes, pion-exchange
currents can make sizeable contributions. Contributions from heavy-meson exchange are

smaller, because nucleons don't like to overlap at small separations. The dominant meson-exchange effects occur in the isovector current operator, while small (ambiguous) contributions can occur to the charge, isoscalar current, and axial vector current operators.

ACKNOWLEDGEMENTS

This work was performed under the auspices of the U.S. Department of Energy. The author would like to thank his collaborators, G. L. Payne, B. F. Gibson, C. R. Chen, and E. L. Tomusiak for their assistance in preparing the lectures.

REFERENCES

[1]   H. A. Bethe, and R. F. Bacher, Rev. Mod. Phys. **8,** 82 (1936).

[2]   J. L. Friar, B. F. Gibson, and G. L. Payne, Ann. Rev. Nucl. Part. Sci. **34,** 403 (1984).

[3]   G. L. Payne, J. L. Friar, B. F. Gibson, and I. R. Afnan, Phys. Rev. **C22,** 823 (1980).

[4]   J. L. Friar, E. L. Tomusiak, B. F. Gibson, and G. L. Payne, Phys. Rev. **C24,** 677 (1981).

[5]   R. A. Malfliet, and J. A. Tjon, Nucl. Phys. **A127,** 161 (1969).

[6]   L. L. Foldy, and W. Tobocman, Phys. Rev. **105,** 1099 (1957).

[7]   W. Glöckle, Nucl. Phys. **A141,** 620 (1970).

[8]   G. L. Payne, W. H. Klink, W. N. Polyzou, J. L. Friar, and B. F. Gibson, Phys. Rev. **C30,** 1132 (1984).

[9]   L. D. Faddeev, Zh. Eksp. Teor. Fiz. **39,** 1459 (1960).

[10] H. P. Noyes, in ''Three Body Problem in Nuclear and Particle Physics'', edited by J. S. C. McKee and P. M. Rolph (North-Holland, Amsterdam, 1970), p. 2.

[11] W. Glöckle, Z. Phys. **271,** 31 (1974).

[12] P. M. Prenter, ''Splines and Variational Methods'', (Wiley, New York, 1975).

[13] C. De Boor, and B. Swartz, SIAM (J. Num. Anal.) **10,** 582 (1973).

[14] C. R. Chen, G. L. Payne, J. L. Friar, and B. F. Gibson, Phys. Rev. **C31,** 2266 (1985).

[15] R. V. Reid Jr., Ann. Phys. (N.Y.) **50,** 411 (1968).

[16] R. B. Wiringa, R. A. Smith, T. A. Ainsworth, Phys. Rev. **C29,** 1207 (1984).

[17] R. de Tourreil, and D. W. L. Sprung, Nucl. Phys. **A201,** 193 (1973).

[18] M. LaCombe, B. Loiseau, J. M. Richard, R. Vinh Mau, J. Côte, et al., Phys. Rev. **C21,** 861 (1980).

[19] C. Hajduk, and P. U. Sauer, Nucl. Phys. **A369,** 321 (1981).

[20] J. L. Friar, B. F. Gibson, C. R. Chen, and G. L. Payne, Phys. Lett. **161B,** 241 (1985)

[21] J. L. Friar, Phys. Rev. **C20,** 325 (1979).

[22] J. L. Friar, and B. F. Gibson, Phys. Rev. **C18,** 908 (1978).

[23] G. L. Payne, J. L. Friar, and B. F. Gibson, Phys. Rev. **C22,** 832 (1980).

[24] J. L. Friar, Nucl. Phys. **A156,** 43 (1970); M. Fabre de la Ripelle, Fizica **4,** 1 (1972).

[25] M. H. Kalos, Phys. Rev. **128,** 1791 (1962).

[26] J. L. Friar, B. F. Gibson, and G. L. Payne, Phys. Rev. **C30,** 441 (1984).

[27] D. O. Riska, and G. E. Brown, Phys. Lett. **38B,** 193 (1972).

[28] M. G. Olsson, E. T. Osypowski, and E. H. Monsay, Phys. Rev. **D17**, 2938 (1978).

[29] A. J. F. Siegert, Phys. Rev. **52,** 787 (1937).

[30] J. Kirk, private communication, points out that the inclination of the satellite orbit is changed in a significant (though very small) way by the tides. Computer codes which accurately calculate satellite motion must include this effect.

[31] B. M. Axilrod, and E. Teller, J. Chem. Phys. **11,** 299 (1943).

[32] R. J. Bell, and I. J. Zucker, in "Rare Gas Solids, Vol. I'', ed. by M. L. Klein and J. A. Venables (Academic Press, London, 1976), p. 122. The effects of three-atom forces are rather important in rare gas solids.

[33] J. L. Friar, and S. A. Coon, Phys. Rev. C (to be published).

[34] H. Primakoff, and T. Holstein, Phys. Rev. **55,** 1218 (1939).

[35] S. A. Coon, M. D. Scadron, P. C. McNamee, B. R. Barrett, D. W. E. Blatt, and B. H. J. McKellar, Nucl. Phys. **A317,** 242 (1979).

[36] H. T. Coelho, T. K. Das, and M. R. Robilotta, Phys. Rev. **C28,** 1812 (1983).

[37] J. Carlson, V. R. Pandharipande, and R. B. Wiringa, Nucl. Phys. **A401,** 59 (1983).

[38] C. Hajduk, and P. U. Sauer, Nucl. Phys. **A322,** 329 (1979).

[39] C. Hajduk, P. U. Sauer, and W. Strueve, Nucl. Phys. **A405,** 581 (1983), and to be published.

[40] J. L. Friar, B. F. Gibson, E. L. Tomusiak, and G. L. Payne, Phys. Rev. **C24,** 665 (1981).

[41] J. L. Friar, B. F. Gibson, Payne, G. L., and C. R. Chen, Phys. Rev. **C30,** 1121 (1984); and Phys. Rev. **C33** (to appear).

[42] A. C. Phillips, Rep. Prog. Phys. **40,** 905 (1977).

[43] I. Slaus, Y. Akaishi, and H. Tanaka, Phys. Rev. Lett. **48,** 993 (1982).

[44] A. Bömelburg, W. Glöckle, and W. Meier, in "Few Body Problems in Physics,'' Vol. II, ed. by B. Zeitnitz (North-Holland, Amsterdam, 1984), p. 483.

[45] C. R. Chen, G. L. Payne, J. L. Friar, and B. F. Gibson, Phys. Rev. Lett. **55,** 374 (1985).

[46] R. B. Wiringa, J. L. Friar, B. F. Gibson, G. L. Payne, and C. R. Chen, Phys. Lett. **143B,** 273 (1984).

[47] J. L. Friar, AIP Conference Proceedings **97,** 378 (1983).

[48] J. L. Friar, in "Electron and Pion Interactions with Nuclei at Intermediate Energies,'' W. Bertozzi, S. Costa, and C. Schaerf, eds., (Harwood, N.Y., 1980), p. 143.

[49] J. L. Friar, Phys. Rev. **C27,** 2078 (1983).

[50] D. K. Campbell, in "Heavy Ions and Mesons in Nuclear Physics,'' Les Houches, Course XXX, R. Balian, ed. (North Holland, Amsterdam, 1977), p. 553, is an excellent introduction to this topic.

[51] J. L. Friar, E. L. Tomusiak, and J. Dubach, Phys. Rev. **C25,** 1659 (1982).

[52] J. Thakur, and L. L. Foldy, Phys. Rev. **C8,** 1957 (1973), contains an excellent discussion of these results, with references to previous work.

[53] E. L. Tomusiak, M. Kimura, J. L. Friar, B. F. Gibson, G. L. Payne, and J. Dubach, Phys. Rev. **C32,** 2075 (1985)

[54] C. Bargholtz, Phys. Lett. **112B,** 193 (1982).

[55] F. P. Juster, et al., contributed paper at conference "Nuclear Physics with Electromagnetic Probes,'' (Paris, July 1-5, 1985), p. 156.

[56] J. Torre, and B. Goulard, Phys. Rev. **C28,** 529 (1983).

SUBNUCLEAR DEGREES OF FREEDOM IN PHOTOABSORPTION

AND SCATTERING

Hartmuth Arenhövel

Institut für Kernphysik
Johannes Gutenberg-Universität
6500 Mainz, Federal Republic of Germany

1. INTRODUCTION

One of the main fields of interest in medium energy nuclear physics is the study of subnuclear or non-nucleonic degrees of freedom (d.o.f.) in nuclei, like meson and isobar or quark-gluon degrees of freedom. It bridges the gap between classical or low energy nuclear physics as understood in terms of nucleon-only degrees of freedom and elementary particle or high energy physics. Hereby one may distinguish roughly two major objectives.

One is to understand the rôle of such subnuclear d.o.f. in nuclear structure. How can we, for example, understand the NN-interaction from an underlying more fundamental theory of hadrons? What is the real nature of the unphysical hard core? To what extent will internal nucleon properties be changed by binding effects, i.e., in which kinematic regions are, e.g., quark-gluon d.o.f. important? Do we have evidence for such d.o.f. in nuclear properties like in electromagnetic and weak currents?

The other objective is to gain information on the properties of "elementary" particles by using the nucleus as a laboratory and studying such an elementary particle in the vicinity of a nuclear medium. A classical example is the field of hypernuclear physics where one studies the behaviour of strange particles bound in a nucleus. Pion physics is another vast field where the pion serves as a probe and a study object as well.

In the present lectures I will concentrate on the study of subnuclear d.o.f. in nuclei using the electromagnetic probes of photoabsorption and photon scattering. In these electromagnetic processes subnuclear d.o.f. manifest themselves by the presence of so-called exchange operators like exchange current and exchange two-photon operators, which effectively describe underlying hidden d.o.f..

The necessity for such exchange currents had already been realized in the early days of nuclear physics when phenomenological exchange forces were introduced by Majorana and Heisenberg[1] . The underlying physical mechanism was put on a firmer basis by Yukawa[2] 50 years ago who explained it as the mediation of the pion field. At the same time the nature of the exchange currents became apparent as the electromagnetic interaction with the underlying subnuclear d.o.f., in this case the pion field. Later, it was furthermore realized that the concept of inert particles has only limited value and that in certain processes isobar d.o.f., which are dominated by Δ–d.o.f., should be considered.

Thus, a new picture of the nucleus emerged, in which the nucleons are no longer considered as being inert but of having internal d.o.f. effectively described, e.g., by allowing virtually excited isobars to be present in the nucleus, by the so-called isobar configurations (IC)[3,4]. Meson d.o.f. are still treated through operators, e.g., by meson exchange current (MEC) operators, usually in lowest order. At present we see the first steps of introducing quark d.o.f. into nuclear theory.

In my talk, subnuclear d.o.f. will refer mainly to mesonic and isobar d.o.f. and not to quark-gluon d.o.f. if not stated otherwise. The reason for not considering quark-gluon d.o.f. is twofold:

(i) The question "How far can we go with conventional hadronic d.o.f. (in an effective theory) before we are forced to accept a quark model description" has not yet been answered satisfactorily. In other words, we do not know at which level or in which kinematic region quark-gluon d.o.f. have to enter the *dynamical* description of nuclear structure. I am not referring to static properties like magnetic moments and vertex or electromagnetic form factors of the nucleons.

(ii) Nuclear structure belongs to the non-perturbative domain of QCD with its difficult confinement mechanism, where at present we have again only effective theories of various types of quark models, bag, constituent, etc.. Thus, we cannot expect at present a clear cut answer to the above posed question from such effective theories, which without doubt are necessary and useful for the development of a deeper understanding of hadronic systems.

The problems one encounters even at the less ambitious level of meson and isobar d.o.f. have been pointed out by D. H. Wilkinson[5] in his introduction to the panel discussion of the Heidelberg Conference 1984, where he emphasizes "the great difficulty that we have encountered in making explicit and reliable demonstrations of mesonic and isobaric contributions to nuclear properties, the necessary rules of the game being that every reasonable effort must be made to find a nucleons-only explanation before claiming the demonstration of a non-nucleonic effect".

In this spirit, I will outline the rôle of subnuclear d.o.f. in electromagnetic processes. I will briefly review in section 2 the general form of the electromagnetic interaction, the definition of current and two-photon operators and their specific form for a non-relativistic system of interacting nucleons. Furthermore, the gauge conditions for these operators will be discussed. Section 3 is devoted to the photon scattering amplitude, where the specific forms of resonance and two-photon amplitudes are given. The derivation of the low energy expansion with the aid of the gauge conditions is sketched. Optical theorem and dispersion relations are briefly mentioned. The generalized polarizabilities are introduced in section 4 and the contributions from resonance and two-photon amplitude are discussed. Low energy theorems for the electric polarizabilities are derived and the decomposition of the low energy expansion of the scattering amplitude of section 3 in polarizabilities is given in detail. At the end of section 4 the use of dispersion relations for the polarizabilities is outlined and their importance is stressed for the calculation of the scattering cross section at finite angles.

In section 5 the explicit forms of π–exchange current and TPA are given and the construction of a consistent exchange current for the Paris potential is outlined. Then exchange contributions to various electromagnetic processes using a deuteron target are discussed: electrodisintegration near threshold, photodisintegration below and above π–production threshold, and elastic photon scattering.

Real Δ–excitation in photon scattering is considered in section 6, first in the elementary approach. Then corrections from Δ–propagation , nuclear medium effects and nucleonic contributions are discussed. Furthermore, the importance of inelastic scattering contributions at backward angles is stressed. Finally, a summary is given in section 7.

2. THE ELECTROMAGNETIC INTERACTION

Since the electromagnetic interaction is relatively weak, it is usually sufficient to consider only the lowest order contributions of the perturbation expansion. That means for the one- and two-photon processes of photoabsorption and scattering we need to know the electromagnetic interaction only up to second order in the electromagnetic field operator $A_\mu(x)$,

$$\hat{H}_{em}(A_\mu) = \int d^4x \, \hat{j}^\mu(x)A_\mu(x) + \frac{1}{2}\int d^4x d^4y \, A_\mu(x)\hat{B}^{\mu\nu}(x,y)A_\nu(y) . \tag{2.1}$$

Here

$$\hat{j}^\mu = \frac{\delta\hat{H}(A_\mu)}{\delta A_\mu}\bigg|_{A_\mu=0} \tag{2.2}$$

is the current density operator and

$$\hat{B}^{\mu\nu}(x,y) = \frac{\delta^2\hat{H}(A_\mu)}{\delta A_\mu(x)\delta A_\nu(y)}\bigg|_{A_\mu=0} \tag{2.3}$$

describes first order contributions to two-photon processes, the so-called seagull terms (see Fig. 1).

Then, the one-photon processes, γ–emission and absorption, are determined by the current matrix element $<P_f|\varepsilon_\mu\cdot j^\mu(0)|P_i>$, where ε_μ denotes the photon polarization vector and P_i and P_f denote the total 4-momenta of initial and final states, respectively.

The total photoabsorption cross section for unpolarized radiation is then given by

$$\sigma_{tot}(\omega) = \frac{2\pi^2 e^2}{\omega}\sum_\lambda\sum_f \delta^{(4)}(P_f - P_i - k)|<P_f|\hat{J}_\lambda(0)|P_i>|^2 . \tag{2.4}$$

Similarly, the photon scattering cross section has the form

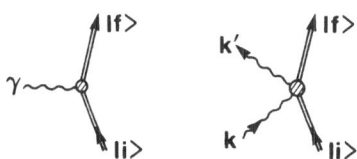

Fig. 1   Diagrams for current and two-photon operators.

$$\frac{d\sigma}{d\Omega} = \frac{k'}{k} |T_{\lambda'\lambda}^{fi}(\vec{k}',\vec{k})|^2 \,, \tag{2.5}$$

where the T-matrix is defined by

$$T_{\lambda'\lambda}^{fi} = \varepsilon_\mu^*(\vec{k}'\lambda')[\, i\!\int\! d^4x\, e^{ik'x} <P_f| T(\hat{j}^\mu(x),\hat{j}^\nu(0)) | P_i>$$

$$-\!\int\! d^4x\, e^{ikx} <P_f| \hat{B}^{\mu\nu}(x,0) | P_i> \,]\varepsilon_\nu(\vec{k},\lambda) \,. \tag{2.6}$$

Here, incoming and scattered photon momenta and polarization vectors are denoted by k_μ, $\varepsilon_\mu(\vec{k},\lambda)$ and k'_μ, $\varepsilon_\mu(\vec{k}',\lambda')$ respectively. The second term is called the seagull or two-photon amplitude (TPA). A more detailed discussion of the scattering amplitude is given in the following section.

In order to derive the explicit forms of the electromagnetic operators I will now consider a non-relativistic system of nucleons interacting through two-body forces in the presence of an electromagnetic field $A_\mu(\vec{x},t)$

$$\hat{H}(A_\mu) = \hat{H}_0 + \hat{H}_{em}(A_\mu) \,, \tag{2.7}$$

where

$$\hat{H}_0 = \sum_i p_i^2/2M + \frac{1}{2}\sum_{i\neq j} V_{ij}$$

$$= \hat{T} + \hat{V} \,. \tag{2.8}$$

Since we now use the Schrödinger picture the electromagnetic interaction will in general have the following form up to second order in the electromagnetic field

$$H_{em}(A_\mu) = \int d^3x\, \hat{j}_\mu(\vec{x})\, A^\mu(\vec{x},t)$$

$$+ \frac{1}{2}\sum_{\mu\nu}\int d^3x\, d^3y\, A^\mu(\vec{x},t) B_{\mu\nu}(\vec{x},\vec{y}) A^\nu(\vec{y},t) \,, \tag{2.9}$$

which defines the current density operator

$$\hat{j}_\mu(\vec{x}) = \left.\frac{\delta\hat{H}_{em}}{\delta A_\nu(\vec{x})}\right|_{A_\mu=0} \tag{2.10}$$

and the two-photon operator

$$\hat{B}_{\mu\nu}(\vec{x},\vec{y}) = \left.\frac{\delta^2\hat{H}_{em}}{\delta A^\mu(\vec{x})\delta A^\nu(\vec{y})}\right|_{A_\mu=0} \,. \tag{2.11}$$

In the non-relativistic framework no explicit time dependence will occur in these operators, and a δ–function $\delta(x_0-y_0)$ has been separated out from $\hat{B}_{\mu\nu}$. Furthermore, the two-photon operator has the symmetry property

$$\hat{B}_{\mu\nu}(\vec{x},\vec{y}) = \hat{B}_{\nu\mu}(\vec{y},\vec{x}) \tag{2.12}$$

and the time components vanish

$$\hat{B}_{0\nu}(\vec{x},\vec{y}) = \hat{B}_{\mu 0}(\vec{x},\vec{y}) = 0 \,, \tag{2.13}$$

because the electromagnetic interaction is linear in the time component A_0 as long as the forces are not energy dependent.

Before considering the specific forms of these operators I would like to discuss the requirements of invariance under gauge transformations $A_\mu \to A_\mu - \partial_\mu \lambda(\vec{x},t)$ since the electromagnetic field strengths are invariant under such transformations. Invariance means that a gauge transformation of the fields which changes the Hamiltonian

$$\hat{H}(A_\mu) \to \hat{H}'(A_\mu) = \hat{H}(A_\mu - \partial_\mu \lambda) \tag{2.14}$$

is accompanied by a local phase transformation of the wave function

$$\psi \to \psi' = e^{i\hat{\chi}(\lambda)}\psi \tag{2.15}$$

in such a way that the physical properties of the system remain unchanged. The phase χ is a linear functional of λ and has to satisfy the following equation

$$e^{-i\hat{\chi}} \hat{H}(A_\mu - \partial_\mu \lambda) e^{i\hat{\chi}} + \frac{\partial \hat{\chi}}{\partial t} = \hat{H}(A_\mu) , \tag{2.16}$$

which one obtains from the transformed Schrödinger equation by comparison with the original one.

Keeping only linear terms in λ , one finds

$$\frac{\partial \hat{\chi}}{\partial t} + i[\hat{H}(A_\mu),\hat{\chi}] - \int d^3x \, \hat{j}_\mu(\vec{x})\partial^\mu \lambda(\vec{x},t)$$

$$= \int d^3x d^3y \, \partial^\mu \lambda(\vec{x},t) B_{\mu\nu}(\vec{x},\vec{y}) A^\nu(\vec{y},t) . \tag{2.17}$$

Since this equation should hold for any gauge function λ , one obtains with

$$\hat{\chi} = \int d^3x \, \hat{j}^0(\vec{x})\lambda(\vec{x},t) , \tag{2.18}$$

the final condition

$$i[\hat{H}(A_\mu),\hat{\rho}(\vec{x})] + \vec{\nabla}\cdot\hat{\vec{j}}(\vec{x}) = \int d^3y \nabla_{x,i}\hat{B}_{ij}(\vec{x},\vec{y})A_j(\vec{y},t) , \tag{2.19}$$

from which the gauge conditions for the current density and two-photon operators follow by comparing the various orders in A_μ , i.e., the continuity equation for the current

$$\vec{\nabla}\cdot\hat{\vec{j}}(\vec{x}) + i[\hat{H}_0,\hat{\rho}(\vec{x})] = 0 , \tag{2.20}$$

and for the two-photon operator

$$\nabla_{x,i}\hat{B}_{ij}(\vec{x},\vec{y}) = i[\hat{\rho}(\vec{x}),\hat{j}_j(\vec{y})] . \tag{2.21}$$

The latter condition links the two-photon amplitude to charge and current density operators and is very important for maintaining the theory gauge invariant[6] .

Now, I first will assume point particles and strictly local interactions, i.e., no momentum dependent nor exchange forces. Then, the electromagnetic interaction is obtained by the minimal substitution

$$\hat{H}_0 \to \hat{H}_0 + \sum_j e_j A_0(\vec{r}_j,t) , \tag{2.22}$$

$$\hat{\vec{p}}_j \to \hat{\vec{p}}_j - e_j\hat{\vec{A}}(\vec{r}_j,t) , \tag{2.23}$$

which maintains gauge invariance as is well known. With the Hamiltonian of Eq. 2.8 one finds

$$\vec{\hat{\rho}}_{[1]}(\vec{x}) = \sum_j e_j\delta(\vec{x}-\vec{r}_j) \tag{2.24}$$

$$\hat{\vec{j}}_{[1]}(\vec{x}) = -\frac{\delta\hat{T}(\vec{A})}{\delta\vec{A}(\vec{x})}\Bigg|_{\vec{A}\equiv 0} = \frac{1}{2M}\sum_j e_j\{\hat{\vec{p}}_j,\delta(\vec{x}-\vec{r}_j)\} \tag{2.25}$$

$$\hat{B}_{kl}^{kin}(\vec{x},\vec{y}) = \frac{\delta_{kl}}{M}\sum_j e_j^2 \delta(\vec{x}-\vec{r}_j)\delta(\vec{y}-\vec{r}_j) \, . \tag{2.26}$$

One can also obtain the spin current, if one remembers the origin of the kinetic energy in the non-relativistic reduction of the Dirac equation (the spin is a genuine relativistic phenomenon!). There one has

$$\frac{1}{2M}(\vec{\sigma}\cdot\hat{\vec{p}})^2 = \frac{1}{2M}\hat{\vec{p}}^2 + \frac{i}{2M}\vec{\sigma}\cdot(\hat{\vec{p}}\times\hat{\vec{p}}) \, , \tag{2.27}$$

from which the spin current originates by minimal substitution

$$\hat{\vec{j}}_{[1]}^s(\vec{x}) = \frac{i}{2M}\sum_j e_j\vec{\sigma}_j \times [\hat{\vec{p}}_j,\delta(\vec{x}-\vec{r}_j)]$$

$$= \vec{\nabla}_x \times \sum_j \frac{e_j}{2M}\, \vec{\sigma}_j\delta(\vec{x}-\vec{r}_j) \, . \tag{2.28}$$

In this way one obtains the spin current of the normal Dirac moment. One has to add the spin current of the anomalous moment, which is a manifestation of internal nucleon structure, in replacing e_i by μ_i. Note that the spin current is divergence free and, thus, is not restricted by the gauge condition (Eq. 2.20).

It is useful to separate the current into an intrinsic and a cm-current by introducing relative and total momenta

$$\hat{\vec{p}}_j' = \hat{\vec{p}}_j - \hat{\vec{P}}/A \, , \tag{2.29}$$

where

$$\hat{\vec{P}} = \sum_j \hat{\vec{p}}_j \, . \tag{2.30}$$

Then one finds

$$\hat{\vec{j}}_{[1]}^c(\vec{x}) = \hat{\vec{j}}^{cm,c}(\vec{x}) + \hat{\vec{j}}^{in,c}(\vec{x}) \, , \tag{2.31}$$

with the intrinsic current

$$\hat{\vec{j}}^{in,c}(\vec{x}) = \frac{1}{2M}\sum_j e_j\{\hat{\vec{p}}_j',\delta(\vec{x}-\vec{r}_j)\} \tag{2.32}$$

and the cm-current

$$\hat{\vec{j}}^{cm,c}(\vec{x}) = \frac{1}{2M_A}\{\hat{\vec{P}},\vec{\rho}_{[1]}(\vec{x})\} \, . \tag{2.33}$$

For the explicit expressions of Eqs. 2.24-25 one easily finds that the following relations hold for the total one-body current

$$\vec{\nabla}\cdot\hat{\vec{j}}_{[1]}(\vec{x}) + i[\hat{T},\hat{\rho}_{[1]}(\vec{x})] = 0 \, , \tag{2.34}$$

$$\frac{\partial}{\partial x_k}\hat{B}_{kl}^{kin}(\vec{x},\vec{y}) = i[\hat{\rho}_{[1]}(\vec{x}),\hat{j}_{[1]l}(\vec{y})] \, . \tag{2.35}$$

Similarly for intrinsic and cm-currents

$$\vec{\nabla}\cdot\hat{\vec{j}}^{cm,c}(\vec{x}) + i[\hat{T}^{cm},\hat{\rho}_{[1]}(\vec{x})] = 0 \, , \tag{2.36}$$

$$\vec{\nabla} \cdot \hat{\vec{j}}^{\text{in,c}}(\vec{x}) + i[\hat{T}^{\text{in}}, \hat{\rho}_{[1]}(\vec{x})] = 0 \,, \tag{2.37}$$

$$\frac{\partial}{\partial x_k} \hat{B}_{kl}^{\text{kin}}(\vec{x}, \vec{y}) = i[\hat{\rho}_{[1]}(\vec{x}), \hat{j}_l^{\text{in,c}}(\vec{y})] + \frac{1}{M_A} \frac{\partial}{\partial x_l} \hat{\rho}_{[1]}(\vec{x}) \hat{\rho}_{[1]}(\vec{y}) \,. \tag{2.38}$$

To obtain Eq. 2.38 I have used

$$[\hat{\rho}(\vec{x}), \{\hat{\vec{P}}, \hat{\rho}(\vec{y})\}] = -i\{\vec{\nabla}_x \hat{\rho}(\vec{x}), \hat{\rho}(\vec{y})\} \,. \tag{2.39}$$

The kinetic energies of cm-motion and intrinsic motion are denoted by $\hat{T}^{\text{cm}}$ and $\hat{T}^{\text{in}}$, respectively.

Since V has been assumed to be local, i.e.,

$$[\hat{V}, \hat{\rho}(\vec{x})] = 0 \,, \tag{2.40}$$

the gauge conditions of Eqs. 2.20-2.21 are fulfilled. This is still true if we weaken the assumption of point particles and allow for an extended nucleon structure in the charge and current density operators

$$\hat{\rho}(\vec{x}) = \sum_j g_j(\vec{x} - \hat{\vec{r}}_j) \tag{2.41}$$

$$\hat{\vec{j}}(\vec{x}) = \frac{1}{2M} \sum_j \left[\{\hat{\vec{P}}_j, g_j(\vec{x} - \hat{\vec{r}}_j)\} + i\hat{\vec{\rho}}_j \times [\hat{\vec{P}}_j, h_j(\vec{x} - \hat{\vec{r}}_j)] \right] \tag{2.42}$$

where g_j and h_j describe the internal charge and magnetization density of the j-th nucleon. The gauge condition (Eq. 2.21) requires then for the two-photon operator the form

$$\hat{B}_{ij}^{\text{kin}}(\vec{x}, \vec{y}) = \frac{\delta_{ij}}{M} \sum_l g_l(\vec{x} - \hat{\vec{r}}_l) g_l(\vec{y} - \hat{\vec{r}}_l) \,. \tag{2.43}$$

This can be generalized in order to allow for internal nucleon dynamics by introducing a more general nucleon two-photon operator

$$\hat{B}_{ij}^{\text{kin}}(\vec{x}, \vec{y}) = \sum_l \hat{b}_{ij}^l (\vec{x}, \vec{y}) \,. \tag{2.44}$$

But then the gauge condition (Eq. 2.21) is in general not fulfilled automatically and the one-body current needs a corresponding modification.

If, however, the NN-interaction contains momentum dependent and/or exchange forces, then the commutator of V with the charge density operator does not vanish any longer and additional exchange currents and exchange two-photon operators are required in order to fulfill gauge invariance

$$\hat{\vec{j}}(\vec{x}) = \hat{\vec{j}}_{[1]}(\vec{x}) + \hat{\vec{j}}_{[2]}^{\text{ex}}(\vec{x}) \,, \tag{2.45}$$

$$\hat{B}_{kl}(\vec{x}, \vec{y}) = \hat{B}_{kl}^{\text{kin}}(\vec{x}, \vec{y}) + \hat{B}_{kl}^{\text{ex}}(\vec{x}, \vec{y}) \,. \tag{2.46}$$

Note that these additional exchange operators have two-body character and are restricted by the following gauge conditions

$$\vec{\nabla} \cdot \hat{\vec{j}}_{[2]}^{\text{ex}}(\vec{x}) = i[\hat{\rho}_{[1]}(\vec{x}), \hat{V}] \,, \tag{2.47}$$

$$\frac{\partial}{\partial x_k} \hat{B}_{kl}^{ex}(\vec{x},\vec{y}) = i[\hat{\rho}_{[1]}(\vec{x}), \hat{j}_{[2]l}^{ex}(\vec{y})] .$$ (2.48)

But these relations are not sufficient to determine the exchange operators as is well known. Only for the low energy theorems can they be utilized to determine the most important exchange contributions as will be discussed in the next section.

In the above relations (Eqs. 2.47-2.48) it has been assumed that the charge density is not affected by exchange effects (Siegert's hypothesis[7]), which is true only in the strict non-relativistic limit with instantaneous interactions. But in general, if one includes retardation in the interaction, one has to consider also exchange contributions to the charge density and then, the gauge conditions have to be modified accordingly. In particular, the time components of $\hat{B}_{\mu\nu}$ need not vanish any more.

3. THE PHOTON SCATTERING AMPLITUDE

I will now discuss in some detail the photon scattering amplitude which has been defined in the previous section in Eq. 2.6. As mentioned there, one has in general two contributions because this process is of second order in the electromagnetic field variables. A second order contribution from the current interaction with virtual excitation of intermediate states is called the resonance amplitude because it may lead to a resonant behaviour of the scattering amplitude. The other contribution is of first order in the electromagnetic interaction arising from the two-photon operator (see Eq. 2.6). It is often called the seagull or two-photon amplitude (TPA). The corresponding diagrams are shown in Fig. 2.

It is important to note that photon scattering in general cannot be described by a form factor as is possible in electron scattering in the one-photon-exchange approximation. The reason for this is that because of the two-step nature of the process the momentum is transferred to the target in two steps with in general intermediate propagation of the system. Therefore, absorption and emission in general take place at two different points in space-time, and because of this non-locality a form factor description fails.

A second remark is concerned with the fact that only the total scattering amplitude has a physical meaning and that the separation into resonance and two-photon amplitude is not a physical one but to some extent arbitrary. For example, in the long wavelength limit

$$\vec{A}(\vec{x},t) = \vec{a} e^{-i\omega t} \qquad (\vec{a} = const.)$$ (3.1)

one can transform $\vec{A}$ away using the gauge function

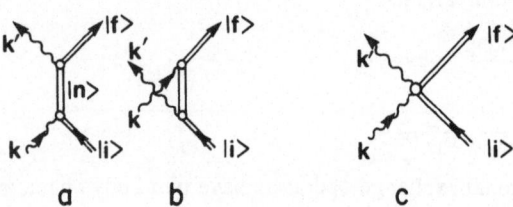

Fig. 2   Photon scattering diagrams: a) and b) direct and crossed term for resonance amplitude, c) two-photon (seagull) amplitude.

$$\lambda(\vec{x},t) = \vec{x}\cdot\vec{a}\,e^{-i\omega t} \tag{3.2}$$

yielding the well known dipole interaction

$$H_{em} = -\vec{E}\cdot\hat{\vec{D}}, \tag{3.3}$$

where $\vec{E}$ denotes the electric field and $\hat{\vec{D}}$ the dipole operator of the system[8,9]. This interaction now is linear in the electromagnetic field and, therefore, no two-photon amplitude occurs.

In addition, the form of the resonance and two-photon amplitude will depend on which degrees of freedom are treated explicitly. This is because the two-photon amplitude will contain contributions from hidden degrees of freedom in terms of effective operators which otherwise would occur in the resonance amplitude. For example, the Compton amplitude for a Dirac-particle has no seagull term. Projecting out the negative energy contributions leads to a seagull term which reduces in the non-relativistic limit to the classical Thomson amplitude.

Keeping this in mind, one may write the scattering amplitude in the form

$$T^{fi}_{\lambda'\lambda}(\vec{k}',\vec{k}) = B^{fi}_{\lambda'\lambda}(\vec{k}',\vec{k}) + R^{fi}_{\lambda'\lambda}(\vec{k}',\vec{k}) \tag{3.4}$$

where the resonance amplitude is given by

$$R^{fi}_{\lambda'\lambda}(\vec{k}',\vec{k}) = \sum_n \Bigg\{ \int d^3P_n \delta(\vec{P}_n - \vec{P}_i - \vec{k}) N_{ni}(k_0) <P_f|J^*_{\lambda'}(0)|P_n> <P_n|J_\lambda(0)|P_i>$$

$$+ \int d^3P_n \delta(\vec{P}_n - \vec{P}_i + \vec{k}') N_{ni}(-k'_0) <P_f|J_\lambda(0)|P_n> <P_n|J^*_{\lambda'}(0)|P_i> \Bigg\}. \tag{3.5}$$

Here, P_n denotes the total 4-momentum including the cm-motion of a specific state of the system. Incoming and scattered photons are denoted by $\vec{k}$ and $\vec{k}'$, respectively with corresponding polarization vectors $\vec{\varepsilon}_\lambda$ and $\vec{\varepsilon}'_{\lambda'}$. The second term of Eq. 3.5 describing the crossed process b) of Fig. 2 is obtained from the first by the substitution $k_\mu \leftrightarrow -k'^\mu$, $\vec{\varepsilon}_\lambda \leftrightarrow \vec{\varepsilon}'^*_{\lambda'}$. The energy denominators are given by

$$N^{-1}_{ni}(k_0) = P_{n,0} - P_{i,0} - k_0 - i\varepsilon . \tag{3.6}$$

For a non-relativistic description one has

$$P_{n,0} = \varepsilon_n + \frac{\vec{P}^2_n}{2M_A}, \tag{3.7}$$

where ε_n denotes the intrinsic excitation energy and M_A the total mass of the target nucleus. Using the Breit-frame from now on, one obtains for elastic scattering ($k'_0 = k_0$)

$$N^{-1}_{ni}(k_0) = \varepsilon_{ni} + \frac{\vec{k}'\cdot\vec{k}}{2M_A} - k_0 - i\varepsilon . \tag{3.8}$$

The current matrix elements have the general structure

$$<P_a|\hat{J}_\lambda(0)|P_b> = <P_a|\vec{\varepsilon}_\lambda\cdot\hat{\vec{j}}(0)|P_b>$$

$$= <a|\bar{J}_\lambda(q_{ab},P_{ab})|b>, \tag{3.9}$$

where

$$q_{ab} = P_a - P_b,$$

$$P_{ab} = P_a + P_b .$$ (3.10)

The state $|a>$ refers to intrinsic motion only (no cm-motion) and correspondingly the current operator $\tilde{J}_\lambda(q,P)$ acts on intrinsic variables only. Non-relativistically one has (see Eq. 2.31)

$$\tilde{J}_\lambda(q,P) = \vec{\varepsilon}_\lambda \cdot [\hat{\vec{J}}_{in}(\vec{q}) + \frac{\hat{\rho}(\vec{q})}{2M_A} \vec{P}] ,$$ (3.11)

where $\hat{\vec{J}}_{in}$ describes the intrinsic current with respect to the cm-system and the second term represents the cm-convection current. In the following I will introduce the short-hand notation

$$J_{ab}(\vec{q},\vec{p}) := <P_a | \vec{\varepsilon}_\lambda \cdot \hat{\vec{j}}(0) | P_b>$$

$$J'_{ab}(\vec{q},\vec{p}) := <P_a | \vec{\varepsilon}'^*_\lambda \cdot \hat{\vec{j}}(0) | P_b> .$$

Then for elastic scattering the resonance amplitude has the following form

$$R^{ii}_{\lambda'\lambda}(\vec{k}',\vec{k}) = \sum_n \{N_{ni}(k_0)J'_{in}(-\vec{k}',\vec{k})J_{ni}(\vec{k},\vec{k}') + \text{cross term}\} .$$ (3.12)

Using Eq. 3.11 and keeping only the first order cm-contributions of the energy denominators

$$N_{ni}(k_0) = N^0_{ni}(k_0) - (N^0_{ni})^2 \frac{\vec{k}'\cdot\vec{k}}{2M_A}$$ (3.13)

with

$$N^0_{ni}(k_0) = (\varepsilon_{ni}-k_0-i\varepsilon)^{-1} ,$$ (3.14)

one can write the resonance amplitude in the following form

$$R^{ii}_{\lambda'\lambda}(\vec{k}',\vec{k}) = <i | \int d^3x d^3y \; e^{-\vec{k}'\cdot\vec{x} + i\vec{k}\cdot\vec{y}} \; \varepsilon'^*_{l'} \hat{R}_{l'l}(\vec{x},\vec{y},k',k)\varepsilon_l | i>$$ (3.15)

where

$$\hat{R}_{l'l}(\vec{x},\vec{y};k',k) = \hat{R}^{in}_{l'l} + \hat{R}^{cm}_{l'l}$$ (3.16)

$$\hat{R}^{in}_{l'l} = \hat{j}^{in}_{l'}(\vec{x})(H^{in}_0-\varepsilon_i-k_0-i\varepsilon)^{-1} \hat{j}^{in}_l(\vec{y}) + \text{cross term}$$ (3.17)

$$\hat{R}^{cm}_{l'l} = \frac{1}{2M_A} \Big[-\vec{k}'\cdot\vec{k} \, \hat{j}^{in}_{l'}(\vec{x})(H^{in}_0-\varepsilon_i-k_0-i\varepsilon)^{-2} \hat{j}^{in}_l(\vec{y})$$

$$+ k_{l'}\hat{\rho}(\vec{x})(H^{in}_0-\varepsilon_i-k_0-i\varepsilon)^{-1} \hat{j}^{in}_l(\vec{y})$$

$$+ \hat{j}^{in}_{l'}(\vec{x})(H^{in}_0-\varepsilon_i-k_0-i\varepsilon)^{-1} \hat{\rho}(\vec{y})k'_l$$

$$+ \text{cross term} \Big] .$$ (3.18)

Here "cross term" means interchange of $(\vec{x},l')$ with $(\vec{y},l)$ and k_μ with $-k'_\mu$.

According to Eqs. 2.6, 2.12 and 2.13, the two-photon amplitude is given by

$$B_{\lambda'\lambda}^{fi}(\vec{k}',\vec{k}) = - \int d^3x \, e^{-i\vec{k}'\cdot\vec{x}} <P_f | \epsilon'_{\lambda',l'} \hat{B}_{l'l}(x,0) \epsilon_{\lambda,l} | P_i > . \tag{3.19}$$

Separating out the cm-motion one obtains in the non-relativistic regime

$$B_{\lambda'\lambda}^{fi}(\vec{k}',\vec{k}) = -<f| \int d^3x \, d^3y \, e^{-i\vec{k}'\cdot\vec{x}} \, e^{-i\vec{k}\cdot\vec{y}} \sum_{l',l} \epsilon'^{*}_{\lambda',l'} B_{l'l}(\vec{x},\vec{y}) \epsilon_{\lambda,l} \, |i> . \tag{3.20}$$

For the one-body operator of Eq. 2.26 one easily finds

$$B_{\lambda'\lambda}^{fi,kin}(\vec{k}',\vec{k}) = - \frac{\vec{\epsilon}'^{*}_{\lambda'}\cdot\vec{\epsilon}_{\lambda}}{M} <f| \sum_j e_j^2 e^{i(\vec{k}-\vec{k}')\cdot\vec{r}_j} |i> , \tag{3.21}$$

which has the appearance of a form factor. It reflects the assumed point-like nature of the nucleon. This, however, is no longer true if one considers internal nucleon dynamics. Then the operator becomes non-local and looses its simple form factor character. Furthermore, one has to remember the origin of this two-photon operator arising from the intermediate anti-particle contribution, which reduces to the above form in the static limit only. Any nonstatic effects will destroy this simple local form.

I will now exploit the gauge conditions in order to derive the low energy expansion of the scattering amplitude up to second order in the photon momenta following closely the discussion of Friar[6]. To this end one needs the low energy expansions of the current and two-photon amplitude. I will first consider the current in Eq. 3.11 up to second order in $\vec{q}$. One finds for charge and intrinsic current density

$$\hat{\rho}(\vec{q}) = Z + i\vec{q}\cdot\hat{\vec{D}} - \frac{1}{2} q_{l'} \hat{Q}_{l'l} q_l \tag{3.22}$$

where

$$\hat{\vec{D}} = \int d^3x \, \vec{x}\hat{\rho}(\hat{x}) \tag{3.23}$$

$$\hat{Q}_{l'l} = \int d^3x \, x_{l'} \, x_l \hat{\rho}(\vec{x}) , \tag{3.24}$$

and

$$\hat{j}_l^{in}(\vec{q}) = \hat{j}_l^{in}(0) + q_{l'} \nabla_{l'} \hat{j}_l^{in}(0) - \frac{1}{2} \hat{\Omega}_l(\vec{q}) . \tag{3.25}$$

Here

$$\hat{\Omega}_l(\vec{q}) = \int d^3x \, (\vec{q}\cdot\vec{x})^2 \hat{j}_l^{in}(\vec{x}) \tag{3.26}$$

is of second order in $\vec{q}$ and contains E1 retardation and higher multipoles. Inserting these expansions into the continuity equation

$$\vec{q}\cdot\hat{\vec{j}}^{in}(\vec{q}) = - [H_0^{in},\hat{\rho}(\vec{q})] , \tag{3.27}$$

and comparing the coefficients on both sides, one finds

$$\hat{\vec{j}}^{in}(0) = i[\hat{H}_0^{in},\hat{\vec{D}}] , \tag{3.28}$$

$$\nabla_l \hat{j}_{l'}^{in}(0) + \nabla_{l'} \hat{j}_l^{in}(0) = - [\hat{H}_0^{in},\hat{Q}_{l'l}] . \tag{3.29}$$

That means only the symmetric part of $\nabla_{l'} \hat{j}_l^{in}(0)$ with respect to interchange of l' with l is determined by the charge quadrupole, whereas the antisymmetric part

$$q_{l'}(\nabla_{l'} \hat{j}_l^{in}(0) - \nabla_l \hat{j}_{l'}^{in}(0)) = -2i(\vec{q} \times \hat{\vec{M}})_l , \tag{3.30}$$

the magnetic moment contribution

$$\hat{M} = \frac{1}{2} \int d^3x \; \vec{x} \times \vec{j}^{in}(\vec{x}) \tag{3.31}$$

remains undetermined. A systematic extension to higher electric multipoles has been given recently by Friar and Fallieros[10].

Collecting the various terms one obtains then for the expansion of the current matrix elements

$$J_{ab}(\vec{q},\vec{p}) = J_{ab}^{(0)} + J_{ab}^{(1)} + J_{ab}^{(2)} , \tag{3.32}$$

where

$$J_{ab}^{(0)} = i[\hat{H}_0^{in}, \vec{\epsilon} \cdot \hat{\vec{D}}]_{ab} \tag{3.33}$$

$$J_{ab}^{(1)} = -i\vec{\epsilon} \cdot (\vec{q} \times \hat{\vec{M}}_{ab}) - \frac{1}{2}[\hat{H}_0^{in}, q_l Q_{ll'} \epsilon_{l'}] + \frac{Z}{2M_A} \vec{\epsilon} \cdot \vec{P} \delta_{ab} \tag{3.34}$$

$$J_{ab}^{(2)} = -\frac{1}{2} \vec{\epsilon} \cdot \vec{\Omega}_{ab} + \frac{i}{2M_A} \vec{\epsilon} \cdot \vec{P} \; \vec{q} \cdot \vec{D}_{ab} . \tag{3.35}$$

I would like to point out that the Siegert operator (Eq. 3.33) includes also possible exchange current contributions. Inserting this expression in Eq. 3.12, expanding also the energy denominators and keeping consistently all terms up to second order in $\vec{k}$ and $\vec{k}'$ and up to first order in $1/M_A$, one arrives at the low energy expansion of the resonance amplitude of which I will give the zeroth order only

$$R_{\lambda'\lambda}^{ii}(0,0) = <i | [\vec{\epsilon}_{\lambda'}^* \cdot \hat{\vec{D}}, [\hat{H}_0^{in}, \vec{\epsilon}_\lambda \cdot \hat{\vec{D}}]] | i>$$

$$= \vec{\epsilon}_{\lambda'}^* \cdot \vec{\epsilon}_\lambda \frac{NZ}{A} \frac{e^2}{M} + <i | [\vec{\epsilon}_{\lambda'}^* \cdot \hat{\vec{D}}, [\hat{V}, \vec{\epsilon}_\lambda \cdot \hat{\vec{D}}]] | i> . \tag{3.36}$$

For the low energy expansion of the two-photon amplitude the starting point will be the gauge condition in the form

$$k'_l \hat{B}_{l'l}(\vec{k}',\vec{k}) = - [\hat{\rho}(-\vec{k}'), \hat{j}_l^{in}(\vec{k})] - \frac{\vec{k}'_l}{M_A} \hat{\rho}(-\vec{k}')\hat{\rho}(\vec{k}) , \tag{3.37}$$

which is the Fourier-transform of Eq. 2.38. The expansion of $\hat{B}_{l'l}(\vec{k}',\vec{k})$ up to second order is given by

$$\hat{B}_{l'l}(\vec{k}',\vec{k}) = \hat{B}_{l'l}(0,0) + (\vec{k}' \cdot \vec{\nabla}' + \vec{k} \cdot \vec{\nabla})\hat{B}_{l'l}(0,0)$$

$$+ \frac{1}{2}[(\vec{k}' \cdot \vec{\nabla}')^2 + (\vec{k} \cdot \vec{\nabla})^2 + 2(\vec{k}' \cdot \vec{\nabla}')(\vec{k} \cdot \vec{\nabla})]\hat{B}_{l'l}(0,0) \tag{3.38}$$

It is a nice consequence of gauge invariance that all terms except for a part of the last second order term can be determined from the gauge condition (Eq. 3.36) as has been pointed out first by Friar[6]. To this end we expand both sides of Eq. 3.37 up to second order in $\vec{k}'$, and by comparison one finds for the first order

$$\hat{B}_{l'l}(0,\vec{k}) = i[\hat{D}_{l'}, \hat{j}_l(\vec{k})] - \frac{Z}{M_A} \rho(\vec{k})\delta_{l'l} , \tag{3.39}$$

and for the second order

$$k'_{l'}(\vec{k}'\cdot\vec{\nabla}')\hat{B}_{l'l}(0,\vec{k}) = \frac{1}{2}\left[[k'_m\hat{Q}_{mn}k'_n, \hat{j}_l(\vec{k})] + i\frac{k'_l}{M_A}\{\vec{k}'\cdot\hat{\vec{D}},\hat{\rho}(\vec{k})\} \right] .$$

(3.40)

Expanding now (Eq. 3.39) up to second order in $\vec{k}$ one obtains

$$\hat{B}_{l'l}(0,0) = - [\hat{D}_{l'},[\hat{H}_0^{in},\hat{D}_l]] - \frac{Z^2}{M_A}\delta_{l'l} ,$$

(3.41)

$$\vec{k}\cdot\vec{\nabla}\hat{B}_{l'l}(0,0) = [\hat{D}_{l'},(\vec{k}\times\vec{M})_l - \frac{i}{2}[\hat{H}_0^{in}, k_m\hat{Q}_{ml}]] - i\frac{Z}{M_A}\vec{k}\cdot\hat{\vec{D}}\delta_{l'l} ,$$

(3.42)

$$(\vec{k}\cdot\vec{\nabla})^2\,\hat{B}_{l'l}(0,0) = - i[\hat{D}_l,\hat{\Omega}_l] + \frac{Z}{M_A}k_m\hat{Q}_{mn}k_n\delta_{l'l} .$$

(3.43)

The corresponding terms for $\vec{k}'$ are obtained from the crossing symmetry

$$\hat{B}_{l'l}(\vec{k}',\vec{k}) = \hat{B}_{ll'}(-\vec{k},-\vec{k}') .$$

(3.44)

Thus it remains to determine the mixed gradient terms of the second order in Eq. 3.38. Applying $\vec{\nabla}\cdot\vec{k}$ to Eq. 3.40, one finds

$$k'_{l'}(\vec{k}'\cdot\vec{\nabla}')(\vec{k}\cdot\vec{\nabla})\hat{B}_{l'l}(0,0) = \frac{1}{2}[k'_m\hat{Q}_{mn}k'_n, -i(\vec{k}\times\vec{M})_l - \frac{1}{2}[H_0^{in}, k_m\hat{Q}_{ml}]]$$

$$- \frac{k'_l}{2M_A}\{\vec{k}'\cdot\hat{\vec{D}}, \vec{k}\cdot\hat{\vec{D}}\}$$

(3.45)

As one can see, it is not possible to obtain completely the mixed gradient term, because this expression is of second order in $\vec{k}'$ and, therefore, only the symmetric part of $(\partial/\partial k'_m)\,(\partial/\partial k_m)\,B_{l'l}\,(0,0)$ with respect to the interchange of m' and l' is determined from this equation. This is analogous to the fact that for the current operator the gauge condition fixes only the quadrupole part of the term linear in $\vec{q}$, which is the symmetric part of $\nabla_{l'}\,j_l^{in}(0)$ with respect to the interchange of l' with l as we have seen above.

I will not list here the complete expansion of the two-photon amplitude up to second order (see Ref. 6). The important result is that due to cancellations between resonance and two-photon amplitudes the total amplitude is determined by a few structure constants. In particular, one notes from Eq. 3.36 and 3.41 the cancellation of the double commutator leaving only the classical Thomson amplitude in the zero energy limit. The complete expression for a non-relativistic system of nucleons keeping terms up to $1/M_A$ is given by[6].

$$T_{\lambda'\lambda}^{ii} = e^2(t_{scalar} + t_{vector} + t_{tensor})$$

(3.46)

$$t_{scalar} = \left[-\frac{Z^2}{M_A} + (\frac{Z}{3M_A}<r^2> + \alpha_E^{(0)})k^2 \right] \vec{\varepsilon}'\cdot\vec{\varepsilon}$$

$$+ [\chi_p^{(0)} + \chi_D^{(0)} - \frac{1}{2M_A}<\vec{D}^2>$$

$$- \frac{1}{3M_A}<\vec{M}^2>\hat{\vec{k}}'\cdot\hat{\vec{k}}](\vec{\varepsilon}'\times\vec{k}')\cdot(\vec{\varepsilon}\times\vec{k})$$

(3.47)

$$t_{vector} = -ik \left[\mu^2((\vec{\varepsilon}'\times\hat{\vec{k}}')\times(\vec{\varepsilon}\times\hat{\vec{k}}))\cdot\langle\vec{S}\rangle + \frac{\mu Z}{M_A}((\vec{\varepsilon}'\times\hat{\vec{k}}')\cdot\langle\vec{S}\rangle\,\vec{\varepsilon}\cdot\hat{\vec{k}} - (\vec{\varepsilon}\times\hat{\vec{k}})\cdot\langle\vec{S}\rangle\,\vec{\varepsilon}'\cdot\hat{\vec{k}}) \right] , \quad (3.48)$$

$$t_{tensor} = \alpha_E^{(2)} \, k^2 \, S(\vec{\varepsilon}',\vec{\varepsilon})$$

$$+ (\chi_P^{(2)} + \chi_D^{(2)} + \frac{3}{8M_A}D^{(2)} - \frac{\mu^2}{2M_A}\hat{\vec{k}}'\cdot\hat{\vec{k}}) \, S(\vec{\varepsilon}'\times\hat{\vec{k}}',\vec{\varepsilon}\times\hat{\vec{k}})$$

$$+ \frac{ZQ}{4M_A}[S(\hat{\vec{k}}',\hat{\vec{k}}') + S(\hat{\vec{k}},\hat{\vec{k}})]\,\vec{\varepsilon}'\cdot\vec{\varepsilon}$$

$$+ \frac{1}{2}\,\mu Q\,[\hat{\vec{k}}'\cdot\hat{\vec{k}}\,S(\vec{\varepsilon}',\vec{\varepsilon}) - \vec{\varepsilon}'\cdot\vec{\varepsilon}\,S(\hat{\vec{k}}',\hat{\vec{k}}')] , \quad (3.49)$$

$$S(\vec{a},\vec{b}) = \{\vec{a}\cdot\vec{S}, \vec{b}\cdot\vec{S}\} - \frac{2}{3}\,(\vec{a}\cdot\vec{b})\,\vec{S}^2 , \quad (3.50)$$

where μ and Q are related to the ground state magnetic dipole and electric quadrupole moment, respectively.

$$\vec{M} = \mu\,\vec{S} \quad (3.51)$$

$$Q^{[2]} = Q\,S^{[2]}, \quad S^{[2]} = [S^{[1]}\times S^{[1]}]^{[2]} . \quad (3.52)$$

The ground state spin operator is denoted by $\vec{S}$.

The additional structure constants, the static electric polarizability α and the magnetic susceptibility χ are model dependent and contain information on conventional nuclear structure and on subnuclear dynamics as well. They are related to energy weighted sum rules $\Sigma_{-2}(E1)$ and $\Sigma_{-2}(M1)$. In detail one has

$$\alpha_E^{(0)} = \frac{2}{3}\langle i|\vec{D}\cdot\frac{\hat{Q}_i}{H_0 - \varepsilon_i}\,\vec{D}|i\rangle \quad (3.53)$$

$$\alpha_E^{(2)} = \langle I\| [D^{[1]}\,\frac{\hat{Q}_i}{H_0 - \varepsilon_i}\,D^{[1]}]^{[2]}\,\|I\rangle / C_2(I) \quad (3.54)$$

$$C_2(I) = \langle I\| S^{[2]}\,\|I\rangle \quad (3.55)$$

for the scalar and tensor parts of the electric polarizability and

$$\chi_P^{(0)} = \frac{2}{3}\langle i|\vec{M}\cdot\frac{\hat{Q}_i}{H_0 - \varepsilon_i}\,\vec{M}|i\rangle \quad (3.56)$$

$$\chi_P^{(2)} = \langle I\| [M^{[1]}\,\frac{\hat{Q}_i}{H_0 - \varepsilon_i}\,M^{[1]}]^{[2]}\,\|I\rangle / C_2(I) \quad (3.57)$$

for the paramagnetic susceptibility. $\hat{Q}_i$ projects off the ground state. Both contributions arise from the resonance amplitude whereas the diamagnetic susceptibility comes from the TPA;

$$\chi_D^{(0)} = -\frac{1}{\sqrt{3}}\langle i|\hat{\chi}_D^{[0]}|i\rangle \quad (3.58)$$

$$\chi_D^{(2)} = \frac{1}{2}\,\langle I\| \hat{\chi}_D^{[2]}\,\|I\rangle / C_2(I) , \quad (3.59)$$

where $\hat{\chi}_D^{[J]}$ ($J = 0, 2$) are the scalar and tensor parts of the totally antisymmetric part of the mixed gradient term

$$\hat{\chi}_{D,rs} = \frac{1}{4}\,\varepsilon_{rij}\,\varepsilon_{smn}\frac{\partial^2\hat{B}_{im}(\vec{k}',\vec{k})}{\partial k'_j\,\partial k_n}\bigg|_{\vec{k}=\vec{k}'=0}. \tag{3.60}$$

From the kinetic part of the TPA (Eq. 2.26) one obtains

$$\chi_D^{(0),kin} = -\frac{1}{6M}<i|\sum_\alpha e_\alpha^2\,r_\alpha^2|i>. \tag{3.61}$$

Additional contributions arise from the exchange TPA.

At the end of this section I would like to briefly mention the optical theorem, which connects the imaginary part of the coherent elastic forward scattering amplitude to the total absorption cross section

$$\sigma_{tot}(k) = \frac{4\pi}{k}\,Im\,T_{\lambda\lambda}^{ii}\,(\vec{k},\vec{k}). \tag{3.62}$$

This means that any absorptive process contributes to the scattering amplitude and leads to an imaginary part. It is also important with respect to the two-photon amplitude, because it tells us that the underlying physical mechanism will contain absorptive processes above a certain threshold energy. For example, the TPA corresponding to Eq. 2.26 originates from projecting out antiparticle degrees of freedom. Thus, the corresponding absorptive process will be particle-antiparticle pair creation with a threshold energy $E = 2M$. The absorptive reaction connected with the exchange TPA will be discussed in section 5.

The optical theorem is also important if the forward scattering amplitude obeys dispersion relations, e.g., a once-subtracted dispersion relation

$$Re\,(T_{ii}(\omega)-T_{ii}(0)) = \frac{2\omega^2}{\pi}\,P\int_0^\infty d\omega'\,\frac{Im\,T_{ii}(\omega')}{\omega'(\omega'^2-\omega^2)}$$

$$= \frac{\omega^2}{2\pi^2}\,P\int_0^\infty d\omega'\,\frac{\sigma_{tot}(\omega')}{\omega'^2-\omega^2} \tag{3.63}$$

which allows one to calculate the complete forward scattering amplitude from the total absorption cross section and provides also the sum rule of Gell-Mann, Goldberger and Thirring[11].

On the other hand, it will not be sufficient for calculating the scattering cross section at finite scattering angles. How this can be done will be discussed in the following section.

4. GENERALIZED POLARIZABILITIES

Very useful quantities for the description of the scattering cross section are the generalized polarizabilities, which are introduced by the following expansion of the scattering amplitude[12,13]

$$T_{\lambda\lambda}^{fi}(\vec{k}',\vec{k}) = (-)^{1+\lambda'+I_f-M_i}\sum_{\substack{L,L',J\\M,M'}}(-)^{L+L'}(2J+1)\begin{pmatrix}I_f & J & I_i\\-M_f & m & M_i\end{pmatrix}\begin{pmatrix}L & L' & J\\M & M' & -m\end{pmatrix}$$

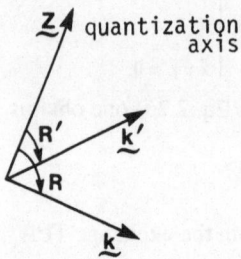

Fig. 3   Scattering geometry and the rotations R and R′ carrying the quantization axis
into $\vec{k}$ and $\vec{k}'$, respectively.

$$\times P_{J,fi}^{L'L\lambda'\lambda}(k',k)D_{M,\lambda}^{L}(R)D_{M',-\lambda'}^{L'}(R'),$$

(4.1)

Here, initial and final spins are denoted by I_i and I_f with projections M_i and M_f, respectively. The rotations R and R′ denote the rotation of the chosen quantization axis into $\vec{k}$ and $\vec{k}'$, respectively (see Fig. 3). The polarizabilities represent a classification of the scattering amplitude according to the various angular momentum and parity transfers from incoming and scattered photons and according to the total momentum transfer J to the target. In detail one has

$$P_{J,fi}^{L'L\lambda'\lambda}(k',k) = \sum_{\nu,\nu'=0,1} \lambda'^{\nu'}\lambda^{\nu} P_{J}^{fi}(M^{\nu'}L',M^{\nu}L,k',k),$$

(4.2)

i.e.,

$$P_{J}^{fi}(M^{\nu'}L',M^{\nu}L,k',k) = \frac{1}{4} \sum_{\lambda,\lambda'=-1,1} \lambda'^{\nu'}\lambda^{\nu} P_{J,fi}^{L'L\lambda'\lambda}(k',k),$$

(4.3)

where ν indicates the type of multipole transition (M^0 = E (electric),M^1 =M(magnetic)). Parity conservation requires the vanishing of a given polarizability

$$P_{J}^{fi}(M^{\nu'}L',M^{\nu}L) = 0 \qquad \text{if } (-)^{L'+\nu'+L+\nu}\neq\pi_i\pi_f,$$

(4.4)

where initial and final state parity are denoted by π_i and π_f, respectively.

This concept was first introduced by Fano[14] for pure E1 radiation. In this case one has only scalar, vector and tensor polarizabilities corresponding to J = 0, 1 and 2, respectively. For a spin zero nucleus only the scalar polarizability contributes to elastic scattering because of the angular momentum selection rule. But vector and tensor polarizabilities may be seen in inelastic (Raman-) scattering as has been demonstrated in the giant resonance region for vibrational and deformed nuclei[15,16].

Subsequently, the polarizabilities have been generalized[12,13] for arbitrary electric and magnetic multipoles by explicitly introducing the multipole expansion for the current

$$\vec{\epsilon}_\lambda\cdot\hat{\vec{j}}^{in}(\vec{k}) = -\sqrt{2\pi} \sum_{L,M,\nu=0,1} \lambda^{\nu} \hat{L} T_{\nu,M}^{[L]} D_{M,\nu}^{L}(R)$$

(4.5)

in the resonance amplitude and by coupling of the different angular momentum transfers. But the cm-contributions had been neglected.

However, in general the polarizabilities may formally be defined by inverting Eq. 4.1 yielding

$$P_{J,fi}^{L'L\lambda'\lambda}(k,k) = (-)^{I_f+L+L'}\frac{\hat{L}^2\hat{L}'^2}{(8\pi^2)^2} \sum_{MM'M_fM_im} (-)^{M_i} \begin{pmatrix} I_f & J & I_i \\ -M_f & m & M_i \end{pmatrix} \begin{pmatrix} L & L' & J \\ M & M' & -m \end{pmatrix}$$

$$\int dR dR' \, D_{m,\lambda}^{L^*}(R) D_{M',-\lambda'}^{L'^*}(R') \, T_{\lambda'\lambda M_f M_i}^{fi} \,. \tag{4.6}$$

This expression makes apparent that the polarizabilities describe the decomposition of the scattering amplitude according to the angular momentum transfer L and L' of incoming and scattered photon, respectively, and to the total angular momentum J (see Fig. 4). They allow one therefore to separate the geometrical properties as described by the various angular momentum and parity transfers from the dynamical ones given by the strengths of the various polarizabilities.

Explicitly, one obtains for the intrinsic part of the resonance amplitude (see Eq. 3.17)

$$P_J^{res,in}(M^vL',M^vL,k',k) = 2\pi\hat{L}\hat{L}'(-)^{L+I_f+I_i}$$

$$\times \sum_n \left[\left\{ \begin{matrix} L & L' & J \\ I_f & I_i & I_n \end{matrix} \right\} \frac{<I_f\|T_v^{[L']}(k')\|I_n><I_n\|T_v^{[L]}(k)\|I_i>}{\varepsilon_{ni}-k_0-i\varepsilon} \right.$$

$$\left. + (-)^{L+L'+J} \left\{ \begin{matrix} L' & L & J \\ I_f & I_i & I_n \end{matrix} \right\} \frac{<I_f\|T_v^{[L]}(k)\|I_n><I_n\|T_v^{[L']}(k')\|I_i>}{\varepsilon_{ni}+k'_0-i\varepsilon} \right] \tag{4.7}$$

and for the cm-part, which is little more involved

$$P_J^{res,cm}(M^vL',M^vL,k',k) = \frac{k}{2M_A}(-)^{J+L}\hat{L}^2\hat{L}'^2 \sum_{K'K} (-)^K \left\{ \begin{matrix} K & K' & J \\ L' & L & 1 \end{matrix} \right\}$$

$$\times \left[\sqrt{2} \begin{pmatrix} 1 & K' & L' \\ -1 & 0 & 1 \end{pmatrix} \begin{pmatrix} 1 & K & L \\ 0 & 1 & -1 \end{pmatrix} \pi(K'+L'+v'+1)(\pi(K+L+v+1) \right.$$

$$\left. \times \tilde{P}_J(CK',EK) + \pi(K+L+v)\tilde{P}_J(CK',MK)) \right.$$

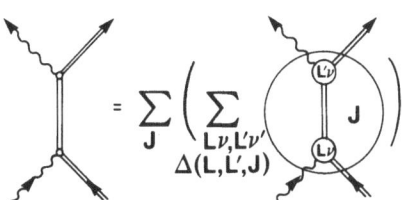

Fig. 4   Decomposition of the scattering amplitude into polarizabilities according to angular momentum and parity transfer.

$$+\sqrt{2}\begin{bmatrix} 1 & K & L \\ 1 & 0 & -1 \end{bmatrix}\begin{bmatrix} 1 & K' & L' \\ 0 & -1 & 1 \end{bmatrix}\pi(K+L+v+1)(\pi(K'+L'+v'+1)$$

$$\times\tilde{P}_J(EK',CK)+\pi(K'+L'+v')\tilde{P}_J(MK',CK))$$

$$-k\begin{bmatrix} 1 & K & L \\ 0 & 1 & -1 \end{bmatrix}\begin{bmatrix} 1 & K' & L' \\ 0 & -1 & 1 \end{bmatrix}\left[\sum_{\mu',\mu=0,1}\pi(K'+L'+v'+\mu'+1)\right.$$

$$\left.\left.\pi(K+L+v+\mu+1)\tilde{P}'_J(M^{\mu'}K',M^{\mu}K)\right]\right], \tag{4.8}$$

where I have introduced

$$\tilde{P}_J(O'K',OK)=2\pi\hat{L}\hat{L}'(-)^{L+I_f+I_i}\sum_n\left[\begin{Bmatrix} K & K' & J \\ I_f & I_i & I_n \end{Bmatrix}\frac{<I_f\|O'^{[K']}\|I_n><I_n\|O^{[K]}\|I_i>}{\varepsilon_{ni}-k_0-i\varepsilon}\right.$$

$$\left.-(-)^{K+K'+J}\begin{Bmatrix} K' & K & J \\ I_f & I_i & I_n \end{Bmatrix}\frac{<I_f\|O^{[K]}\|I_n><I_n\|O'^{[K']}\|I_i>}{\varepsilon_{ni}+k_0-i\varepsilon}\right] \tag{4.9}$$

$$\tilde{P}'_J(O'K',OK)=\frac{\partial}{\partial k_0}\tilde{P}_J(O'K',OK) \tag{4.10}$$

and

$$\pi(n)=\frac{1}{2}(1+(-)^n). \tag{4.11}$$

Correspondingly, one finds as the contribution of the two-photon amplitude to the polarizabilities

$$P_J^{TPA}(M^{v'}L',M^vL,k',k)=2\pi(-)^{L+J}\frac{\hat{L}\hat{L}'}{\hat{J}}<I_f\|\sum_{l'l}\int d^3x d^3y$$

$$\times[A_{l'}^{[L']}(M^{v'};\vec{k}',\vec{x})\times A_l^{[L]}(M^v;\vec{k},\vec{y})]^{[J]}\hat{B}_{l'l}(\vec{x},\vec{y})\|I_i>. \tag{4.12}$$

For the one-nucleon two-photon operator of Eq. 2.44 one obtains

$$\hat{P}_{L'L\lambda'\lambda}^{[J]TPA,kin}=\frac{\sqrt{4\pi}}{\hat{J}}\sum_{\alpha,ll'j,LL'J}i^{L-L'}(-)^{l+l'}\hat{j}^2\hat{l}\hat{L}^2\hat{L}'^2\hat{L}^2\hat{L}'^2 j_{L'}(kr_\alpha)j_L(kr_\alpha)\begin{bmatrix} L' & l' & L \\ 0 & -\lambda' & \lambda' \end{bmatrix}$$

$$\times\begin{bmatrix} L & L' & J \\ 0 & 0 & 0 \end{bmatrix}\begin{bmatrix} L & l & L \\ 0 & \lambda & -\lambda \end{bmatrix}\begin{Bmatrix} L & L' & J \\ l & l' & j \\ L & L' & J \end{Bmatrix}[Y^{[J]}(\hat{r}_\alpha)P_{l'l\lambda'\lambda}^{[j]}(\alpha)]^{[J]} \tag{4.13}$$

where $\hat{P}_{l'l\lambda'\lambda}^{[j]}(\alpha)$ denote the nucleon polarizability operators. This expression reflects the transformation of angular momentum with respect to the nucleon center-of-mass to the nuclear c.o.m. For the limiting case of a point nucleon one has

$$\hat{P}^{[j]}_{l'l\lambda\lambda}(\alpha) = -\delta_{j,0}\delta_{l,1}\delta_{l',1}\sqrt{3}\frac{e_\alpha^2}{M} \; . \tag{4.14}$$

Eq. 4.13 will be important for the discussion of photon scattering in the Δ–region .

The following crossing symmetries hold for the polarizabilities

$$P_J(M^{v'} L',M^v L,k',k) = (-)^J P_J(M^{v'} L,M^{v'} L',-k,-k') \; . \tag{4.15}$$

and in addition

$$P_J^{TPA}(M^v L,M^{v'} L',k,k') = (-)^J P_J^{TPA}(M^{v'} L',M^v L,k',k) \; . \tag{4.16}$$

because of the crossing symmetry (Eq. 2.12). From the latter property follows the selection rule

$$P_J^{TPA}(M^v L,M^v L,k,k) = 0 \tag{4.17}$$

for odd J. For elastic scattering a further symmetry relation holds if time reversal invariance is valid

$$P_J(M^v L,M^{v'} L') = P_J(M^{v'} L',M^v L) \; . \tag{4.18}$$

It is possible to derive low energy theorems also for the polarizabilities analogous to the scattering amplitude[17]. Using, for example, the low energy limit of the electric multipoles of the intrinsic current

$$\hat{T}_e^{[L]}(k) \xrightarrow{k\to0} -k^{L-1} \sqrt{\frac{L+1}{L}} \, [\hat{H}_0^{in}, \tilde{C}^{[L]}] \; , \tag{4.19}$$

where

$$\tilde{C}^{[L]} = \frac{i^L}{(2L+1)!!} \int d^3x \, x^L \hat{\rho}(\vec{x}) Y^{[L]}(\hat{x}) \tag{4.20}$$

is the low energy limit of the Coulomb multipole, one finds for the lowest order term of the resonance contribution to the scalar polarizability

$$\hat{P}_0^{res}(EL,EL) \xrightarrow{k\to0} +2\pi k^{2L-2} (-)^{L+1} \frac{L+1}{L} (2L+1)\Big[[\tilde{C}^{[L]},\times [\hat{H}_0^{in},\tilde{C}^{[L]}]]^{[0]}$$

$$+ \frac{1}{\sqrt{(2L-1)}\hat{L} M_A} [\tilde{C}^{[L-1]} \times \tilde{C}^{[L-1]}]^{(0)} (1-\delta_{L,1})\Big] \; . \tag{4.21}$$

The first term is the contribution of the intrinsic part (Eq. 4.7) and the second the cm-contribution (Eq. 4.8), which vanishes for L = 1. For the TPA contribution one find similarly

$$\hat{P}_0^{TPA}(EL,EL) \xrightarrow{k\to0} -2\pi k^{2L-2} (-)^{L+1} \frac{L+1}{L} (2L+1)\Big[[\tilde{C}^{[L]} \times [\hat{H}_0^{in}, \tilde{C}^{[L]}]]^{[0]}$$

$$+ \frac{1}{\sqrt{(2L-1)}\hat{L} M_A} [\tilde{C}^{[L-1]} \times \tilde{C}^{[L-1]}]^{[0]}\Big] \; . \tag{4.22}$$

That means one finds exact cancellation of the lowest order terms of the electric polarizabilities except for E1. This is a generalization of what has been derived previously for E2[6]. For the low energy limit of the dipole polarizability one finds

$$P_0(E1,E1) \xrightarrow{k\to0} -\frac{Z^2e^2}{M_A} \sqrt{3} \, \hat{1} \; , \tag{4.23}$$

i.e., the Thomson limit again. It is also possible to go one step further beyond the low energy limit. That is discussed in detail in Ref. 17. Here, I will list only the generalized polarizabilities for the low energy expansion of Eq. 3.45[18].

Scalar polarizabilities:

$$P_0(E1,E1) = -e^2 \frac{\sqrt{3}\hat{I}}{M_A} [Z^2 - (\frac{Z}{3} <r^2> + M_A \alpha_E^{(0)} - \frac{I(I+1)}{6} \mu^2)k^2] \tag{4.24}$$

$$P_0(M1,M1) = e^2 \sqrt{3}\hat{I} \, (\chi_P^{(0)} + \chi_D^{(0)} - \frac{<\vec{D}^2>}{2M_A})k^2 \tag{4.25}$$

$$P_0(E2,E2) = 0 \tag{4.26}$$

$$P_0(M2,M2) = -e^2 \frac{\sqrt{5}}{12M_A} \hat{I} \, I(I+1)\mu^2 k^2 \tag{4.27}$$

Vector polarizabilities:

$$P_1(E1,E1) = e^2 \sqrt{\frac{2}{3}} \hat{I}\sqrt{I(I+1)} \frac{\mu Z}{M_A} k \tag{4.28}$$

$$P_1(M1,M1) = e^2 \sqrt{\frac{2}{3}} \hat{I}\sqrt{I(I+1)} \, \mu^2 k \tag{4.29}$$

$$P_1(E1,M2) = -e^2 \frac{1}{3} \sqrt{\frac{5}{3}} \hat{I}\sqrt{I(I+1)} \frac{\mu Z}{M_A} k \tag{4.30}$$

Tensor polarizabilities:

$$P_2(E1,E1) = e^2 \frac{C_2(I)}{\sqrt{5}} (\frac{ZQ}{5M_a} - 2\alpha_E^{(2)} - \frac{\mu^2}{4M_A})k^2 \tag{4.31}$$

$$P_2(M1,M1) = -e^2 \frac{2C_2(I)}{\sqrt{5}} (\chi_P^{(2)} + \chi_D^{(2)} + \frac{3}{8M_A} D^{(2)})k^2 \tag{4.32}$$

$$P_2(E1,M2) = e^2 \sqrt{\frac{3}{5}} \frac{C_2(I)}{2M_A} (\frac{ZQ}{3} + \frac{\mu^2}{2})k^2 \tag{4.33}$$

$$P_2(M1,E2) = -e^2 \sqrt{\frac{3}{5}} \frac{C_2(I)}{2} \mu Q k^2 \tag{4.34}$$

$$P_2(M2,M2) = -e^2 \sqrt{\frac{7}{15}} \frac{C_2(I)}{4M_A} \mu^2 k^2 \tag{4.35}$$

with

$$D^{(2)} = <I|| [D^{[1]} \times D^{[1]}]^{[2]} ||I>/C_2(I) . \tag{4.36}$$

It is interesting to note that the existence of a non-vanishing ground state magnetic moment implies a vector polarizability, i.e., an optical activity. This is not surprising since the magnetic moment implies a dependence of the scattering process on the circular polarization.

Using the optical theorem, one may also express the total absorption cross section in terms of the imaginary parts of the polarizabilities. For unpolarized photons and nuclei only the scalar polarizabilities contribute. One has[19]

$$\bar{\sigma}_{tot} = \sum_L (\sigma(EL) + \sigma(ML))$$

$$= \frac{4\pi}{k} \frac{1}{\hat{I}_i} \sum_L \frac{(-)^{L+1}}{\hat{L}} \, Im \, (P_0(EL,EL) + P_0(ML,ML)) \, . \tag{4.37}$$

which gives for the partial cross sections

$$\sigma(M^vL) = \frac{4\pi}{\hat{I}_i \hat{L} k} (-)^{L+1} \, Im \, P_0(M^vL, M^vL) \, . \tag{4.38}$$

This relation says that the imaginary part of the polarizabilities can be simply obtained from the partial multipole contributions to the total cross section. The problem arises for the real part, where for each photon energy an explicit sum over the whole excitation spectrum has to be evaluated. Here, dispersion relations are of great help and, in fact, have been successfully used for the calculation of the generalized polarizabilities[19,20].

In order to insure the proper low energy behaviour, one has to use appropriately subtracted dispersion relations. For example, in the previous discussion of low energy theorems we have seen that the scalar electric polarizability $P_0(EL,EL)$ behaves for $k \to 0$ like k^{2L}. In this case then one uses the dispersion relation

$$Re \, (P_0(EL,EL,k) - P_0(E1,E1,0)\delta_{L1}) = \frac{2k^{2L}}{\pi} \, P\!\!\int_0^\infty dk' \, \frac{Im \, P_0(EL,EL,k')}{k'^{2L-1}(k'^2 - k^2)}$$

$$= (-)^{L+1} \hat{I}_i \hat{L} \, \frac{k^{2L}}{2\pi^2} \, P\!\!\int_0^\infty dk' \, \frac{\sigma(EL,k')}{k'^{2L-2}(k'^2 - k^2)} \, . \tag{4.39}$$

More details for the general case can be found in Ref. 20. For illustration I show in Fig. 5 a recent calculation of $P_0(E1,E1)$ by Fein[21], where the explicit evaluation is compared to the dispersion relation result for ^{12}C in and above the giant resonance region. The

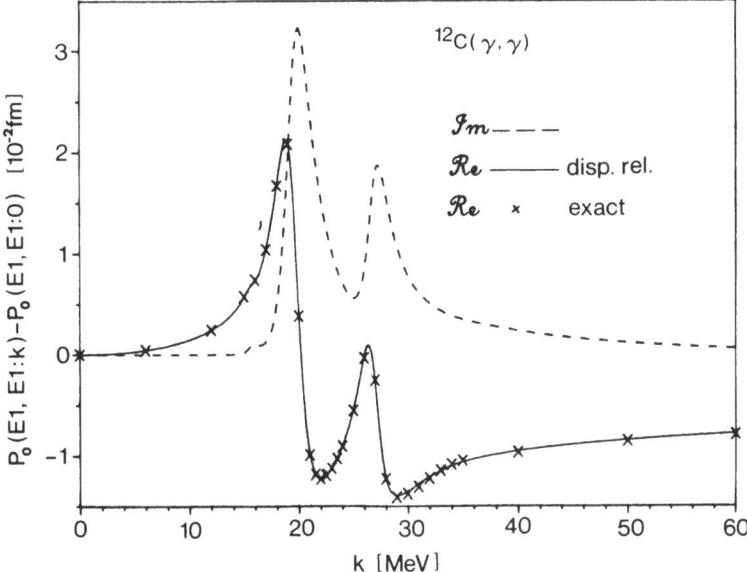

Fig. 5   Real and imaginary part for scalar E1-polarizability of ^{12}C. Full curve represents real part as calculated from dispersion relation, crosses represent explicit evaluation of (4.7) from Fein[21].

agreement is impressive. In this case the dispersion methods meant a gain in computing time of about a factor 50.

The important feature of having dispersion relations for the generalized polarizabilities is that they allow one to calculate the scattering cross section for arbitrary angles from the total absorption cross section, if the partial multipole contributions are known.

5. EXCHANGE CONTRIBUTIONS TO CURRENT AND TWO-PHOTON AMPLITUDE

As mentioned in section 2 the presence of exchange and/or momentum dependent forces requires additional currents and two-photon amplitudes in order to satisfy the corresponding gauge conditions. They are manifestations of the physical mechanism underlying the NN-interaction and internal nucleon structure, i.e., manifestations of subnuclear degrees of freedom, like meson and isobar degrees of freedom in the conventional framework of nuclear physics, or quark-gluon d.o.f. in QCD. This feature makes the study of exchange currents and exchange TPA so interesting, because they provide additional information on these subnuclear d.o.f. and, thus, serve as testing grounds for our present knowledge of the nature of the hadronic interactions.

However, as has been discussed before, the gauge conditions are not sufficient to fix the exchange operators connected to the NN-interaction and internal nucleon structure. Thus, for phenomenological potentials a wide range of arbitrariness is left open for possible exchange operators in order to satisfy the gauge conditions.

On the other hand, if the underlying mechanism for the interaction is known, then one is able to construct uniquely the corresponding exchange operators and one may test them in experiments.

The best known and most reliable exchange operators arise from π–meson exchange, which describes the long range part of the NN-interaction. The diagrams for current and TPA are shown in Figs. 6 and 7. The corresponding operators for the current $\vec{J}_{(2)}^{\pi-MEC}$ are:[22]

(i) pair current

$$e\frac{f^2}{m^2}\sum_{jj'}(\vec{\tau}_j \times \vec{\tau}_{j'})_3\, \vec{\sigma}_j\cdot\vec{\nabla}_j\, \sigma_{j'l}\delta_{j'}(\vec{x})J^1(\vec{r}_j-\vec{r}_{j'}) \qquad (5.1)$$

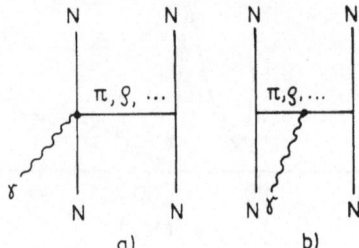

Fig. 6 π- and ρ-exchange current diagrams. (a) pair current, (b) meson current.

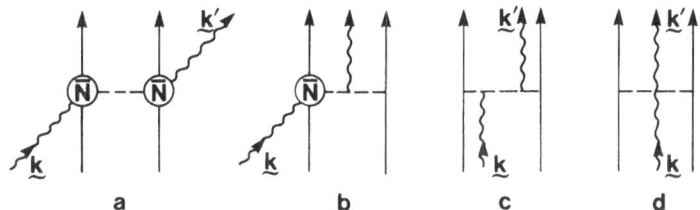

Fig. 7   Diagrams of π-two-photon operator: (a) pair-pair current, (b) pair-π current, (c) π–π current, and (d) π-two photon operator.

(ii) π–current

$$-e^2 \frac{f^2}{m^2} \sum_{jj'} (\vec{\tau}_j \times \vec{\tau}_{j'})_3 \frac{1}{2} \vec{\sigma}_j \cdot \vec{\nabla}_j \, \vec{\sigma}_{j'} \cdot \vec{\nabla}_{j'} \, \vec{J}^2(\vec{r}_j, \vec{y}, \vec{r}_{j'}) , \tag{5.2}$$

where

$$\delta_j(\vec{x}) = \delta(\vec{x} - \vec{r}_j) , \tag{5.3}$$

and for the π–exchange two-photon operator $B_{kl}^{\pi-\text{MEC 23,24}}$

(i) pair-pair current

$$-e^2 \frac{f^2}{m^2} \sum_{jj'} T_{jj'}^0 \sigma_{j,k} \delta_j(\vec{x}) \, \sigma_{j',l} \delta_{j'}(\vec{y}) J^1(\vec{r}_j - \vec{r}_{j'}), \ T_{jj'}^0 = \vec{\tau}_j \cdot \vec{\tau}_{j'} - \tau_j^3 \tau_{j'}^3 , \tag{5.4}$$

(ii) pair–π–current

$$e^2 \frac{f^2}{m^2} \sum_{jj'} T_{jj'}^0 \left[\sigma_{j,k} \delta_j(\vec{x}) \vec{\sigma}_{j'} \cdot \vec{\nabla}_{j'} \, J_l^2(\vec{r}_j, \vec{y}, \vec{r}_{j'}) + \begin{bmatrix} l \leftrightarrow k \\ \vec{y} \leftrightarrow \vec{x} \end{bmatrix} \right] , \tag{5.5}$$

(iii) π–π–current

$$\frac{e^2}{2} \frac{f^2}{m^2} \sum_{jj'} T_{jj'}^0 \vec{\sigma}_j \cdot \vec{\nabla}_j \, \vec{\sigma}_{j'} \cdot \vec{\nabla}_{j'} \left[J_{kl}^3(\vec{r}_j, \vec{x}, \vec{y}, \vec{r}_{j'}) + \begin{bmatrix} l \leftrightarrow k \\ \vec{y} \leftrightarrow \vec{x} \end{bmatrix} \right] , \tag{5.6}$$

(iv) π–two–photon operator

$$e^2 \frac{f^2}{m^2} \delta_{kl} \delta(\vec{x} - \vec{y}) \sum_{jj'} T_{jj'}^0 \, \vec{\sigma}_j \cdot \vec{\nabla}_j \, \vec{\sigma}_{j'} \cdot \vec{\nabla}_{j'} \, J^1(\vec{x} - \vec{r}_j) \, J^1(\vec{x} - \vec{r}_{j'}) , \tag{5.7}$$

where

$$J^1(\vec{r}) = \frac{e^{-mr}}{4\pi r} , \tag{5.8}$$

$$\vec{J}^2(\vec{r}_j, \vec{y}, \vec{r}_{j'}) = J^1(\vec{r}_j - \vec{y}) \overset{\leftrightarrow}{\vec{\nabla}}_y J^1(\vec{y} - \vec{r}_{j'}) , \tag{5.9}$$

$$J^3_{kl}(\vec{r},\vec{x},\vec{y},\vec{r}') = J^1(\vec{r}-\vec{x}) \overset{\leftrightarrow}{\nabla}_{x,k} J^1(\vec{x}-\vec{y}) \overset{\leftrightarrow}{\nabla}_{y,l} J^1(\vec{y}-\vec{r}') .$$

(5.10)

These operators fulfill the gauge conditions

$$\vec{\nabla} \cdot \hat{\vec{j}}^{\pi MEC}(\vec{x}) + i[V^\pi, \hat{\rho}_{[1]}(\vec{x})] = 0$$

(5.11)

$$\frac{\partial}{\partial x_k} \hat{B}^{\pi MEC}_{kl}(\vec{x},\vec{y}) = i[\hat{\rho}_{[1]}(\vec{x}), \hat{j}^{\pi MEC}_l(\vec{y})] ,$$

(5.12)

with the OPE-potential

$$V^\pi = -\frac{f^2}{m^2} \vec{\tau}_1 \cdot \vec{\tau}_2 \, \vec{\sigma}_1 \cdot \vec{\nabla}_1 \vec{\sigma}_2 \cdot \vec{\nabla}_2 J^1 .$$

(5.13)

The π–exchange TPA which one obtains from Eqs. 5.4 - 5.7 can be written in the following form[19,20].

$$B^{\pi MEC}_{fi,\lambda'\lambda} = \frac{f^2 e^2}{m^2} \sum_{j \neq j'} <f| e^{i(\vec{k}-\vec{k}')\cdot\vec{R}_{j'j}} T^0_{j'j} \, \hat{b}^{\pi MEC}_{\lambda'\lambda j'j} (\vec{r}_{j'j}) | i>$$

(5.14)

$$\vec{R}_{j'j} = \frac{1}{2}(\vec{r}_{j'} + \vec{r}_j) ,$$

(5.15)

$$\vec{r}_{j'j} = \vec{r}_{j'} - \vec{r}_j ,$$

(5.16)

where $\hat{b}^{\pi-MEC}_{\lambda'\lambda j'j}$ represents the exchange TPA for a nucleon pair relative to the pair cm-coordinate $\vec{R}_{j'j}$. It depends only on the relative pair coordinate $\vec{r}_{j'j}$. Thus, the total nuclear exchange TPA is obtained by multiplying each individual pair exchange TPA with the pair cm-form factor and then summing over all pairs.

Correspondingly, the contribution to the polarizabilities are expressed in terms of the pair polarizabilities like Eq. 4.13, where on has to replace the one-body polarizability operators by the two-body ones, the coordinates $\vec{r}_\alpha$ by the two-body cm-coordinates $\vec{R}_{j'j}$ and the sum over α by the sum over all two particle contributions $j' \neq j$. Further details can be found in Ref. 19-20.

Besides the π–exchange operators, which describe the long range part, one has to consider additional medium and short range exchange operators for current and TPA in correspondence to the medium and short range exchange pieces of realistic NN-potentials. Very often these are mocked up by including into the current ρ–exchange in addition to π–exchange only, with some regularization procedure. But then one violates the gauge conditions in the medium and short range regions. That consistency is not unimportant and has first been demonstrated for the OBE potential model of Bryan and Gersten[25] in deuteron electrodisintegration near threshold[26].

Recently, a consistent exchange current has been constructed for the parametrized Paris potential[27] independently by Riska[28] and Buchmann et al.[29]. I will briefly describe the construction of this exchange current. The parametrized Paris potential can be brought into the following form:

$$V = V^0 + \vec{\tau}_1 \cdot \vec{\tau}_2 \, V^\tau ,$$

(5.17)

where both the isospin-independent part V^0 and the isospin-dependent V^τ contain central, spin-spin, tensor, spin-orbit and quadratic spin-orbit components with a quadratic momentum dependence for the central and spin-spin parts

$$V^{0/\tau} = \sum_\alpha V_\alpha^{0/\tau}(r,p^2)\,\Omega_\alpha \tag{5.18}$$

where

$$\Omega_0 = 1, \; \Omega_\sigma = \vec{\sigma}_1 \cdot \vec{\sigma}_2, \; \Omega_T = S_{12}$$

$$\Omega_{LS} = \vec{L} \cdot \vec{S}, \; \Omega_{S02} = \frac{1}{2}(\vec{\sigma}_1 \cdot \vec{L}\, \vec{\sigma}_2 \cdot \vec{L} + \vec{\sigma}_2 \cdot \vec{L}\, \vec{\sigma}_1 \cdot \vec{L}) \tag{5.19}$$

and

$$V(r,p^2) = V^a(r) + \{\frac{p^2}{M}, V^b(r)\} \ . \tag{5.20}$$

Assuming the charge density as given by the non-relativistic point particle density $\rho_{(1)}$, it then remains to find a two-body exchange current $\vec{j}_{(2)}^{MEC}$, which fulfills the gauge condition (Eq. 2.47). Contributions to the commutator in Eq. 2.47 arise from:

(i) the momentum dependence and
(ii) the isospin dependence of the potential.

The current belonging to the momentum dependence is easily constructed by the standard method of minimal coupling substituting $\vec{p}$ by $\vec{p}-e\vec{A}$. The isospin dependence poses in general a problem for phenomenological potentials because in principle it comes from the exchange of charged particles, which are not known except for the longest range tail, being the one-pion-exchange part.

Fortunately, the Paris potential in its parametrized form resembles the superposition of one-boson-exchanges of different kinds having fictitious masses and coupling strengths. For example, the spin and isospin dependent central and tensor potential part can be split into a $\pi-$ and $\rho-$like potential

$$\vec{\tau}_1 \cdot \vec{\tau}_2(\vec{\sigma}_1 \cdot \vec{\sigma}_2 V_\sigma + S_{12} V_T) = V(\pi-\text{like}) + V(\rho-\text{like}) \ . \tag{5.21}$$

Knowing the $\pi-$ and $\rho-$MEC one can immediately write down the corresponding gauge invariant MEC. Details can be found in Refs. 28, 29.

It is clear that this procedure is not unique, as is well known, since the gauge condition leaves the transverse current undetermined. However, we consider it as a reasonable method because it incorporates as much as possible the underlying physical mechanisms which are best known in the long range region. The most uncertain part concerns the short range region which is purely phenomenological in the Paris potential. Hopefully, in the future a better knowledge of quark-gluon d.o.f. will give a more reliable theory.

In addition to these meson exchange operators the contributions of internal nucleon d.o.f., like $\Delta-$isobar d.o.f., to the electromagnetic process have to be considered. This can be done in terms of effective operators[30] in the nucleonic space or by allowing isobars to be present in the nuclear wave function in terms of so-called isobar configurations (IC)[3,4] . I will not go into details but show only the corresponding diagrams for current and TPA in Figs. 8 and 9.

Before discussing various applications I would like to point out that the exchange operators in Eqs. 5.1 - 5.6 are of lowest order in the static limit in the spirit of static meson exchange potentials. However, we know that these have limited validity only. With increasing energy retardation effects and finally inelasticities become important. Thus, the $\pi-$exchange currents are in principle linked to the Born-terms of pion photoproduction above threshold as the $\pi-\Delta-$current is related to the resonance term. As long as pion d.o.f. are not treated explicitly these pion-production processes will manifest

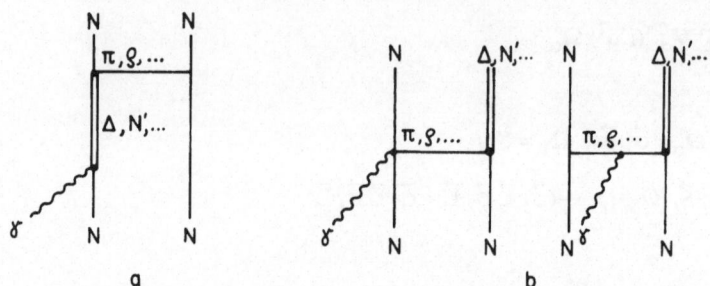

Fig. 8   Diagram for Δ-IC current (a) and Δ-IC-MEC current (b).

themselves in the exchange TPA through imaginary contributions above the pion-production threshold. That remains to be studied in the future.

Let me now turn to the discussion of exchange effects. In the spirit of what has been said in the introduction we need a reliable nucleon-only description before we may claim evidence for subnuclear d.o.f. Therefore, the lightest nuclei are certainly favoured since only for these can an exact treatment within the nucleon-only theory be done.

In fact, very well known examples for the evidence of pionic d.o.f. are the M1-transition in deuteron break-up near threshold[31] and the magnetic form factors of ^{3}H and ^{3}He , where the best nucleon-only description fails badly to describe the experimental results[32]. Also E1-transitions, e.g., in deuteron photodisintegration, of the enhancement of the TRK-sum rule[33] show strong evidence for the failure of nucleon-only description. The reason why these are not considered as evidence of π d.o.f. lies in the fact that the low energy theorem for E1 transitions, the Siegert operator (see Eq. 3.33), masks the evidence for non-nucleonic effects by allowing one to calculate the MEC contributions without explicit knowledge of the MEC-operators. However, as has been shown by various authors[34], evaluation of the nucleon-only current results in an E1-strength far too small and leads in fact to a decrease of the TRK-sum rule below its classical value[33]. This is demonstrated in Fig. 10 for deuteron photodisintegration, which is dominated by E1. One readily sees that the Siegert operators incorporate the largest part of the exchange current. More sensitive to exchange contributions beyond the Siegert operators is the asymmetry with respect to linearly polarized photons as is shown in Fig. 11 and in fact only including them leads to a satisfactory agreement with experiment.

Turning now to d(e,e')pn near threshold, I show first in Fig. 12 an inelastic form factor at moderate momentum transfer[36]. Very nicely seen is the $^1S_0$ peak at threshold and the failure of the nucleon-only description. Only with the pionic MEC contribution of about 50% of the total strength is a quantitative agreement obtained. There is also very little model dependence with respect to various realistic NN-potentials.

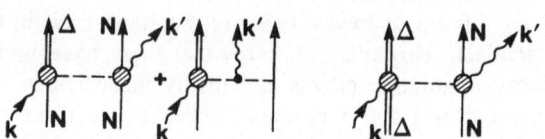

Fig. 9   Δ-IC contributions to two-photon operator.

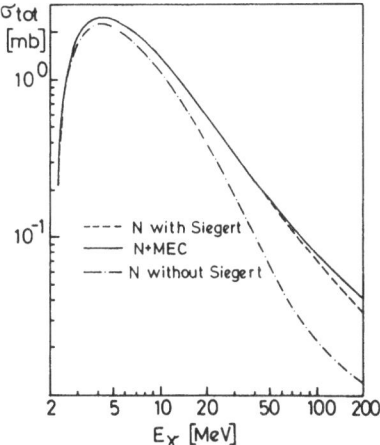

Fig. 10   Total deuteron photodisintegration cross section for conventional one-body contributions with and without Siegert operator and with additional explicit exchange effects (N + MEC).

Fig. 13 shows the averaged cross section ($E_{np} = 0$–3 MeV) at higher momentum transfers. One readily sees that approaching the region of momentum transfer where the normal (nucleon-only) cross section has a minimum, a systematic underestimation of the theory develops. This minimum occurs because the dominant $M1(^1S_0)$ transition form factor passes through zero at $q^2 = 12$ fm^{-2} beyond which normal and MEC contribution interfere destructively.

It is therefore evident that in this specific kinematic region the cross section is much more sensitively dependent on the details and uncertainties of the theoretical description.

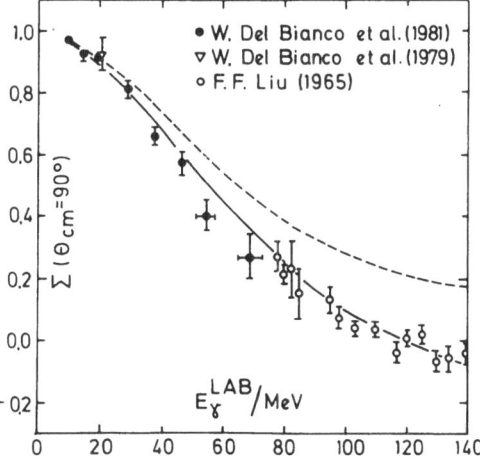

Fig. 11   Linearly polarized photon asymmetry at $\theta_{cm} = 90°$ for $d(\vec{\gamma},p)n$. Data from Ref. 35.

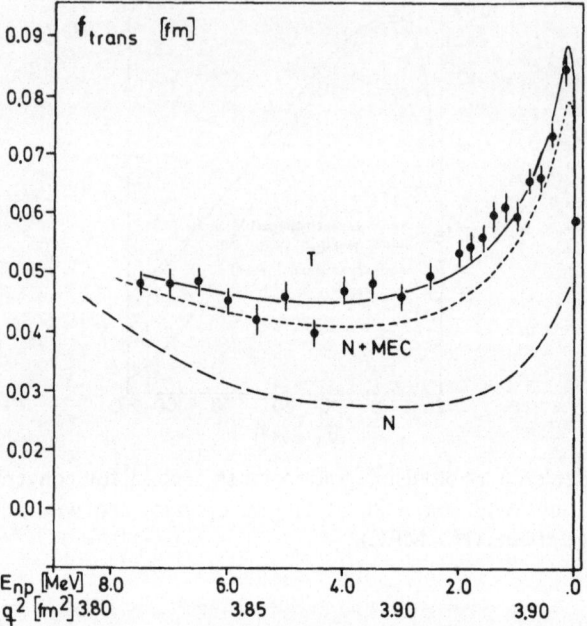

Fig. 12   Inelastic transverse form factor, f_{trans}, as function of relative n-p cm-energy
around $(\bar{q}_{cm})^2 = 4 \text{ fm}^{-2}$. Data from Ref. 36. The long-dashed curve is obtained
using the Reid soft-core potential[37] without exchange effects, while the short-
dashed curve includes π-MEC and the full curve both π-MEC and IC[36].

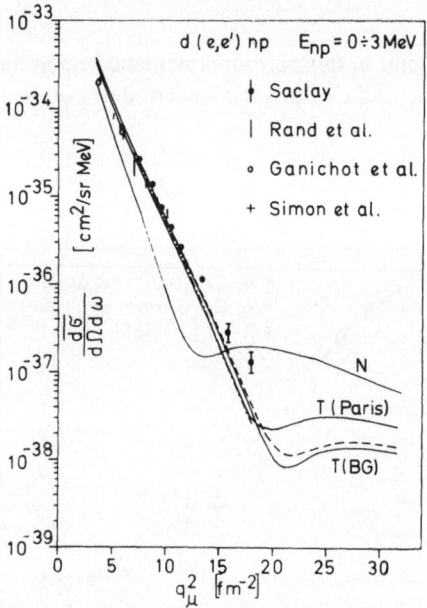

Fig. 13   Comparison of experimental data[38] for d(e,e')pn with various theoretical contributions: one-body part (N) for the Paris potential[27], N+π–MEC for Paris
potential (dashed) and Reid soft-core (dash-dot). Full curves (T) include also
ρ-exchange and neutron electric form factor for Paris and Bryan-Gersten[25]
potentials.

Such uncertainties come from

(i) contribution of medium and short range MEC,
(ii) contribution of isobar d.o.f.,
(iii) choice of elementary electromagnetic form factors for the
     MEC, and,
(iv) relativistic corrections.

This I have discussed recently[39]. I only show in Fig. 14 the effect of consistency of the exchange current by comparing an older calculation[26] with $\pi-$ and $\rho-$exchange with the consistent Paris exchange current[29]. At higher momentum transfer the effect is quite substantial but not sufficient to obtain a good agreement with recent high momentum data[40].

In these examples the exchange contributions were dominated by $\pi-$exchange . But isobar d.o.f. are not totally negligible. For example, the total isobar contribution to d(e,e′)pn near threshold amounts to about 20% (see Ref. 26). However, in contrast to the model independence of lowest order $\pi-$MEC , isobar contributions show a larger uncertainty and model dependence because of their not always well defined rôle in the NN-potential used, and different treatments of $\Delta-$dynamics like static approach, impulse approximation or coupled channel calculation. At low energies the impulse approximation for the calculation of isobar configurations is quite reliable[41]. But in the region of real $\Delta-$excitation it is expected that a coupled channel approach will be important and I will illustrate this for the two-body break-up process d(γ,p)n in this energy region. The following results are largely based on the thesis work of W. Leidemann[42,43].

Fig. 15 shows in the upper part the various contributions of $\Delta-$IC , $\Delta-$MEC, MEC and the normal part, the latter without and with $\Delta-$degrees in a coupled channel (CC) approach. The dominance of $\Delta-$IC , essentially the $^5D_2$ (NΔ)- -partial wave, is readily seen. But also MEC contribute significantly even at higher energies. The lower part in Fig. 15 shows a comparison of the impulse approximation versus a coupled channel

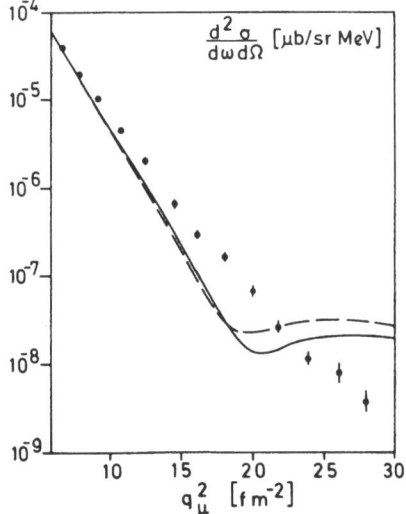

Fig. 14   d(e,e′)pn cross section at $\theta_e = 155°$. Data from Ref. 40. Calculations include one-body, isobar and exchange currents. Full curve with consistent Paris exchange current and dashed curve with π- and ρ-exchange from Ref. 26.

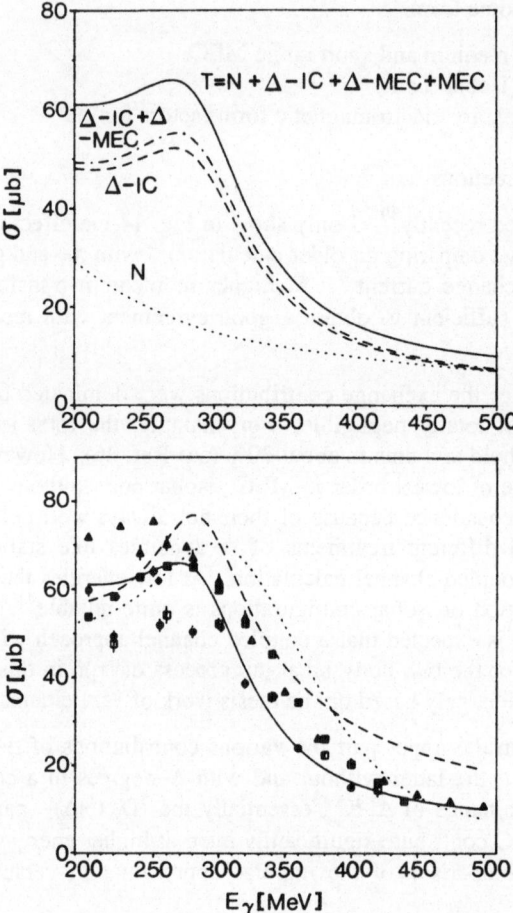

Fig. 15   Total cross section for d(γ,p)n. Data from Ref. 44. Upper part shows various
           contributions; lower part: theoretical curves with RSC in IA (dashed) and CC
           for Δ–IC (solid) and from Ref. 45 (dotted).

approach and the potential model dependence. Strikingly seen is the strong reduction in
the coupled channel calculation. With respect to the experimental results[44] a quantitative
agreement of the theory has not yet been reached. More detailed results on angular distributions, photon asymmetry and nucleon polarization can be found in Refs. 42, 43 and
46.

Turning now to photon scattering let me first consider the structure constants of the
low energy expansion (Eq. 3.46) for the nucleon and the deuteron. In the framework of a
non-relativistic constituent quark model Drechsel and Russo[47] obtained the following
results

$$\alpha_E^{(0)} + \frac{1}{M_N} <r^2> = 13.7\cdot10^{-3}\ \text{fm}^3 \qquad \{\text{Exp.} = (10.8\pm4.2)\cdot10^{-3}\ \text{fm}^3\} \qquad (5.22)$$

$$\chi_p^{(0)} + \chi_D^{(0)} - \frac{1}{2M_N} <\vec{D}^2> = 1.5 \cdot 10^{-3} \text{ fm}^3 \qquad \{\text{Exp.} = (2 \pm 2) \cdot 10^{-3} \text{ fm}^3\} \qquad (5.23)$$

which are in decent agreement with experiment[48]. For the deuteron Weyrauch and myself obtained (in units of fm^3, without nucleon contributions)

$$\alpha_E^{(0)} = 0.561 \text{ without MEC}$$

$$= 0.639 \text{ with MEC} \qquad (5.24)$$

$$\frac{1}{3M_d} <r^2> = 0.196 \cdot 10^{-2}$$

$$= -2\chi_D^{(0),kin} \qquad (5.25)$$

$$\chi_p^{(0)} = 0.665 \cdot 10^{-1} \text{ without MEC}$$

$$= 0.750 \cdot 10^{-1} \text{ with MEC}$$

$$= 0.795 \cdot 10^{-1} \text{ with MEC and IC } (\Delta) \qquad (5.26)$$

$$\chi_D^{(0),MEC} = 0.65 \cdot 10^{-4}. \qquad (5.27)$$

These values are in agreement with earlier sum rule results[49] and a recent analysis[50]. Thus, one sees that electric polarizability and paramagnetic susceptibility are enhanced by about 15 to 20% by exchange effects. However, for the diamagnetic susceptibility exchange contributions are considerably smaller.

Beyond the low energy region one needs the generalized polarizabilities. As an example, I show for the deuteron in Figs. 16.1, 16.2 the resonance contribution to scalar, vector and tensor polarizabilities for E1 and M1 as well (from Ref. 19). At low energy,

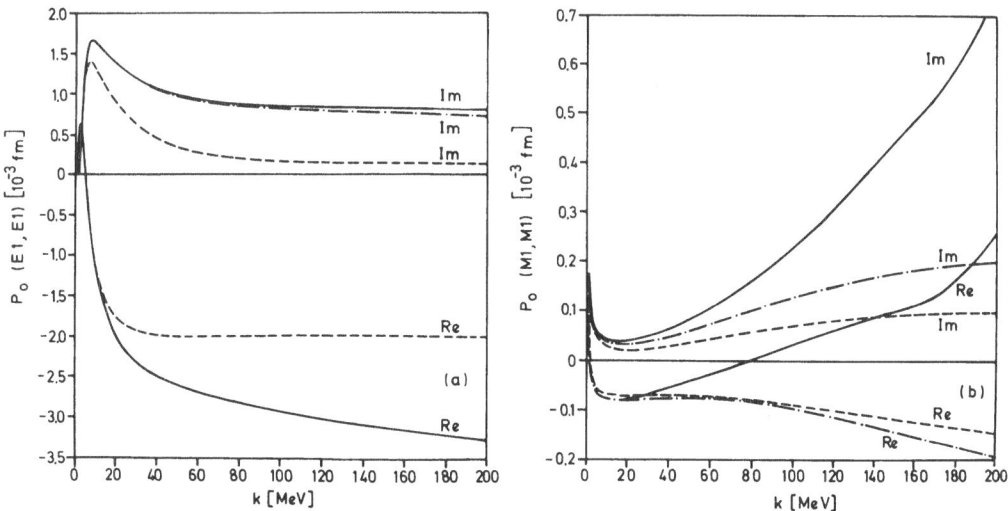

Fig. 16.1 Scalar E1 and M1 polarizabilities for elastic photon-deuteron scattering. Dashed curves are without MEC (no Siegert operators) and IC, dash-dot curves include MEC and full curves include MEC and IC. For the real part of E1 scalar polarizability dash-dot and full curves coincide.

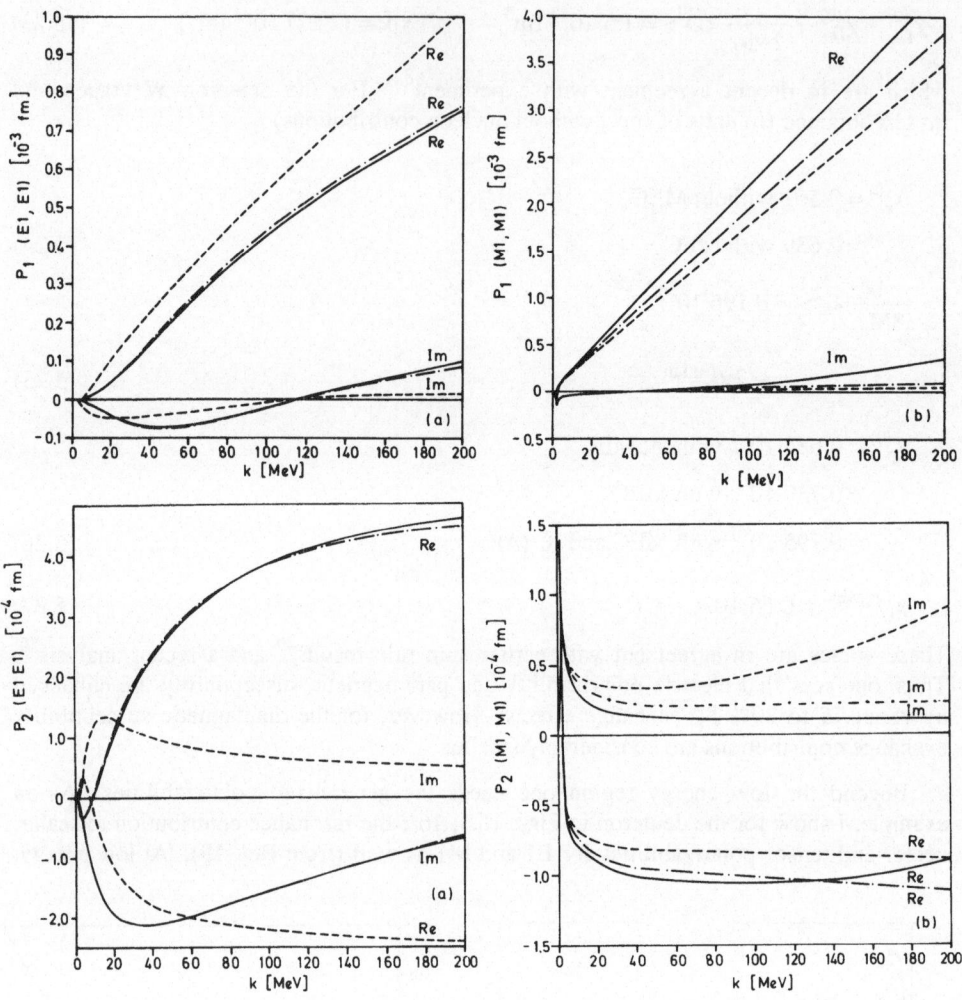

Fig. 16.2 Vector and tensor E1 and M1 polarizabilities for elastic photon-deuteron scattering. Details are described in the caption of Fig. 16.1.

say up to 50 MeV the scalar E1-polarizability is dominating whereas M1 is considerably smaller except near the threshold region. However, while E1 is appreciably reduced at higher energies M1 shows a steady increase which is dramatically enhanced due to the Δ-isobar excitation. Both E1 and M1 show a strong effect on MEC the major part of which is contained in the Siegert operator for E1. Contrary to M1 the IC effects are small for E1 which is not surprising because the dominating NΔ configuration is essentially excited via M1 transition.

The vector polarizability characterizing the optical activity of the deuteron becomes quite sizable with increasing energies. Here M1 is clearly dominating over E1. It is interesting to note that exchange effects are much smaller in comparison to the scalar

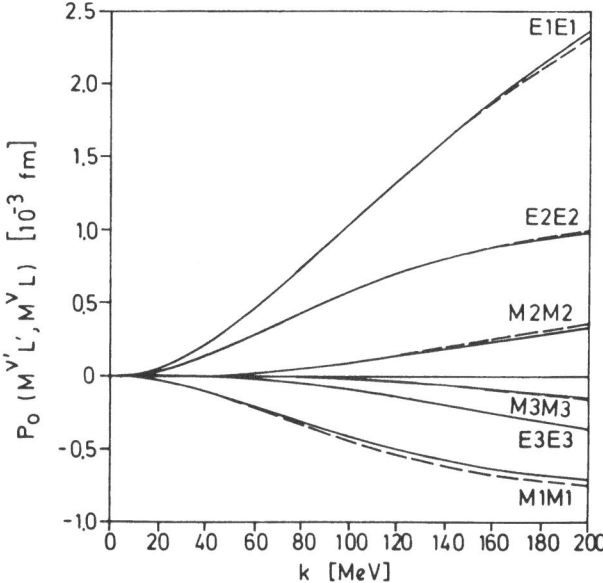

Fig. 17   Scalar polarizability of the two-photon amplitude (beyond low-energy limit) for kinetic part only (dashed curves) and with exchange part (full curves).

polarizabilities. The non-vanishing tensor polarizability in Fig. 16 reflects the fact that the deuteron is optically anisotropic because of its slight deformation due to the small D-component. Even though the tensor polarizability is much smaller than scalar and vector polarizabilities it is remarkable to note the very strong MEC effect in E1 leading even to a sign change.

The scalar and tensor polarizabilities of the two-photon amplitude are shown in Figs. 17 and 18 for the kinetic part and the exchange part as well. The vector polarizability vanishes because of Eq. 4.16. One readily sees that the exchange effects beyond the zero-energy limit (Eq. 4.22) are small for the scalar polarizabilities in Fig. 17. In contrast to this one observes in Fig. 18 a strong enhancement for the pure E1 and M1 tensor polarizabilities. However, these relatively large effects in the tensor polarizabilities do not show up in the averaged scattering cross section where the scalar and vector polarizabilities dominate.

Even though the separate pieces of the TPA have not an independent physical meaning it is interesting from a theoretical point of view to interpret the fourth diagram of Fig. 7 corresponding to the Thomson-amplitude of an exchanged pion as the form factor of the pion excess in a nucleus. Weyrauch and myself[51] have shown that it is not the form factor of the exchanged pions but rather a transition form factor representing the difference of the virtual pion density in a nucleus minus the pion density of the free nucleons, and that means, the pion excess density. Its monopole part is shown in Fig. 19 for the deuteron.

For heavier nuclei we have little reliable information on exchange effects in photon scattering. There have been attempts to simulate the exchange TPA by a simple form factor[52,53]. But in a similar analysis Alberico and Molinari[54] reach the conclusion that such an extrapolation of the low energy theorem of the TPA is not sufficient. Probably a

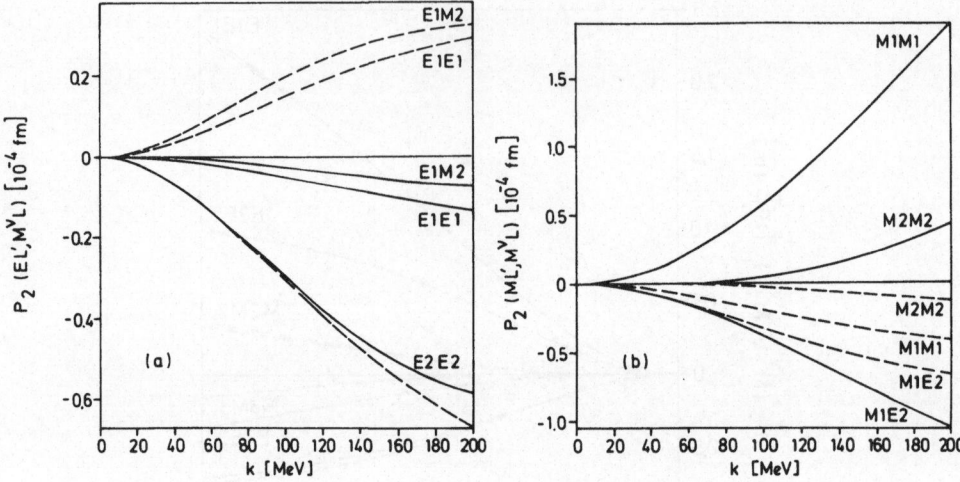

Fig. 18   Tensor polarizabilities of the two-photon amplitude (beyond low-energy limit)
for various multipole contributions. Notation as in Fig. 17.

quasi-deuteron approach for the exchange TPA will be promising.

6. PHOTOABSORPTION AND SCATTERING IN THE Δ–REGION

As already mentioned subnucleonic degrees of freedom can also be studied by looking at the internally excited nucleon states, the nucleon resonances or isobars. A very
prominent example is the Δ–excitation in pion and photo-reactions[55]. The interesting
questions are how the Δ is affected by the presence of a nuclear medium, how its

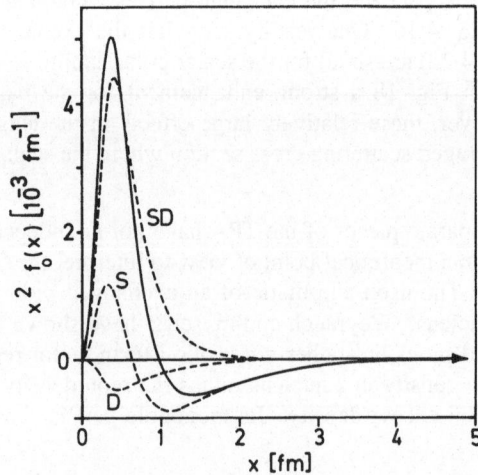

Fig. 19   Monopole part of π-exchange transition (π-excess) density for the deuteron as
measured from the center-of-mass (full curve) and separate contributions of S-
and D-waves and S-D interference (dashed curves), from Ref. 51.

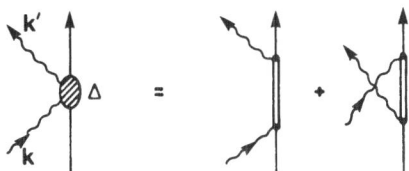

Fig. 20   One-nucleon Δ-resonance model for photon scattering in the Δ-region.

resonance position and width are changed and whether nuclear structure effects influence the Δ-propagation.

The simplest model assumes a one-body excitation mechanism $\gamma+N \rightarrow \Delta \rightarrow N+\gamma$ (Fig. 20) neglecting background contributions from nucleonic excitations and π-production, Δ-propagation and medium corrections on the Δ as well. The elementary scattering amplitude is given by

$$t_{\lambda'\lambda} = c_- \, \vec{\varepsilon}_{\lambda'}^{\,\prime *} \cdot (\vec{\sigma}_{N\Delta} \times \vec{k}') \, \vec{\varepsilon}_\lambda \cdot (\vec{\sigma}_{\Delta N} \times \vec{k})$$
$$+ \, c_+ \, \vec{\varepsilon}_\lambda \cdot (\vec{\sigma}_{N\Delta} \times \vec{k}) \, \vec{\varepsilon}_{\lambda'}^{\,\prime *} \cdot (\vec{\sigma}_{\Delta N} \times \vec{k}') \, , \qquad (6.1)$$

where

$$c_- = c^2 (M_\Delta - M_N - k - i\Gamma_\Delta/2)^{-1} \qquad (6.2)$$

$$c_+ = c^2 (M_\Delta - M_N + k' + i\Gamma_\Delta/2)^{-1} \qquad (6.3)$$

$$c = G_{M1}^{\Delta N} (M_\Delta + M_N)/(4 M_N \, M_\Delta) \, . \qquad (6.4)$$

I have denoted the momenta of incoming and scattered photons by $\vec{k}$ and $\vec{k}'$, respectively, the corresponding polarization vectors by $\vec{\varepsilon}_\lambda$ and $\vec{\varepsilon}'_\lambda$ and the $\Delta \rightarrow N$ spin transition matrix by $\vec{\sigma}_{N\Delta}$. The magnetic transition strength $G_{M1}^{\Delta N}$, the resonance position M_Δ and the resonance width Γ_Δ are fitted to the experimental photoabsorption cross section per nucleon in the Δ-resonance region. Rewriting (Eq. 6.1) in a form corresponding to the defining equation (Eq. 4.1), one finds for the elementary scalar and vector polarizability operators

$$\hat{P}_\Delta^{[0]}(M1,M1) = \frac{2}{\sqrt{3}} \, c_0 \, k^2 \qquad (6.5)$$

$$\hat{P}_\Delta^{[1]}(M1,M1) = \sqrt{\frac{2}{3}} \, c_1 \, k^2 \, \sigma^{[1]} \qquad (6.6)$$

$$c_{0/1} = c_- \pm c_+ \, . \qquad (6.7)$$

Inserting these operators into Eq. 4.13 and specializing to a spin-zero nucleus, it turns out that only the elementary polarizability scalar contributes. In detail, one finds for the elastic scattering amplitude[18]

$$T^{\ddot{i}}_{\lambda'\lambda}(\Delta) = -\sum_L \frac{(-)^L}{\hat{L}} d^L_{\lambda'\lambda}(\theta) \, (P_0(EL,EL) + P_0(ML,ML))$$

$$= \frac{2}{3} c_0 k^2 \vec{\varepsilon}' \cdot \vec{\varepsilon} \; f^0(\vec{k} - \vec{k}') \tag{6.8}$$

with the polarizabilities

$$P_0(EL,EL) = \frac{1}{2} (-)^{L+1} \frac{\hat{L}^3}{\sqrt{3}} <i| \sum_\alpha \hat{p}^{[0]}_\Delta j^2_L(kr_\alpha)|i>$$

$$= (-)^{L+1} \frac{\hat{L}^3}{3} c_0 k^2 \int d^3r \, \rho_N(r) j^2_L(kr) \tag{6.9}$$

$$P_0(ML,ML) = \frac{1}{2}(-)^{L+1} \frac{\hat{L}}{\sqrt{3}} <i| \sum_\alpha \hat{p}^{[0]}_\Delta [(L+1)j^2_{L-1}(kr_\alpha)+Lj^2_{L+1}(kr_\alpha)]|i>$$

$$= (-)^{L+1} \frac{\hat{L}}{3} c_0 k^2 \int d^3r \, \rho_N(r) \, [(L+1)j^2_{L-1}(kr) + Lj^2_{L+1}(kr)] \, . \tag{6.10}$$

Here we have introduced the nucleon mass density $\rho_N(r)$ and its form factor $f^0(q)$. Furthermore, the quantization axis in Fig. 3 is taken parallel to the incoming photon direction $\vec{k}$, thus, θ in Eq. 6.8 is the scattering angle. The scattering cross section is then given by a mass density form factor $f^0(q)$

$$\frac{d\sigma}{d\Omega} = \frac{8\pi}{9} k^4 \, |c_0|^2 \, f^0(q)(1 + \cos^2\theta) \, , \tag{6.11}$$

$$f^0(q) = |<0|| \sum_{\alpha=1}^A j_0(qr_\alpha) \, Y^{[0]}(\hat{r}_\alpha) \, ||0>|^2 \, , \tag{6.12}$$

$$c_0 = c_- + c_+ \, . \tag{6.13}$$

The appearance of the form factor is a consequence of the neglect of Δ–propagation and of other contributions making the scattering process strictly local.

The total absorption cross section of this elementary model for Δ–excitation has the form

$$\sigma_{tot}(\Delta) = \frac{8\pi}{3} \, k \, A \, Im \, c_0$$

$$= \sum_L (S(EL) + S(ML)) \, , \tag{6.14}$$

where I have introduced the multipole strengths according to Eq. 4.38

$$S(M^\nu L) = \frac{4\pi}{k} \frac{(-)^{L+1}}{\hat{L}} \, Im \, P_0(M^\nu L, M^\nu L), \quad \nu = 0, 1 \, . \tag{6.15}$$

These are shown in Fig. 21 for ^{208}Pb . Remarkable is the increase up to L = 7 reflecting the strong influence of the form factor because of the high momentum transfer of a photon absorbed in the Δ–region .

The scattering cross section is thus governed by the nuclear density form factor. However, this simple model fails to describe a recent experiment by Hayward and Ziegler[56] (see Fig. 22). This might not be too surprising since at this rather large angle of 115° the momentum transfer is large and, thus, the form factor drops rapidly so that

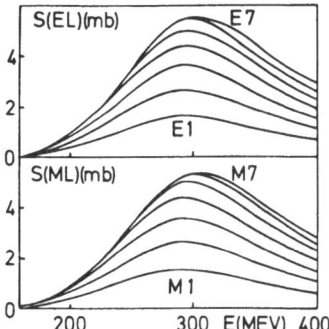

Fig. 21   Multipole strengths for photoabsorption in the Δ-region for ^{208}Pb.

other effects, which safely may be neglected in the forward direction, can become important. Such effects are:

(i)   nucleonic contributions which might effectively be
       described by the quasi-deuteron model,
(ii) Thomson and exchange two-photon amplitude,
(iii) Δ–propagation and medium corrections, and
(iv) background contributions from Born-terms of
       π–photoproduction.

The first two points have been discussed recently[18], even though in an approximate way, and it was found that they give significant contributions at higher momentum transfer in the right direction. A more careful treatment is necessary.

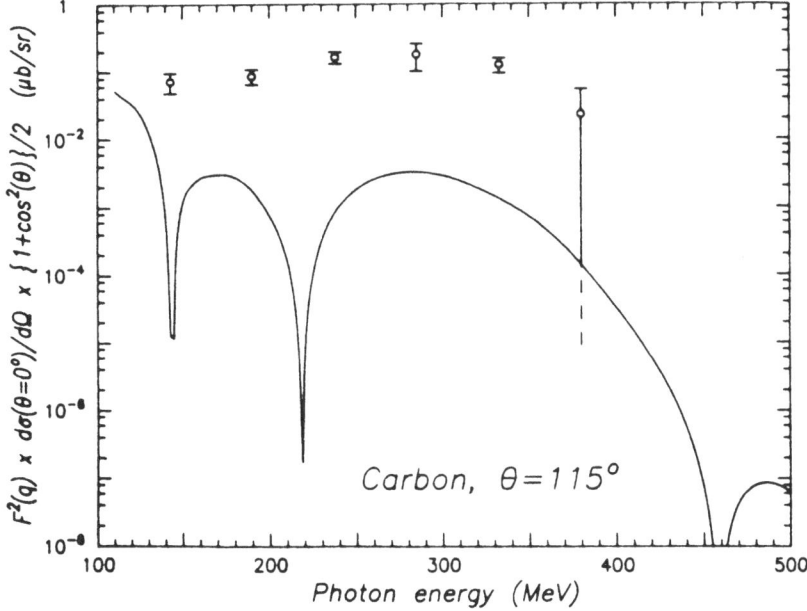

Fig. 22   The prediction of the simple model for ^{12}C compared with experimental data
          (from Ref. 56).

Medium corrections and nuclear structure effects have been studied in the Δ–hole model by Koch, Moniz and Ohtsuka[57]. They use as scattering amplitude

$$T_\Delta = \bar{F}^+_{\gamma N\Delta}\, G_{\Delta h}(E)\, \bar{F}_{\gamma N\Delta}\,,$$

where the vertex operator is the coherent sum of the various Δ–hole excitation operators dressed by inclusion of background terms, and $G_{\Delta h}(E)$ is the full many-body Δ–hole Green function. It contains different pieces of medium corrections as is discussed in detail in Ref. 57. A comparison of this model with the elementary approach is shown in Fig. 23. The importance of a proper treatment of Δ–propagation is clearly seen. But still the experimental data is considerably higher.

However, Hayward and Ziegler[56] already remarked that a possible explanation for the large discrepancy between the elementary theory and experiment could be the inclusion of inelastic contributions in the experimental data because the energy resolution of their photon detector was only 10 per cent. Therefore, we recently have estimated the inelastic contributions to photon scattering in this energy region using the same simple model[58], in which the inelastic cross section can be expressed by transition form factors of mass and spin density. Using simple 1p-1h excitations for the final states, one finds important inelastic contributions at larger scattering angles, i.e., at higher momentum transfer. The reason for this is that due to the high momentum transfers, e.g., $q = 2.55$ fm^{-1} at $k = 300$ MeV and $\theta = 115°$, the elastic form factor suppresses the elastic scattering cross section considerably. Then transitions to higher final state spins described by higher multipole transition form factors are favoured. In fact, the dominant inelastic contributions come from $I = 2$ and 3 for ^{12}C and I around 8 for ^{208}Pb with still sizable contributions from higher spins.

This calculation has been improved recently by Vesper et al.[59] for ^{12}C by again considering Δ–propagation and medium corrections in the Δ–hole approach of Koch et al.[57].

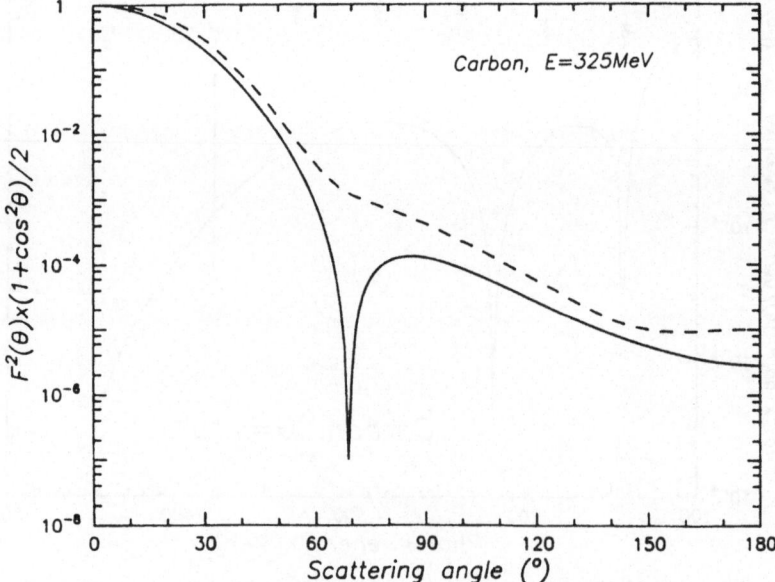

Fig. 23   Comparison of the simple model (full curve) with a Δ-hole calculation (dashed curve) from Ref. 57 for ^{12}C.

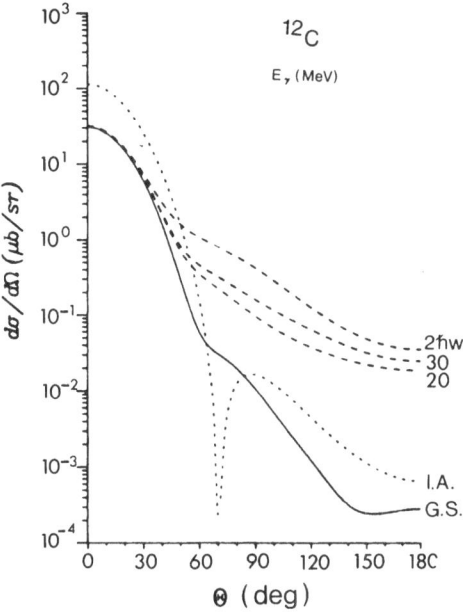

Fig. 24   Angular distribution for photon scattering off ^{12}C at photon energy E = 325 MeV. Solid curve: elastic scattering, dashed curve: elastic and inelastic scattering up to ΔE = 20, 30 and 42 MeV (2ℏω), dotted curve: elastic scattering in IA (Ref. 59).

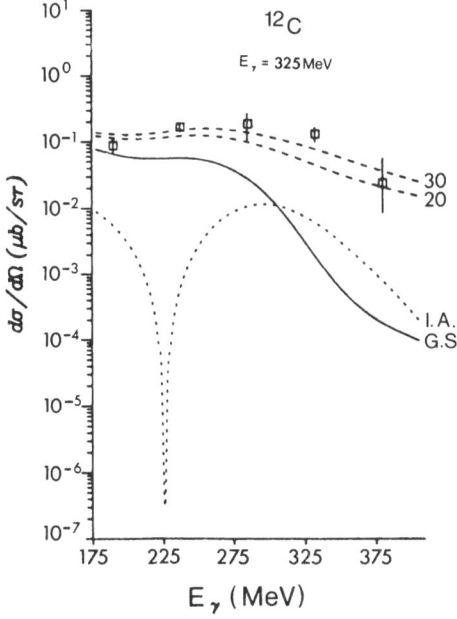

Fig. 25   Differential cross sections dσ/dΩ for photon scattering off ^{12}C at constant scattering angle θ = 115°. Solid curve: elastic scattering, dashed curve: elastic and inelastic scattering up to ΔE = 20 MeV, dashed-dotted curve: same for ΔE = 30 MeV, dotted curve: elastic scattering in IA (Ref. 59).

From their results I show in Fig. 24 an angular distribution and in Fig. 25 the comparison with the experimental data. The angular distributions are in qualitative agreement with the elementary approach in Ref. 58. However, at backward angles both elastic and inelastic cross sections are considerably larger than in the elementary approach and a good agreement with the data is achieved.

Future improvements of the theory should be concerned with a better final state description for inelastic scattering, inclusion of the quasi-deuteron tail as well as the two-photon amplitude. On the experimental side, the separation of elastic and inelastic scattering is highly desirable and more data on angular distributions, particularly in the forward direction is needed.

7. SUMMARY

In the present lectures I have discussed the rôle of subnuclear degrees of freedom in electromagnetic exchange operators. They arise from meson and isobar degrees of freedom and give important contributions to one-photon processes such as photoabsorption and electron scattering and to two-photon processes, such as photon scattering, where the two-photon operator contributes.

Gauge conditions for charge and current densities and for the two-photon operator are shown to give strong constraints on the form of the corresponding exchange operators. Except for the low-energy theorems for the current in the form of the Siegert operators and for the two-photon amplitude these gauge conditions, however, are not sufficient to fix the explicit form of the exchange operators. One needs specific models.

A detailed discussion of contributions from resonance and the two-photon amplitude to the photon scattering amplitude has been given. Specific emphasis has been put on the expansion of the scattering amplitude in terms of generalized polarizabilities. Low-energy theorems have been derived with the help of the gauge conditions. Up to second order in the photon energy the scattering amplitude is decomposed into scalar, vector and tensor polarizabilities, which are determined by a few structure constants like, e.g., total charge, magnetic dipole and charge quadrupole moments, electric polarizability and magnetic susceptibility. Furthermore, it has been shown that the lowest order term of the electric scalar polarizabilities $P_0(EL,EL)$ is proportional to k^{2L}, except for $L = 1$, for which the lowest order term is the Thomson amplitude.

The optical theorem and the use of dispersion relations for the forward scattering amplitude are discussed. Then a generalization to the polarizabilities is given linking their imaginary parts to the corresponding multipole contributions to the total cross section. The importance of dispersion relations for the polarizabilities is stressed, which simplify their calculation considerably.

For the long-range and most important pion exchange the explicit form of the exchange current and the exchange two-photon operators have been given. The construction of a consistent exchange current for the parametrized Paris potential has been sketched briefly. Then exchange effects for the deuteron have been discussed for electrodisintegration near threshold and for photodisintegration below π–production threshold and in the region of real Δ–excitation, where a proper treatment of the Δ–dynamics is important. At low momentum and energy transfer the π–exchange contributions give in general a fair agreement with experimental data. For higher momentum and energy transfer larger uncertainties arise because of the less well understood medium and short range behaviour of the NN-interaction and a lack of a proper treatment of retardation and relativistic effects. Probably, an explicit treatment of quark-gluon degrees of freedom will be necessary for a more reliable description.

The exchange effects in photon scattering have been demonstrated for elastic photon-deuteron scattering. The scalar polarizability is dominated by E1, where the exchange effects are mostly covered by the Siegert operator. The vector polarizability is dominated by M1 contributions. The largest explicit exchange effects show up in the tensor polarizability. From the Thomson scattering diagram off an exchanged pion the pion excess density in the deuteron has been deduced.

Finally, Δ-excitation in a nucleus by photon scattering has been considered, first in an elementary static approach. Then the importance of Δ-propagation, medium corrections and nuclear structure effects are discussed. Furthermore, it has been shown that at larger scattering angles corresponding to high momentum transfer large inelastic contributions exceeding the elastic scattering cross section are to be expected. This will certainly be an interesting field to be studied in the future both experimentally and theoretically.

REFERENCES

[1]   W. Heisenberg, Z Phys. **77,** 1 (1932); E. Majorana, Z. Phys. **82,** 137 (1933).

[2]   H. Yukawa, Proc. Phys.- Math. Soc. Jap. **17,** 48 (1935).

[3]   A. M. Green, Rep. Progr. Phys. **39,** 1109 (1976).

[4]   H. J. Weber and H. Arenhövel, Phys. Rep. **C36,** 277 (1978).

[5]   D. H. Wilkinson, Nucl. Phys. **A434,** 629 (1985).

[6]   J. L. Friar, Ann. Phys. (N.Y.) **95,** 170 (1975).

[7]   A. J. F. Siegert, Phys. Rev. **52,** 787 (1937).

[8]   M. Goeppert-Mayer, Ann. Phys. **9,** 273 (1931).

[9]   P. I. Richards, Phys. Rev. **73,** 254 (1948).

[10]   J. L. Friar and S. Fallieros, Phys. Rev. **C29,** 1645 (1984).

[11]   M. Gell-Mann, M. L. Goldberger and W. E. Thirring, Phys. Rev. **95,** 1612 (1954).

[12]   H. Arenhövel and W. Greiner, Prog. Nucl. Phys. **10,** 167 (1969).

[13]   R. Silbar and H. Überall, Nucl. Phys. **A109,** 146 (1968).

[14]   U. Fano, NBS Technical Note 83 (1960), reprinted in Photonuclear Reactions ed. E. G. Fuller and E. Hayward (Dowden, Hutchinson & Ross, 1976) p. 338.

[15]   H. Arenhövel, Proc. Int. Conf. on Photonuclear Reactions and Applications, B. L. Berman, ed. (Univ. of California, Livermore 1973) p. 449.

[16]   R. Moreh, Proc. Int. School of Intermediate Energy Nuclear Physics, R. Bergere, S. Costa and C. Schaerf, eds. (World Scientific, Singapore, 1982) p. 1.

[17]   H. Arenhövel and M. Weyrauch, to be published.

[18]   H. Arenhövel, Proc. Workshop on Perspectives in Nuclear Physics at Intermediate Energies, S. Boffi, C. Ciofi degli Atti, M. M. Giannini, eds. (World Scientific, Singapore 1984) p. 97.

[19]   M. Weyrauch and H. Arenhövel, Nucl. Phys. **A408,** 425 (1983).

[20]   M. Weyrauch, Ph.D. thesis, Mainz (1984).

[21]   E. Fein, private communication.

[22]   P. Stichel and E. Werner, Nucl. Phys. **A145,** 257 (1970).

[23]   J. L. Friar, Phys. Rev. Lett. **36,** 510 (1976).

[24]   H. Arenhövel, Z. Phys. **A297,** 129 (1980).

[25]   R. A. Bryan and A. Gersten, Phys. Rev. **D6,** 341 (1972).

[26] W. Leidemann and H. Arenhövel, Nucl. Phys. **A393,** 385 (1983).

[27] M. Lacombe *et al.,* Phys. Rev. **C21,** 861 (1980).

[28] D. O. Riska, preprint, University of Helsinki, HU-TFT-84-48.

[29] A. Buchmann, W. Leidemann and H. Arenhövel, Nucl. Phys. **A443,** 726 (1985).

[30] M. Chemtob and M. Rho, Nucl. Phys. **A163,** 1 (1971).

[31] J. Hockert *et al.,* Nucl. Phys. **A217,** 14 (1973); J. A. Lock and L. L. Foldy, Ann. of Phys. **93,** 276 (1975); W. Fabian and H. Arenhövel, Nucl. Phys. **A258,** 461 (1976); B. Mosconi and P. Ricci, Nuovo Cim. **36A,** 67 (1976); B. Sommer, Nucl. Phys. **A308,** 263 (1978).

[32] P. Sauer, invited talk at ELSA-Workshop, Bonn (1984), Lecture Notes in Physics, **234,** (Berlin-Heidelberg, 1985) p. 276.

[33] H. Arenhövel, Nuovo Cim. **76A,** 256 (1983).

[34] J. M. Laget, Nucl. Phys. **A312,** 265 (1978); W.-Y.P. Hwang and G. A. Miller, Phys. Rev. **C22,** 968 (1980); H. Arenhövel, Z. Phys. **A302,** 25 (1981); M. Gari and H. Hebach, Phys. Rep. **72C,** 1 (1981).

[35] F. F. Liu, Phys. Rev. **B138,** 1443 (1965); W. Del Bianco *et al.,* Nucl. Phys. **A343,** 121 (1980); W. Del Bianco *et al.,* Phys. Rev. Lett. **47,** 118 (1981).

[36] G. G. Simon *et al.,* Nucl. Phys. **A324,** 277 (1979).

[37] R. V. Reid, Ann. of Phys. **50,** 411 (1968).

[38] R. E. Rand *et al.,* Phys. Rev. Lett. **18,** 469 (1967); D. Ganichot *et al.,* Nucl. Phys. **A178,** 545 (1972); G. G. Simon *et al.,* Phys. Rev. Lett. **37,** 739 (1976); M. Bernheim *et al.,* Phys. Rev. Lett. **46,** 402 (1981).

[39] H. Arenhövel, Proc. of CEBAF 1985 Summer Workshop (to be published).

[40] S. Auffret *et al.,* Phys. Rev. Lett. **55,** 1362 (1985).

[41] H. Arenhövel, Z. Phys. **A275,** 189 (1975).

[42] W. Leidemann and H. Arenhövel, Can. J. Phys. **62,** 1036 (1984).

[43] W. Leidemann, doctoral thesis, Mainz (1985).

[44] R. L. Anderson *et al.,* Phys. Rev. Lett. **22,** 651 (1969) ( ); D. I. Sober *et al.,* Phys. Rev. Lett. **22,** 430 (1969) (Δ); P. Dougan *et al.,* Z. Phys. **A280,** 341 (1977) (Δ); R. Rose *et al.,* Z. Phys. **202,** 364 (1967) ($\square$); A. M. Smith *et al.,* J. Phys. **A1,** 553 (1968) ($\bigcirc$); J. Buon *et al.,* Phys. Lett. **26B,** 595 (1968) ($\bullet$); J. Arends *et al.,* Nucl. Phys. **A412,** 509 (1984) ($\blacksquare$).

[45] J. M. Laget, private communication.

[46] W. Leidemann and H. Arenhövel (to be published).

[47] D. Drechsel and A. Russo, Phys. Lett. **137B,** 297 (1984).

[48] V. A. Petrun'kin, Sov. J. Part. Nucl. **12,** 278 (1981).

[49] H. Arenhövel and W. Fabian, Nucl. Phys. **A292,** 429 (1977).

[50] J. L. Friar *et al.,* Phys. Rev. **C27,** 1364 (1983).

[51] M. Weyrauch and H. Arenhövel, Phys. Lett. **134B,** 21 (1984).

[52] R. Leicht *et al.,* Nucl. Phys. **A362,** 111 (1981).

[53] M. Sanzone Arenhövel *et al.,* Proc. Int. School of Intermediate Energy Nuclear Physics, R. Bergere, S. Costa and C. Schaerf, eds. (World Scientific, Singapore 1982) p. 291.

[54] W. Alberico and A. Molinari, Z. Phys. **A309,** 143 (1982).

[55] F. Lenz and E. J. Moniz, Phys. Rev. **C12,** 909 (1975); E. Oset and W. Weise, Nucl. Phys. **A319,** 477 (1979); K. Klingenbeck and M. G. Huber, J. Nucl. Phys. **G6,** 961 (1980).

[56] E. Hayward and B.Ziegler, Nucl. Phys. **A414,** 333 (1984).

[57] J. H. Koch, E. J. Moniz and N. Ohtsuka, Ann. Phys. **154,** 99 (1984).

[58] H. Arenhövel, M. Weyrauch and P.-G. Reinhard, Phys. Lett. **155B,** 22 (1985).

[59] J. Vesper *et al.,* Phys. Lett. **159B,** 233 (1985).

SCATTERING AND ABSORPTION OF PHOTONS BY NUCLEI AND NUCLEONS

B. Ziegler

Max Planck Institut für Chemie
D6500, Mainz, West Germany

1. INTRODUCTION

Photon scattering cross sections are determined by counting the number of photons scattered into a certain solid angle $d\Omega$ at a given scattering angle Θ. The scattering objects must possess electric charges or magnetic moments or both. The objects under investigation can in principle be anything: a small droplet[1], an atom, a nucleus or a nucleon. Here, we shall confine ourselves to elastic coherent scattering from nuclei and nucleons in energy ranges that are adequate to reveal the particles' internal structure. Since the physical phenomena involved are similar for nucleons and nuclei, and most experiments and theories deal with the case of nuclei, the formalism will be described for nuclei. An attempt will be made to apply the results for the case of nucleons.

The information contained in the angular distribution pattern essentially can be regarded as a kind of outside projection of the particle's internal structure and dynamics. As far as a mapping of the charge or magnetization densities is concerned, the necessary scale is uniquely provided by the wavelength $\lambda = \hbar c/E$ of the incident wave, and the scattering angle Θ. There is no distortion of the incoming or outgoing waves by Coulomb effects, a fact that favors photon scattering experiments over charged particle, and even electron scattering. However, since electron scattering has been cultivated for a number of decades, both experimentally and theoretically, the absence of Coulomb effects would hardly justify more than a few cross checks on electron scattering results. Although it will not be mentioned everywhere explicitly, it is one of the purposes of this lecture to show basic differences of electron and photon scattering and what kind of new information can be obtained using real photons.

After a short outline of necessary definitions[2,3], the starting point for the theoretical description will be scattering from a point charge. With increasing complexity of the scatterer, the phenomena also become more complex. Actually, there is no well defined region in photon energy or in momentum transferred to the nucleus where a single elementary process dominates. One is never far from resonances and, on the other hand, always has to deal with continuous amplitudes interfering with resonances. An attempt will be made to give a complete description of the processes consistent with general accepted theorems and rules. However, it is not intended to derive the formulas nor to claim to be original. The goal will be to find a workable formalism that can be used to interpret experimental results consistently over an energy range from zero to well above the characteristic resonances. The analysis will be confined to the strongest multipolarities, E1, M1, and E2.

Photon absorption cross sections measured over a wide energy range serve a twofold purpose. One is an evaluation of sum rules that can be derived from very general assumptions. The other is the correspondence of the total absorption cross section with the forward scattering cross section. Both aspects are essential and very much relied on in an attempt to interpret the whole nuclear response to photons.

Published experiments will be discussed at the end to show the most important techniques and also in order to get an idea which technique might be best suited for the higher energy scattering experiments that have to come in the future.

2. SCATTERING AMPLITUDE, MOSTLY FORWARD

The elastic scattering cross section can be written as the square of the modulus of the complex scattering amplitude:

$$\frac{d\sigma}{d\Omega}(E,\Theta) = |R(E,\Theta)|^2 . \tag{1}$$

In general, there are scattering amplitudes of different electromagnetic character, λ, electric or magnetic, and of different multipolarities L, at the same energy, that are added coherently:

$$R(E,\Theta) = \sum_{\lambda,L} R^{\lambda,L}(E,\Theta) . \tag{2}$$

The total cross section for the removal of photons from the primary beam is given by the optical theorem:

$$\sigma_{\gamma,tot}(E) = 4\pi\lambda \ Im\{R(E,\Theta{=}0)\} . \tag{3}$$

Causality requires the validity of a pair of equations that connect the real with the imaginary part of the scattering amplitude, known as dispersion relations[4,5]. With $D(E) = Re\{R(E,\Theta{=}0)\}$ and $A(E) = Im\{R(E,\Theta{=}0)\}$, they can be written

$$D(E) = \frac{2}{\pi} P \int \frac{E' A(E')}{E^2-E'^2} dE' , \tag{4a}$$

and

$$A(E) = \frac{2}{\pi} P \int \frac{E D(E')}{E'^2-E^2} dE' . \tag{4b}$$

If the real part of R at energy zero is subtracted from Eq. 4a, the integral converges faster, or does not diverge any more, as a function of energy, and $A(E)/E$ can be replaced by $\sigma_{\gamma,tot}(E)$ using Eq. 3. Then one has

$$D(E) = D(0) + \frac{E^2}{2\pi^2\hbar c} P \int \frac{\sigma_{\gamma,tot}(E')}{E'^2-E^2} dE' . \tag{5}$$

Combining Eqs. 1, 3, and 5, an expression is obtained that allows the calculation of the scattering cross section, if the photon absorption cross section $\sigma_{\gamma,tot}(E)$ is known between zero and infinite energies:

$$\frac{d\sigma}{d\Omega}(E,\Theta{=}0) = \frac{\sigma_{\gamma,tot}^2(E)}{(4\pi\lambda)^2} + \left\{ D(0) + \frac{E^2}{2\pi^2\hbar c} P \int \frac{\sigma_{\gamma,tot}(E')}{E'^2-E^2} dE' \right\}^2 . \tag{6}$$

This formula was first derived by Fuller and Hayward[2] . It was used to calculate nuclear photon scattering cross sections. Recently[6] , an attempt was made to work backward and to try to find a function $\sigma_{\gamma,tot}(E)$ satisfying Eq. 6, for a given finite set of measured scattering cross sections. This is not possible in general, since there is no information obtained on the phase of the scattering amplitude, if only scattering cross sections are measured. It could be shown in Ref. 6, how $\sigma_{\gamma,tot}(E)$ in an energy region above the measured scattering cross sections can be used as independent input to the dispersion integral in order to constrain the absorption cross section deduced from scattering.

Leicht's "magic" angle[7] is an example, how to get information on the absorption cross section by choosing a special scattering cross section (see section 3). Another rather direct way of obtaining absorption data is the remark by Fuller and Hayward[2] that in the neighbourhood of an isolated peak absorption cross section, the principal value integral vanishes, the scattering amplitude is pure imaginary, and the absorption cross section is given by the first term in Eq. 6:

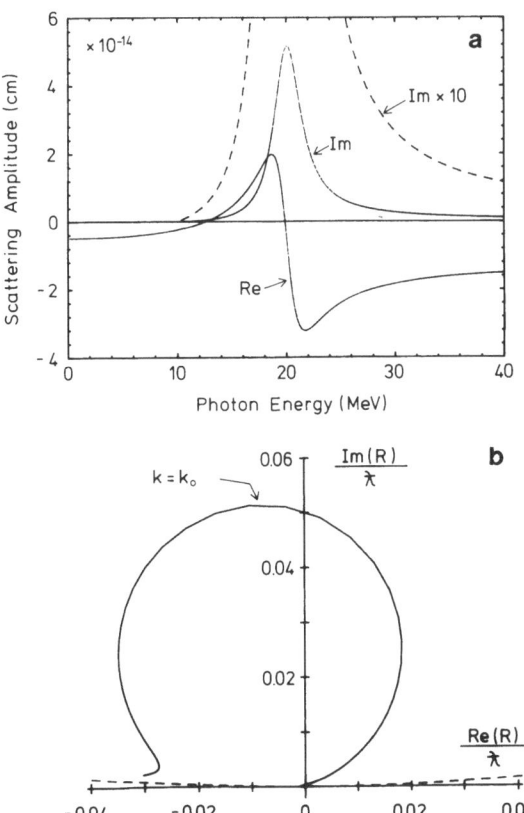

Fig. 1   The scattering amplitude (a) of a fictitious lead nucleus with a resonance at 20 MeV, that contains one classical sum. The Argand diagram (b) shows how strongly the process is absorptive: the peak scattering cross section is only 0.05^2 times the value for a pure elastic scattering process at the same energy. Dashed is the unitarity circle.

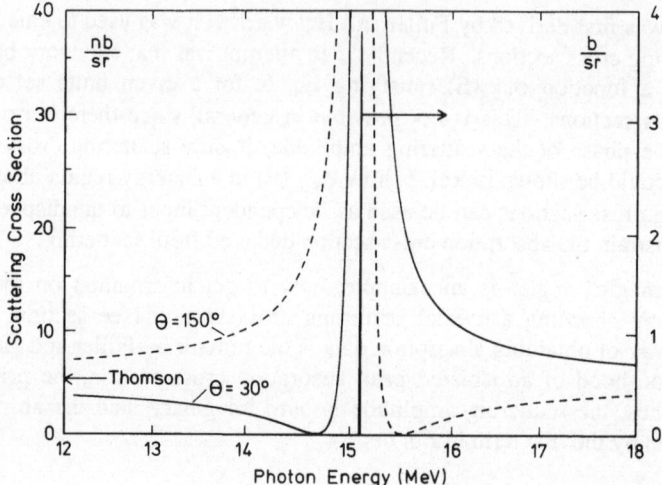

Fig. 2   A very narrow resonance, the 15.1 MeV resonance in ^{12}C, and its interference
with the Thomson amplitude. Far from resonance, the resonance scattering
amplitude is real and entirely determined by the integrated scattering cross section (2mb·MeV) of the line. Neglected is the giant resonance amplitude, inelastic scattering to the 4.4 MeV level, and a small absorptive (γ,α) cross section.

$$\frac{d\sigma}{d\Omega}(E_0, \Theta = 0) = \frac{\sigma^2_{\gamma,\text{tot}}(E_0)}{(4\pi\lambda_0)^2} . \tag{6a}$$

Coming back to Eq. 6, it may be noted that in $\sigma_{\gamma,\text{tot}}(E)$ alone there is as well no direct
phase information. It is the knowledge of D(0) that has to be used as an independent
input. Fortunately, there is absolutely no doubt about the value of D(0), and its electric
dipole character.

An interesting possibility was mentioned by Schumacher[8]. He proposes to measure
interference terms between the nuclear scattering amplitude and Delbrück-amplitudes for
small scattering angles where Delbrück is well known, in order to get information on the
nuclear phase.

It seems to be generally accepted that for each multipole λ,L there is a separate relation according to Eq. 6 connecting the multipoles' absorption with the multipoles' forward scattering cross section.

2.1 Example 1: An Isolated Lorentzian

This is a study to show the most simple case for a resonance at 20 MeV with threshold at 10 MeV and a width of 3 MeV. Fig. 1 shows the real and imaginary parts of the
scattering amplitude. In the Argand diagram, real and imaginary parts are made dimensionless by dividing both by λ . They are plotted in the complex plane. The unitarity circle is the locus of pure scattering amplitudes not accompanied by absorptive cross sections at the same energy.

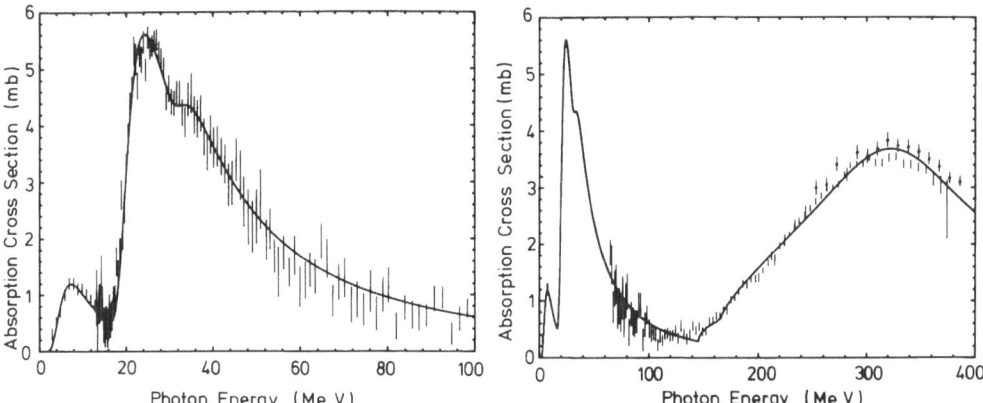

Fig. 3   Measured ^{9}Be absorption cross section; (γ,n) below 10 MeV ;[9] attenuation technique[10] ; hadron detection[11] above 250 MeV, with points. The curves are
explained in the text.

Fig. 2 shows the other extreme, the 15.1 MeV, M1 level in ^{12}C , and its interference
with the Thomson amplitude. The giant resonance amplitude is not taken into account.
Typically, the real part of the Thomson scattering amplitude is strongly modified even at
a distance of several MeV from the resonance that has a width of only 30 eV.

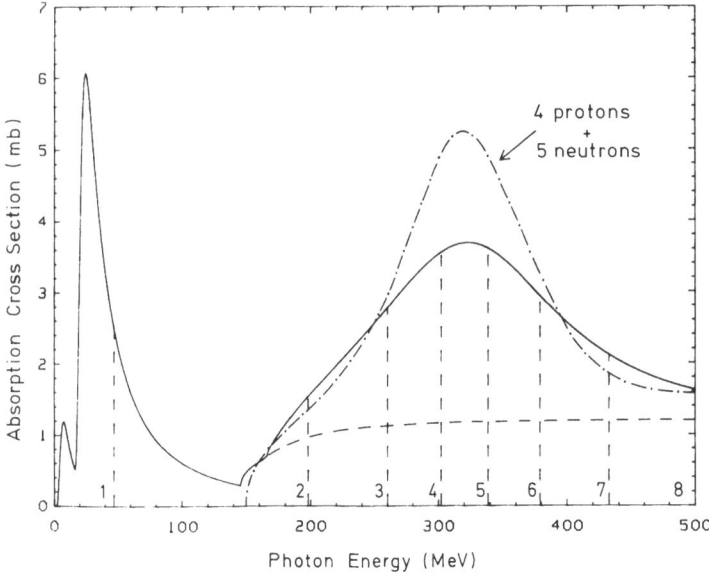

Fig. 4   The curve fitted to the ^{9}Be absorption cross sections. The vertical dashed lines
show the increasing number of classical sums. The almost horizontal dashed
line separates resonance (M1), from direct reaction (Born-term, E1) cross sections. Dashed-dotted is the cross section of the incoherent sum of 4 free protons and 5 free neutrons.

2.2 Example 2: ^{9}Be

Beryllium total absorption cross sections are well known[9,10,11] up to ≈ 400 MeV. In Fig. 3, the experimental points are shown together with fitted curves. For the nuclear part, below pion threshold, the type of cross section used is a direct reaction

$$\sigma_{\gamma,tot}(E) = \sum_{i=1}^{3} T(E-E_{th,i})\, M_i \lambda^2 \tag{7}$$

where the energy dependence is simply determined by a transmission coefficient[12] $T(E-E_{th,i})$ and λ^2 . There are at most three pieces with essentially six parameters. The three thresholds $E_{th,i}$ are 2, 16 and 29 MeV, coinciding (accidentally?) with the (γ,n) , the (γ,p) , and the (γ,2p) thresholds. The strengths, M_i , correspond to 0.15, 1.4, and 0.35 classical sums, adding up to a total strength of 1.9 classical sums for the nuclear part.

The mesonic part is formally treated as described in Ref. 13. The parameters have been adjusted to the new experimental results published recently. In Fig. 4, the fitted curve is shown once more together with some partitioning lines showing the considerable number of classical sums in the Δ–resonance region and the almost equal resonance and direct reaction parts in the same region.

In comparing the ^{9}Be cross section with that for the incoherent sum of 9 nucleons, Fig. 4 suggests a certain reduction of the bound particles' cross section. However, it seems to be appropriate to consider first kinematic reasons for differences, such as the effect of different threshold energies and nuclear sizes. Taking these effects into account, the reduction of the free compared to the bound particles' cross sections may even be larger[13]. However, the tentative inclusion of the more recent experimental values tends to go into the opposite direction. In order to be able to draw final conclusions, there is certainly a need for absorption data in the whole resonance region.

The absorption cross sections of Fig. 3 have been inserted into Eqs. 3 and 5. The resulting imaginary and real parts of the ^{9}Be scattering amplitudes are plotted in Fig. 5. The imaginary part resembles the energy weighted absorption cross section, whereas the real part crosses the zero four times. In addition to the zeros close to an absorption peak, there are two more zeros close to absorption minima.

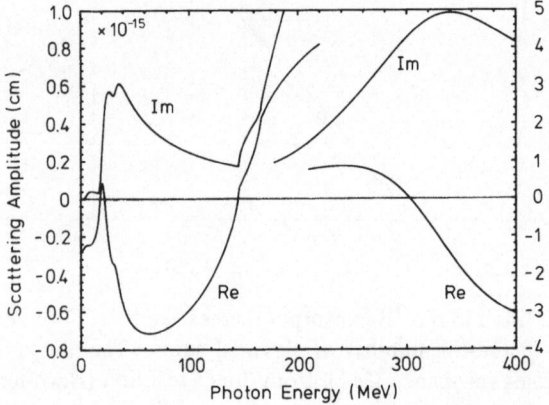

Fig. 5   The forward scattering amplitudes calculated from the absorption cross sections of Fig. 4, using the dispersion integral Eq. 6.

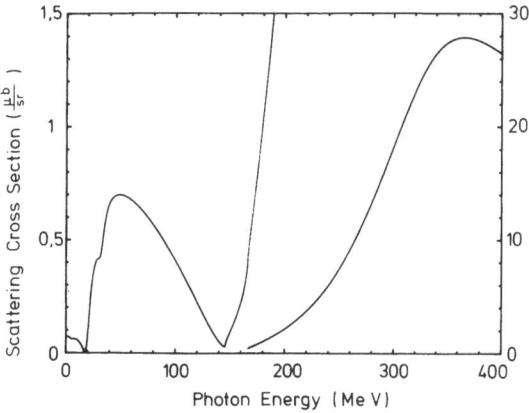

Fig. 6   The forward scattering cross sections for ^{9}Be, as determined from the absorption cross section of Fig. 4. By far the largest peak is the Δ-resonance.

The dominant feature in the whole energy range really is the Δ−resonance , where the
forward scattering cross section, (see Fig. 6), exceeds the nuclear scattering peak by a
factor of forty. Every time a new level of absorption opens, (atoms → nuclei around 10
MeV; nuclei → nucleons at 140 MeV), the scattering cross section passes through a pronounced minimum.

3. MULTIPOLES, LINEAR POLARIZATION, AND ANGULAR DISTRIBUTIONS

In addition to the total photon absorption cross section, which shows in its energy
dependence the main phenomena of the photon-nucleus interaction quite clearly, the
measurement of scattering cross sections provides information on multipole strengths and
particle sizes. This information in turn may help in clarifying the discrepancies that persist in a theoretical description of total cross sections for the light elements. For example, a recent calculation[14] of the total absorption cross section for ^{16}O still differs from
experiment by a factor of two in the giant dipole region. Fig. 7 shows this comparison.
The experimental values[10] have been corrected by -3.5 mb as prescribed by Fuller[15] , in
order to be consistent with all other experiments that rely on different techniques.

Analogous to Eq. 2, the nuclear absorption cross section can be written as sum of different multipoles

$$\sigma_{\gamma,tot}(E) = \sum_{\lambda,L} \sigma^{\lambda L}(E), \tag{8}$$

with λ=E , or M; L=1, or 2. Using sum rules and theoretical predictions of the resonance
energies[16], one can show that the strength of the dipole absorption is roughly one order of
magnitude larger than that of electric quadrupole or magnetic dipole absorption, and that
higher multipolarities give only negligible contributions. As a consequence, only E1, M1
and E2 absorption will produce measurable effects in elastic photon scattering. However, it should be mentioned that there are cases where many higher multipoles contribute if the center of mass of absorptive substructures (for example "quasideuterons")
does not coincide with the c.m. of the whole nucleus. This situation will be discussed in
section 6.3.

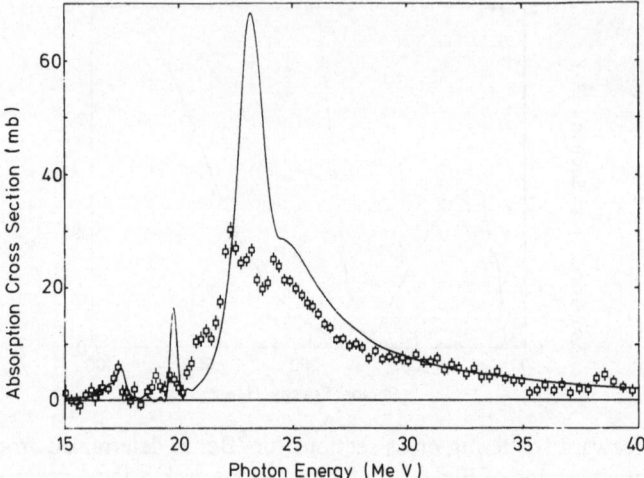

Fig. 7   The total absorption cross sections for ^{16}O. Points are from Ref. 10, with corrections ($\approx$3mb) as deduced in Ref. 15. The curve is from Ref. 14.

In general, the absorption cross section consists of E1 strengths at all energies above the particle emission threshold, so that resonances of higher multipolarity always overlap with the electric dipole, giving rise to interferences:

$$\frac{d\sigma}{d\Omega} = |R^{E1} + R^{E2} + R^{M1}|^2 = |R^{E1}|^2 + |R^{E2}|^2 + |R^{M1}|^2$$

$$+ 2\text{Re}\{R^{E1}R^{E2*}\} + 2\text{Re}\{R^{E1}R^{M1*}\} + 2\text{Re}\{R^{E2}R^{M1*}\} . \qquad (9)$$

Compared to the E1-E2 or E1-M1 interference, the pure M1 and pure E2 and the E2-M1 interference in most cases can be neglected.

Photons of the same momentum k can differ by their polarization. For each value of k, there are two linearly independent polarizations, and any other polarization is a linear combination of these basis polarizations. In general, one has partially polarized states which are statistical mixtures of pure states. They can be represented by a density operator of the 2-dimensional space of the polarization states.

As basis polarization, one can take the two circular polarizations or the two orthogonal rectilinear polarizations. The latter will be used here because atomic scattering cross sections (Rayleigh- and Delbrück-scattering) usually are given in that basis, and also the small number of experiments done with polarized photons[17,18] have been performed using linear polarization. In photon scattering, the polarizations are decomposed in components parallel ($^{\parallel}$) and perpendicular ($\leftrightarrow$) to the scattering plane which is defined by the directions of the incident and of the scattered photons, $\vec{k}$ and $\vec{k}'$, respectively. The polarization P, effective in the scattering of a photon beam with intensity I, is then given by

$$P = \frac{I^{\parallel} - I^{\leftrightarrow}}{I^{\parallel} + I^{\leftrightarrow}} . \qquad (10)$$

The "intrinsic" polarization of the photon beam, p, is given by the density matrix. In the following, it is assumed that the beam is prepared in such a way, that this matrix is diagonal, thus one has P=p, P varies between -1 and +1:

$P = +1$, $I^{\parallel} = I$ completely parallel polarized,

$P = 0$, $I^{\parallel} = I^{\leftrightarrow} = I/2$ unpolarized, and

$P = -1$, $I^{\leftrightarrow} = I$ completely perpendicular polarized .

The scattering cross section for incident photons with polarization P can, therefore, be expressed as the sum of the parallel and perpendicular cross sections:

$$\frac{d\sigma}{d\Omega} = \frac{1}{2}\{(1+P)\frac{d\sigma^{\parallel}}{d\Omega} + (1-P)\frac{d\sigma^{\leftrightarrow}}{d\Omega}\} . \tag{11}$$

For unpolarized photon, P=0, we have

$$\frac{d\sigma}{d\Omega} = \frac{1}{2}\{\frac{d\sigma^{\parallel}}{d\Omega} + \frac{d\sigma^{\leftrightarrow}}{d\Omega}\} . \tag{12}$$

Considering different multipolarities, one has

$$\frac{d\sigma^{\parallel}}{d\Omega}(E,\Theta) = |\sum_{\lambda,L} R_{\lambda L}(E,\Theta=0)g_{\lambda L}^{\parallel}(\Theta)|^2 , \tag{13}$$

and the same for $\leftrightarrow$. The angular distribution factors, $g_{\lambda L}$, normalized to 1 at 0°, are listed in Table I for the most important cases. Table II shows the angular distributions of the resulting interference terms.

A special situation occurs at $\Theta=120°$ for parallel polarization. Leicht[7] found that this scattering cross section is independent of the multipole composition of the absorption if only E1 and E2 transitions contribute - a fact otherwise only found at zero degrees:

$$\frac{d\sigma^{\parallel}}{d\Omega}(E,120°) = \frac{1}{4}\frac{d\sigma}{d\Omega}(E,0) . \tag{14}$$

Thus, it is possible - at least for light nuclei, where form factor effects are negligible and all M1 strength is concentrated in bound levels - to measure the forward scattering cross section by measuring the scattering cross section at 120° , for photons linearly polarized parallel to the scattering plane. A further discussion of the use of linear polarization can be found in Refs. 7, 19, 20.

4. CIRCULAR POLARIZATION

With recent advances in the technique of producing strong beams of longitudinally polarized electrons[21] , it seems to be easier to produce rather strongly circular polarized photons in the upper half of a bremsspectrum than linearly polarized ones. Therefore, it may be appropriate to make some qualitative remarks on the scattering of circularly polarized photons. For circularly polarized photons, the cross section is different for the two polarization directions if, and only if, the target possesses a definite "handedness". Such an optical activity can be imposed on the target by external magnetic fields or it can be, in principle, also a property of the target material itself. In nuclear physics, the latter case usually is called a "parity experiment" and should not be discussed here.

The effect of spin polarized targets with s=1/2 is most easily seen for electrons, polarized along the photon beam axis. The Compton cross section[22,23] depends strongly on the direction of polarization. It is much smaller and even almost vanishes for a certain scattering angle for the spins parallel to the beam polarization. Compton scattering from polarized electrons is well known as an efficient analyzer for circular polarization. The

Table I   Angular distribution factors

λ, L, and P	$g_{\lambda L}{}^P(\Theta)$
E1$^\parallel$ or M1$^\leftrightarrow$	$\cos\Theta$
E1$^\leftrightarrow$ or M1$^\parallel$	1
E2$^\parallel$ or M2$^\leftrightarrow$	$2\cos^2\Theta - 1$
E2$^\leftrightarrow$ or M2$^\parallel$	$\cos\Theta$
E3$^\parallel$ or M3$^\leftrightarrow$	$(15\cos^3\Theta - 11\cos\Theta)/4$
E3$^\leftrightarrow$ or M3$^\parallel$	$(5\cos^2\Theta - 1)/4$

quantity being measured is the asymmetry

$$A = \frac{d\sigma^r/d\Omega(E,\Theta) - d\sigma^l/d\Omega(E,\Theta)}{d\sigma^r/d\Omega(E,\Theta) + d\sigma^l/d\Omega(E,\Theta)} , \tag{15}$$

where l and r denote the two directions of the beam polarization or, for a fixed beam polarization, the two directions of target polarization. A is different from zero only if there is a target as well as a beam-polarization with parallel components.

For the case of protons, the non relativistic low energy expansion of the photon-nucleon interaction can be written[24]

$$Re\{R(E\rightarrow 0,\Theta)\} = \frac{-e^2}{Mc^2}\left[(1 - \frac{1}{3}k^2\langle r^2\rangle) - \hat{\alpha}E^2\right]\vec{\epsilon}_f\vec{\epsilon}_i$$

$$+ E^2(\chi_P + \chi_D)\vec{\epsilon}_f\times\vec{k}_f \cdot \vec{\epsilon}_i\times\vec{k}_i . \tag{16}$$

$\langle r^2\rangle$ is the mean square radius and $\hat{\alpha}=\alpha/r_0$, χ_P and χ_D are the second order structure constants: the static electric polarizability, the paramagnetic susceptibility, and the diamagnetic susceptibility. The region of validity of Eq. 16 is expected to extend up to 100 MeV (see Fig. 7 in section 5.3).

The asymmetry A, defined by Eq. 15 is only sensitive to the spin magnetic term, which changes sign by reversing $\vec{\epsilon}_i$, or the direction of target "magnetization". A measurement of A between 50 and 100 MeV photon energy is selectively sensitive to magnetic structure constants.

For nuclei, the low energy expansion was given in full completeness in Ref. 25. There, one can also find vector polarizabilities that lead to a dependence of the scattering amplitude on the degree of circular polarization. The effect is similar to the atomic case, where, for example, the absorption of unpolarized photons by polarized iron (or circular polarized photons by unpolarized spins) depends on the degree of circular polarization. Measuring the absorption of magnetized versus unmagnetized iron[26] allows to

Table II   Interference Terms

P	g_{E1E2}	g_{E1M1}
0	$\cos^3\Theta$	$\cos\Theta$
+1	$\cos\Theta(2\cos^2\Theta-1)$	$\cos\Theta$
-1	$\cos\Theta$	$\cos\Theta$

distinguish spin magnetization from orbital magnetization. Since the change of the absorption is the same for both directions of circular polarization, compared to the unmagnetized case, the effect can be used for separating spin magnetic effects. It is to be distinguished from optical activity which requires a different sensitivity for left- and right-handed photons[27].

5. SCATTERING AMPLITUDES

In order to conserve the photon number, the lowest order amplitudes are of the second order. One in general has two types, resonant and non-resonant ones. They are equally important throughout the whole energy range considered here, but there are not more than these two types. Postponing exchange current effects to section 6, the following sections deal with properties of non-resonant and resonant amplitudes.

5.1 The Thomson Limit

For very low energy E and a scatterer with radius R, for which kR>>1 holds, the scattering amplitude is real, and given by the Thomson amplitude

$$R(E \to 0,\Theta) = -r_0 g_{E1}(\Theta) \tag{17}$$

with $r_0 = e^2/Mc^2$, e the particle's charge, and M its mass. Eq. 17 is valid exactly to all orders in e^2 in quantum electrodynamics[28]. It can also be derived classically[29]. In this classical point-particle approach, the current $\vec{j}$, with which the external field $\vec{A}$ interacts, is set up by the same field $\vec{A}$. Therefore, this term is proportional to A^2, and

    a) the amplitude is proportional to e^2, independent of the sign of the electric charge, and,

    b) it is independent of the momentum the point particle had prior to the interaction.

5.2 The A^2- Term for the Particles of Finite Size: Seagull Term

To the extent that one can neglect the size of a nucleon, the non-resonant amplitude R_{nr} for a nucleus is the coherent superposition of the constituents' amplitudes

$$R_{nr}(E,\Theta) = -r_0 <0| \sum_{j=1}^{Z} e^{i(\vec{k}-\vec{k}')\cdot\vec{r}_j} |0> g_{E1}(\Theta) . \tag{18}$$

For $\Theta=0$, R_{nr} is given by $-Zr_0$, independent of the photon energy. For $E \to 0$, the

Thomson-limit (Eq. 17) for a nucleus with Z charges and A masses, $-Zr_0/A$, is not fulfilled, since in Eq. 18 only the mass M of a single nucleon enters, and, furthermore, there is no mention of the neutrons at all, because the summation runs over charged particles only. Eq. 18 describes the scattering from Z free protons far above all nuclear resonances. There is, however, no energy region where this term clearly shows up because above the nuclear excitations, there is absorption by correlated pairs and meson production.

If one introduces a form factor[24,30] , for the proton charge distribution $\rho(r)$, defined by:

$$F_Z(q) = \frac{1}{q} \int \rho(r)\sin(qr)\, r\, dr \tag{19}$$

Eq. 18 can be written

$$R_{nr}(E,\Theta) = -Zr_0 F_Z(q) g_{E1}(\Theta) . \tag{20}$$

q is the momentum-transfer to the nucleus

$$q = |\vec{k}-\vec{k'}| = \frac{2E}{\hbar c} \sin(\Theta/2) . \tag{21}$$

R_{nr} is energy independent, and, as well as the expression for the point particle, is not dependent on the scattering particles' momentum. $F_Z(q)$ describes the spatial distribution of the protons in the nucleus. There is no absorptive part of the same order of magnitude as this term - it is entirely real - and $F_Z(q)$ should be considered as representing the higher multipoles. This fact also implies certain conditions on higher multipole absorption, as described later in section 5.2.2.

The concept of a form factor not dependent on E relies on the point-picture of the nuclear constituents. As it has been restated recently[31] , in a relativistic theory the non-resonant term (Eq. 18) actually is a sum over pair terms, that are off shell by $2Mc^2$ in the intermediate state. The A^2–term is an exact expression for the limiting case q<<2Mc. That certainly is a good approximation up to q≈200 MeV/c .

R_{nr} (as well as other photon scattering amplitudes) is proportional to the square of the scattering charge. It is impossible to tell the sign of a charge by photon scattering and - in contrast to electron scattering - charge distributions of opposite charges do not cancel. This fact is important for the effect that exchange currents have on scattering cross sections.

5.3 Resonance Terms

A composite system like a nucleus always must show dipole resonant absorption and scattering, since the non-resonant amplitude alone differs for E→0 from the Thomson limit by a factor A/Z. The resonant dipole amplitude has to provide the correct transition from rigidly bound particles at the Thomson-limit to virtually free protons, "far above resonances". Higher multipoles also have to fulfill requirements[32] as E→0. Apart from vanishing at E=0, the lowest order with which they can approach zero is E^{2L} . This condition is of the same general validity as the Thomson limit itself, and also can be deduced from classical electrodynamics. The whole complex of conditions on the real part of the scattering amplitude at E=0 and for E→0 is called Low Energy Theorem. It can be shown that complying with these low energy requirements on the scattering amplitude is equivalent to fulfilling the conditions, imposed on the absorption cross section by multipole sum rules.

5.3.1 <u>Dipole</u> It was shown by Überall[30], how the correct low energy limit is obtained. The resonance terms R_{res} are given by

$$R_{res}(E,\Theta) = \frac{1}{M^2} \{\sum_n <0| \sum_j e_j \vec{p}_j \cdot \vec{\varepsilon}' e^{i\vec{k}'\cdot\vec{r}_j} |n> \cdot \frac{1}{E_{ni}-E+i\varepsilon}$$

$$\cdot <n| \sum_j e_j \vec{p}_j \cdot \vec{\varepsilon} e^{i\vec{k}\cdot\vec{r}_j} |0> + c.c.\} g_{E1}(\Theta) . \tag{22}$$

The low energy limit is found to be (exchange currents are postponed to section 6)

$$R_{res}(E=0,\Theta) = r_0 \frac{ZN}{A} g_{E1}(\Theta) \tag{23}$$

where the summation over all excited states was done by invoking closure $\sum_n |n><n|=1$.

The sum of the non-resonant and the resonant amplitudes is

$$R_{nr}(E=0,\Theta) + R_{res}(E=0,\Theta) = -Zr_0 g_{E1}(\Theta) + \frac{NZ}{A} r_0 g_{E1}(\Theta) = -\frac{Z^2}{A} r_0 g_{E1}(\Theta) \tag{24}$$

where $F_Z(q=0)=1$ has been used in the non-resonant part. The Thomson limit now agrees with the sum of the two terms. The Thomson limit for a nucleus apparently is a more complicated construct than the high energy limit (Eq. 20), that is not altered by the resonant term, since resonance terms vanish for large energies:

$$R_{res}(E \rightarrow \infty,\Theta) = 0 . \tag{25}$$

In principle, the scattering amplitude can be written differently, either as a sum of the high energy limit and a resonance part, that vanishes at high energies, as given above, or as a sum of the Thomson amplitude and a resonance amplitude, the latter being zero[3] at E=0. The two descriptions are identical, if one neglects the form factor $F_Z(q)$. Since it is required to apply the form factor only to the high energy limit, Eq. 20, the correct dipole amplitude is obtained by summing Eqs. 20 and 23, where neither term is the Thomson amplitude.

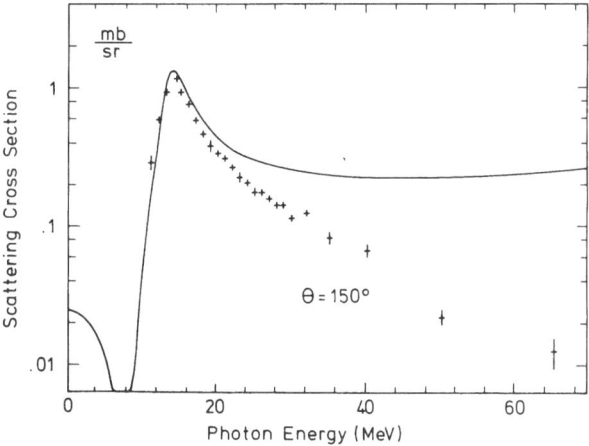

Fig. 8   The curve is calculated assuming a pure dipole angular distribution without form factor ($F_Z\equiv1$). This assumption leads to a big discrepancy with experimental points.[33]

It can be easily seen, using the unsubtracted dispersion relation Eq. 4a for E→0 that the value of the resonant scattering amplitude at E=0, as given in Eq. 23, is equal to the absorption cross section integrated over the resonance. This amplitude is therefore finite:

$$R_{res}(E=0,\Theta=0) = \frac{1}{2\pi^2\hbar c} \int \sigma_{\gamma,tot}^{E1}(E)dE \equiv \frac{\sigma_0^{E1}}{2\pi^2\hbar c} . \tag{26}$$

The integrated E1 cross section σ_0^{E1} therefore has to be equal to

$$\sigma_0^{E1} = \frac{2\pi^2 e^2 \hbar}{Mc} \frac{NZ}{A} . \tag{27}$$

This expression is the Bethe-Levinger sum rule, for a nucleus without exchange current effects.

To calculate the real part of the scattering amplitude from a total E1 absorption cross section that includes one dipole sum, Eq. 27, it is convenient to use the dispersion relation. D(E=0) then is the sum, according to Eq. 24, of the correct low energy limit of the resonance amplitude, $(NZ/A)r_0 g^{E1}(\Theta)$, and the A^2-term carrying a form factor, $-ZF_Z(q) g_{E1}(\Theta)$:

$$Re\{R^{E1}(E,\Theta)\} = \left\{ (-ZF_Z(q) + \frac{NZ}{A})r_0 \right.$$
$$\left. + \frac{E^2}{2\pi^2\hbar c} P \int \frac{\sigma^{E1}(E')}{E'^2-E^2}dE' \right\} g_{E1}(\Theta) . \tag{28}$$

The high energy limit of this expression is the non-resonant term, that alone survives at energies far above resonances. Experimentally, the need for a form factor $F_Z(q)\neq 1$ was clearly demonstrated in the 30-60 MeV region for lead, see Fig. 8, Eq. 28, together with the sum rule, Eq. 27, is the final and complete description of E1 scattering without exchange currents.

As shown in Eq. 16, the factor multiplying the term quadratic in E of the resonance scattering amplitude for E→0, is the static electric polarizability of the scatterer. The

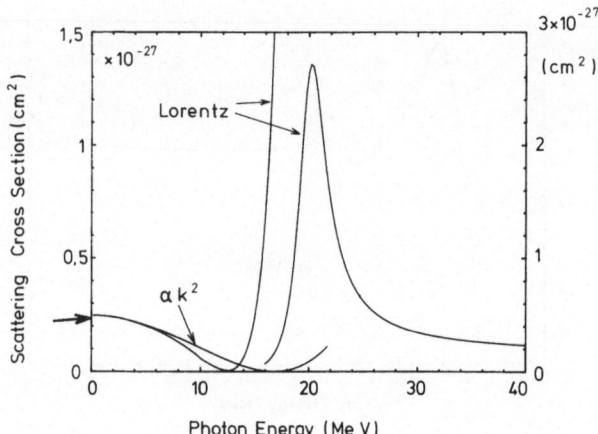

Fig. 9   The scattering cross section of the Lorentzian of Fig. 1 is shown together with the low energy expansion, $\frac{d\sigma}{d\Omega} (E,\Theta=0) = (r_0-\alpha E^2)^2$.

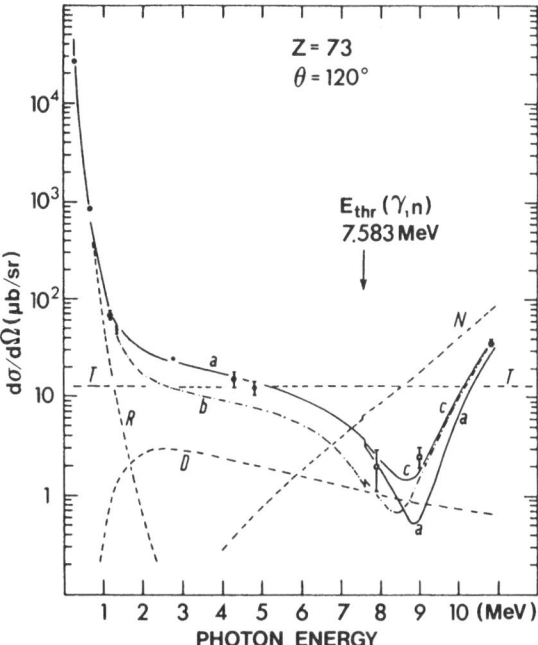

Fig. 10   Elastic scattering cross section for Z=73. R. Rayleigh, T. Thomson, D. Delbrück, N nuclear resonance scattering. a) is the sum of all partial amplitudes. Data from Refs. 34, 35, and 36.

retardation term $(1/3)k^2<r^2>$, also multiplying E^2 , can safely be neglected for nuclei in the several MeV region. Using the dispersion integral Eq. 28, one obtains:

$$\alpha = \frac{\hbar c}{2\pi^2} \int \frac{\sigma^{E1}(E)}{E^2} \, dE \equiv \frac{\hbar c}{2\pi^2} \, \sigma^{E1}_{-2} . \qquad (29)$$

As an example for the range of validity of the first term of the expansion of Eq. 16, the scattering cross section of the Lorentzian of Fig. 1 is plotted in Fig. 9, together with the low energy approximation

$$\frac{d\sigma}{d\Omega}(E \to 0, \Theta = 0) = (-r_0 + \alpha k^2)^2 . \qquad (30)$$

Eq. 30 is a good approximation up to about 1/3 of the resonance energy. Fig. 10 shows the situation for a true nuclear scattering cross section in this energy range[34] .

5.3.2 Higher Multipoles, Mainly E2 The most important point about higher multipoles may be that, once the Thomson limit has been reached correctly, not only the E1 absorption integral σ^{E1}_0 has been given a certain value by Eq. 27, but also all other multipoles are constrained by the requirement of vanishing powers of E, smaller than E^{2L} [32,37].

Following Hayward[38], the relevant relations can be derived as follows: The introduction of the form factor $F_Z(q)$ to the non-resonant term modifies the angular distribution of $d\sigma(E,\Theta)/d\Omega$ for all photon energies, except E=0. The value of the Thomson cross section is preserved, independent of the shape of the nucleus. However, the low energy theorem and classical electrodynamics state that also those E2 amplitudes proportional to

E^2 have to be zero for $E \to 0$. On the other hand, the general form of the dispersion integral shows that there really is always a E^2-proportional term, if there is a non-vanishing E2-resonant absorption. Therefore, in order to satisfy the Low Energy Theorem, there has to be a cancellation of this unphysical term by another, also proportional to E^2. This other term automatically appears in a power series expansion of the E2-component of the non-resonant term as soon as one applies a form factor to it. In this way, the size of the nucleus (form factor) and the excitation of E2 strength are strongly correlated.

The E2 amplitude, introduced by the form factor that describes a charge distribution with RMS radius $<r^2>$, is given model-independently by

$$-Zr_0[\frac{E}{\hbar c}]^2 \frac{<r^2>}{6} g_{E2}(\Theta) . \tag{31}$$

This amplitude must be canceled exactly by the low energy limit of the real part of the E2-resonance amplitude

$$Re\{R^{E2}(E \to 0),\Theta)\} = \lim_{E \to 0} \left\{ \frac{E^2}{2\pi^2\hbar c} P \int \frac{\sigma^{E2}(E')}{E'^2-E^2} dE' \right\} \tag{32}$$

or, by comparing Eqs. 31 and 32, the requirement

$$\sigma_{-2}^{E2} \equiv \int \frac{\sigma^{E2}(E)}{E^2} dE = Zr_0\pi^2 \frac{<r^2>}{3\hbar c} \tag{33}$$

has to be fulfilled. This expression, the E2 sum rule, can also be derived without recourse to the low energy theorem[39]. It is also possible to introduce the low energy limit of the E2-dispersion integral explicitly by subtracting a second time the $E \to 0$ limit of the integral (the first subtraction occurred from Eq. 4a to Eq. 5)

$$Re\{R^{E2}(E,\Theta)\} = \{E^2 \frac{\sigma_{-2}^{E2}}{2\pi^2\hbar c} + \frac{E^4}{2\pi^2\hbar c} P \int \frac{\sigma^{E2}(E')}{E'^2(E'^2-E^2)} dE'\}g_{E2}(\Theta) . \tag{34}$$

Of course, $\sigma^{E2}(E)$ has to be chosen such that σ_{-2}^{E2} fulfils the sum rule Eq. 33. In modifying $\sigma^{E2}(E)$ in an attempt to fit a measurement, one always has to comply with this boundary condition.

Discussing Eq. 34, it is found that the second term, containing the twice subtracted dispersion integral, as well as a Lorentzian multiplied by[3] E^2, diverges for $E \to \infty$, proportional to E^2. Neither expression should be used unless the E^2-proportional E2-term of the form factor expansion is also dropped.

For M1 nuclear absorption, similar expressions are valid. They have been worked out in detail in Ref. 38, and should be used with similar precautions concerning the E proportional M1 term in the form factor expansion. In general, for each multipole, there is a dispersion integral

$$Re\{R^{\lambda L}(E,0)\} = R^{\lambda L}(0,0) + \frac{E^2}{2\pi^2\hbar c} P \int \frac{\sigma^{\lambda L}(E')}{E'^2-E^2} dE' . \tag{35}$$

In order to satisfy the low energy theorem, $R^{\lambda L}$ has to be zero except for E1, and the function $\sigma^{\lambda L}(E)$ has to be chosen in accordance with the λL sum rule.

For the limit $E \to \infty$ (or better $E \gg E_{res}$), $Re\{R^{\lambda L}(E,\Theta)\}$ is bound to assume a constant value, that is proportional to the absorption cross section integrated over energy. Thus, these amplitudes, extending to infinite energy, may be interpreted as retardation terms to the E1 seagull term, Eq. 20, describing coherent scattering from Z free protons.

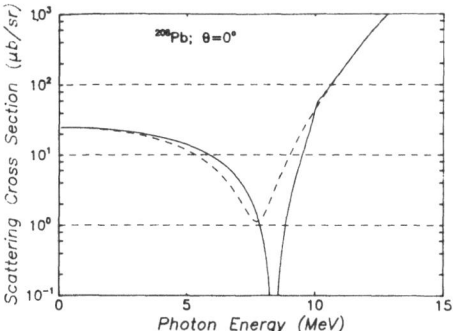

Fig. 11   Scattering cross section for a Lorentzian with the particle threshold at 10 MeV
(full line). Dashed is the same Lorentzian assuming the threshold at zero
energy. Despite using the same resonance parameters, the polarizabilities, α,
read from the slope with which the curves approach zero, are different.

5.3.3 <u>Particle Thresholds</u>   If resonances occur close to thresholds for particle emission
(or meson production), then the resonance widths Γ are formally taken to be energy
dependent[12], $\Gamma(E) = T(E)\cdot\gamma$. The width γ is the true width that is approached asymptotically for large E. The transmission factor T(E) always is <1 and is given by the probability for the particle to tunnel through the centrifugal and Coulomb barrier. T(E) is fully
determined by the angular momentum of the outgoing particle and the size of the
nucleus.

The scattering cross section in the neighbourhood of a threshold can easily be calculated by applying the dispersion integral, Eq. 28, to the absorption cross section that has
been modified by a threshold. The result is shown in Fig. 11, which is a realistic case for
^{208}Pb . The scattering cross section indeed changes very little by removing the absorption strength below threshold, because this area resembles only a small part of the
integrand of the dispersion integral. Scattering by bound levels increases the energy-
averaged cross-section below threshold for a nucleus with a high level density like ^{238}U

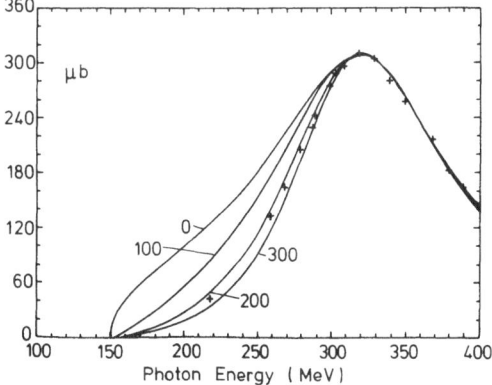

Fig. 12   The π^0 photo-production cross section[41], and a relativistic resonance fit[13] with
various channel radii R_c. The numbers give $1/R_c$ in MeV/c.

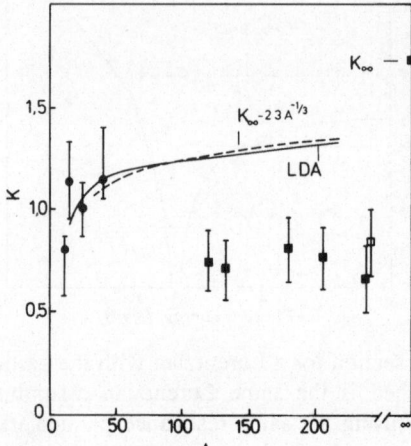

Fig. 13   The experimental[10,43], and theoretical[42] values of the enhancement factor κ,
         due to exchange currents.

by only a factor of two[40] , because the ground state radiation width is small and the
scattering is predominantly inelastic. On the other hand, it could be shown[2] for a low
level-density nucleus like ^{208}Pb that the energy-averaged scattering from bound levels is
in the order of the peak cross section in the giant dipole resonance.

As another example of how absorption cross sections can be modified drastically, the
π^0-production cross section on the proton[41] is shown in Fig. 12, together with relativistic
resonance curves, calculated for different values of the channel radius for this reaction[13].
It is surprising how well the simple assumption - a smooth logarithmic derivative at the
nucleon's "boundary" - also works in this case. The channel radius, read off Fig. 12, is
0.8 fm. Interpreted in the classical way, this distance should be taken as the sum of the
strong interaction radii of the proton and the π^0 .

6. EXCHANGE CURRENTS

For nuclei, there is a very important contribution to the total photon interaction, due
to the presence of charged mesons that provide the nuclear binding force, and that appear
in nuclei, in addition to those mesons which form the free nucleons' meson clouds. The
interaction of photons with these electrically charged exchange currents multiplies the
absorption integral Eq. 27 roughly by two. Fig. 13, taken from Ref. 42 summarizes the
present status of experiment and theory. The usual notation is[44]

$$\int \sigma^{E1}(E)dE \equiv \sigma_0^{E1} = \sigma_{TRK}(1+\kappa) \tag{36}$$

with σ_{TRK} given by Eq. 27. As was pointed out in Ref. 45, the role of the mesons in
nuclei is dual: on the one hand, they give rise to two-body currents, to be added to the
conventional nucleonic one; on the other hand, they correlate pairs of nucleons which
will consequently act as a unique entity in the nuclear response to the external field.

The first mechanism enhances resonance excitations; the giant dipole resonance strength is enhanced by 20-40%[46] . Second, a new kind of reaction takes place, where proton-neutron pairs form substructures that absorb the photon leaving the remaining nucleus almost unchanged. This effect, called the quasideuteron[47], leading to an emission of an n-p pair with roughly the kinematics of the deuteron photon disintegration, adds another 0.5 to 0.8 classical sums, σ_{TRK}, to the absorption integral. Since it is easy to separate experimentally the two contributions to κ , they will be given different symbols, $\kappa = (\kappa_{res} + \kappa_{QD})$.

These and the following considerations are certainly rather simplified in order to arrive at a point where one has some expressions to use for a comparison with experiment. The situation is still changing quite rapidly, as the appearance of new papers on fundamentals[48] or special aspects[49] indicate.

6.1 The Nonresonant Exchange Term

Quite intuitively, one might describe the exchange currents using expressions very similar to those used to describe the normal currents. In order to satisfy the low energy theorem, the real part of the exchange resonance amplitudes at E=0, $\kappa_{res} r_0 NZ/A$ has to be canceled by a seagull amplitude[50] , of equal magnitude and opposite sign. This term shall be considered analogous to the term Eq. 20 that describes scattering from individual protons in the nucleus. It is assumed to be a term describing coherent scattering from charged mesons. Since mesons, or nucleon pairs, may be spatially distributed differently from protons, a second form factor, $F_{ex}(q)$, has been introduced in addition to $F_Z(q)$ for the protons:

$$R_{nr}(E=0,\Theta=0) = r_0\{(1+\kappa_{res}) \frac{NZ}{A} - ZF_Z(q) - \kappa_{res} \frac{NZ}{A} F_{ex}(q)\} . \qquad (37)$$

Despite the fact, that the numerical analysis of an exact theory[49,51] is still lacking, there is no doubt about the existence of exchange seagull terms. There is, however, no quantitative answer yet as to which part of the whole exchange contribution, $\kappa r_0 NZ/A$, is canceled by a term analogous to Eq. 20, with a point interaction on charged mesons. In Eq. 37, it is assumed that only κ_{res} is canceled by this type of seagull term.

For the exchange form factor $F_{ex}(q)$ one may use[25]

$$F_{ex}(q) = F_Z^2(q/2) , \qquad (38)$$

based on the assumption of non-interacting particles. A more refined derivation of the angular dependence of the exchange seagull-term was given in Ref. 49; however, it has not yet been evaluated numerically.

6.2 Exchange Enhancement of Resonances

It is well known that the Lorentzians, fitting experimental absorption photon cross sections in the giant dipole region, contain more than one classical sum[46]. The increase of the real part of the resonance amplitude at E=0, given by $\kappa_{res} r_0 NZ/A$, is compensated by an equivalent part of the exchange seagull term. The resonance term has the usual $g_{\lambda L}(\Theta)$ angular distribution, whereas the seagull term in addition is multiplied by $F_{ex}(q)$, as given in expression Eq. 37.

The same applies for higher resonances, especially E2, for which an enhancement κ_{E2} is expected. The low energy theorem requires, analogous to Eq. 33,

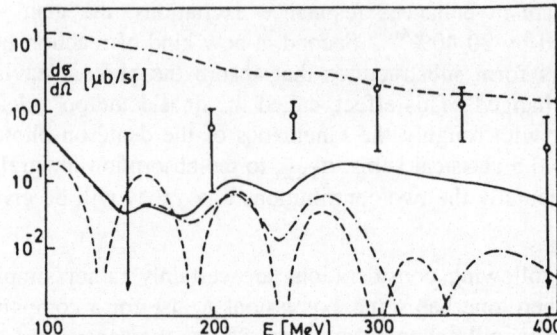

Fig. 14   The differential scattering cross sections for $\Theta=115^0$, for ^{208}Pb.[25] The experimental points are taken from Ref. 54. The full curve is calculated with $F_{ex}(q)$
as given by Eq. 38; for the dashed-dotted curve $F_{ex}\equiv1$ has been assumed.

$$\sigma_{-2}^{E2} \equiv \int \frac{\sigma_{ex}^{E2}(E)}{E^2} \, dE = \kappa_{E2}r_0 \, \frac{NZ}{A} \, \pi^2 \, \frac{\langle r_{ex}^2 \rangle}{3\hbar c} . \tag{39}$$

σ_{ex}^{E2} also contains the E2-cross sections from the quasideuteron absorption. $\langle r_{ex}^2 \rangle$ is the
mean squared radius of the exchange current distribution. In this way, the E2 exchange
cross sections are linked to the spatial distribution of the exchange currents.

6.3 Absorptive Substructures

The main absorption effect, introduced by exchange currents, is the quasideuteron
effect. It is intuitively clear that for the associated scattering process, the same form factor $F_{ex}(q)$ has to be applied to the scattering amplitude, that was given to the exchange
seagull term. Since the three parts of the scattering amplitude, real and imaginary parts
of the quasideuteron absorption and the seagull term, that cancels the real part for E=0 all
have the same angular distribution, there is no need to introduce explicitly the seagull
term. Instead one can simply use Eq. 35 with D(0)=0, and apply $F_{ex}(q)$. κ_{QD} does not

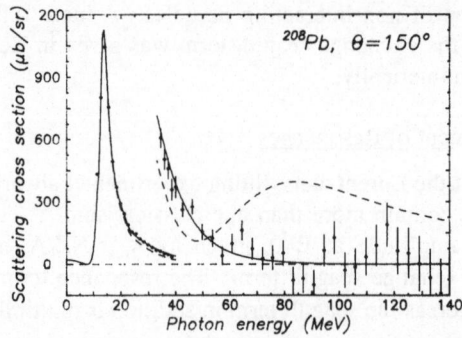

Fig. 15   The preliminary experimental results[53] for the scattering cross section from
^{208}Pb; full line: the scattering amplitude in the quasideuteron region is multiplied by $F_{ex}(q)$ given by Eq. 38; dashed line, the same with $F_{ex}(q)\equiv1$.

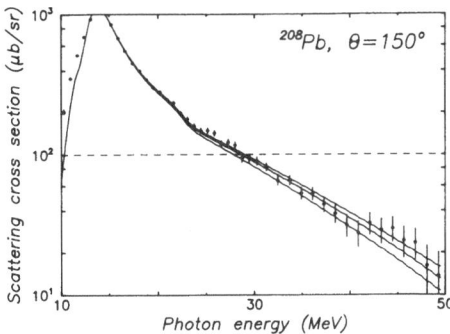

Fig. 16   The preliminary experimental results[53] are compared to three curves, calculated for three different Fermi 2-parameter charge distributions. Top line: half
density radius is $c=6.8$ fm, and skin thickness is $z=0.54$ fm; middle line: $c=6.6$
fm, and $z=0.75$ fm; bottom line: $c=6.6$ fm, and $z=1.0$ fm. The corresponding
r.m.s. radii are 5.64, 5.76 and 6.07 fm, respectively.

appear in $R(E{\rightarrow}0,\Theta)$, and Eq. 37 is the final expression for the nonresonant part of the
scattering amplitude including exchange effects. The fact that κ_{QD} does not appear in the
amplitude at zero energy, agrees with Christillin[49] , who showed how the real part of the
exchange amplitude from virtual meson effects (quasideuteron), and real meson production (quenching by the Pauli effect in a bound system) cancel at $E=0$.

The use of a form factor, $F_{ex}(q)$, instead of the exact expression, which is a function
of q, and also E, is even more questionable than the use of $F_Z(q)$. It should be used more
as a reminder that there is a spatial effect the magnitude of which can be estimated by a
reasonable value for $F_{ex}(q)$ that, however, still waits for a theoretical and/or numerical
analysis.

Experimentally, a large influence of $F_Z(q)$ on the scattering cross section was shown
for ^{208}Pb in Fig. 14 taken from Ref. 33. The need for an exchange form factor $F_{ex}(q)$ was
demonstrated by comparing theoretical curves[25] with the result for lead in the 300 MeV
region[25] (see Fig. 14). Since to both curves in this figure, an inelastic contribution has to
be added, the curve with $F_{ex}=1$ can be excluded experimentally.

Describing the angular distribution given by $F_{ex}(q)$ in terms of a multipole expansion,
introduces strong contributions from higher multipoles, increasing with energy and
nuclear radius; whereas, in the center of mass of the absorptive substructures, the reaction still takes place predominantly by E1 absorption. Typical pictures for these multipole strengths as a function of photon energy are known from the Δ−resonance, where
the nucleons constitute the absorptive substructures[52], (see for example Fig. 19, below).

In a recent measurement[53] the statistical accuracy could be increased to the point,
where also at lower energies the need for $F_{ex}(q)$, applied to the quasideuteron part of the
absorption strength, becomes evident (see Fig. 15). In this way, absorption by collective
effects (giant resonances) and by absorptive substructures ("quasideuteron") becomes
clearly distinguishable experimentally.

In Fig. 16, the results of the same experiment are compared to calculations assuming
somewhat different sizes of the proton distributions. The data allow the determination of

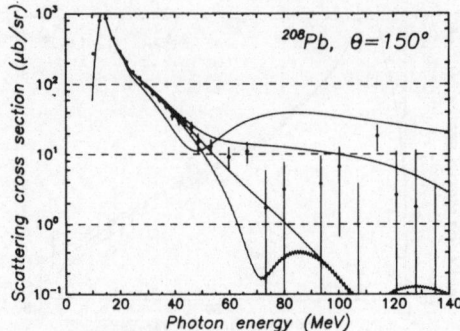

Fig. 17   The preliminary experimental results[53] and calculated curves for $F_z(q)$, given
by a charge distribution for which a half density radius c=6.6 fm and a skin
thickness of z=0.75 fm were assumed. For the exchange form factor, $F_{ex}(q)$,
four different values have been assumed. These are for the four lines from top
to bottom: a) $F_{ex}\equiv 1$; b) c=2 fm; c) c=4.5 fm; d) c=6.6 fm; z is always 0.75
fm; scattering angle is $\Theta=150°$.

$<r^2>$ with a statistical accuracy of only a few percent. The energy region shown in Fig.
17 is not sensitive to the parameters of $F_{ex}(q)$. In Fig. 17 the full energy region is shown,
together with curves for three different radii for $F_{ex}(q)$. Albeit the statistical accuracy of
the experimental result in the upper energy region is worse, improving the statistics by
factor of two will be sufficient to obtain a first number for the two particle $<r_{ex}^2>$. This
quantity is not accessible by charged particle scattering.

6.4 The Delta-Region

Absorption in the Δ–region also takes place by absorptive substructures, the nucleons
themselves. As a first approximation, the E1 and M1 parts of the cross sections shown in
Fig. 4 can be used to calculate forward scattering amplitudes that are to be multiplied by

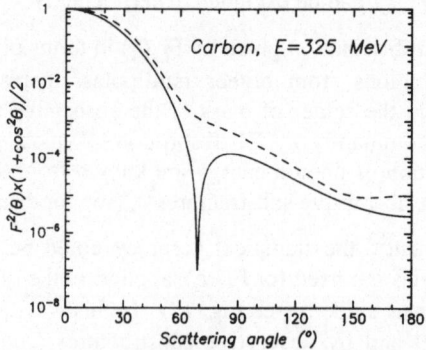

Fig. 18   The dotted line is the simple approach $F_z^2(q) \cdot d\sigma/d\Omega(E,0) \cdot (1+\cos^2\Theta)/2$, where
in $d\sigma/d\Omega$ the measured absorption cross section have been used[54]. The full line
is a particle hole calculation neglecting quasideuteron and seagull terms[55].
$F_z(q)$ is the charge distribution form factor for ^{12}C.

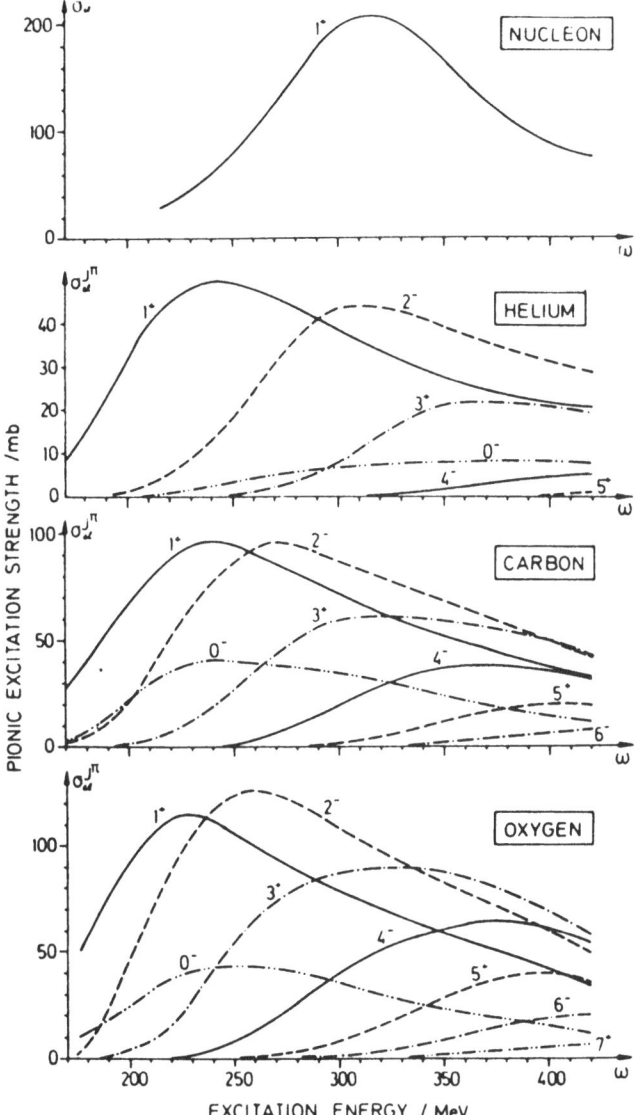

Fig. 19   The multipole strengths for nuclear excitations in the Δ-region[52]. The direct
reaction at these energies essentially proceeds through E1 pion production on
the nucleon, thereby providing for nuclei also the natural parity "states".

a form factor, $F_m(q) \approx F_Z(q)$, describing the mass distribution of the nucleus. This procedure may be allowed for calculating real parts of scattering amplitudes below meson
threshold. In the Δ−resonance region itself, the form factor picture is expected to fail
since the seagull term can no longer be regarded as a point-like interaction compared
with λ . However, a comparison[54] of a simple form factor approach with a full particle-
hole calculation[55] shows remarkable similarities over five orders of magnitude in the
cross section (see Fig. 18). Part of the difference may even stem from the neglect of
quasideuteron and seagull amplitudes in the particle-hole calculation. The curves shown

in Fig. 19 can be regarded as a multipole expansion of the form factor. The E1 direct reaction (Born term) pion production is comparable in magnitude (see Fig. 4) with the resonance term and - when occurring in an extended nucleus - gives similar multipole decompositions, also with natural parity.

Measurements of the ^{12}C and ^{208}Pb cross section give results which are much larger than the particle hole calculation as well as the simple ansatz of Fig. 18. It has been shown that the inclusion of an inelastic scattering component[56] could account for the discrepancy, if inelastic processes up to 30 MeV energy-transfer to the nucleus, were included. Fig. 20 shows the situation. Experimentally, inelastic processes with an average excitation energy of 5% of the photon energy are included in the results. ΔE_{exc} is 10 MeV at 200 and 15 MeV at 300 MeV. These lower values cannot quite account for the whole difference.

At the end of these paragraphs on scattering amplitudes, the summary is given in Table III. Again, it should be remembered, that the F's mean form factors literally only as long as exact calculations are not available that avoid the factorization in functions of E and q.

7. SCATTERING FROM NUCLEONS

In the future, the most important experiments will be on nucleons, either free or bound. One will have to apply the formalism developed for the nucleus to free nucleons and, furthermore, one then can ask whether or not the structure of the free protons has changed by binding them. Photon scattering may be a complementary experiment to electron, μ–meson , or pion scattering, with certain advantages as to the cleanness of the interaction and drawbacks for the smallness of the cross sections. For such a program,

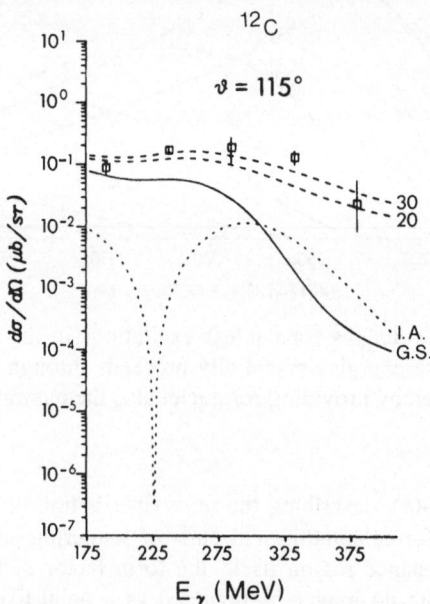

Fig. 20   The points at energies E(MeV) are measured elastic cross sections[54]. They contain inelastic processes, due to the finite energy resolution, of a maximum excitation energy of - on the average - 0.05 E. The curves include inelastic scattering[56], up to an excitation energy of 20 or 30 MeV.

Table III   Form factors, applied to different parts of the scattering amplitude. F_z, F_{ex} and F_m describe the proton, exchange current and mass distributions, respectively.

		Re(R)	IM(R)
Nucleon currents	Giant resonances	1	1
	Seagull term	$F_z(q)$	x
Exchange currents	Giant resonances	1	1
	Quasideuteron	$F_{ex}(q)$	$F_{ex}(q)$
	Seagull term	$F_{ex}(q)$	x
Δ-resonance and Born π-production		$F_m(q)$	$F_m(q)$

the knowledge of the nuclear effects, i.e. nuclear resonances, exchange currents, etc. is of prime importance, since for the real part of the scattering amplitude, all energies contribute.

For unpolarized light, there are two physical quantities appearing at low energies, the second order (i.e. multiplying terms proportion to E^2) structure constants α and β, describing the response of the nucleon to an electric and magnetic field. Up to now, no explicit mention of quarks seemed to be necessary. Of course, an understanding of α and β requires an explicit quark treatment[31,57]. An excellent review of the present status of experiment and theory is given in Refs. 57 and 58. In the following, only these two constants and their measurement for free and bound protons shall be considered.

7.1 Photon Absorption and Forward Scattering Cross Section of the Free Proton

Fig. 21 shows the present knowledge of the photon absorption by protons. The three main resonances can be identified by the channel in which pion scattering shows a

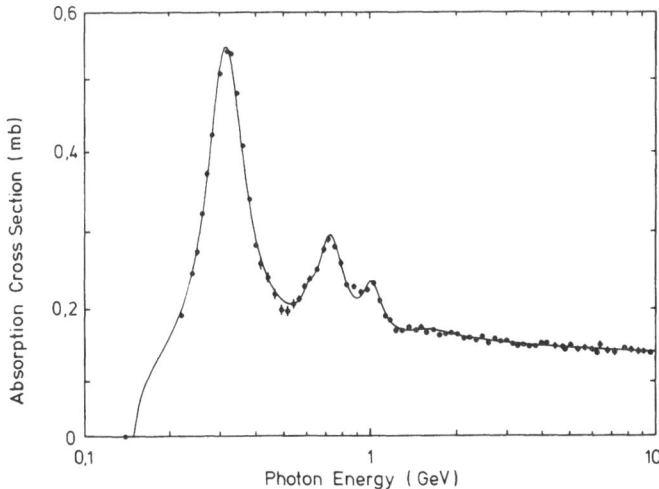

Fig. 21   The total absorption cross section $\sigma_{\gamma,tot}$ for the proton[59]. In the threshold region, π-production cross sections[41] have been added. The curve[60] is a fit to all available experimental numbers. In each point, many experimental data have been combined.

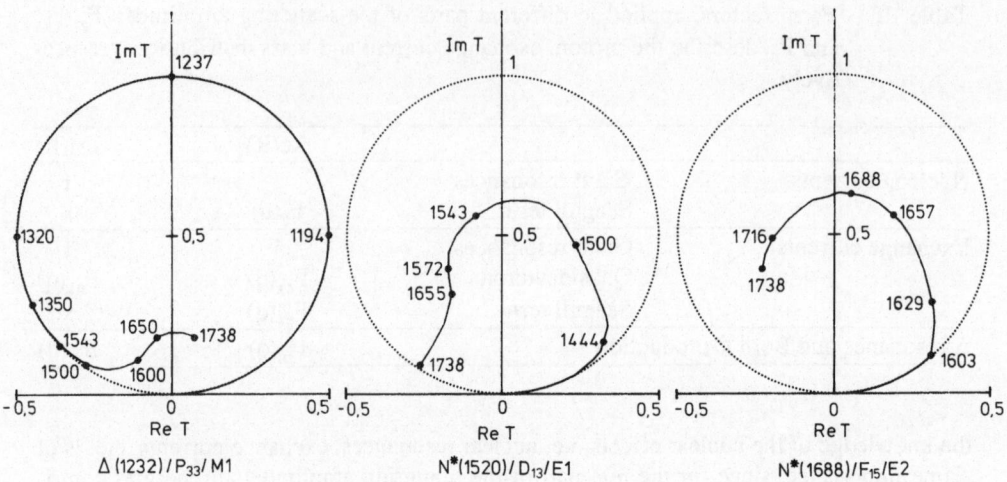

Fig. 22   The Argand diagrams of π-scattering amplitudes for the three main nucleon
           resonances; taken from Ref. 61.

resonance behaviour. Argand diagrams for these three channels are shown in Fig. 22. In
the nuclear physics terminology they are M1, E1, and E2 in character.

The forward scattering cross section can be easily calculated, using Eq. 35. The
result looks qualitatively very similar to Fig. 9 and was given quantitatively in Ref. 62. It
is shown as Fig. 23. Characteristically, there is at low energies a departure from r_0^2 proportional to $(\alpha+\beta)E^2$ with the electric polarizability α and the magnetic susceptibility β.
From Fig. 9, identifying the resonance with the Δ–resonance, one may judge the validity
of only the term quadratic in E up to 60 MeV. There are attempts[57] to calculate also
terms to higher order in E. It seems, however, better justified to use the resonance formulas instead[13] that take care of the whole energy region.

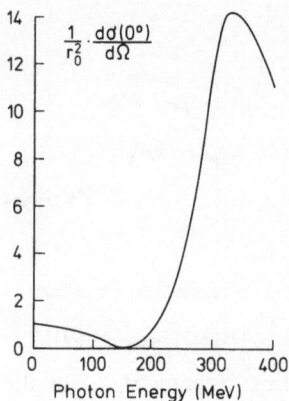

Fig. 23   The proton forward-scattering cross section from theory[62].

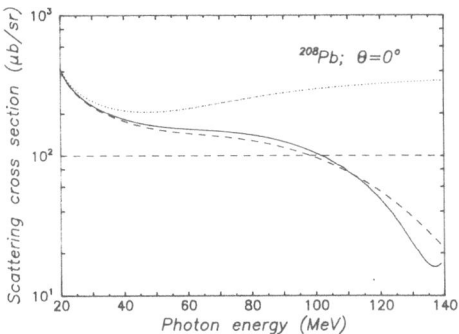

Fig. 24   The forward-scattering cross section for ^{208}Pb. Dotted: the static electric polarizability α is assumed to be zero; full line: the full dispersion integral up to 30 GeV, using experimental nucleon absorption cross sections; dashed: α taken from theory, Ref. 31. In this last case the dispersion integral was extended only to pion threshold[6].

At about meson threshold the magnitude of $(\alpha+\beta)E^2$ is equal to r_0^2. The forward cross section for the free proton is zero at that energy and so is the seagull term, Eq. 18, as coherent sum over bound proton scattering cross sections. It is, therefore, extremely important to consider also the αE^2–term (the "modification of the Thomson cross section") in calculations of the nuclear scattering below pion threshold. This can be done in two ways:

1) Extend the dispersion integral up to "infinity" (for the nuclear case it turned out to be sufficient to go up to 20 GeV) with an attempt (see Figs. 21 and 22) to specify the multipole resonances. The assumption, thereby, is that bound nucleons are the same as free ones.

2) Extend the dispersion integral only over the nuclear region up to meson threshold and rely on a theory for the electric polarizability α of bound protons.

The first way was used in the figures shown in the preceding paragraphs, where experiments were compared to theory. The second way was used in Ref. 6, with the theory for α taken from Ref. 31. A comparison of the effect of both on the forward scattering cross section of ^{208}Pb, is shown in Fig. 24. The quark theory for bound nucleons results in a somewhat larger α, than for the free proton.

7.2 Angular Distributions

Below 60 MeV the second order term in E is the only one to be considered. The theory is not yet complicated and the angular distributions can be written[63] in terms of α and β

$$\frac{d\sigma}{d\Omega} = \frac{d\sigma}{d\Omega}\bigg|_{point} - r_0 E^2 \left\{ \frac{\alpha+\beta}{2}\ (1+\cos\Theta)^2 + \frac{\alpha-\beta}{2}\ (1-\cos\Theta)^2 \right\} \qquad (40)$$

where the first term is the charge and magnetic moment scattering from a point particle with charge e, spin 1/2, and anomalous magnetic moment $\lambda(e/2Mc)$[64]. This cross section is a complicated function, yielding terms proportional to E and E^2 in the scattering amplitude, which complicate the simple picture suggested by Eq. 40.

Since the sum $(\alpha + \beta)$ can be derived from σ_{-2}^{E1+M1}, as shown in Eq. 29, the expression, Eq. 40, can be used to derive the difference $(\alpha - \beta)$ by one single measurement, preferably in the backward direction. The statistical errors of existing measurements[58] are still too large for a sensible application of Eq. 40.

The most recent analysis[57] of experimental data gives $\hat{\alpha} = (0.76\pm0.16)\mathrm{fm}^2$, and $\beta = (0.19\pm0.16)\mathrm{fm}^2$. Very surprisingly, β is still compatible with zero.

7.3 Bound Protons

Since the nuclear seagull term is coherent scattering from bound protons, it may be possible to compare free protons (as soon they are measured with sufficient accuracy) with bound ones. From an experimental point of view, and neglecting all difficulties of interpretation, it might even be easier measuring bound protons since the cross sections vary as Z^2.

There will be probably no chance to measure directly the internal spatial distribution of charges or spins of bound protons. All q-dependent factors are dominated by the nucleus' spatial extent. However, the polarizabilities multiply the term quadratic in E. By looking at the E-dependence, at forward angles, where the nuclear form factor is not yet too small, and selecting the angles such that q is fixed, it may be possible to determine the second order structure constants of bound protons. In order to see the sensitivity of these cross sections to $(\alpha + \beta)$, in Fig. 25 the cross sections for $^{208}\mathrm{Pb}$ and $^{12}\mathrm{C}$ for $\Theta=60°$ have been calculated[65] using parameters, that reproduce existing experiments. The effect of a variation of 50% and 100% in $(\alpha + \beta)$ is shown. It may be allowed to speculate about the size of a possible effect by assuming the EMC-effect being due to an increase in quark bag size by 20% for lead[66]. Since the polarizability is proportional to the volume of the particle if nothing else has changed, this radius increase should lead to a measurable 70% effect on α.

8. EXPERIMENTS

There are excellent summaries on modern photon sources[67,68]. Their development is strongly linked to accelerator progress. In some cases it cannot even be avoided that a

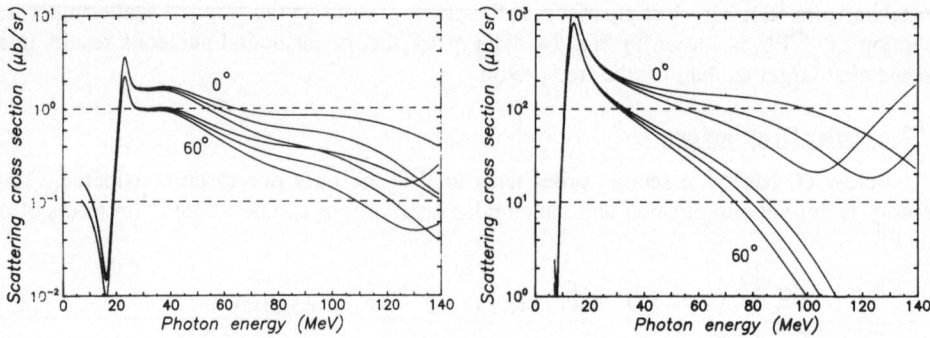

Fig. 25   The effect of the value of the static electric polarizability of the proton on the energy dependence of the scattering cross section for (a) ^{12}C, and (b) ^{208}Pb. The upper three curves are for $\Theta=0$, the lower 3 curves are calculated for the experimentally accessible angle of $\Theta=60°$. $(\alpha+\beta)$ has been assumed to be $(\alpha+\beta)_{free}$, $1.5\cdot(\alpha+\beta)_{free}$, and $2\cdot(\alpha+\beta)_{free}$, respectively, from top to bottom.

photon source is part of an accelerator. In the following it will be sufficient to mention the four different techniques. The only points being looked at in more detail, will be the requirements to measure absolute cross sections and possible future directions for development, both of course reflecting a rather personal point of view.

8.1 Absorption

The attenuation method[69,70] relies on the measurement of a ratio of counts, taken with the same detector. From this ratio the absolute absorption cross section is determined. These are favourable conditions to obtain high accuracy. Systematic errors are of the order of ±0.2% of the atomic cross section (essentially the pair cross section) that has to be subtracted from the measured value. Atomic physics, more precisely the uncertainty of the electron density modifying the field of the naked nucleus, limits the accuracy of this method and, even worse, limits the range in Z that can be used to $Z \leq 20$. More accurate calculations and also measurement of the total pair cross section are needed to extend these measurement to higher-Z elements.

Recently the method has been applied very successfully to monochromatic radiation from particle reactions[71] and neutron capture γ–rays[72]. In the latter case a record has been set for the accuracy of a photonuclear cross section. The total deuteron disintegration cross section between 6 and 11 MeV was measured with only $\approx 1\%$ statistical error.

The attenuation method will certainly be used also in the future, especially for the nucleon and the very lightest elements. At Mainz, it is planned to use a pair spectrometer and rather long cryogenic attenuators (the optimal length of liquid H_2 is 25m), once the MAMI-B accelerator is in operation.

The complementary method[11,59,73], detecting all hadronic events that emerge from a target, relies on the knowledge of counter efficiencies (also for the neutral particles) and on an extrapolation of angular distributions into regions of solid angle, where no counter can be placed. These corrections are calculated using a code in which all the processes are supposed to be correctly described. Restricting the magnitude of these corrections to some ten percent, confines the applicability of this method to low Z elements.

In some cases, it is sufficient to measure one or a few well-defined cross sections[43], in order to find $\sigma_{\gamma,\text{tot}}$. There is no doubt that for a heavy element, with $Z \geq 50$, the sum of all possible (γ,xn) cross sections, with properly identified multiplicity x, is the total cross section up to the energy where quasideuteron absorption dominates. In the quasideuteron region the experimental setup should be able to measure the real deuteron's disintegration as well. Otherwise, one has to rely on preequilibrium calculations that have to tell the probability of a certain number of quasideuteron-associated neutrons.

The fission cross section[74] plays a special role. In some cases, especially for the actinides in the 100 MeV region, it is taken as the total photon absorption cross section, including secondary or ternary fission after a particle emission process. The crucial parameter is the energy given to the primary emitted particle. If it is more then some 30 MeV, the nucleus cools down too quickly and the sum of the fission probabilities of the target and all daughter nuclei does not approach 100%.

In all methods, except the attenuation method, the primary photon flux has to be known. This requirement also holds for scattering experiments and will be discussed briefly there.

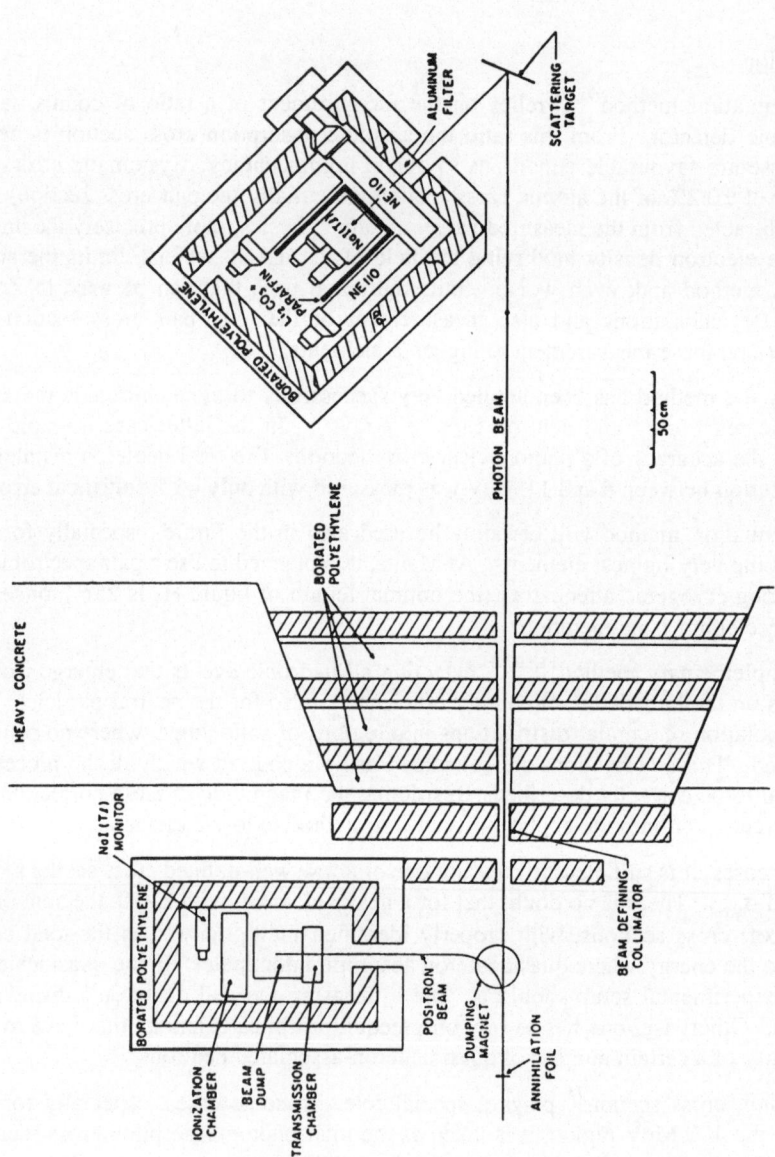

Fig. 26   The experimental setup for photon scattering using positron annihilation in flight[38]. Absolute cross sections were obtained by placing the detector assembly in the direct beam. A linear response of the beam monitors over 6 decades of intensity is essential.

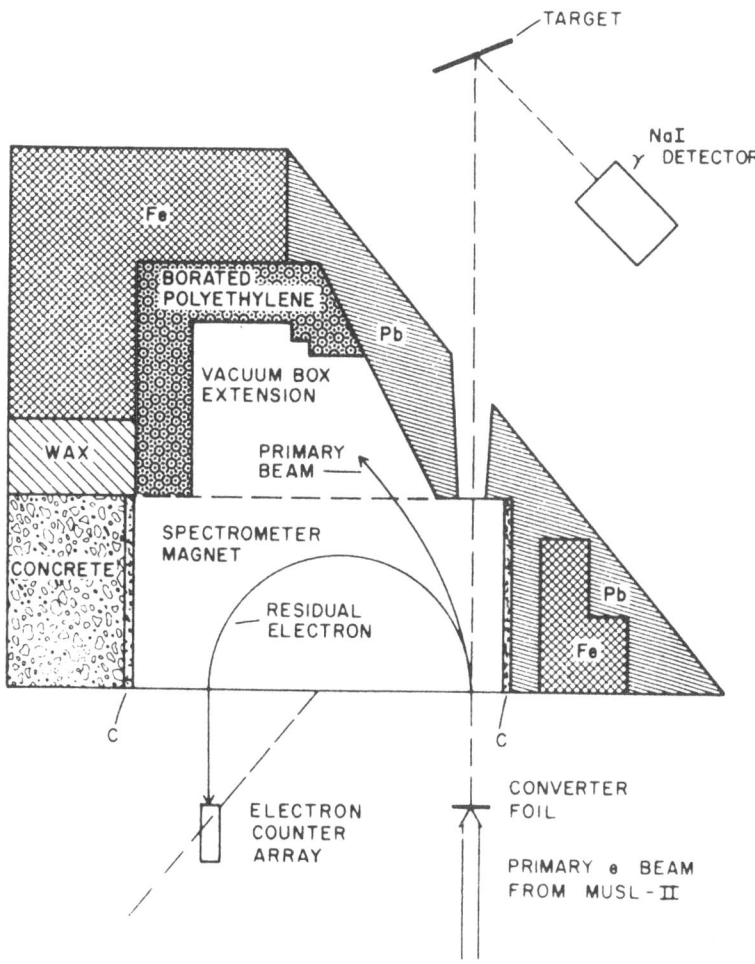

Fig. 27   The experimental setup of P. Axel[77]: the tagged photon technique.

8.2 Scattering

The first photon scattering experiments in the giant dipole region and above were performed in the fifties[2,75]. The photon source was bremsstrahlung radiation from betatrons. The detectors were rather primitive, compared to present standards. The same detector was placed alternately in the direct and the scattered beam. The cross section then was a ratio of counts, taken with the same detector, and the art of the experimentalist consisted in monitoring a 10^6 variation in intensity, in order not to damage the photon detector in the direct beam.

The same technique of absolute measurement later was used with monochromatic beams from positron annihilation. The use of monoenergetic positron beams for the production of "monochromatic" photons was invented by Tzara[76]. It turned out to be an extremely useful technique that produced the bulk of all (γ,n) cross sections[46]. Photon scattering was measured at the NBS (see Fig. 26) and at Mainz[33]. Both installations have been disassembled.

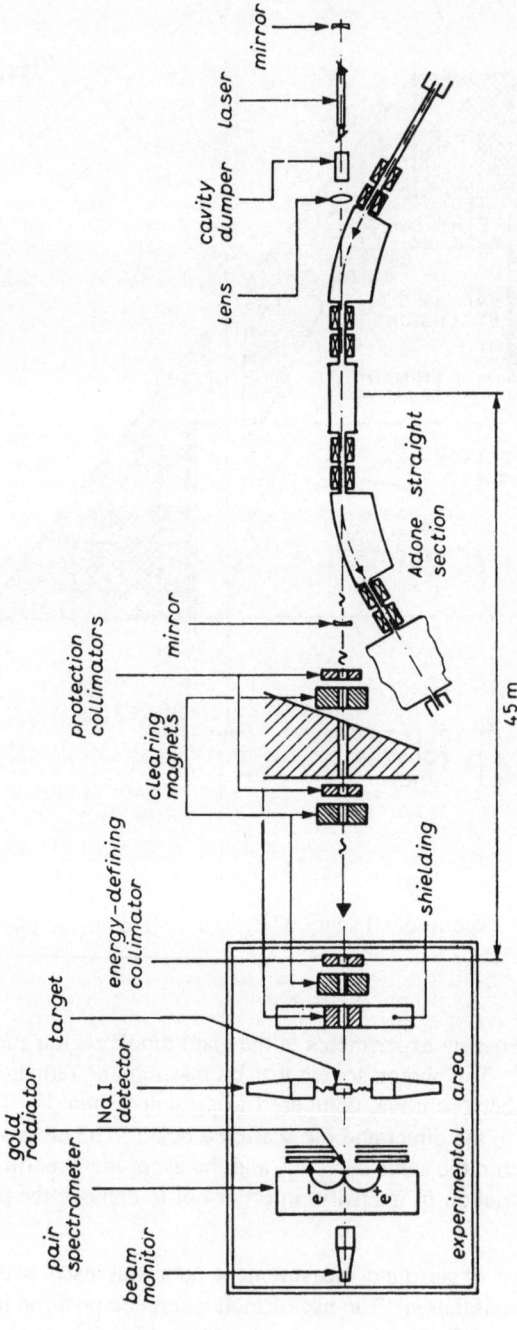

Fig. 28   The experimental setup "LADON"[78]: laser backscattering from relativistic electrons.

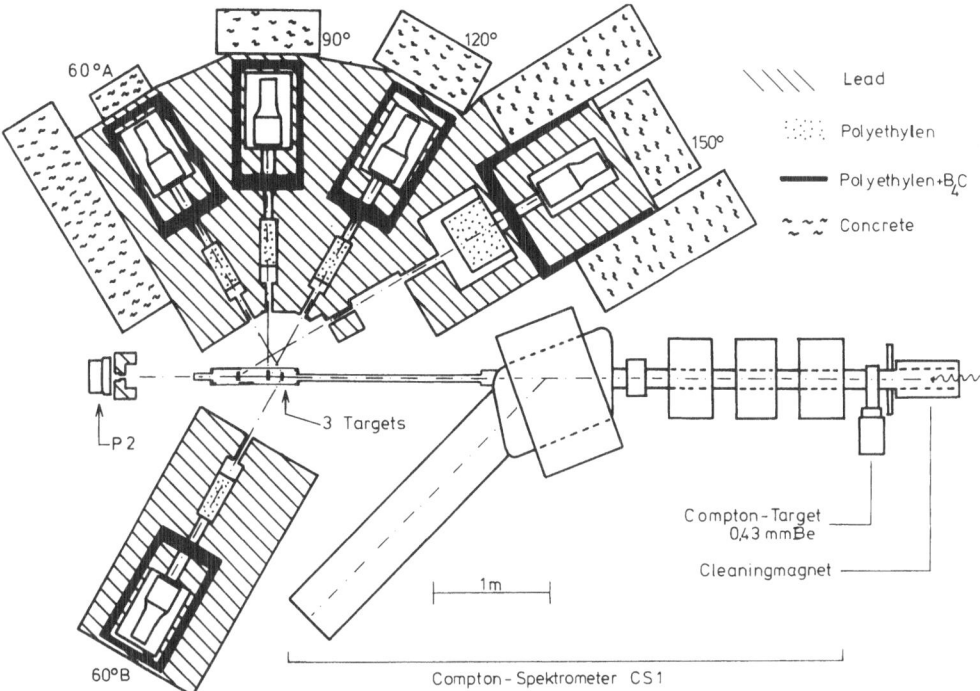

Fig. 29   The measurement of the elastic plus inelastic scattering cross section, using bremsstrahlung, five NaI-crystals, and a Compton spectrometer[53].

The tagged photon technique[68] was pioneered by Peter Axel. The original setup[77] (see Fig. 27), is one of the most successful experiments in photonuclear physics. With some refinements, it is still being used for photon scattering experiments[6]. The technique has been made feasible by the development of c.w. accelerators and is about to replace the positron beams. The investment for a tagging facility is much smaller. One has not to waste 10^8 electrons for one photon, and it is possible to measure a large number of photon energies simultaneously. The quality of a set-up, regarding absolute cross sections, is mainly determined by a large ratio of tagged photons divided by electron counts ("tagging efficiency") approaching the theoretical value as closely as possible and, even more important, a stable ratio of these counts. Then the essential photon monitors are the electron counts, and no separate photon detectors in the direct beam are necessary. The main pre-requisite for a constant tagging efficiency is a stable primary beam with changes of the angle in time of typically not more than 1 mrad. This technique will certainly be most useful in connection with c.w. electron beams.

The second very promising technique is the Compton-backscattering of Laser light from relativistic electrons. The first installation was built at Frascati[78] with some 10^5 photons per sec in an energy bin comparable to the resolution of a good NaI-crystal, and very impressive degree of polarization, approaching 100%. The set-up (see Fig. 28),

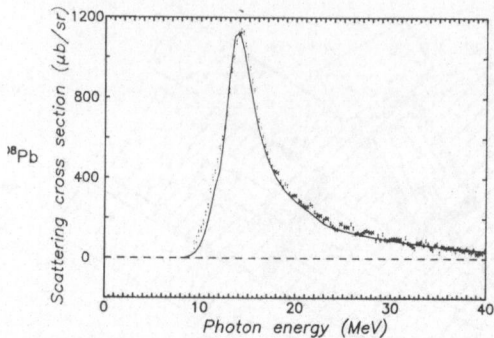

Fig. 30   Preliminary data[53] of bremsstrahlung photon scattering on ^{208}Pb; the target
thickness is 0.05 cm; scattering angle is $\Theta=150°$; a calculated background, proportional to $1/E^2$, of 70 μb/sr at 10 MeV is subtracted. The cross section curve
is calculated from absorption data[43]; the detector is a NaI crystal 10"ϕ×14"; the
measured detector response has been unfolded from the data. The Lorentz-
parameters are: 0.08 classical sums at 11.4 MeV with Γ=1.3 MeV; and 1.13
classical sums at 13.4 MeV with Γ=3.5 MeV.

includes photon spectrometers for photon counting and flux measurement, since no flux
monitor is inherently provided by this system. The stability of the photon flux seems to
be mainly determined by the stability of the high power laser in intensity as well as in
angle and position.

The LEGS-facility[79] (at Brookhaven National Laboratory) will combine the advantages of the tagging and the laser technique. It will have a resolution of the order of 1%,
a built-in photon monitoring, and the polarization of the high energy (100-300 MeV)
photon beam can be chosen at will. This source will be the most promising development
for future photon scattering experiments on nuclei and nucleons.

At the end of this short overview, I shall report on a recent experiment at Mainz[53],
that became feasible only when the second stage of the MAMI-accelerator was completed. It is the most primitive approach to a scattering experiment (see Fig. 29), using a
bremsspectrum, a Compton spectrometer as photon counter, and NaJ crystals, measuring
the spectrum of photons scattered from the thin targets. For targets with small or no inelastic scattering[80], the scattered photon spectrum is converted into a scattering cross section. For backward angles, there is only a small background from wide angle
bremsstrahlung of pair particles created in the thin target. Fig. 30 is the result for ^{208}Pb
for scattering angle $\Theta=150°$. A calculated background, amounting effectively to 70 μb/sr
at 10 MeV and falling proportional to $1/E^2$, has been subtracted.

This picture of ^{208}Pb giant resonance fluorescence reflects the progress over the last
30 years achieved in accelerator and detector technology. It will be the task of the near
future to obtain similar pictures from the proton's excited states.

I want to thank K. P. Schelhaas for his help in preparing this lecture; I also thank
Evans Hayward for reading the manuscript critically.

REFERENCES

[1] G. Mie, Ann. Phys. **25**,377(**1908**).
[2] E. G. Fuller and E. Hayward, Phys. Rev. **101**, 692 (1956).
[3] H. Arenhövel, M. Danos and W. Greiner, Phys. Rev. **157**, 1109 (1967).
[4] M. Gell-Mann, M. L. Goldberger and W. Thirring, Phys. Rev. **95**, 1612 (1954).
[5] M. Danos, Photonuclear Physics, University of Maryland Technical Report **221**, (1961).
[6] D. H. Wright, P. T. Debevec, L. J. Morford and A. M. Nathan, preprint P/85/3/35, University of Illinois, 1985.
[7] R. G. Leicht and B. Ziegler, in: From Nuclei to Particles 1981, LXXIX Corso, Soc. Italiana di Fisica, Bologna.
[8] M. Schumacher, private communication.
[9] R. J. Hughes, R. H. Sambell, E. G. Muirhead and B. M. Spicer, Nucl. Phys. **A238**, 189 (1975).
[10] J. Ahrens, H. Borchert, K. H. Czock, H. B. Eppler, H. Gimm, H. Gundrum, M. Kröning, P. Riehn, G.Sita Ram, A. Zieger and B. Ziegler, Nucl. Phys. **A251**, 479 (1975).
[11] J. Arends, J. Eyrink, A. Hegerath, K. G. Hilger, B. Mecking, G. Nöldecke and H. Rost, Phys. Lett. **89B**, 423 (1981).
[12] J. M. Blatt and V. F. Weisskopf, Theoretical Nuclear Physics, reprinted 1979, Springer Verlag. p. 358.
[13] B. Ziegler, Lecture Notes in Physics **108**, 148 Springer (1979).
[14] S. Krewald and G. Co in: Proceedings of the Bates Users Theory Group, MIT July 1984, p. 80.
[15] E. G. Fuller, Physics Reports, in press.
[16] E. Hayward, in: Lecture Notes in Physics, **61**, eds. S. Costa and C. Schaerf, Springer 1977, p. 374.
[17] U. E. P. Berg, Journal de Physique, Colloque, C4, 359 (1984).
[18] R. M. Laszewski, P. Rullhusen, S. O. Hobbit and S. F. Lebrun, preprint P/84/8/105, University of Illinois, 1984.
[19] B. Ziegler, Proceedings of the Workshop on Intermediate Energy Nuclear Physics, eds. G. Matone and S. Stipcich, Frascati, July 1980.
[20] G. Matone, Proceedings of the School on Intermediate Energy Nuclear Physics, eds. G. Schaerf and S. Costa, Verona 1985, to be published.
[21] C. K. Sinclair, Proc. Intern. Symp. on High Energy Physics with Polarized Beams and Polarized Targets, eds. C. Joseph and J. Soffer, Basel 1981, p.27.
[22] R. D. Evans, The Compton Effect, in: Handbuch der Physik, Vol. XXXIV, S. Flügge ed., Springer 1958.
[23] B. Ziegler, Z. Physik, **C15**, 95 (1982).
[24] T. E. O. Ericson and J. Hüfner, Nucl. Phys. **B57**, 605 (1973).
[25] H. Arenhövel, Proceedings of the Workshop on Perspectives in Nuclear Physics at Intermediate Energies, Trieste 1983, p.97.
[26] R. Moreh et al., Nucl. Phys. **A275**, 445 (1977).
[27] M. Born, Optik, Springer 1932, p. 404.
[28] W. Thirring, Phil. Mag. **41**, 1193 (1950).
[29] W. Heitler, The Quantum Theory of Radiation, Oxford 1954, p. 36.
[30] R. Silbar, C. Werntz and H. Überall, Nucl. Phys. **A107**, 655 (1968).
[31] D. Drechsel and A. Russo, Phys. Rev. **137B**, 294 (1984).
[32] J. L. Friar, Annals of Physics, **95**, 170 (1975).
[33] R. Leicht, M. Hammen, K. P. Schelhaas and B. Ziegler, Nucl. Phys. **A362**, 111 (1981).

[34] M. Schumacher, P. Rullhusen, F. Smend and W. Mückenheim, Nucl. Phys. **A346,** 418 (1980).

[35] S. Kahane and R. Moreh, Nucl. Phys. **A308,** 88 (1978).

[36] T. J. Bowles, R. J. Holt, H. E. Jackson, R. M. Laszewski, R. D. McKeown, A. M. Nathan and J. R. Specht, Phys. Rev. **C24,** 24 (1981).

[37] S. Fallieros, private communication.

[38] W. R. Dodge, E. Hayward, R. G. Leicht, M. McCord and R. Starr, Phys. Rev. **C28,** 8 (1983).

[39] J. S. O'Connell, Proceedings of the Intern. Conf. on Photonuclear Reactions and Applications, ed. B. L. Berman, 1973, p. 71.

[40] U. Zurmühl, P. Rullhusen, F. Smend, M. Schumacher, H. G. Börner and S. A. Kerr, Z. Phys. **A314,** 171 (1983).

[41] Landolt-Börnstein, New Series I/8, Springer Berlin (1973).

[42] F. Fantoni, Proceedings of the Second Workshop on Perspectives in Nuclear Physics at Intermediate Energies, Trieste 1985, to be published.

[43] A. Leprêtre, H. Beil, R. Bergère, P. Carlos, J. Fagot, A. de Miniac and A. Veyssière, Nucl. Phys. **A367,** 237 (1981).

[44] J. S. Levinger and H. A. Bethe, Phys. Rev. **78,** 115 (1950).

[45] W. M. Alberico, M. Ericson and A. Molinari, Ann. Phys. **154,** 356 (1984).

[46] Atlas of Photoneutron Cross Sections, ed. B. L. Berman, UCRL-78482.

[47] J. S. Levinger, Nuclear Photodisintegration, Oxford University Press 1960, p. 97.

[48] W - Y. P. Hwang and G. E. Walker, Annals of Physics **159,** 118 (1985).

[49] P. Christillin, Proceedings of the Second Workshop on Perspectives in Nuclear Physics at Intermediate Energies, Trieste 1985, to be published.

[50] P. Christillin and M. Rosa-Clot, Nuovo Cimento, **28A,** 29 (1975).

[51] H. Arenhövel, Z. Physik, **A297,** 129 (1989).

[52] M. G. Huber and Klingenbeck, Proceedings of the Workshop on Perspectives in Nuclear Physics at Intermediate Energies, Trieste 1983, p. 365.

[53] Göttingen-Mainz collaboration, to be published.

[54] E. Hayward and B. Ziegler, Nucl. Phys. **A414,** 333 (1984).

[55] J. H. Koch and E. J. Moniz, Annals of Physics, **154,** 99 (1984).

[56] J. Vesper, N. Oktsuka, L. Tiator and D. Drechsel, Nuclear Physics with Electromagnetic Probes, Paris 1985, 11th Europhysics Conference, to be published.

[57] V. A. Petrun'kin, Sov. J. Part. Nucl. **12,** 278 (1981).

[58] P. S. Baranov and L. V. Fil'kov, Sov. J. Part. Nucl. **7,** 42 (1976).

[59] T. A. Armstrong *et al.,* Nucl. Phys. **B41,** 445 (1972).

[60] J. Ahrens, Nuclear Physics with Electromagnetic Probes, Paris 1985, 11th Europhysics Conference, to be published.

[61] D. H. Miller, in: High Energy Physics, Vol. IV, ed. E. H. S. Burhop, Academic Press 1969.

[62] L. V. Fil'kov, in: The Nuclear Compton Effect, ed. D. V. Shobel'tsyn, P. N. Lebedev Physics Institute, Moscow 1969, p. 49.

[63] T. E. O. Ericson, in: Interaction Studies in Nuclei, eds. H. Jochim and B. Ziegler, North Holland Publishing Company 1975, p. 577.

[64] J. L. Powell, Phys. Rev. **75,** 32 (1949).

[65] K. P. Schelhaas, Thesis University of Mainz 1984, and private communication.

[66] C. M. Shakin, in: Proceedings of the Bates Users Theory Group, MIT July 1984, p. 251.

[67] H. Beil and R. Bergere, Note CEA-N-2144, July 1980.

[68] L. S. Cardman, Photon Tagging, University of Illinois, preprint P/83/12/168.

[69] H. W. Koch and J. M. Wyckoff, IRE Transactions on Nuclear Science **NS-5,** 127 (1958).

[70]   B. Ziegler, Z. Physik, **152,** 566 (1958).

[71]   Ph. B. Smith, W. Bieriot, H. Groenveld and P. van den Hock, Phys. Rev. **C30,** 2085 (1984).

[72]   R. Moreh, Y. Birenbaum and O. Shakal, Nucl. Phys. **A275,** 445 (1977).

[73]   P. Argan *et al.,* Rapport DPh-N Saclay 2162 (1984).

[74]   J. Ahrens, J. Arends, P. Bourgois, P. Carlos *et al.,* Phys. Lett. **146B,** 303 (1984).

[75]   G. E. Pugh, R. Gomez, D. H. Frisch and G. S. Jones, Phys. Rev. **105,** 983 (1957).

[76]   C. Tzara, Comptes Rendus Acad. Sci. **254,** 56 (1957).

[77]   J. O'Connell, P. Tipler and P. Axel, Phys. Rev. **126,** 228 (1962).

[78]   L. Fedderici *et al.,* Il Nuovo Cimento, **59B,** 247 (1980).

[79]   C. E. Thorn, M. J. Levine and A. M. Sandorfi, Proceedings of the Workshop on the Use of Electron Rings for Nuclear Physics Research, Lund, Sweden (Oct. 1982), p. 204.

[80]   A. M. Nathan and R. Moreh, Physics Letters **91B,** 38 (1980); and A. M. Nathan, private communication.

HIGH RESOLUTION QUASI-FREE (e,e'p) EXPERIMENTS

P. K. A. de Witt Huberts

NIKHEF-K
P.O. Box 4395
1009 AJ Amsterdam/The Netherlands

INTRODUCTION

The purpose of these lectures is to provide an introduction to recent state-of-the-art developments in the field of the quasi-free proton-knock-out process induced by high-energy (≈ 500 MeV) electrons. In the electron scattering (e,e') response function $d^2\sigma/d\Omega de'$ of a nucleus the quasi-free process becomes the dominant feature when the wavelength of the virtual photon becomes of the order of the nucleon dimension $\delta r \approx 0.8$ fm. Since the average internucleon distance amounts to ≈ 2 fm and provided that the mean free path of the knocked-out nucleon is substantially larger than the internucleon distance, the dominant reaction mechanism involves incoherent scattering off individual (quasi-free) nucleons. In this incoherent scattering limit the probability distribution of momentum and energy of the struck nucleon can be reconstructed if a coincidence (e,e'p) experiment is performed. Due to the, in principle, straightforward simplicity of interpretation the quasi-free process has found widespread application in diverse fields varying from the few kilo electron volt (e,2e) process[1] at the level of atomic dimensions, to the deep-inelastic scattering of multi-GeV electrons from point-like constituents (partons) in the nucleon.[2]

In nuclear structure physics (p,2p) experiments were the first to provide evidence for the validity of the nuclear shell model. Since the basic formulation of this model around 1950 a vast effort has been invested in developing a detailed understanding of the remarkable fact that it appears to be a good approximation to describe the densely packed and correlated system of strongly interacting fermions, that constitute the atomic nucleus, as a collection of independent particles moving in a mean field. Precision experiments with the electromagnetic interaction, mediated by virtual photons provided by the electron scattering process, have substantially advanced our understanding of the nuclear many-body system, notably so for the inner regions of the nuclear density.[3,4] The (e,e'p) experiments carried out fairly recently at the Saclay facility with a moderate missing mass resolution $\Delta E_m \approx 1.2$ MeV have, among other things, provided much information on the separation energy and spreading width of deeply-bound hole states in a series of nuclei up to mass 58. Compared to this generation of experiments a major improvement in the quality of (e,e'p) data has been achieved with the high resolution dual-spectrometer setup[5] at NIKHEF-K. Upon application of a novel beam-dispersion matching technique[6] a missing-mass resolution of (100 - 150) keV can now routinely be achieved. This high-resolution feature gives us a spectroscopic tool at hand with a largely improved sensi-tivity and signal-to-noise ratio. Given this new tool the question arises as to which open problems concerning the single-particle structure of nuclei are foremost. One such prob-lem concerns absolute spectroscopic factors and the question concerning the mechanism of (ground state) correlations in the nuclear many-body wave function. Several experi-mental observations[7] provide evidence, albeit indirectly, for the lack of correlations in standard (shell-model) wave functions. Another issue concerns the amount of high-momentum components that may be generated by short-range nucleon-nucleon collisions in the dense nuclear system. There is very little known experimentally about such processes at present. In standard calculations of nuclear structure the constituent nucleons are treated as elementary particles; that is one assumes that the nuclear medium does not affect their intrinsic properties. Recent developments of Dirac-equation based relativistic mean field theory propose a different picture.[8] In this approach the nucleon size is increased in the medium and this leads to a modification of the virtual photon-nucleon coupling. Quasi-elastic (coincidence) reactions, that are least obscured by nuclear-structure effects, may profitably be used to investigate this interesting field where nuclear and elementary particle physics intersect.

These lecture notes are organized as follows. Section 2 contains the basics of the (coincidence) quasi-free reaction formalism in a tutorial rather than rigorous style. Corrections to the plane-wave impulse approximation will be briefly discussed. We also describe the instrumentation for (e,e′p) experiments and summarize the essentials of the data analysis. In addition the complementary aspects of (e,e′p), single-arm (e,e′) and one nucleon transfer reactions (e.g. d,³He) will be pointed out. In Section 3 a selection of novel high-resolution (e,e′p) data and their interpretation is presented. In Section 4 a summary and an outlook for future experiments is given.

2. THE QUASI-FREE KNOCKOUT PROCESS

Consider the quasi-free proton knockout process in the plane-wave impulse approxi-mation limit. In such a process, illustrated in the kinematics diagram of Fig. 1, the energy momentum $(\omega,\vec{q})$ of the virtual photon is absorbed by one particular proton mov-ing with momentum $\vec{p}_{initial}$, while the residual (A-1) system acts as a spectator. In the laboratory coordinate system the requirement of conservation of momentum and total energy yields

$$\vec{q} = \vec{e} - \vec{e}\,' = \vec{p} - \vec{p}_{initial} \tag{1}$$

$$e_o + M_A = e_o' + T_p + m_p + M_{A-1} + p_m^2/2M_{A-1} \tag{2}$$

where in (2) the (A-1) system has been treated non-relativistically, a good approximation

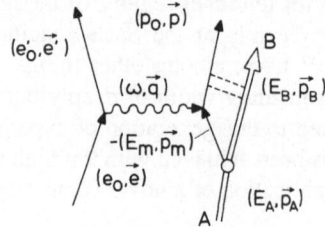

Fig. 1   Diagram of the proton knockout reaction in the impulse approximation. The dashed lines indicate the final state interaction.

for most practical cases. We choose to select the missing mass E_m and missing momentum p_m as variables of the reaction, found by making up the (external) energy and momentum balances,

$$E_m = e_o - e_o' - T_p - p_m^2/2M_{A-1}$$

$$\vec{p}_m = -\vec{p}_{initial} = \vec{p}_{recoil} \quad .$$

The most general form of the cross section, consistent with Lorentz co-variance, can be written as[9]

$$\frac{d\sigma}{dp_o d\Omega_p de_o' d\Omega_e} \propto (\sigma_T + \varepsilon\sigma_L + \varepsilon \cos2\phi \, \sigma_{TT} + \sqrt{\varepsilon(\varepsilon+1)q_\mu^2/2\omega^2} \cos\phi \, \sigma_{TL}) \tag{3}$$

where ϕ is the angle between the knockout proton and the $(\vec{e},\vec{e}')$ scattering plane and ε is the photon polarization $\varepsilon = (1 - 2q^2/q_\mu^2 \, tg^2\theta/2)^{-1}$. In the photon-proton interaction vertex there is coupling with the Coulomb field (σ_L) and the current vector field consisting of the longitudinal current (parallel to $\vec{q}$) and the transverse currents. Interference terms occur between the current components. Lumping the separate terms in the photon-proton vertex together in the off-shell electron-proton cross section σ_{ep}^*, expression (3) may be written as

$$\frac{d\sigma}{dp_o d\Omega_p de_o' d\Omega_e} = K \, \sigma_{ep}^* \, S \, (E_m, \vec{p}_m) \quad . \tag{4}$$

Here K is a kinematic factor (phase space) and $S(E_m, \vec{p}_m)$ is the so-called spectral function defined as the joint probability of having a proton with $(E_m, \vec{p}_m)$ in the nucleus. In order to see what to expect for the global features of $S(E_m, \vec{p}_m)$ let us consider the extreme case of the independent particle shell-model (IPSM). Single-particle orbits are labeled by $\alpha = (n_\alpha l_\alpha j_\alpha)$, orbits are filled up to the Fermi surface (F). Then the spectral function is given by

$$S(E_m, \vec{p}_m) = \frac{1}{4\pi} \sum_{\alpha \le F} (2j_\alpha+1) \, R_\alpha^2(p_m) \, \delta(E_m + \varepsilon_\alpha) \tag{5}$$

where ε_α is the single-particle energy of orbit $\{\alpha\}$. The single-particle wave function information is contained in $R_\alpha(p_m)$, which is the *Fourier transform* of the coordinate space wave function

$$R_\alpha(p_m) = \sqrt{\frac{2}{\pi}} \int_0^\infty j_{l_\alpha} (p_m r) \, R_\alpha(r) \, r^2 dr \tag{6}$$

with j_{l_α} being the spherical Bessel function of order l_α. In this extreme model the spectral function simply consists of regions of strength at specific ε_α values each with a dependence on p_m characteristic of (n_α, l_α). Given the unpolarized cross sections we are considering here there is no direct sensitivity to j_α in the cross sections. In Fig. 2 is shown a set of single-particle orbits evaluated in the harmonic oscillator (H.O.) basis. One should note that the shape of the H.O. wave function is the same in both momentum and coordinate space, the argument should only be changed from $p_m b$ to r/b. Switching on the residual nucleon-nucleon interaction the single-particle orbits mix and couple to complicated (2p-2h), (3p-3h)-states. This leads to a fragmentation in E_m of the original IPSM states. In order to maintain the similarity with Eq. (5) one may expand the spectral function as

$$S(E_m, p_m) = \sum_{\alpha\beta} P_{\alpha\beta}(E_m) \, \phi_\alpha^*(p_m) \, \phi_\beta(p_m)$$

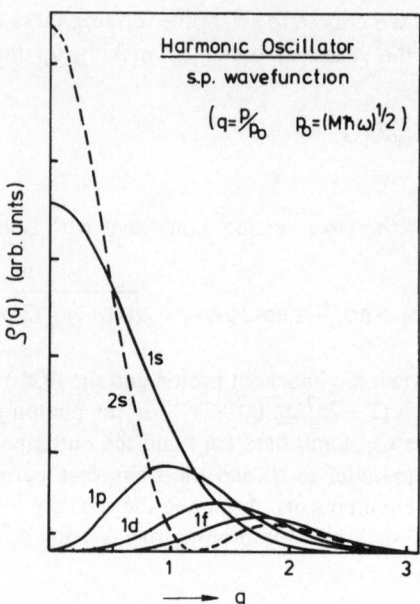

Fig. 2   Single particle harmonic-oscillator wavefunction probabilities in momentum
space for various quantum states.

with $P_{\alpha\beta} = <\psi_A \mid a_\alpha^+ \delta(E-H+E_A)a_\beta \mid \psi_A>$.

With an appropriate choice of single-particle basis wave functions one may diagonalize $P_{\alpha\beta}$ and thus obtain

$$S(E_m,p_m) = \sum_\alpha P_\alpha(E_m) \phi_\alpha^2(p_m) \quad . \tag{7}$$

The spectroscopic factor, well-known from the field of one-nucleon transfer reactions, can be obtained from

$$C^2S = 4\pi \int\limits_0^\infty S(E_m,p_m)p_m^2 dp_m \quad .$$

At present, the practical upper limit of p_m amounts to ⁻300 MeV/c. As is illustrated in Fig. 3 the integral has at this p_m value reached convergence to a good approximation. Furthermore an obvious sumrule holds $4\pi\iint S(E_m,p_m) p_m^2 dp_m dE_m = Z$ (the number of protons in the target nucleus). The inclusive quasi-elastic cross section can be obtained by integrating expression (3) over the variable $d\vec{p} = p^2 d\Omega_p\, dp$. Since

$$\int\limits_0^{2\pi} \cos\phi_p\, d\phi_p = \int\limits_0^{2\pi} \cos2\phi_p d\phi_p = 0$$

the interference terms disappear and we are left with *two* response functions in the inclusive cross section

$$\frac{d\sigma}{d\Omega_e de_o'} = \sigma_{\text{Mott}} [\frac{q_\mu^4}{q^4} R_L(q,\omega) + \epsilon^* R_T(q,\omega)] \tag{8}$$

where

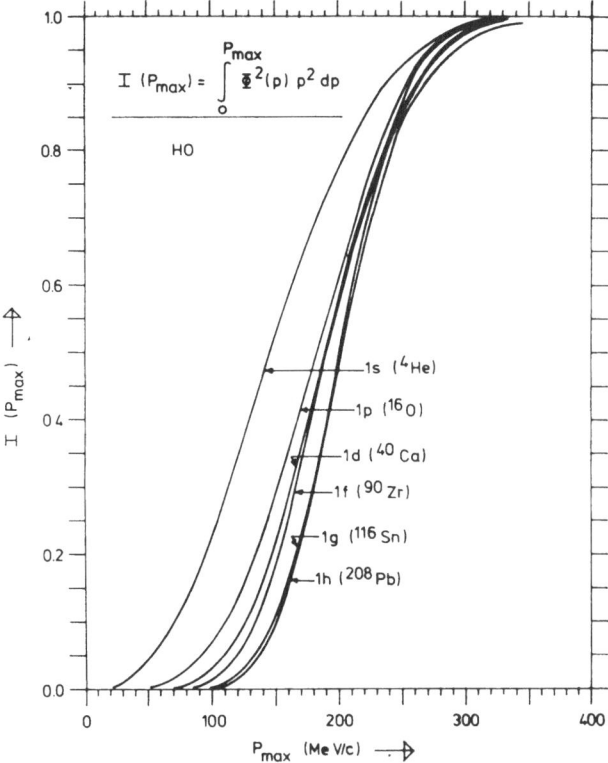

Fig. 3   Convergence of the H.O. spectral function for various nuclei.

$$\epsilon^* = [\frac{q_\mu^2}{2q^2} + tg^2\theta/2] \quad .$$

In particular the transverse term is exclusively due to the photon-nucleon spin interaction

$$\frac{d\sigma}{d\Omega_e de'_o} = \sigma_{Mott} F_D^2(q) [(Z \mu_p^2 + N \mu_n^2)\epsilon^* q^2 R_T^*(q,\omega) + \frac{q_\mu^4}{q^4} R_L^*(q,\omega)] \qquad (9)$$

where $F_D(q)$ is the dipole parametrization of the proton form factor $F_D(q) = (1 + q^2/M_o^2)^{-2}$ with $M_o^2 = 0.71 \, GeV^2$.

Let us return to the exclusive (e,e′X) process and investigate the experimental methods involved in the measurement of the nuclear structure information contained in the spectral function $S(E_m,p_m)$. In practical experiments two (orthogonal) methods are used to cover a region of the (E_m,p_m)-plane, i) parallel kinematics (Fig. 4b) and ii) perpendicular kinematics (Fig. 4a). In most of the experiments to be discussed parallel kinematics were employed. In such kinematics the missing momentum $\vec{p}_m$ is parallel to $\vec{q}$ and a range of p_m values is covered by changing the magnitude of $\vec{q}$ (and its direction) by varying the electron scattering angle $\theta_{e'}$. This is done in such a way that the outgoing proton momentum or better the relative kinetic energy of proton and (A-1) system, is kept constant (T_p = constant). This latter constraint is effective in order to keep the final-state interaction constant. We will discuss this point in some more detail later on.

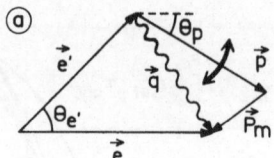

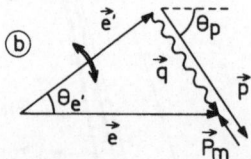

Fig. 4   Perpendicular (a) and parallel (b) kinematics.

A particular property of parallel kinematics is that no interference between the longitudinal- and transverse-current occurs[10]. Therefore the expression for the cross section resembles expression (8) for the inclusive (e,e′) process

$$\frac{d\sigma}{d\Omega_e de_o{}' d\Omega_p dp_o} = \sigma_{Mott} \left\{ \frac{q_\mu^4}{q^4} S_L(p_m, E_m) + \varepsilon^* S_T(p_m, E_m) \right\} \quad . \tag{10}$$

In perpendicular kinematics the missing momentum vector is perpendicular to the momentum transfer. Here p_m is scanned by varying the knocked-out proton direction. In our experiments this mode is used to check on specific aspects of the final state interaction. The momentum of the scattered electron $\vec{e}'$ and the momentum of the knocked-out proton $\vec{p}$ are measured by two magnetic spectrometers, with the best resolution one can achieve, and with large solid angles. The focal planes are equipped with fast wire chamber systems[11] to allow a precise reconstruction of the momentum vectors at the scattering vertex. The NIKHEF-K set-up[5] is illustrated in Fig. 5. Upon application of a special beam-imaging technique called two-arm dispersion matching[6] the resolution contribution due to the finite momentum spread on target (≈ 1 MeV) can be nullified. All this results in a standard missing-mass resolution of $\Delta E_M = 150$ keV and ~100keV in special cases. Since each of the spectrometers spans a momentum bite of *ten percent* simultaneously an area of the space of the observed momenta $(\vec{e}', \vec{p})$ is covered. Inspired by the plane wave impulse approximation we would like to represent the cross section data in terms of a spectral function in the (p_m, E_m) plane. In other words, we would like to carry out the transformation

$$\frac{d\sigma}{de_o{}' dp_o d\Omega_e d\Omega_p} \to \frac{d\sigma}{dE_m dp_m d\Omega_e d\Omega_p} \quad .$$

The weight of the phase-space volume element corresponding to a given elementary bin $(p_m \pm \frac{1}{2} \Delta p_m, E_m \pm \frac{1}{2} \Delta E_m)$ (typically, $\Delta p_m = 10$ MeV/c, $\Delta E_m = 50$ keV) can be calculated by using the Jacobian of the $(e_o{}', p_o) \to (p_m, E_m)$ transformation; however, the treatment of the integration boundaries becomes very cumbersome analytically. A more practical method consists of determining the phasespace volume via a Monte-Carlo method, in which a $(\vec{e}', \vec{p})$ pair is drawn at random in the interval spanned by the spectrometer solid angles and momentum ranges. The resulting weight distribution, typical for the NIKHEF setup, is illustrated in Fig. 6. A fairly large range of $E_m(\sim 30 \text{MeV})$ is spanned with good

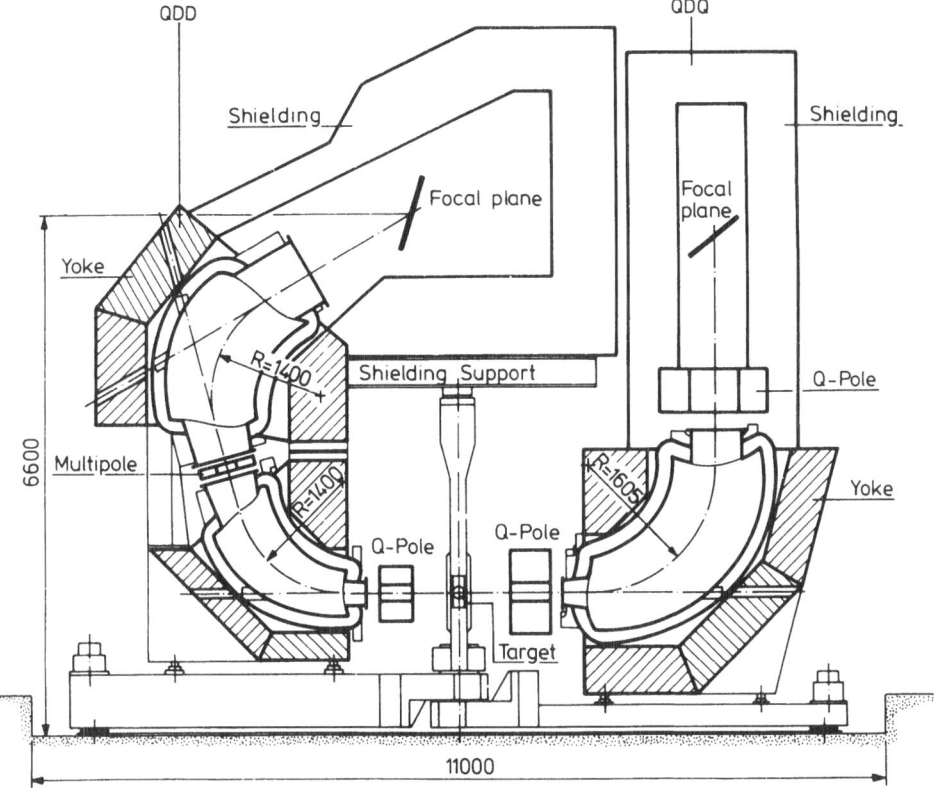

Fig. 5   Two spectrometer setup of NIKHEF-K. The sizes are indicated in mm's.

detection efficiency, the simultaneous momentum interval covered is 50 MeV/c. Several consecutive angle and/or momentum settings are used to cover a useful range of p_m = (0 - 250) MeV/c. We note finally that the precise particle trajectory reconstruction[11] enables us to achieve a coincidence time-resolution of better than 1 ns (fwhm), a prerequisite to obtain an acceptable real to accidental ratio at the larger p_m-values.

Having discussed the ideal world of the plane wave impulse approximation we will now have to deal with some harsh realities of the real world. Although the corrections to

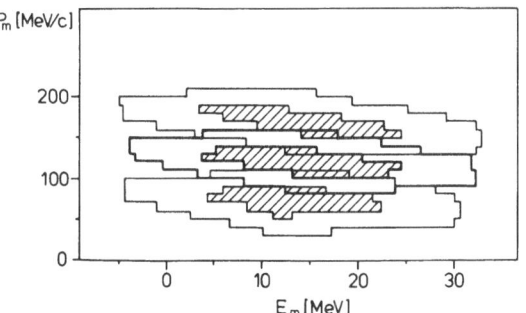

Fig. 6   Phase space weight distribution for three overlapping kinematics. At each set-ting the dashed region indicates > 80% of maximum weight.

the PWIA, required for a quantitative interpretation of (e,e′p) data, are non-negligible, they do not basically impair the physics of the simple plane-wave interpretation. The corrections may be divided in basically two categories: i) QED effects due to radiative processes and Coulomb distortion effects in the incoming and outgoing electron spinors and ii) strong interaction effects (acting only on the outgoing proton) before it is detected in asymptotia.

2.1 QED Corrections

The radiative effects can be handled in a manner similar to the one used in single-arm electron scattering. Most of the data that we will discuss in the following sections have been corrected for these relatively small effects. The Coulomb distortion effects, that should ultimately be calculated with a phase shift code, are at present handled in an effective momentum transfer approximation $q_{eff} = q(1 + Z\alpha/R_{eq}e_o)$. In addition the effect due to the flux-converging action of the nuclear Coulomb field is accounted for[12] by a correction factor $\Delta\sigma_{max}/\sigma_{max} = [(e,e')^{1/2}/p] * Z * 0.3\%$. This is presumably a good approximation for light nuclei, but for heavier nuclei and for momentum distributions that show minima (2p, 2s, 3s, 2d states for instance) a more precise treatment of Coulomb distortions will be necessary. Work in this direction is in progress.

2.2 Final State Interaction

The strong interaction in the final state of the knocked-out proton and recoiling (A-1) system is at present the most important but least understood correction to the PWIA. Formally the expression for the coincidence cross section can be written as

$$\frac{d\sigma}{de_o'd\Omega_e dp_o d\Omega_p} = \sum_{f,i} | \hat{T}_{ep} \cdot M_{fi} |^2$$

with

$$M_{fi} = \int X_p^{(-)*}(\vec{r}_p)e^{i\vec{q}\cdot\vec{r}_p}<\psi_{A-1} | \psi_A > dr_p$$

where $\hat{T}_{ep}$ is the electron-proton scattering amplitude and $X_p^{(-)}$ is the distorted wave function of the outgoing proton. Notably so for parallel kinematics the $\hat{T}_{ep}$ can be factorized from the (distorted) nuclear term to a good approximation. The resulting expression then reads

$$\frac{d\sigma}{de_o'd\Omega_e dp_o d\Omega_p} = K \sigma_{ep}^* S^D (E_m,\vec{P}_m,\vec{p})$$

where the distorted spectral function is defined as

$$S^D(E_m,\vec{P}_m,\vec{p}) = \sum P_\alpha(E_m) | \phi_\alpha^D(\vec{P}_m,\vec{p}) |^2$$

with

$$\phi_\alpha^D(\vec{P}_m,\vec{p}) = \int X_p^{(-)*} (\vec{r}_p) e^{i\vec{q}\cdot\vec{r}_p} \phi_\alpha(\vec{r}_p)d\vec{r}_p$$

In practice the distorted proton wave is generated in an optical model framework. The parameters of the potential well are chosen such that the *elastic* scattering of protons off the (A-1) nucleus is well reproduced. As is well known, the optical potential consists basically of a real part V(r) and an imaginary part W(r). The real part parametrizes essentially the elastic small-angle scattering that gives rise to fairly moderate distortion effects in the spectral function. Qualitatively the effect can be estimated by a local wavelength approximation: Let $\overline{V}$ be some average (attractive) potential in which $\vec{p}$ moves, transform $\vec{p}$ to an effective local momentum $\vec{p}_D$ by using $\vec{p}_D^2/2M = \vec{p}^2/2M - \overline{V}$;

therefore $\vec{p}_D \approx \vec{p}(1-M|\overline{V}|/\vec{p}^2)$. Now, the observed missing momentum is $\vec{p}_m = \vec{q} - \vec{p}_D$, whereas the 'real' $\vec{p}_m^* = \vec{q} - \vec{p}$. Therefore, $\vec{p}_m = \vec{p}_m^* + M|\overline{V}|\vec{p}/\vec{p}^2$. Thus in parallel kinematics the distortion corresponds to a shift of the momentum distribution along the p_m- axis. The reactive part of the p-(A-1) cross section involves inelastic excitation, two- and more-nucleon emission processes. These catastrophic events that deplete the flux in the outgoing channel at a given E_m value are approximated by the absorption due to the imaginary part of the optical potential.

Essentially the flux reduction can be expressed by an absorption factor

$$\eta_\alpha(p) = \frac{\int d\vec{p}_m |\phi_\alpha^D(\vec{p}_m,\vec{p})|^2}{\int d\vec{p}_m |\phi_\alpha(\vec{p}_m)|^2} .$$

In Fig. 7 is shown the effect of distortions on the 2p momentum distribution of ^{90}Zr for a 100 MeV outgoing proton. The calculated absorption factor for $T_p = 100$ MeV is typi- cally $\eta \approx 0.7$ for ^{12}C and amounts to $\eta \approx 0.5$ for the extreme case of knockout from the $3s_{1/2}$ state in ^{208}Pb, i.e. rather moderate effects are involved. The treatment of the absorptive processes by an elastic optical potential is very crude and rests on assumptions that have not yet been tested in detail. In addition it is known that the imaginary poten- tial is quite poorly determined from elastic scattering. Its determination should be made from the reactive or total (p-A) cross section data, a procedure that remains to be ela- borated. It is in the description of the fate of the knocked out proton on its way through the nucleus that theoretical progress is most urgently needed in order to reduce the present uncertainty ($\approx 30\%$) in the extracted absolute spectroscopic factors. In addition the following approach is available. Experimentally, the kinematic freedom of the (e,e'p) reaction allows one to put constraints on the FSI description by measuring a spectral function at different T_p-values. We will discuss examples of this in section 3.

Let us finally discuss the sensitivity in r-space of the (e,e'p) reaction and compare the observations with the radial-sensitivity function of various other reactions. We consider the following reactions that may be used to determine single-particle wave functions. *Electromagnetic:* (e,e'p), the magnetic elastic (e,e') of high multipolarity[13], the differ- ence of the charge scattering (e,e') cross sections from isotone pairs and the *hadronic:* one nucleon transfer (i.e. proton pickup) reaction. The scattering amplitudes or, equivalently, the form factors pertinent to each of the reactions considered, are summar- ized in Table I. For the electromagnetic probes the specific sensitivity in r-space is

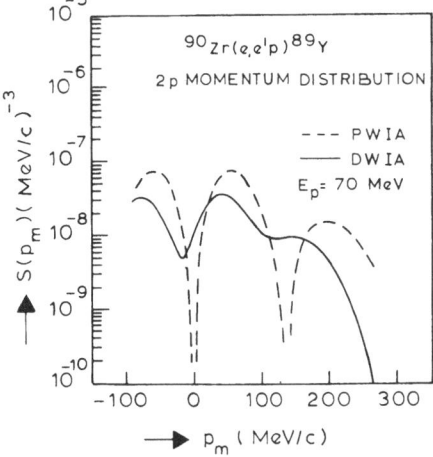

Fig. 7   Final state interaction effect for the ^{90}Zr(e,e'p) reaction.

Table I.   Survey of the scattering amplitude expressions for various reactions, as an
             illustration of the radial sensitivities involved.

Reaction	Scattering amplitude or form factor
High multipole magnetic (e,e′) ($\Lambda = 2j_\alpha$)	$F_{M\Lambda}(q) \propto q\mu_n \int\limits_0^\infty j_{\Lambda-1}(qr) R_\alpha^2(r) r^2 dr$
Charge scattering from isotone pairs $[\Delta\rho = \rho(Z) - \rho(Z-1)]$	$F_{C0}(q) \propto \int\limits_0^\infty j_0(qr) \Delta\rho(r) r^2 dr$
Quasi free (e,e′p)	$\sqrt{S_\alpha(p_m)} \propto \int\limits_0^\infty j_{l\alpha}(p_m r) R_\alpha(r) r^2 dr$
(d,^{3}He) ($T_d = 80$ MeV, $\theta = 10°$)	$\sqrt{\sigma_\alpha(\theta)} \propto \int\limits_0^\infty f_s(r,\theta) R_\alpha(r) r^2 dr$

determined by the value of the momentum transfer. In the majority of cases one avails of
data in a large range of momentum transfer and thus the sensitivity window is modulated
over a region of r-space. In order to illustrate the typical spatial sensitivity of the electron microscope we show in Figs. 8-10 the sensitivity curves of the various reactions.
Note that the sensitivity functional is defined as $\Delta\sigma/\Delta r$ where

$$\sqrt{\sigma} \propto \int\limits_r^\infty f_s(r)R_\alpha(r) r^2 dr \ .$$

The most conspicuous feature of the sensitivity curve for the proton pickup reaction,
illustrated in Fig. 10 for the case of ^{208}Pb(d,^{3}He)^{207}Tl at an angle $\theta_{3_{He}} = 10°$, is the following. Due to the strong absorption acting on both the projectile and the ejectile, the
bound state wave function is sampled in the asymptotic tail. This in essence explains the
extreme model dependence of the spectroscopic factors $C^2S \propto R_\alpha^2$ (r = large) deduced
from these reactions. For instance, adopting mean field Woods-Saxon model wave functions, the dependence of C^2S on the rms radius of the wavefunction is given by
$\Delta C^2S/C^2S \approx (10-15)$ Δrms/rms. Nonetheless the hadronic reactions are very attractive in
that the cross sections are large, the availability of polarized $\vec{d}$-beams offers the

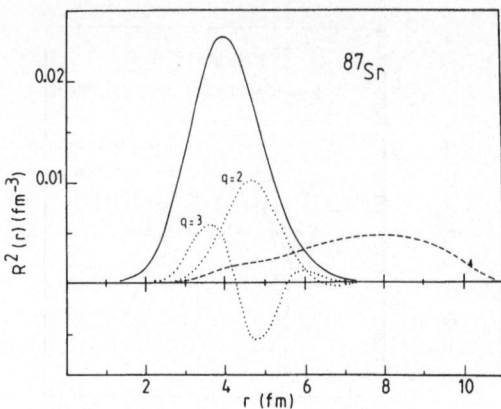

Fig. 8   1g$_{9/2}$-density and radial sensitivity curves for the (e,e′)$_{magn.}$ and the (d,^{3}H) reaction (long-dashed).

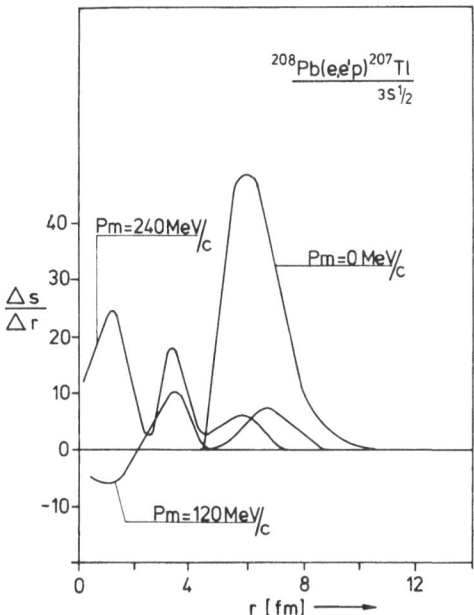

Fig. 9   Radial sensitivity curves for the ^{208}Pb(e,e′p) reaction for the various values of
P_m.

possibility to make the j-assignment for the picked-up proton, and the reaction has good
energy resolution (50-100 keV). Our philosophy therefore is to try and exploit the particular advantages of each of the reactions in a combined analysis of data in order to obtain
a more unambiguous picture of the single-particle structure physics probed with these
reactions. We will give one example of such an endeavor in section 3.

3. EXPERIMENTAL RESULTS

3.1 High-momentum components and scaling in ^{3}He

Since realistic wave functions can be constructed with the Faddeev-method, the
three-nucleon system constitutes a principal case in the search for correlations and high-
momentum components in nuclei. The probability of such high-momentum components
is closely related to the dynamics of the nucleon-nucleon interaction at short distance,
that is poorly understood at present.

The experimental situation in this issue is rather confused. The proton momentum
distribution ρ(k), extracted from ^{3}He(e,e′p) data[14] up to k = 300 MeV/c does not show an
excess probability relative to the Faddeev calculation. In contrast, inclusive quasi-elastic
data at large momentum transfer q at the low-energy tail of the q.e. peak, when interpreted in a y-scaling approach, indicate an appreciable enhancement of momentum components in the range k > 300 MeV/c. It can be shown[15] that the inclusive quasi-elastic
cross section $d\sigma/d\Omega_e d\omega$, which is a function of three momentum transfer q and energy
transfer ω, depends in the limit q → ∞ on a single variable y = y(q,ω). This scaling vari-
able $y \approx (\omega^2 + 2m_n\omega - q^2)/2q$ represents the component in the initial momentum $(\vec{k})$ of the
nucleon parallel to q, $y = \vec{k}\cdot\vec{q}/|q|$. The cross section can then be written as

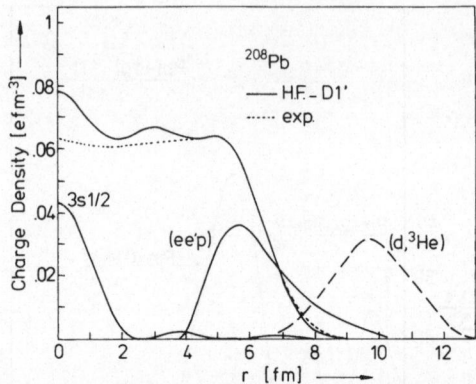

Fig. 10   Experimental and theoretical densities of total charge and of the $3s_{1/2}$ orbital.
Radial sensitivity curves are indicated for the $(d,^3He)$ and the $(e,e'p)$ reaction at
$p_m = 10$ MeV/c.

$$\frac{d\sigma}{d\Omega_e \, d\omega} = (Z\sigma_{ep} + N\sigma_{en})\frac{dy}{d\omega} \, F(y)$$

where the function $F(y)$ represents the probability to find nucleons with momentum y in
the nucleus and σ_{ep} (σ_{en}) is the elementary electron-proton (-neutron) cross section.
However, it should be noted that controversy exists as to whether the scaling interpretation is valid at the relatively small q-values of the experiment[16]. Alternative explanations
of the experimental data without invoking enhanced high momentum components have
been forwarded. Pirner and Vary[17] reproduce the (e,e') data well without invoking
enhanced high momentum components by using a quark-cluster model for ^{3}He, in which
the energy-momentum of the virtual photon is absorbed by three-, six- and nine-quark
clusters.

Using a diagrammatic expansion of the scattering amplitude J. M. Laget[18] has proposed that part of the excess cross section may be due to the knockout of correlated (p-n)
pairs. The proposed mechanism has indeed been empirically supported recently by a
coincident ^{3}He(e,e'd) ^{1}H study, in which deuterons were observed in the direction of
three-momentum transfer $\vec{q}$.

In the absorption of virtual photons two main mechanisms can be envisioned, absorption by either a proton, $(e,e'p)$, or a correlated pn-pair which eventually recombines into a
deuteron, $(e,e'd)$. In the Plane Wave Impulse Approximation (PWIA) the two-body
breakup coincidence cross section for $(e,e'p)$ is given by $d\sigma/de'd\Omega_e d\Omega_p = K_p\sigma_{ep}S_p(k)$,
where K_p is a kinematical factor, σ_{ep} is the elastic electron-proton cross section and $S_p(k)$
the spectral function for the two-body breakup of ^{3}He at recoil momentum k. Likewise
for the $(e,e'd)$ reaction one has $d\sigma/de'd\Omega_e d\Omega_d = K_d\sigma_{ed}S_d(k)$, where σ_{ed} is the elastic
electron-deuteron cross section in which only active pn-pairs with T = 0 are considered.
For the two-body breakup channel one has: $S_p(k) = S_d(k)$. In parallel kinematics the
knocked-out deuterons have momentum d = k + q. For relatively large values of the
momentum transfer recoil deuterons have a largely suppressed probability since
S(d)<<S(k), an effect only partially counter balanced by σ_{ed} being smaller than σ_{ep}. Two
kinematics were used in the experiment[19].

I.   The momentum transfer was kept constant at q = 380 MeV/c and the recoil-
momentum range k = 0 - 200 MeV was covered in four intervals.

II. At fixed recoil momentum (k = 60 MeV/c) the momentum transfer was varied in the range of 350 to 450 MeV/c.

In Fig. 11 the experimental cross sections, averaged over 10 MeV/c momentum bins, are shown (kinematics I). The solid curve represent a PWIA calculation in which the photon couples to a proton or a pn-pair (in either a T = 0 or a T = 1 state) as illustrated by the diagrams shown in the inset. The cross section calculated with the photon-proton coupling diagram alone is largely suppressed (dotted curve) in agreement with the schematic model discussed above. In a diagrammatic approach[18] the first order diagrams in Fig. 11 were included in the cross section calculation (dashed curve). The diagrams include the meson exchange effect (diagram c), proton rescattering (d) and deuteron rescattering (e).

The data follow with fair agreement the fall-off of the calculated cross sections with recoil momentum. At the lowest recoil momenta the data fall below the calculation including the FSI, while at higher recoil momenta the calculation does represent the data fairly well. The data of kinematics II are shown in Fig. 12. Again the calculation including FSI (dashed curve) describes the data well. The q-dependence of the cross section follows that of the elastic electron-deuteron cross section σ_{ed}.

In conclusion the present (e,e'd) coincidence data constitute evidence for the direct coupling of the virtual photon with correlated nucleon pairs. This mechanism has been proposed to account for part of the excess cross section observed in the small-ω region of

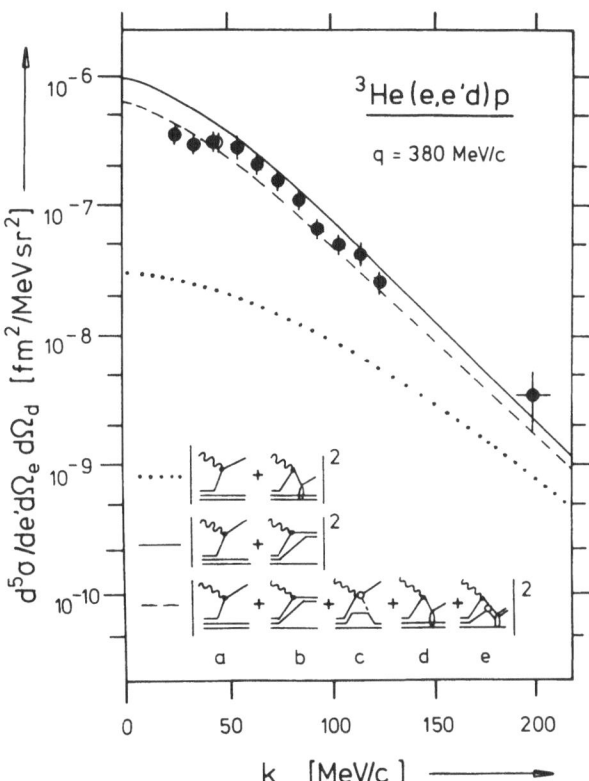

Fig. 11 (e,e'd) cross sections as a function of recoil momentum K at fixed momentum transfer. Calculated curves correspond to the diagrams shown.

the inclusive (e;e′) response function, especially at relatively small momentum transfer (q = 0.45 GeV/c) (see Fig. 13). The implication of the results of the coincidence experiment for the interpretation of the inclusive experiments solely in terms of an excess of high-momentum components in the ^{3}He wave function clearly merits further study.

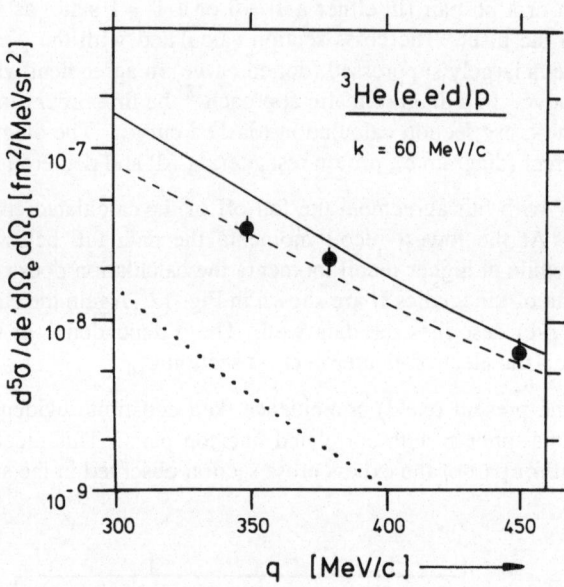

Fig. 12   Cross sections as a function of q at fixed recoil momentum K = 60 MeV/c.

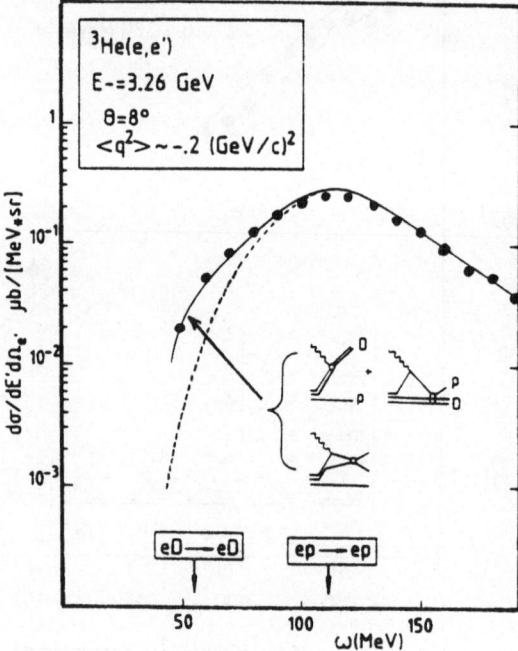

Fig. 13   Inclusive ^{3}He(e,e′) quasi elastic cross section. The dashed curve is calculated in PWIA with a Faddeev momentum distribution. The solid curve includes the photon-(pn) coupling mechanism.

3.2 Occupation Probabilities and Correlations in the 1p-shell

In spite of the impressive success of the shell model in the 1p-shell, the approximate nature of shell-model wave functions is indicated by the need for effective operators in the calculation of, say, electromagnetic observables. The renormalized (effective) operators correct for the lack of correlation in the s.m. wave function. Experimental information on wave-function components beyond the standard $0\hbar\omega$ configuration is of importance since speculations about the role of mesonic degrees of freedom in e.g. the spin-isospin response function of nuclei depend critically upon the assumed wave function in the nucleonic sector[20].

Recently, an interesting and novel theoretical development has been proposed by Glaudemans et al[21] for 1p-shell nuclei. The configuration space is spanned by harmonic oscillator single-particle states (1s, 1p, 1d2s, and 1f2p), *no* inert core is assumed, and spurious states can be completely eliminated. The most notable feature of the calculation is the large amount of correlations, as is evidenced by a $\approx 20\%$ depletion of the standard $0\hbar\omega$ 1p-shell wave function. Experimental confirmation of the validity of the proposed wave functions is clearly called for but is as yet scant.

The occupation probability of shell-model orbits occupied by the least bound nucleons near the Fermi edge could be assessed if experimental methods were available to determine *absolute* spectroscopic factors. Unfortunately the uncertainty inherent in the treatment of the reaction mechanism involved in experiments with strong interaction probes (i.e. in pick-up (d,^{3}He) and knockout reactions) has thus far prevented obtaining sufficiently reliable absolute spectroscopic factors.

An alternative, though less direct, method would be to observe the knockout of particles from normally unoccupied states located above the Fermi edge. In Fig. 14 is shown a high-resolution spectrum of the ^{12}C(e,e′p)^{11}B reaction. The well-known states at $E_x(J^\pi)$ 0 Mev (3/2⁻), 2.12 MeV (1/2⁻) and 5.02 Mev (3/2⁻) in ^{11}B correspond to 1p proton knockout and their momentum distributions show the expected $l = 1$ behaviour. At $E_x = 6.79$ MeV the known $J^\pi = 1/2^+$ state is excited with appreciable strength. This state was clearly observed in all four (e,e′p) spectra[22] measured in a p_m-range of -20 MeV/c to +110 MeV/c. Its momentum distribution, shown in Fig. 15 exhibits a typical $l = 0$

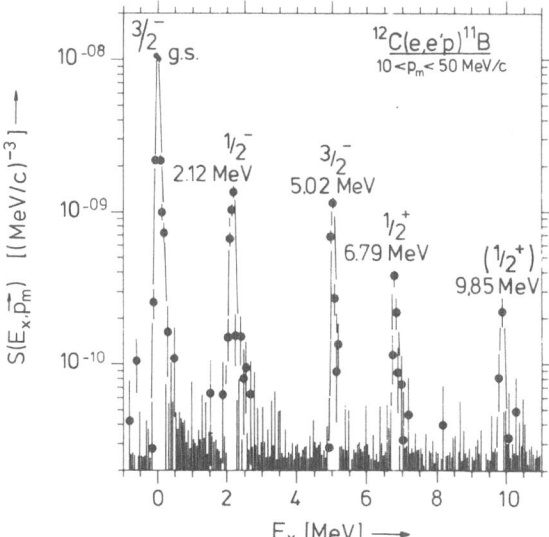

Fig. 14   Excitation spectrum of ^{11}B observed in the (e,e′p) reaction.

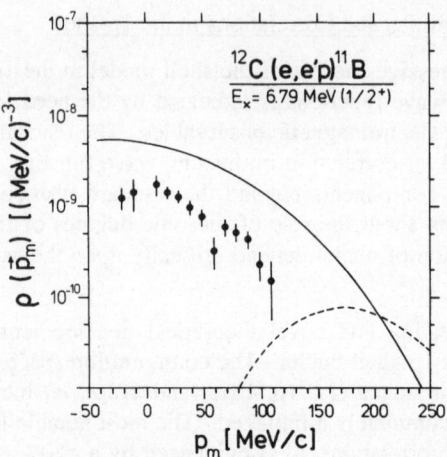

Fig. 15   Momentum distribution of the $j^\pi = 1/2^+$ state.

behaviour, indicating the knockout of a proton from a $s_{1/2}$ orbit. In order to see what configurations may contribute to such a transition from the $J^\pi = 0^+$ ground state of ^{12}C, consider the schematic picture of possible configurations in a $(2\hbar\omega)$ space, shown in Fig. 16. The major connections between (n) and (n+1)$\hbar\omega$ configurations are indicated by arrows. In the analysis of the momentum distribution possible contributions from the unresolved $7/2^-$ state at 6.74 MeV have been estimated as follows. With the extreme assumption of 100% overlap between this state and the $1f_{7/2}$ component of the ^{12}C ground state and using a calculated $1f_{7/2}$ amplitude of 0.10, the $l = 3$ momentum distribution of the $7/2^-$ state can be calculated. As is shown in Fig. 15 (dashed curve) the largest contribution to the $1/2^+$ state cross section occurs at $|\vec{p}_m| = 110$ MeV/c and amounts to 20%. The data have been interpreted in the framework of distorted wave impulse approximation (DWIA) with an optical model description of the FSI. The ground state wave function was evaluated in the shell-model. Recently, it has been shown by Glaudemans et al[21] to be possible to perform shell-model calculations in the mass region A = 4-16 in a large $(3\hbar\omega)$ configuration space with complete elimination of spurious states. The space is spanned by 1d2s and 1f2p configurations, albeit with restrictions on the total number of oscillator quanta involved.

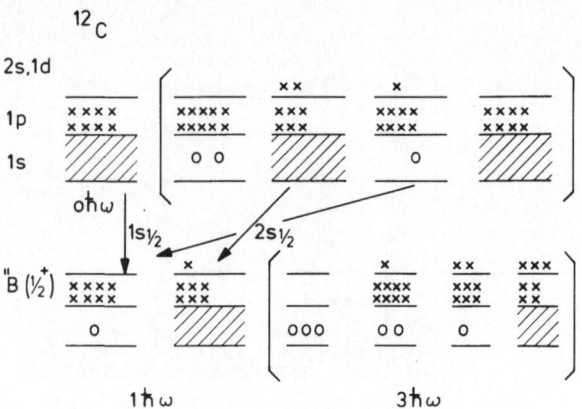

Fig. 16   Diagram showing the major configurations that may contribute to the ^{12}C$_{gs} \rightarrow$ ^{11}B$(1/2^+)$ transition.

The overlap integral of the ^{12}C ground state and the ^{11}B $1/2^+$ wave function can be written as a coherent sum of $1s_{1/2}$ and $2s_{1/2}$ single-particle wave functions

$$<\psi(^{11}B_{1/2}+) \mid \psi(^{12}C_{g.s.})> = A_{1s}\phi(1s) + A_{2s}\phi(2s) \quad .$$

The square of the amplitudes A_{1s} and A_{2s} are the spectroscopic factors C^2S for the orbitals involved. In the spectral function $(\mid<\mid>\mid^2)$ an interference term occurs. For ^{12}C a $2\hbar\omega$ model space has been used, which resulted in the following major components of the ground state wave function:

$$0.80(1s)^4(1p)^8 + 0.33(1s)^3(1p)^8(2s1d)^1 + 0.25(1s)^4(1p)^6(2s1d)^2 \quad .$$

The most notable feature of this calculation is the large depletion of the 1p-shell, which is indicated by the relatively small probability (64%) of the standard $(1s)^4(1p)^8$-component. Furthermore the ^{4}He-core has a remarkably large $1s_{1/2}$ hole content of 10%; the $2s_{1/2}$-orbit has an occupancy of 2.2%. The wave function for the $1/2^+$ state in ^{11}B was evaluated in a smaller ($1\hbar\omega$) space. As shown in Fig. 15 the shape of the calculated momentum distribution is in fair agreement with the data, considering the as yet rather restricted model space used. It is interesting to note that a second thus far unknown $1/2^+$ state at $E_x = 9.85$ MeV has been observed in the experiment in fair agreement with its model counterpart calculated at $E_x = 9.2$ MeV. It is to be noted that the calculated summed strength of the two $1/2^+$ states is quite close to the empirical value.

The analysis discussed so far rests on the assumption that the fairly weakly populated $1/2^+$ states are excited in a one-step direct knockout process. The role of the two-step processes, i.e., the $l = 1$ knockout of a proton, followed by inelastic excitation of the residual nucleus can be investigated by studying the transition to the $5/2^-$-state at $E_x = 4.45$ MeV. In the most restricted shell model space of $(1s)^4(1p)^8$ this state should not be populated since the $1f_{5/2}$-subshell is empty in ^{12}C and the $5/2^-$ state can therefore not be populated in a direct one-step process. In the $^{12}C(e,e'p)^{11}$B reaction parallel kinematics were chosen such that at $p_m = 120$ MeV/c (close to the maximum of a $l = 1$ momentum distribution) protons were detected at a relatively small kinetic energy $T_p = 40$ MeV. This energy was selected because in general the probability of two-step processes tends to be larger at smaller energies[23]. The remarkable feature of the observed spectrum, shown in Fig. 17, is the absence of strength at $E_x = 4.45$ MeV, the location of the $5/2^-$ state.

Recent calculations in an extended ($2\hbar\omega$) configuration space indicate an amplitude of less than 0.05 for the $1f_{5/2}$ component in the ^{12}C ground-state wave function[21]: The calculated overlap $<^{11}B_{5/2^-} \mid ^{12}C_{g.s.}>$ yields a momentum distribution of

$$\rho(\vec{p}_m) = \int_{\delta E_x} S(E_x,\vec{p}_m)dE_x < 2.10^{-11}(MeV/c)^{-3}$$

corresponding to a spectral function $S(E_x,\vec{p}_m) < 7.10^{-11}$ $(MeV/c)^{-3}$ MeV^{-1}, which is consistent with the non-observance of the state. Consequently the absence of strength at $E_x = 4.45$ MeV can be used to set an upper limit to the amount of two-step processes in the $^{12}C(e,e'p)^{11}$B reaction. From the data shown in Fig. 17 we deduce that there is at least a factor of 30 difference between the excitation strength in the region of $E_x = 4.45$ MeV and the one of the $3/2^-$ state at 5.02 MeV. The result can be expressed in terms of a maximum contribution to the momentum distribution $\rho(\vec{p}_m)$ of 7.10^{-11} $(MeV/c)^{-3}$ to be compared with $\rho(\vec{p}_m) = 1.10^{-9}$ $(MeV/c)^{-3}$ for a typical weak transition. Hence, even when weak transitions are involved, the interpretation of the $(e,e'p)$ reaction appears to be free to a large extent from complications due to two-step processes.

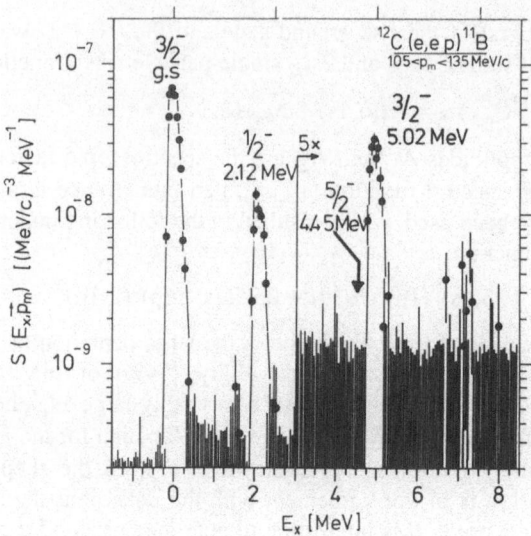

Fig. 17   Spectrum of the ^{12}C(e,e'p) ^{11}B reaction, showing the absence of strength in the region $E_x = 4.45$ MeV.

This observation may be compared with the results of a (p,2p) experiment shown in Fig. 18, in which an incoming proton energy of 100 MeV was used[24] and the energy of the outgoing protons (40 MeV) was the same as in the present (e,e'p) experiment. The $5/2^-$ state was found to be excited in the (p,2p) reaction with one third of the strength of the nearby $3/2^-$ state with an angular distribution characteristic for $l = 1$ knockout. In this hadronic reaction two-step processes are expected to proceed through two main channels:

i.   Inelastic excitation of the target nucleus ^{12}C($0^+ \rightarrow 2^+$) and subsequent $l = 1$ knockout;

ii. $l = 1$ knockout leading to the ^{11}B ground state followed by inelastic excitation of ^{11}B($3/2^- \rightarrow 5/2^-$) by the outgoing protons.

Since the total reaction cross section for proton-nucleus scattering increases with decreasing T_p and two protons contribute to the inelastic excitation in the final state, it has been suggested[23] that the major contribution comes from the latter channel. Adopting the dominance of channel (ii) and assuming the two protons contribute coherently to the inelastic excitation, the probability of populating the $5/2^-$ state in (p,2p) should be a factor four larger than in the (e,e'p) reaction. However, from the present experiment, we infer a difference of at least a factor of ten. A more detailed investigation of the coupled-channel effects in the (p,2p) reaction also involving the entrance channel will be needed for a better understanding of this observation.

3.3 Fragmentation of 1f-hole Strength in ^{90}Zr

The main motivation to study processes in which one nucleon is removed from inner shells is to investigate to what extent the concept of the Independent Particle Shell Model (IPSM) is valid for states well below the Fermi surface. In the study of proton-hole states both knockout - (p,2p), (e,e'p) - and pick-up (d,τ), (τ,α) reactions have been employed. Pick-up reactions are mainly sensitive to the asymptotic tail of the bound-state wave function $\phi_\alpha(r)$, whereas knockout reactions probe the entire wave function. The principal

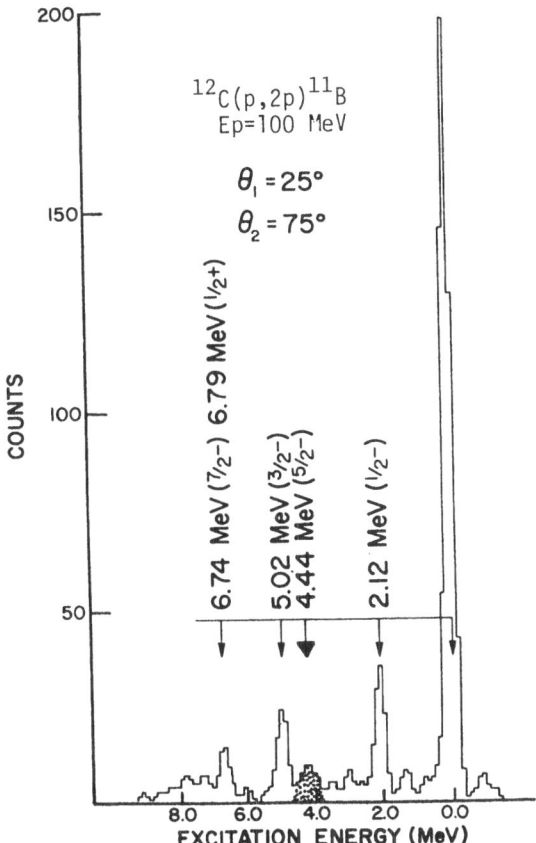

Fig. 18 ^{12}C(p,2p) spectrum with an outgoing proton energy T_p = 40 MeV. The 4.45 MeV state is excited with appreciable strength.

observables in these reactions are the spectroscopic factor C^2S_α and the spreading width Γ.

Nuclei in the vicinity of the neutron shell closure N = 50 are particularly interesting because various different quantum states (1g,2p,1f) are quite close energetically in this region.

Cross sections for the reaction ^{90}Zr(e,e′p)^{89}Y were obtained[25] in parallel kinematics in which the proton ($\vec{p}$) is detected in the direction of the momentum transfer $\vec{q}$ ($|\vec{p}| < |\vec{q}|$). In order to achieve complete coverage of the 0 to 300 MeV/c missing momentum region two proton energies were used: T_p = 70 MeV for the $0 < p_m < 170$ MeV/c region and T_p = 100 MeV for the $140 < p_m < 300$ MeV/c region. The missing energy region covered was $5 < E_m < 28$ MeV ($-3 < E_x < 20$ MeV). A typical missing energy spectrum is shown in Fig. 19.

The population of the low-lying states with J^π 1/2$^-$(g.s.), 9/2$^+$ (0.91 MeV), 3/2$^-$ (1.151 MeV), and 5/2$^-$ (1.75 MeV) is observed. We discuss here the fragmentation of l = 3 (1f$_{5/2}$,1f$_{7/2}$) strength at larger excitation energy. Momentum distributions were obtained by integration over properly chosen intervals ΔE_m.

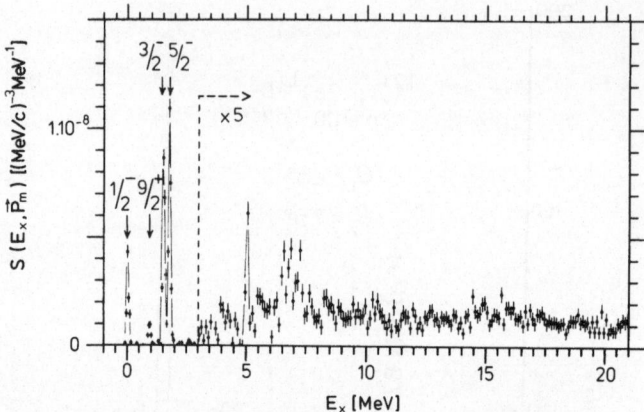

Fig. 19   Excitation energy spectrum of the ^{90}Zr(e,e′p) ^{89}Y reaction at p_m = 125 MeV/c.
Note the break in the vertical scale at E_x = 3.0 MeV.

The FSI effect was calculated in Distorted Wave Impulse Approximation (DWIA) where the knocked-out proton moves in the optical potential of ^{89}Y for which we used a Woods-Saxon type with only volume and surface terms. The potential parameters were taken from a systematic study[26] of the energy dependence of optical potential parameters in the energy region 80<T_p<180 MeV for masses 24<A<208. The bound-state wave functions were calculated in a Woods-Saxon potential, which reproduces the rms radius of ^{90}Zr and the location of the $1f_{7/2}$, $1f_{5/2}$, $2p_{3/2}$, and $2p_{1/2}$ shells as found from Hartree-Fock calculations.

In Fig. 20 the experimental momentum distributions are shown for transitions to the $5/2^-$ level at 1.745 MeV, and to the excitation energy region between 3.0 and 11.3 MeV. The curves (solid for T_p = 70 MeV, dashed for T_p = 100 MeV) were calculated with an incoherent sum of 1f, 2p, 2s and 1d components. Their strengths C^2S were kept equal for the two T_p-sets in the fit. The deduced spectroscopic factor for the $1f_{5/2}$ transition at 1.745 MeV is C^2S_{1f} = 2.6. Fig. 20 shows that for higher excitation energies the spectral function still shows a clear 1f-type momentum distribution at p_m > 50 MeV/c. However, at lower missing momentum a significant low-l admixture is needed to describe the data. The fragmentation $C^2S(E_x)$ of the 1f-strength is shown in Fig. 21. It appears that the 1f-strength extends well beyond 10 MeV excitation energy and that the enhanced structure observed in the missing-energy spectrum between 6 and 8 MeV corresponds to the maximum of the 1f fragmentation. The mean excitation energy $\bar{E}_x = <C^2S(E_x) \cdot E_x>$ and spreading width $\Gamma = <C^2S(E_x)\cdot(E_x-\bar{E}_x)^2>^{1/2}$ are respectively $\bar{E}_x$ = 6.0(4) MeV and Γ = 4.1(1.4) MeV. The total 1f strength from 0 to 20 MeV amounts to 8.9 compared with the total sumrule value of 14. This is in fair agreement with mean-field calculations which predict, after inclusion of short range correlations, a depletion of approximately 30% for shells just below the Fermi surface for the lead region[7].

The total 1f strength 8.9 derived from the (e,e′p) experiment is considerably smaller than the value 18.1 obtained in previous proton pickup experiments[27]. The discrepancy might be partly due to the treatment of strong-interaction effects in the hadronic reaction and the strong dependency of the spectroscopic factor on the radial shape of the bound state wave function in the (d,τ) reaction. In addition we observe, unlike the (d,τ) experiment, 1f strength well beyond 11.3 MeV.

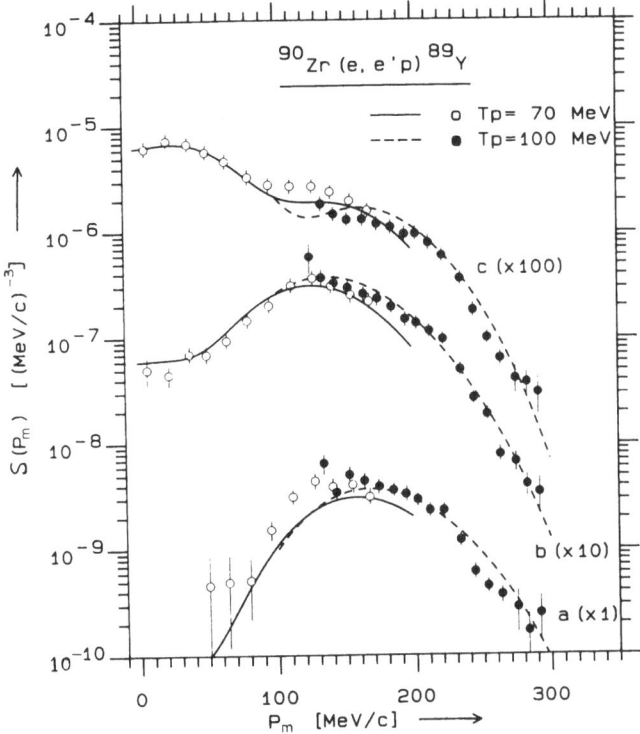

Fig. 20   Momentum distribution for $l=3$ knockout to the state at $E_x = 1.74$ MeV (c)
E_x=6–8Mev (b), and 3 - 11.3 MeV (a). DWBA best-fit curves are shown.

Although the statistical errors in the present experimental results are small and the systematic error is only 10%, we have estimated that the deduced spectroscopic factors are less accurate due to theoretical ambiguities in the treatment of the FSI. Comparison of the $T_p = 70$ MeV data with DWIA calculations using a different set of optical model parameters leads to a difference of 20% in the extracted spectroscopic factors.

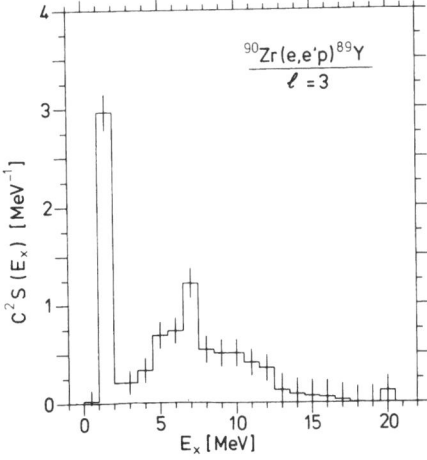

Fig. 21   Fragmentation of 1f-hole strength in ^{90}Zr. The errors indicated are statistical.

In order to obtain more insight into the origins of the optical model ambiguities we have used the (e,e'p) reaction in various different kinematics with the purpose of testing the energy- and density-dependence of the inelastic and absorption processes of the proton on its way out of the nucleus.

3.4 An (e,e'p) Study of the Proton Optical Potential

There exist at least two basic problems with the use of an optical potential, derived from elastic proton-nucleus scattering, to describe the final state interaction in (e,e'p). First the imaginary part which is most important for the FSI, is poorly determined from (p-A) scattering data and second there is no direct sensitivity to the shape of the optical model potentials inside the nucleus. However, the large kinematic freedom at hand with the (e,e'p) reaction allows one to put additional constraints on the optical potential. The basic idea is to measure a spectral function in a given p_m-range and to vary the outgoing proton energy. The prescribed energy dependence of the real and imaginary potentials can thus be checked. In addition, by selecting different quantum states near the Fermi surface, the nuclear density is sampled in a different way by each orbital and one may obtain information on both the energy dependence and the density dependence of the absorption or, phrased differently, the mean free path of protons in nuclear matter. We will discuss two applications in the following sections.

3.4.1 The Reaction $^{90}Zr(e,e'p)^{89}Y$ The experiment[28] consisted of a determination of the spectral function in a fixed range of missing momentum 120MeV/c$<p_m<$160 MeV/c, while the outgoing proton energy varied from 60 MeV to 165 MeV. The p_m-range chosen allowed one to study simultaneously knockout from the $2p_{3/2}$, $2p_{1/2}$, $1g_{9/2}$ and $1f_{5/2}$ orbit, leading to the four low-lying states in ^{89}Y (see Fig. 19).

The $l = 3$ ($1f_{5/2}$) and $l = 4$ ($1g_{9/2}$) transitions are mainly sensitive to the optical potential at the nuclear surface (around r = 5 fm) whereas the $l = 1$ transitions receive a substantial contribution from the first maximum of the 2p wave function at r = 1.5 fm.

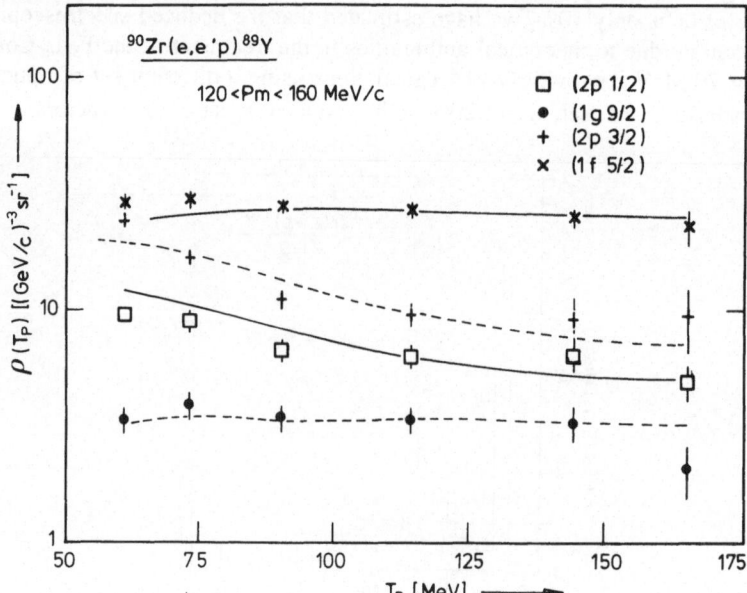

Fig. 22   Dependence on the proton energy of spectral functions for various surface states in ^{90}Zr.

Preliminary results for the T_p-dependence of the spectral function for the four states are shown in Fig. 22. The data are compared with a DWIA calculation which contained the following ingredients:

    i. 1f, 1g and 2p bound state (BS) wave functions as established from our earlier (e,e′p) data measured in the complete $0<p_m<300$ MeV/c missing momentum range.

    ii. Optical potentials with real and imaginary volume and spin-orbit terms as deduced from the parametrization for 80 to 180 MeV protons by Schwandt et al.[26].

    iii. A non-locality correction for both the bound-state and scattering state wave functions.

For ease of comparison we have normalized the DWIA curves (i.e. varied the individual spectroscopic factors) to the $T_p = 114$ MeV experimental point. The 1g data and calculation agree over the whole T_p range. This implies that the optical potential used has the correct energy dependence at the nuclear surface. For the $1f_{5/2}$ orbit, that peaks more inside the nucleus, deviations are observed at low energies. Here the absence of a surface term in the optical potential may play a role. More pronounced effects are observed for the $l = 1$ transitions. DWIA curves neither reproduce the trend of the experimental data for the individual transitions, nor do they reproduce the energy dependence of the ratio of $2p_{3/2}$ to $2p_{1/2}$ strength. These observations possibly indicate the inadequacy of the optical potential with Woods-Saxon shape in the nuclear interior. It is interesting to note that microscopic calculations[30] of the optical potential (of either the Dirac type or the non-relativistic type) predict such non-standard forms. Further detailed analysis is underway[29] to define the origins of the observed deviations from simple optical model predictions.

3.4.2 <u>The Reaction</u> $^{12}C(e,e′p)^{11}B_{g.s.}$ The spectral function for $l = 1$ knockout of the $1p_{3/2}$ proton leading to the ^{11}B ground state has been measured at an outgoing proton energy $T_p = 70$ MeV. At the maximum of the momentum distribution at $p_m \approx 100$ MeV/c cross sections were measured at three values of T_p: 40, 70 and 100 MeV. The observed energy dependence, shown in Fig. 23 is not reproduced by the cross section that was calculated with the optical potential determined by Jackson et al.[31] from a fit to $(p-^{12}C)$ elastic scattering data for a range of proton energies. The reason for the discrepancy may be found in the inadequacy of a $(p-^{12}C)$ optical potential to describe the $(p-^{11}B)$ final state interaction. In contrast the calculation with a potential, determined by Schwandt[26] from a systematic mass (A = 24 - 208) and energy dependence study of

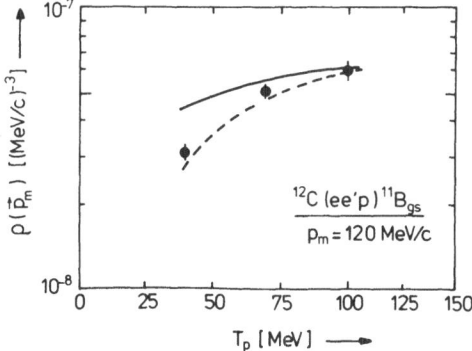

Fig. 23 Energy dependence of the $l=1$ spectral function for the reaction $^{12}C(e,e′p^{11}B_{g.s.})$. Two optical model calculations for the FSI are shown. Dashed line corresponds to Schwandt's potential[26].

elastic proton scattering, yields a good fit. Note that the calculated curves were normal-ized at the $T_p = 100$ MeV point. This example shows that the T_p-dependence of the spectral function reduces the ambiguity of the available optical potential prescriptions, and thus may be used to constrain the uncertainty of the absolute spectroscopic factors that can be deduced from (e,e'p) data.

3.5 Occupation Probability of the $3S_{\frac{1}{2}}$ -Orbit in the Pb-Region

The notion that the least-bound orbits at the Fermi-surface may be depleted to a much larger extent than is usually purported by shell model practitioners has been supported by evidence from various different sources. If the occupation probability of a quantum state with total angular momentum j in a closed-shell nucleus is appreciably smaller than (2j+1), this should most clearly show up in electron scattering form factors of large mul-tipolarity. This is so because transition operator renormalization due to configuration mixing comes mainly from the depletion of the orbit contributing to the observable. Indeed, the observed magnetic and electric form factors of multipolarity $\lambda = 10, 12$ and 14 in the lead region indicate a $\approx 30\%$ depletion of orbits. A similar observation holds for the M9 form factor of elastic scattering from ^{209}Bi. Additional information may be extracted from the difference of the charge density ρ of ^{206}Pb and ^{205}Tl, $\Delta\rho = \rho(206) - \rho(205)$. Such information from the Saclay experiment[32] was the first evi-dence that the very notion of a single-particle orbit makes sense even in the interior of a heavy nucleus. Upon adding one proton to ^{205}Tl the original density changes slightly through the monopole core polarization mechanism ($\rho(205) \rightarrow \rho^*(205)$). Furthermore, one has

$$\rho(205) = \sum_i P_i \, | \alpha_i^{-1} > \otimes Pb(206, J_i) |^2$$

with

$$\sum_i P_i = 1 \ .$$

The core polarization effect $\Delta\rho^{cp} = \rho(205) - \rho^*(205)$ can be calculated to a good approximation with Hartree-Fock theory provided that pairing effects are taken into account. The level ordering of the spherical shell model prescribes that $\alpha = 3S\ 1/2$ should be a prime component of the ^{205}Tl ground state with $J^\pi = 1/2^+$. This component has a characteristic signature in the form factor at $q \cong 2\text{fm}^{-1}$. It has been found[32] that by assuming $P_{2d} = 0.30$ and $P_{3S} = 0.70$ in the expression for $\rho(205)$, given above, fair agree-ment with the data in both r-space and q-space can be obtained. However, at present is is not clear how to derive an absolute occupation probability for the $3S_{1/2}$ orbit from the difference measurement $\Delta\rho$, since appreciable correlations may occur in both the ^{206}Pb and ^{205}Tl- ground state. In order to investigate this issue also in the doubly-magic nucleus ^{208}Pb, a high resolution (e,e'p) experiment on ^{208}Pb and ^{206}Pb has been per-formed recently[33]. Good resolution is a prerequisite to separate the ground state from the nearby 3/2+ state in the two nuclei, as is shown in Figs. 24 and 25.

Data were obtained for three values of missing momentum at which different parts of the $3S_{1/2}$ radial wave function are sampled (see Fig. 9). Special attention has been paid to obtain accurate relative cross sections, resulting in a systematic error of 1.5%. Results for the ratio of cross sections or equivalently the spectroscopic factors for the transition to the ground state are shown in Fig. 26. The dashed line is the ratio of calculated $3S_{1/2}$ distorted spectral functions, assuming equal spectroscopic factors in ^{206}Pb and ^{208}Pb, where the $A^{1/3}$ dependence of the optical potential parameters and the proper separation energies have been taken into account. This result shows that for the (e,e'p) reaction the

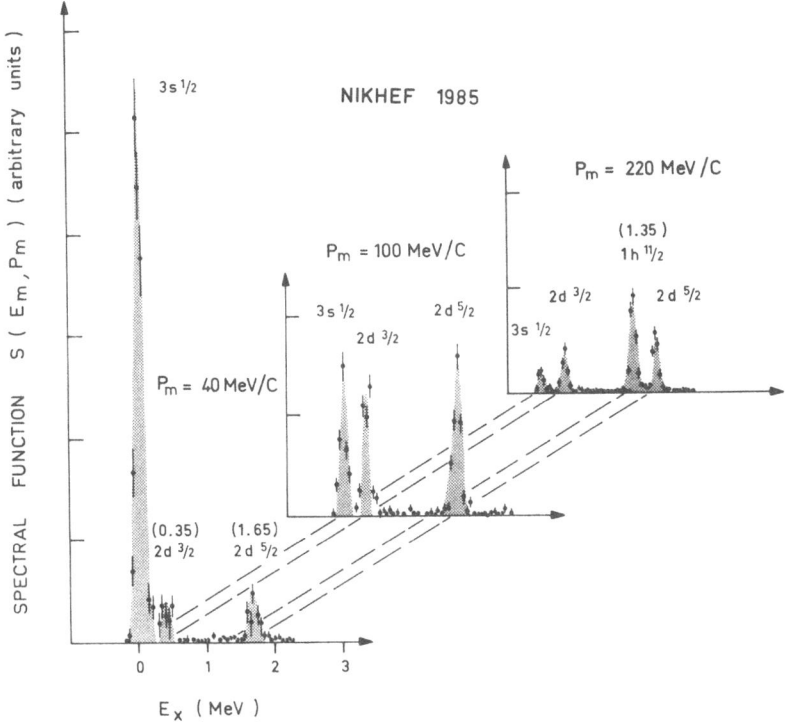

Fig. 24   Missing mass spectrum of the reaction ^{208}Pb(e,e′p) ^{207}Tl for three values of the missing momentum p_m.

model dependence of this particular representation is rather small and constitutes, in consequence, a negligible source of errors. The result for the averaged ratio of spectroscopic factors is $C^2S(206)/C^2S(208) = 0.65\pm0.03$, where the error is statistical.

Since our final goal is to establish the relative occupation probability of the $3S_{1/2}$ orbit in ^{206}Pb and ^{208}Pb, it is necessary to search for $l = 0$ knockout strength residing in excited states of ^{205,207}Tl. As is illustrated in Fig. 25, the 1/2+-state at $E_x = 1.21$ MeV in ^{205}Tl has appreciable strength in this spectrum at the small value of $p_m = 20$ MeV/c, where the $l = 0$ strength is strongly emphasized. Due to shell closure the fragmentation of $l = 0$ strength is much smaller in ^{207}Tl. Integrating the strength up to $E_x \approx 10$ MeV in both final nuclei the ratio of occupation probabilities $P(206)/P(208) = 0.85\pm0.08$, i.e. the $3S_{1/2}$ occupation probabilities is somewhat smaller in ^{206}Pb.

Let us clarify what remains to be done before absolute spectroscopic factors can be obtained, by making the following remarks. It will be necessary to avail upon redundant information to better pin down the various uncertainties of the analysis. This is being done in a collaboration with groups from Tubingen[34] and Indiana[35] where information from the pickup reactions ^{206}Pb($\vec{d}$,^{3}He) and ^{208,206}Pb(d,^{3}He) at two different energies will be used in a synthesizing analysis with (e,e′) and (e,e′p) data. It should be reiterated that provided the necessary refinements can be incorporated in the treatment of the final state interaction, the goal of absolute spectroscopic factors, derived from solely the (e,e′p) data, will come within reach. In this context we note that the use of data, involving both the reactive (σ_R) and the elastic (σ_E) cross sections, leads to a quite good determination[36]

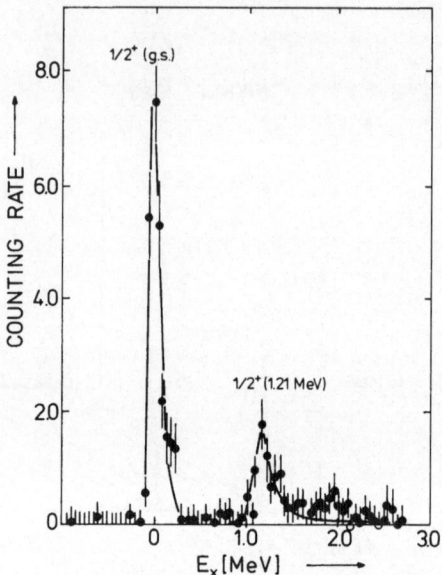

Fig. 25   Spectrum of ^{205}Tl populated in ^{206}Pb(e,e'p) ^{205}Tl reaction. At the small p_m=20MeV/c value chosen the l=0 knockout process is strongly emphasized.

of the imaginary part of the optical potential, the part of primary importance for a realistic calculation of the FSI. The above discussion of the physics has remained within the confines of traditional nuclear physics, in that the primary paradigm, (nucleons in nuclei maintain their free-space identity) is respected. In contrast, an intrigueing explanation of the (ρ(206)–ρ(205)) data has been proposed[8] in the framework of relativistic, Dirac-equation, based, mean-field theory. We will discuss this topic in the next section.

3.6 Medium Modification of Nucleons

In standard models of nuclear structure, nucleons are treated as elementary particles, i.e., one assumes that the nuclear medium does not affect their intrinsic properties (size,

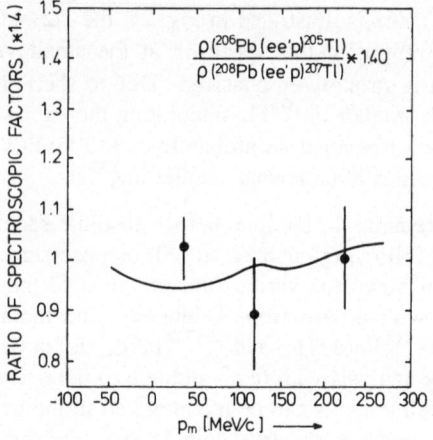

Fig. 26   Ratio of ground state spectral functions for ^{206}Pb(e,e'p) and ^{208}Pb(e,e'p) and a calculated distorted-wave prediction.

effective mass, magnetic moment). Since the nucleus is a relatively loosely-bound system one intuitively expects such medium effects to be of minor importance since the
nucleons are kinematically nearly on-shell[37].

In the framework of the recently developed relativistic mean-field theories, the situation may be drastically different. In the $(\sigma-\omega)$ model, which is a schematic representation of the Dirac-equation based relativistic theories, the average potential, felt by the
nucleon, arises from a strong, attractive, scalar field $V_s(\approx-420\text{ MeV})$ and a repulsive vector field $V_0(\approx320\text{ MeV})$. The effect of these fields can be incorporated in the bound-state
equation by modifying the mass and energy:

$$M \rightarrow M^* = M + V_s \text{ and } E \rightarrow E^* = E - V_0 \text{ .}$$

The nucleon spinors therefore change in the medium and as a consequence, also the coupling of the virtual photon to the nucleon current j_μ is modified. In the $(\sigma-\omega)$ model the
nuclear current for a proton assumes the following schematic form,

$$j_0 = \rho = F\, e^{i\vec{q}\cdot\vec{r}}$$

$$\vec{j} \sim \frac{F}{2M^*}(1+k_p\frac{M^*}{M})i(\vec{q}\times\vec{\sigma})e^{i\vec{q}\cdot\vec{r}}$$

where in the change operator ρ, the Darwin-Foldy term and in the current operator, the
convection current term have been neglected for simplicity. The form factor F may be
taken to be either the free-space form factor or a medium modified form factor. The
former is used in the standard relativistic approach in which a strong interaction vertex
form factor is introduced in order to make the quantum hadrodynamic field theory renormalizable. Shakin and co-workers[8] go one step further in giving a perception for the
change of size of a nucleon dependent on the nuclear density. The corresponding change
in the nuclear form factor $F_{(q)}$ can be appreciable, i.e. one finds, using the dipole form
factors $F_D(q) = (1+q^2/0.71)^{-2}$ at $q = 1$ fm^{-1}, that the cross section changes by as much as
16% for a 10% relative change of radius of the nucleon. This is essentially the mechanism by which the $\Delta\rho$-data, discussed in the previous section, can be explained, without a
30% renormalization of the occupation probability. In addition, the long-standing problem of the missing strength observed[38] in the Coulomb sumrule for the quasi-elastic
scattering process, is purported to be explicable by the same mechanism of inflated
nucleons. However, the transverse response function should then also be explained in the
same framework. Such a complete analysis is however seldom performed. One important issue in the interpretation of the transverse $(e,e')_{Q.E.}$ response function concerns the
role of secondary processes (e.g. two nucleon knockout, meson exchange currents)
beyond the one nucleon knockout part. The coincidence $(e,e'p)$ reaction may be used to
unambiguously select the one-nucleon knockout process because the initial and the final
state of the knocked-out proton can be selected energetically.

In order to investigate this particular aspect of the reaction mechanism and the q-
dependence of the transverse and longitudinal response function, a high-precision experiment has been recently performed at NIKHEF-K in which the $^{12}C(e,e'p)\,^{11}B_{g.s.}$ reaction
was studied in parallel kinematics. In these kinematics the electron-nucleon cross section
can be written as

$$\sigma_{en} = \sigma_{Mott}\left\{\frac{q_\mu^4}{q^4}R_L + \left[\frac{q_\mu^2}{2q^2} + \tan^2\frac{\theta}{2}\right]R_T\right\}$$

By choosing different combinations of incoming energy and electron scattering angle, R_T
and R_L can be determined by a Rosenbluth separation. Two energies were used, 314 and

444 MeV. A range of three-momentum transfer from 270 MeV/c to 460 MeV/c was covered and the kinematic factor multiplying R_T ranged from 0.6 - 3. The prerequisite for a clean interpretation of such data is to keep the final proton energy constant in order to nullify the effect of the FSI in the L/T separation. Special care was taken to ensure high precision and stability by repeated measurements of the elastic ^{12}C cross section, interleaved with the coincidence runs. A stability of 1.010 ± 0.015 was obtained for the ratio of two coincidence runs performed a few days apart. The result for the ratio of the backward angle to forward angle cross sections (not yet separated in longitudinal and transverse pieces) is shown in Fig. 27. The solid curves shown, were calculated in a model[39] in which both the charge radius and magnetic radius of the proton were increased by 10 and 20% ($R^*/R = 1.10$ and 1.20) and the dashed curve is the result of a calculation[40] with the ($\sigma-\omega$) model. The data analyzed thus far have not yet reached the final precision of $\approx 3\%$ and therefore no definite signature of medium modified nucleons can be assessed yet. However, I expect that this type of experiment performed in a more extensive range of kinematics and for several nuclei will be a crucial asset in the investigation of the intriguing properties of the nucleus predicted in relativistic mean-field theories.

4. SUMMARY AND OUTLOOK

I have attempted to illustrate by means of a few selected examples of recent experimental works, the potential of the quasifree proton-knockout reaction, provided that good resolution is available. The nuclear structure questions addressed concern deviations from the simple single-particle picture embodied in the shell model such as the amount of correlations, the related question of high momentum components and the limits of single-particle motion deep inside the nuclear density. The intriguing and more fundamental question whether nucleons retain their free-space properties in nuclei, a question also put forward in the context of the modification of the quark distribution of nucleons in nuclei, may be fruitfully studied with high-precision (e,e'p) experiments. It has been pointed out that it will be conducive to deeper insight into the single-particle aspects of nuclei to analyze the information from both electro-magnetic (e,e') and (e,e'p)) reactions and hadronic (i.e.(d,^{3}He),(p,2p)) reactions in a synthesizing way. Projects conceived in this spirit are underway. The examples discussed have all been obtained with a one-

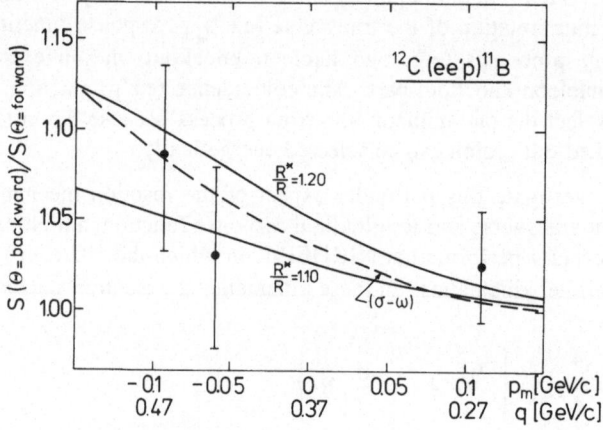

Fig. 27   Ratio of backward angle and forward angle spectral functions in a range of q-values and the prediction of various models.

percent duty-factor beam. Given the high-resolution a rich field has been opened for future research. It is nonetheless clear that two technological developments have to be pursued, increase of duty factor to 100% and increase of the energy to, say, one GeV. The energy increase will open up the kinematic window much further and largely increased cross sections will result. The spatial resolution will improve, still remaining compatible with the nucleon-meson language in which most present-day structure theories are phrased. Clearly the vast increase in duty factor combined with developments in detection techniques (large solid angle devices) will be very important in providing new vistas in electromagnetic nuclear physics.

REFERENCES

[1]   P. Martel et al., Journ. Low Temp., Physics **23**, 285 (1976).
[2]   A. Bodek, Phys. Rev. **D20**, 1471 (1979).
[3]   B. Frois et al., Nucl. Phys. **A396**, 409c (1983).
[4]   B. Frois et al., Nucl. Phys. **A434**, 47c (1985).
[5]   C. de Vries et al., Nucl. Instr. and Meth. **223**, 1 (1984).
[6]   L. Lapikás and P.K.A. de Witt Huberts, Journ. de Physique Colloque, C4, Tome 45, C4-57 (1984).
[7]   V. R. Pandharipande, C. N. Papanicolas and J. Wambach, Phys. Rev. Lett. **53**, 1133 (1984).
[8]   L. S. Celenza et al., Phys. Rev. **C31**, 946 (1985).
[9]   T. W. Donnelly in 'Symmetry in Nuclear Structure', NATO ASI Series B; Physics, Vol. 93, Plenum Press.
[10] T. de Forest, Jr., Ann. Phys. **45**, 365 (1967).
[11] J.H.J. Distelbrink et al., Nucl. Instru. and Meth. **220**, 433 (1984).
[12] J. Knoll, Nucl. Phys. **A223**, 462 (1974).
[13] S. Platchkov et al., Phys. Rev. C, Vol. 25, 2318 (1982).
[14] E. Jans et al., Phys. Rev. Lett. **49**, 974c (1982).
[15] I. Sick et al., Phys. Rev. Lett. **45**, 871 (1980).
[16] C. Ciofi degli Atti et al., Phys. Lett. **127B**, 303 (1983).
[17] N. J. Pirner and J. P. Vary, Phys. Rev. Lett. **46**, 1376 (1981).
[18] J. M. Laget, Phys. Lett. **151B**, 325 (1985).
[19] P.W.M. Keizer et al., Phys. Lett. B, Vol. **157B**, 255 (1985).
[20] W. Weise, Nucl. Phys. **A396**, 373c (1983).
[21] A.G.M. van Hees and P.W.M. Glaudemans, Z. Physik **A315**, 223 (1985).
[22] G. v.d. Steenhoven et al., Phys. Lett. B, **156B** 146 (1985).
[23] G. E. Walker, Proceedings of the Third Workshop of the Bates' Users Theory Group, MIT, July 23-24 (1984). Ed. G. H. Rawitscher.
[24] D. W. Devins et al., Aust. J. Phys. **32**, 323 (1979).
[25] J. W. den Herder et al., Phys. Lett. **B,** to be published.
[26] P. Schwandt et al., Phys. Rev. **C26**, 55 (1982).
[27] A. Stuirbrink et al., Z. Physik **A297**, 307 (1980).
[28] L. Lapikás, private communication.
[29] H. P. Blok et al., Contribution to International Symposium on Medium Energy Nucleon and Anti-nucleon Scattering, Bad Honnef (FRG) (1985).
[30] B. C. Clark et al., Phys. Rev. Lett. **50**, 1644 (1983).
[31] D. F. Jackson and I. Abdul-Jalil, J. Phys. **G6**, 481 (1980).
[32] B. Frois et al., Nucl. Phys. **A396**, 409c (1983).
[33] E. N. M. Quint, to be published in Phys. Lett. B.

[34]   G. J. Wagner, private communication.
[35]   H. Nann, private communication.
[36]   R. Dymark, Phys. Lett. **155B**, 5 (1985).
[37]   T. de Forest, Jr., Nucl. Phys. **A392**, 232 (1983).
[38]   Z. E. Meziani et al., Phys. Rev. Lett. **54**, 1233 (1985).
[39]   P. J. Mulders, Phys. Rev. Lett **54**, 2560 (1985).
[40]   T. de Forest, Jr., Proc. of 3rd Bates' Users Theory Group Workshop, MIT (1984),
        Ed. G. H. Rawitscher.

MESON (PHOTO- AND) ELECTRO-PRODUCTION AND THE STRUCTURE

OF NUCLEI AT SHORT DISTANCES

J. M. Laget

Service de Physique Nucleaire-Haute Energie,
CEN Saclay,
91191 Gif-sur-Yvette cedex, France

1. INTRODUCTION

Over more than half a century, the complementary use of hadronic and electromagnetic probes has led us to a very accurate description of the atomic nucleus in terms of inert nucleons bound together by effective forces. While this simple and economical picture has emerged naturally from the analysis of hadron induced reactions, it is only the advent, fifteen years ago, of high intensity and high energy electron accelerators which has allowed us to put it on solid grounds.

The measurements of elastic and inelastic form factors at high momentum transfer[1] has made possible the accurate determination of the charge and magnetization densities and has led to a good knowledge of the shape of the nuclei. The analysis of quasi elastic electron scattering (when the outgoing electron is detected in coincidence with the struck nucleon) has made possible the straightforward study of the shell structure of the nuclei[2]. The analysis [3] of total photo-absorption cross sections [4] and of deep inelastic electron scattering cross sections[5,6] has clearly confirmed that, to a good approximation nuclei are made of nucleons embedded in a mean effective potential. All these results have led to strong constraints on the self-consistent mean field description of nuclei.

The effective interaction which generates this mean field is constructed by eliminating any explicit reference to the internal degree of freedom of the nucleons and differs from the actual interaction between two free nucleons. On the one hand, the Pauli exclusion principle prevents nucleons from coming close together in a nucleus, and their strong short range repulsion is weakened. On the other, hand the complexity of the many-body problem requires one to parametrize also some aspects of the two-nucleon interaction. In a nucleus they feel a softer long range attraction than if they were free; this can be used in self-consistent methods to describe the ground state and the first excited states of the atomic nucleus.

While it is fair to say that the one-body properties of nuclei are now well under control, the use of probes of higher and higher energy has revealed departures from this simple scheme. At short distance the measured charge distributions (Fig. 1) exhibit smaller

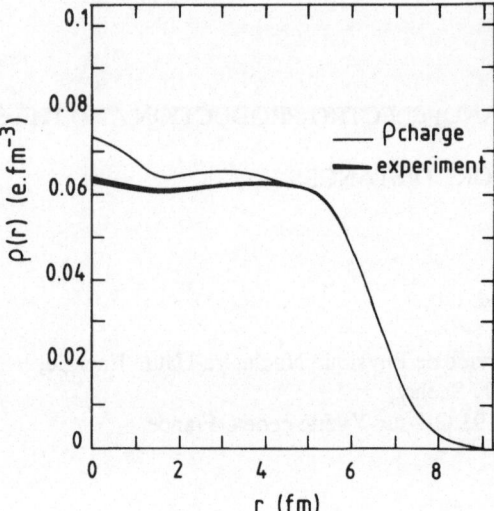

Fig. 1   The charge density of ^{208}Pb. The experimental values, measured at Saclay[1], are represented by the full line, of which the thickness represents the size of the error. The dashed line is the theoretical charge density computed in the framework of mean field theory.[7]

oscillations than predicted by the theory[7]. This may well be the hint that all the long range correlations have not been included properly in the mean field descriptions. The spectroscopic factors determined in the A (e,e′p) reactions[2] or by comparing the charge density of two neighbouring nuclei[8] are systematically lower (30%) than unity. This is the hint that correlations shift a sizeable part of the one-particle one-hole strength in the two-particle two-hole continuum. The total photoabsorption cross section per nucleon (Fig. 2) is the same for a wide range of nuclei (from ^{9}Be to U). At first sight, the difference between its shape and the shape of the free nucleon cross section is primarily due to the mean properties of the nucleus (size, Fermi motion, binding energy, etc.). However, a more careful analysis reveals that a sizeable part is due to the absorption of the photon by two correlated nucleons. This is a dominant mechanism below the pion threshold, and it involves correlations between the nucleons.

Therefore, we have been naturally led to go beyond this simple picture of the nucleus, in terms of inert nucleons, and to consider also the other degrees of freedom which have been explicitly excluded when building up the effective interaction between two nucleons. Although considerable progress had been made in this new field, we are still faced with two open questions.

On the one hand, the increase of the momentum transferred to the nucleus allowed us to probe its spatial structure over distances comparable to or smaller than the nucleon size, where short range correlations between two or several nucleons are important. We must admit that they are badly known, and the first goal of modern nuclear physics is to accurately determine them.

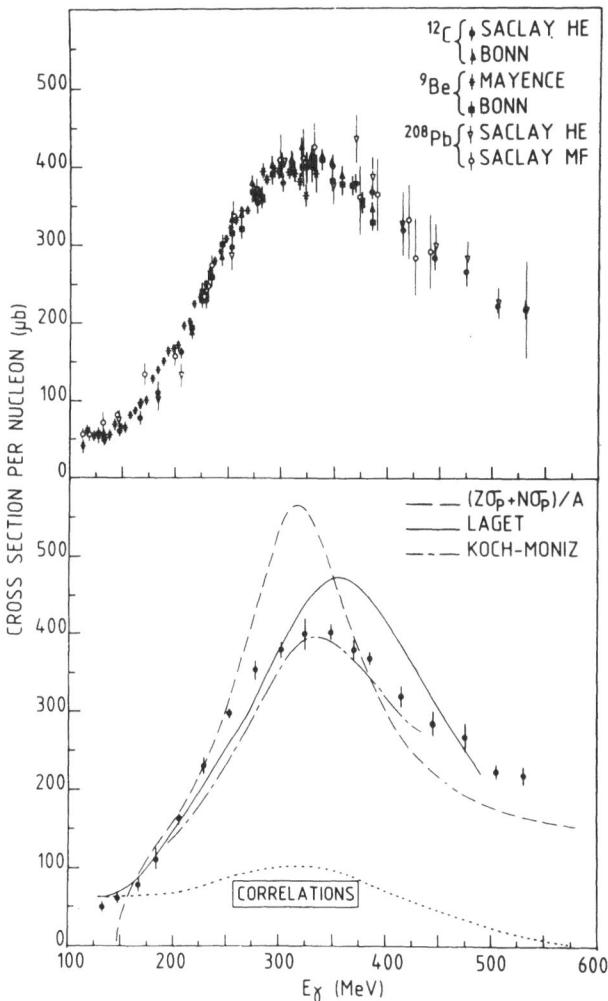

Fig. 2   The upper part of the figure shows the "universal curve" of the total photo-absorption cross section on nuclei, plotted against the incoming photon energy. Here the photo-absorption cross section of a wide variety of nuclei (from ^{9}Be to ^{208}Pb) has been divided by the corresponding number of nucleons. In the lower part, the total photo-absorption of ^{12}C is compared to the free nucleon cross section (dashed line) and two theoretical predictions. The first one (dot-dash line) is based on the Δ-hole model[93]. The second one (full line) is based on a diagrammatic expansion of the amplitude[3] and is discussed in Section 5. The dotted curve line curve represents the two nucleon current contributions.

On the other hand, nuclear physics has now evolved from the study of the many-nucleon problem to the study of the interplay of the degrees of freedom of such a complex system and the internal degrees of freedom of its hadronic constituents. For instance, when the available energy is increasing the nucleon can be deformed, and its first excited state, the Δ, can be created inside the nucleus in the vicinity of another nucleon. Its subsequent propagation is the only way to determine the N–Δ interaction, the knowledge of which is as important as the knowledge of the N-N interaction[9,10]. The second goal of modern nuclear physics is, therefore, to use the nucleus as a laboratory, in order to study the internal structure of hadrons in an environment which cannot be achieved in the scattering of free nucleons.

These two topics will be the axes of the research program of the new electron accelerators which are under construction or planned in various countries[11,12]. Photo- and electroproduction of real mesons on nuclei, or of virtual mesons in nuclei, have been, still are, and will be a powerful tool to study them. Let me illustrate this point with the simplest nuclear system, the two-nucleon system.

Their interaction (Fig. 3) is very well described[13,14] at large distances by the pion exchange potential, and at intermediate distances by the exchange of two correlated pions with a total isospin T = 0 (which are often parametrized in the OBE potential[14] by the

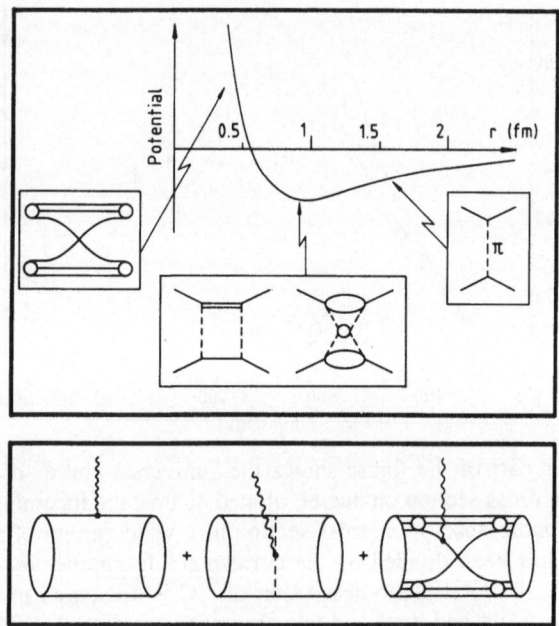

Fig. 3   The nucleon-nucleon potential and the dominant driving terms. They must be iterated to obtain the full T-matrix. At long distance the pion exchange mechanism dominates. At in intermediate distance a Δ can be created between the exchange of two pions. A short distance heavy mesons can be exchanged, but also the subnuclear degrees of freedom are expected to play a role: the Quark Interchange Mechanism is one possible example. When the electromagnetic probe interacts with the nucleon current in nuclei, gauge invariance requires also its interaction with each charged particle which is exchanged in the driving terms.

σ meson). Between the exchange of these two pions, one of the nucleons, or both, can be transformed into a Δ. Below the pion production threshold virtual Δ's enter the description of the nucleon-nucleon interactions, but above they can be created freely during a collision between two nucleons. The problems of the nucleon-nucleon interaction and of the nucleon-delta interaction should be solved at the same time in a coupled channel formalism. This is the first place where the internal degrees of freedom of hadrons enter nuclear physics.

At small distances the exchange of vector mesons (ρ,ω) plays a role, but it is also here that the quark structure of the nucleon is expected to enter into the game. This is the second place where the internal structure of hadrons plays a role, but the relative importance of these two mechanisms, the double counting problem and the relevance of the description of the nucleus in terms of quarks, are still open questions[15,16].

Coincidence experiments performed with the electromagnetic probe are precisely the most straightforward way to answer these questions.

On the one hand, a (real or virtual) photon interacts weakly and is not absorbed at the nuclear surface, as are hadronic probes such as the pion for instance. The photon sees the entire nuclear volume and can create an unstable particle, like the Δ, in the very center of the nucleus, making possible the study of its interaction in the final state.

On the other hand, the photon's coupling to a nuclear system is well under control, since it must satisfy the gauge invariance principle and since it is weak enough to be treated as a small perturbation. For instance, the electromagnetic probe has allowed us to disentangle the long range part of the nucleon-nucleon interaction, which is mediated by the exchange of a charged pion, and the intermediate part, which is mediated by the exchange of two correlated pions of which the total charge vanishes[13,14]. This sensitivity of a (real or virtual) photon to the local variations of the charge and the magnetization densities should be systematically exploited nowadays to study the short range part of the nucleon-nucleon interaction inside the nucleus.

At short distances the problem is to disentangle the mechanisms which have to do with the internal structure of the nucleon (quark interchange for instance) and the contribution due to meson exchange, which dominates at large distances but still contributes here. The best way to do it is to take advantage of the third property of virtual photons, the possibility of varying independently their energy ω , their squared mass q^2 and their degree of longitudinal polarization ε.

This is illustrated in Fig. 4, where the basic feature of the absorption of a photon by a nucleus appears clearly. The spectrum of the electrons inelastically scattered by 8° on ^{3}He is plotted in the upper part. The experimental data have been obtained at SLAC (Ref. 17). In spite of the high energy, $E_- = 3.26$ GeV, of the incoming electron beam, the momentum transfer is small (the squared mass of the virtual photon varies little around $q^2 = -.2$ (GeV/c)^2. The energy transfer is high enough to make it possible to excite the Δ resonance, which is responsible for most of the pions which are electroproduced on a quasi-free nucleon (the pions created through the non resonant part of the electroproduction operator[18] have also been taken into account). The range of momentum and energy transfer is really that which is already allowed by the present generation of high intensity electron machines. As an example, the transverse and longitudinal response functions recently determined at Saclay[19] for $q^2 = -.2$ (GeV/c)^2 are also shown in Fig. 2. Unfortunately, the maximum energy, 720 MeV, of the Saclay Linac is not high enough to allow their determination under the Δ peak. Obviously the energy of the incoming electron beam should be significantly increased.

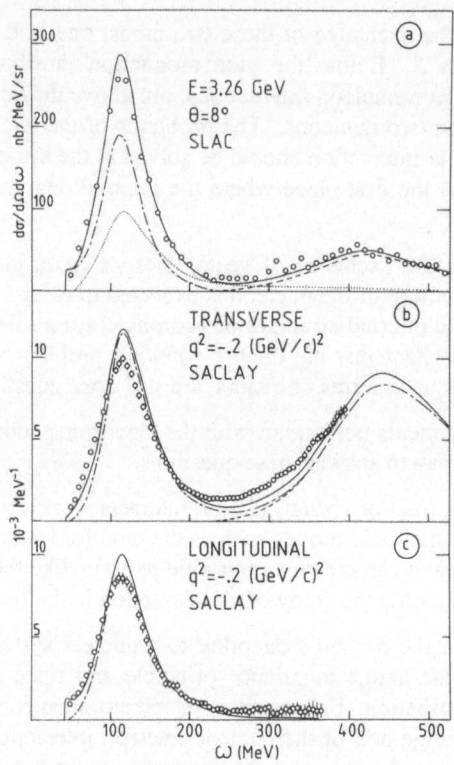

Fig. 4   The contribution to the spectrum[17] of the electron inelastically scattered on ^{3}He
of the two-body (dot-dash), the three-body (dot) break-up channels, and of the
pion electroproduction channel (dashed) are shown separately in a). The
transverse and longitudinal response functions[19] are shown in b and c, when
$q^2 = -.2(GeV/c)^2$. The dot-dashed curves correspond to the usual plane-wave
treatment. The full line curves include the final state interactions and meson
exchange currents[22].

Besides pion electroproduction on quasi-free nucleons, the incoming electron may
also scatter elastically on a quasi-free nucleon. The top of the peak, which appears for
small values of the energy ω of the virtual photon corresponds to the scattering of the
electron on a nucleon at rest in the nucleus. Its width is due to the nucleon Fermi motion.
The use of a good three-body wave function (the solution[20] of the Faddeev equations in
momentum space for the Reid potential[21]) makes it possible to compute separately the
contributions of the two-body and the three-body channels. They add up to give a fair
agreement with the experimental data. While the shape of the quasi-elastic peak is
directly related to the behaviour of the nucleon wave function, the shape of the quasi-free
pion production peak, which appears in the high energy part of the spectrum in Fig. 4, is
due to the internal degrees of freedom of the nucleons: one nucleon is changed into its
first excited state, the Δ, and the others nucleons are spectators.

Between the quasi-elastic scattering and the quasi-free pion electroproduction peaks, the excess of the cross section is well accounted for by the tail of three-body break-up channel, which is mainly due to the meson exchange mechanisms. The pion (or the ρ) which is created on one nucleon is reabsorbed by an other, breaking up the residual nuclear system. This mechanism involves the short range correlation function between two nucleons, which is automatically contained in the three-body realistic wave function[20] which I use. A full account of this calculation is given in Ref. 22.

It is worthwhile to point out that the transverse and longitudinal response functions exhibit a different behaviour. The pion production mechanisms and the exchange currents dominate the transverse response function, but they do not affect the longitudinal response function. Therefore, the best way to go beyond the study of the Δ and pion degrees of freedom in nuclei is to accurately determine the longitudinal response function far from the quasi elastic peak. Its tails are very sensitive to the high momentum components of the nuclear wave function and, therefore, to the mechanisms which occur at short distance.

These spectra are good examples of the interplay between the many body aspects (here the three-nucleon problem) and the internal degrees of freedom of the nucleon (the creation and the propagation of real or virtual pions, the excitation of the Δ, etc.). But they are integrated quantities which tell us only how the proton is absorbed by the nucleus. To go further we must perform more exclusive experiments, in order to single out each channel (pion photoproduction, photodisintegration, etc.) and study the various aspects of the nuclear dynamics.

This is precisely the topic of these lectures, where I review the present status and the future prospects of the studies of very inelastic electronuclear reactions, when both high energy and high momentum are transferred to the nucleus. I will not deal with "coherent" meson electro- or photoproduction reactions, which leave the nucleus in its ground state or its first excited states. So far they have been primarily used as a spectroscopic tool with its own sensitivity (for a review see Refs. 9, 23, 34).

I will treat on the same footing real and virtual mesons. Real meson production allows us to study the propagation of baryonic and mesonic resonances in nuclei and to put constraints on their interaction with the nucleon. Virtual meson creation is an alternative way to deal with exchange currents. On the one hand, gauge invariance links the mechanisms which bind the nucleons together and the mechanism where a virtual meson is created off a nucleon and reabsorbed by an other. On the other hand, the creation of the Δ- resonance, followed by the reabsorption of its decaying pion has to do with the absorptive part of the $N\Delta$ interaction, which is coupled to the intermediate range part of the nucleon-nucleon interaction.

The first lecture deals with the elementary operators, which describe meson photo- and electroproduction on free nucleons. The second one deals with real meson photo- and electroproduction on few-body systems. Since most of the material of these two lectures is already included in an earlier review[9], I discuss here only the main features, and I update it by presenting the last developments.

The third lecture deals with the coupling of the electromagnetic probe to the virtual mesons in nuclei. The emphasis is put on the few-body systems, since their nuclear wave functions are known and since they are simple enough to allow for elaborate calculations. The case of heavy nuclei is also discussed. In the last lecture, I will try to look for evidence of the limits and the breakdown of the description of nuclei in terms of nucleons and mesons and to forecast the new developments.

2. MESON ELECTROPRODUCTION ON FREE NUCLEONS

Below $E_\gamma = 1$ GeV two channels dominate the pion photoproduction reaction on a free nucleon: the $\gamma N \to N\pi$ and the $\gamma N \to N\pi\pi$ reactions, of which typical examples are shown in Fig. 5. While the Born terms are responsible for a large background, which varies slowly with energy but dominates near the single pion photoproduction threshold, the creation of a few baryonic resonances in the s-channel leads to strong variations in the cross section. Among them the $\Delta(1236)$ clearly appears near $E_\gamma \sim 300$ MeV, but the resonances with higher masses $E_\gamma \sim 600–700$ MeV contribute little.

The two-pion production threshold occurs near $E_\gamma \sim 300$MeV but the cross section of the reaction $\gamma N \to N\pi\pi$ begins to be significant near $E_\gamma \sim 400$ MeV and exhibits a rapid rise from $E_\gamma \sim 500$ MeV to $E_\gamma \sim 600$ MeV, where it reaches a broad maximum. This behaviour is characteristic of the threshold creation (in a relative s-wave) of a π–Δ pair[27].

The vector meson (ρ and ω) photoproduction threshold is $E_\gamma = 1110$ MeV and the $\gamma N \to N\rho$ or the $\gamma N \to N\omega$ reactions contribute also to the two- and the three-pion photoproduction channels.

The strange K meson photoproduction threshold are respectively $E_\gamma = 911$ MeV or $E_\gamma = 1046$ MeV, depending whether the target nucleon is changed into Λ or a Σ.

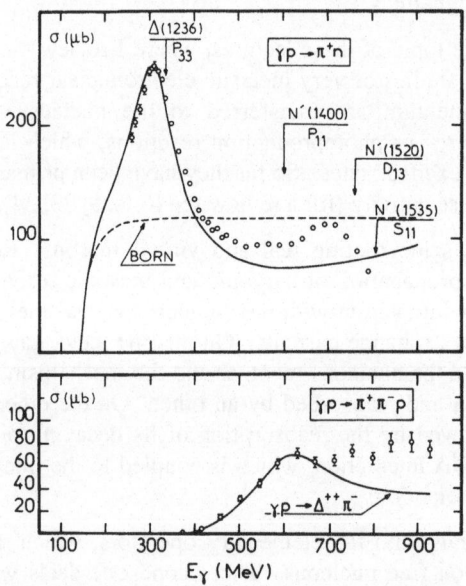

Fig. 5   In the upper part of the figure the cross section of the $\gamma p \to n\pi^+$ is depicted. The experimental points come from Ref. 24 and the curves are the predictions of the model described in Ref. 25. The resonances are labelled by their quantum numbers (mass, isospin I, spin J and orbital angular momentum L) as $L_{2I,2J}$. In the lower part the cross section of the reaction $\gamma p \to p\pi^+\pi^-$ is depicted. The experimental points come from Ref. 26. The curves correspond to the cross section of the reaction $\gamma p \to \Delta^{++}\pi^-$.

Among these channels, the single pion photoproduction reactions have been the most extensively studied. In the Δ- resonance region multipoles have been accurately determined, and their analysis in the framework of dispersive relations has led to a fair understanding of the basic mechanisms. Above the Δ resonance this method is difficult to use, and it is more economical to expand the amplitude in terms of few relevant diagrams. This is also a good way to deal with two-pion photoproduction reactions. Even in the Δ–resonance region, this method has also proved to be efficient, especially when dealing with pion photoproduction on nuclei.

The basic ingredients are the effective Lagrangians which describe the coupling of the photon with the various mesons, the nucleons and the baryonic resonances. They are chosen in such a way as to satisfy the constraints due to the low energy theorem and PCAC. For instance, I use the pseudo-vector (PV) Lagrangian to describe the pion-nucleon coupling. I use the non relativistic reduction of the amplitude corresponding to each relevant diagram in which all terms, up to and including order $1/m^2$, are retained. At this order the corresponding operators are Lorentz and gauge invariant, and are valid in any frame of reference. This is a great advantage when dealing with a nucleon moving in a nucleus. A detailed discussion is given in Refs. 9, 25, and I reproduce here the discussion of Ref. 28, which deals with the extension of the model to the virtual photon case. It gives the basic ingredient which I will need when discussing the gauge invariance of the electronuclear amplitudes. At the end of this section, I will briefly review the status of our knowledge of vector and strange meson electroproduction channels.

2.1 The Nucleon Current

The space component of the nucleon current $J=(J_0,\vec{J})$ is the extension of the virtual photon case of the current, which has already been given in Ref. 29:

$$\vec{J} = -i\frac{\vec{p}-\vec{k}/2}{m}F_1(q^2)+\frac{F_2(q^2)+F_1(q^2)}{2m}\ \vec{\sigma}\times\vec{k}+O\left[\frac{1}{m^3}\right] \tag{1}$$

The time component is obtained in the same way:

$$J_0 = -iF_1(q^2)\left[1+\frac{\vec{p}^2+\vec{p}'^2}{4m^2}\right]+i\frac{\vec{k}^2}{8m^2}\ \{\,2[F_2(q^2)+F_1(q^2)]-F_1(q^2)\}$$

$$+\text{S.O.} + O\left[\frac{1}{m^3}\right] \tag{2}$$

where $\vec{p}$ and $\vec{p}' = \vec{k}+\vec{p}$ are respectively the nucleon momentum before and after the photon absorption. This expression is obtained when the nucleon spinors are normalized to $\bar{u}u = 1$, and differs from the more commonly used[30], where the term $(\vec{p}^2 + \vec{p}'^2)/4m^2$ disappears, and which is obtained with the norm $\bar{u}u = m/E$. Of course the same norm of spinor should be used in the phase space factor and both prescriptions lead to the same result. This point is important when dealing with terms of order $1/m^2$. All the calculations reported here have been made with the conventions of Björken and Drell[31], and using their relativistic expression of the phase space. The kinematics is relativistic, and the expansion at order $1/m^2$ is only made in the vertex operators.

Using the relations between the Dirac and Sachs form factors:

$$G_M(q^2) = F_1(q^2) + F_2(q^2)$$

$$G_E(q^2) = F_1(q^2) + \frac{q^2}{4m^2}F_2(q^2)$$

$$F_1^n(0) = 0, \quad F_1^p(0) = e, \quad F_2^n(0) = e \times \kappa_n, \quad F_2^p(0) = e \times \kappa_p \tag{3}$$

it can be reexpressed as:

$$J_0 = -iG_E(q^2)\left[1 + \frac{\vec{k}^2}{8m^2}\right] + S.O. + O\left[\frac{1}{m^3}\right] \tag{4}$$

when the target nucleon is at rest ($\vec{p} = 0, \vec{p}' = \vec{k}$). The four-momentum q, the three-momentum $\vec{k}$ and the energy ω of the virtual photon are related through:

$$q^2 = \omega^2 - \vec{k}^2 . \tag{5}$$

The spin orbit terms (SO), of order $1/m^2$, vanish when the nucleon is at rest. Therefore, at the top of the quasi-elastic peak, a good approximation to the charge operator is:

$$J_0 = -iG_E(q^2)\left[1 + \frac{\vec{k}^2}{8m^2}\right] \tag{6}$$

For a free nucleon, this operator satisfies the gauge invariance principle, and it is easy to show that:

$$\omega J_0 - \vec{k}\cdot\vec{J} = 0 + O\left[\frac{1}{m^3}\right] \tag{7}$$

since

$$i\omega J_0 = \frac{\vec{p}'^2 - \vec{p}^2}{2m} F_1(q^2) + O\left[\frac{1}{m^3}\right] \tag{8}$$

and

$$i\vec{k}\cdot\vec{J} = \frac{\vec{k}\cdot(\vec{p} - \vec{k}/2)}{m} F_1(q^2) + O\left[\frac{1}{m^3}\right]$$

$$= \frac{\vec{p}'^2 - \vec{p}^2}{2m} F_1(q^2) + O\left[\frac{1}{m^3}\right] \tag{9}$$

For bound (and off-shell) nucleons, this current violates gauge invariance since Eq. 8 is no longer valid (since $\omega = p'_0 - p_0 \neq \frac{\vec{p}'^2 - \vec{p}^2}{2m}$); we shall see that the way to restore it is to consider the exchange currents also.

2.2 Pion Electroproduction off a Nucleon

The space component of the pion electroproduction current is a straightforward extension[18] of the Blomqvist-Laget (B-L) amplitude[25], which reproduces fairly well the pion photoproduction reactions on free nucleons.

The resonant Δ formation part is determined by multiplying the corresponding B-L amplitude by the form factor

$$F_\Delta(q^2) = \frac{(1 + q^2/6)}{\left[1 - \frac{q^2}{0.71}\right]^2} . \tag{10}$$

It takes the form:

$$J^\mu = F_\Delta(q^2)\, C_\pi C_\gamma G_3 G_1\, \frac{e^{i\varphi}}{Q^2 - M_\Delta^2 + i M_\Delta \Gamma}$$

$$\times \left\{ [\,\vec{S} \cdot (\vec{q}_\pi - \frac{E_\pi}{M_\Delta}\,\vec{p}_\Delta\,)]\cdot[\,\vec{S}^\dagger \times (\vec{k} - \frac{M_\Delta - m}{m}\,\vec{p}_i\,)] \right\} \tag{11}$$

with the same notation as in Ref. 25 (but with $q = (\omega, \vec{k})$ and $q_\pi = (E_\pi, \vec{q}_\pi)$ being respectively the four momenta of the photon and the pion). The phase $\phi(Q)$ is determined in such a way that the full amplitude is unitary when the Born terms are added. This aspect of the B-L operator has not been fully acknowledged. For instance the non-unitary version has been criticized in Refs. 32, 33. However, it is the use of this non-unitary version of the model in nuclear calculations[34,35] which might be questioned, and the improvements proposed in Refs. 32, 33 are already contained in the original B-L article (Eqs. 20, 29-31).

The Born part of the amplitude (see Fig. 6) is computed with the pseudovector pion nucleon coupling. This choice is not only required by the low energy theorems and the PCAC constraints (see Ref. 9), but it also leads to a good agreement with the measured electric dipoles E_{0+} from threshold up to $E_\gamma \sim 500\ \mathrm{MeV}$ (Ref. 25). For π^+ production the space part of the current takes the form:

$$\vec{J}^{\pi^+} = i\frac{g_0\sqrt{2}}{2m} \Bigg\{ F_\pi(q^2)\, \frac{2\vec{q}_\pi - \vec{k}}{(q_\pi - q)^2 - m_\pi^2}\, \vec{\sigma}\cdot(\vec{k} - \vec{q}_\pi) - F_1^P(q^2)\, \frac{2\vec{p}_i + \vec{k}}{2E_a(p_a^0 - E_a)}\, \vec{\sigma}\cdot\vec{q}_\pi \tag{12}$$

$$- iG_M^P(q^2)\, \frac{\vec{\sigma}\cdot\vec{q}_\pi\, \vec{\sigma}\times\vec{k}}{2E_a(p_a^0 - E_a)} - iG_M^n(q^2)\, \frac{\vec{\sigma}\times\vec{k}\, \vec{\sigma}\cdot\vec{q}_\pi}{2E_b(p_b^0 - E_b)} - F_A(q^2)\vec{\sigma} + F_1^P(q^2)\, \frac{m\omega\vec{\sigma}}{E_a(p_a^0 + E_a)} \Bigg\} .$$

The denominator in each term tells us what is the corresponding diagram: pion exchange in the t-channel, nucleon exchange in the s-channel (charge and magnetic coupling) or in the u-channel, the contact term and the pair term which is suppressed by a factor m_π/m in PV coupling (as compared to PS coupling).

For π^- production the space part of the current takes the form:

$$\vec{J}^{\pi^-} = i\frac{g_0\sqrt{2}}{2m} \Bigg\{ -F_\pi(q^2)\, \frac{2\vec{q}_\pi - \vec{k}}{(q_\pi - q)^2 - m_\pi^2}\, \vec{\sigma}\cdot(\vec{k} - \vec{q}_\pi)$$

$$- F_1^P(q^2)\, \frac{2\vec{p}_f - \vec{k}}{2E_b(p_b^0 - E_b)}\, \vec{\sigma}\cdot\vec{q}_\pi - iG_M^n(q^2)\, \frac{\vec{\sigma}\cdot\vec{q}_\pi\, \vec{\sigma}\times\vec{k}}{2E_a(p_a^0 - E_a)}$$

$$- iG_M^P(q^2)\, \frac{\vec{\sigma}\times\vec{k}\, \vec{\sigma}\cdot\vec{q}_\pi}{2E_b(p_b^0 - E_b)} + F_A(q^2)\vec{\sigma} + F_1^P(q^2)\, \frac{m\omega\vec{\sigma}}{E_b(p_b^0 + E_b)} \Bigg\} . \tag{13}$$

The Δ does not contribute to the time component of the current, which is dominated by the Born terms

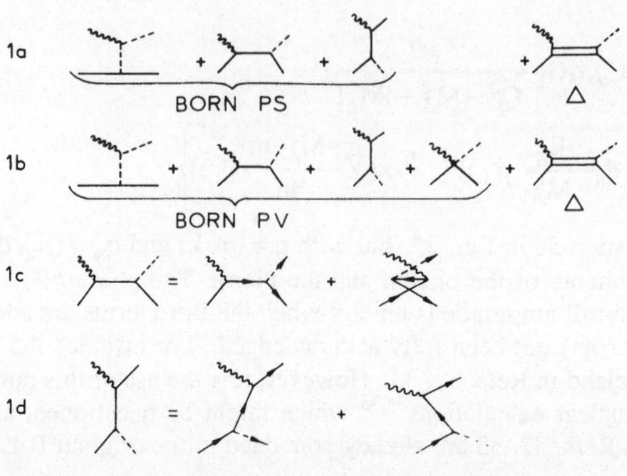

Fig. 6   The amplitude of the $\gamma N \to N\pi$ reaction is expanded in terms of the relevant diagrams in a and b, for respectively pseudo-scalar (PS) or pseudo-vector (PV) πNN coupling. In c and d, the Feynman diagrams, involving a nucleon in the intermediate state, are expanded in terms of times ordered diagrams, in order to exhibit the pair terms. In PV coupling they are suppressed, with respect to PS coupling by a factor m_π/m.

$$J_0^{\pi^+} = i\frac{g_0\sqrt{2}}{2m}\left\{ -F_1^p(q^2)\frac{m\vec{\sigma}\cdot\vec{q}_\pi}{E_a(p_a^0-E_a)} + F_1^p(q^2)\frac{E_\pi\vec{\sigma}\cdot\vec{k}}{2E_a(p_a^0-E_a)}\right.$$

$$\left. +\frac{\vec{\sigma}\cdot(\vec{p}_i+\vec{p}_f)}{2m}[F_1^p(q^2)-F_A(q^2)] + F_\pi(q^2)\frac{2E_\pi-\omega}{(q_\pi-q)^2-m_\pi^2}\,\vec{\sigma}\cdot(\vec{k}-\vec{q}_\pi)\right\} \qquad (14)$$

$$J_0^{\pi^-} = i\frac{g_0\sqrt{2}}{2m}\left\{ -F_1^p(q^2)\frac{m\vec{\sigma}\cdot\vec{q}_\pi}{E_b(p_b^0-E_b)} - F_1^p(q^2)\frac{E_\pi\vec{\sigma}\cdot\vec{k}}{2E_b(p_b^0-E_b)}\right.$$

$$\left. -\frac{\vec{\sigma}\cdot(\vec{p}_i+\vec{p}_f)}{2m}[F_1^p(q^2) - F_A(q^2)] - F_\pi(q^2)\frac{2E_\pi-\omega}{(q_\pi-q)^2-m_\pi^2}\,\vec{\sigma}\cdot(\vec{k}-\vec{q}_\pi)\right\}. \qquad (15)$$

It is worthwhile to note that the time component of the contact term cancels a part of the nucleon Born terms when $F_1^p(q^2) = F_A(q^2)$.

This current satisfies gauge invariance up to order $1/m^2$ (for on-shell nucleons) provided $F_\pi = F_1^p = F_A$, otherwise additional terms should be added. The relations

$$2E_a(p_a^0 - E_a) = 2m\left[\omega - \frac{\vec{k}^2}{2m} - \frac{\vec{p}_i \cdot \vec{k}}{m}\right], \quad 2E_b(p_b^0 - E_b) = 2m\left[-\omega - \frac{\vec{k}^2}{2m} + \frac{\vec{p}_f \cdot \vec{k}}{m}\right]$$

$$2E_a(p_a^0 + E_a) = 2m(2m + \omega), \quad 2E_b(p_b^0 + E_b) = 2m(2m - \omega) \tag{16}$$

are useful to demonstrate it. In the numerical application presented here, we shall use

$$F_\pi(q^2) = F_A(q^2) = F_1^p(q^2) = \frac{G_M^p(q^2)}{\mu_p} = \frac{G_M^n(q^2)}{\mu_n} = \frac{e}{(1 - \frac{q^2}{0.71})^2}$$

$$F_1^n(q^2) \equiv 0 \tag{17}$$

which is a reasonable choice at low momentum transfer.

This type of current has already been used by Tiator and Drechsel[36]. However, the current conservation equation ($\omega J_0 - \vec{k} \cdot \vec{J} = 0$) requires that the space part J of the current vanishes when ω vanishes. Therefore, we have replaced E_π by ω in the antinucleon contribution to the Born terms of the B-L amplitude.

For neutral pions, the extension of the B-L model (as improved in Ref. 37) has been carried out along the same lines and is not given here.

2.3 Cross Sections

Following Dombey[38], the cross section of the inclusive reaction N(e, e')πN is related to the transverse σ_T and the longitudinal σ_L virtual photoabsorption cross section by

$$\frac{d\sigma}{d\Omega_e d\omega} = \Gamma_v[\sigma_T + \varepsilon \sigma_L] \tag{18}$$

where the flux of the virtual photon is

$$\Gamma_v = \frac{\alpha}{2\pi^2} \frac{E'}{E} \frac{k}{1 - \varepsilon} \frac{1}{-q^2} \tag{19}$$

and the degree of polarization of the virtual photon is

$$\varepsilon = \left[1 - 2\frac{k^2}{q^2} \tan^2\theta/2\right]^{-1}. \tag{20}$$

The energy of the incident electron is E and the energy and the polar angle of the scattered electron are respectively E' and θ. The fine structure constant is α. The transverse cross section is related to the integral, over the pion angles, of the square of the transverse part of the current, averaged over the initial polarizations and summed over the final polarizations:

$$\sigma_T = A \int d\Omega_\pi [(J_x^* J_x) + (J_y^* J_y)] \tag{21}$$

where $A = |\vec{q}_\pi| \, [m/4\pi Q]^2/4 \, |\vec{k}|$ is the relativistic phase space factor (Q being the invariant mass of the πN pair). Eq. 21 reduces to the total photoabsorption cross section[25] when $q^2 \to 0$.

The longitudinal cross section is related in the same way to the third component of the current via

$$\sigma_L = 2A \int d\Omega_\pi \frac{-q^2}{\omega^2} (J_z^* J_z) \tag{22}$$

and vanishes at the photon point.

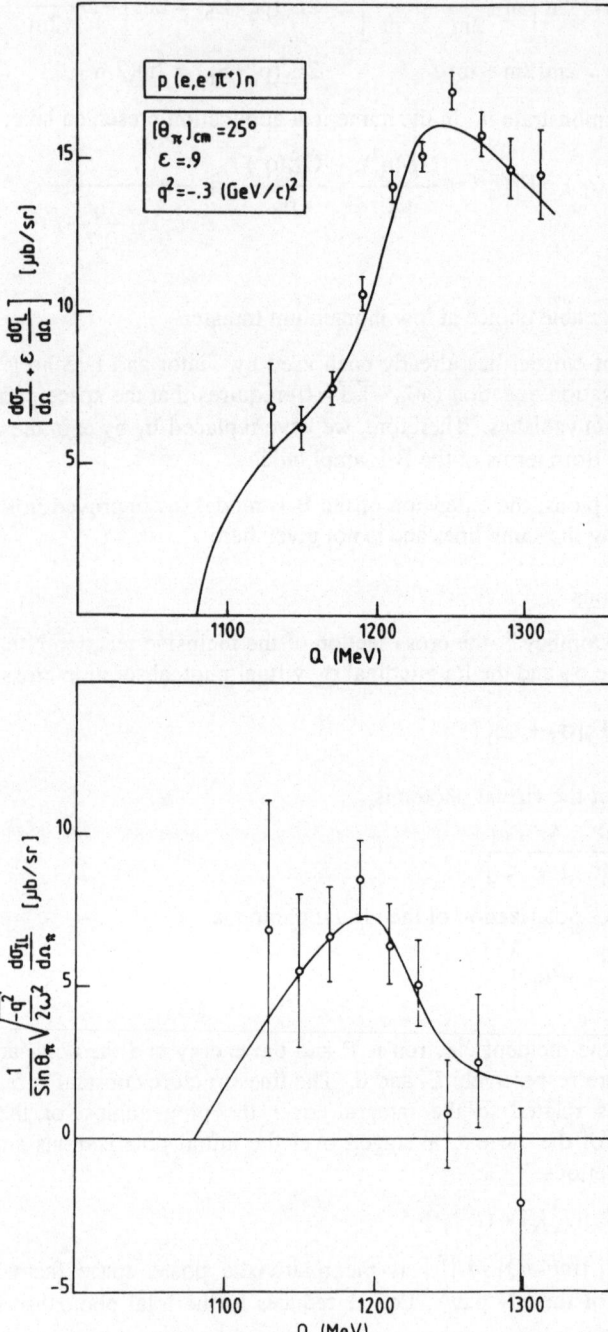

Fig. 7   The excitation function, at $\theta_\pi = 25°$, of the reaction $p(e,e'\pi^+)$ is plotted against
the invariant mass Q of the emitted π–N pair[39,105]. The squared four momen-
tum transfer is $q^2 = -0.3$ $(GeV/c)^2$. the sum of the longitudinal and the
transverse cross sections and the transverse-longitudinal interference cross section are shown separately.

When the pion is detected in coincidence with the scattered electron, two more terms have to be considered: the interference between the two transverse components, and the longitudinal components of the current. The cross section of the exclusive reaction N(e, e′π)N is therefore,

$$
\frac{d\sigma}{d\Omega_e \cdot d\omega d\Omega_\pi} = \Gamma_v \left[\frac{d\sigma_T}{d\Omega_\pi} + \varepsilon \frac{d\sigma_L}{d\Omega_\pi} + \varepsilon \cos 2\phi \, \frac{d\sigma_{TT}}{d\Omega_\pi} \right.
$$

$$
\left. -\cos\phi \sqrt{\frac{-q^2 \varepsilon(\varepsilon+1)}{2\omega^2}} \frac{d\sigma_{TL}}{d\Omega_\pi} \right] \tag{23}
$$

where

$$
\frac{d\sigma_{TT}}{d\Omega_\pi} = A[(J_x^* J_x) - (J_y^* J_y)]
$$

$$
\frac{d\sigma_{TL}}{d\Omega_\pi} = 2A[(J_x^* J_z) + (J_z^* J_x)] \tag{24}
$$

and ϕ is the angle between the electron scattering plane and the pion production plane. It vanishes when the pion is emitted on the same side as the scattered electron, with respect to the direction of the virtual photon.

Current conservation $(\omega J_0 = k J_z)$ has been used to eliminate the time component in Eqs. 18, 22 and 23. This is the usual choice in particle physics. If the third component J_z of the current had been eliminated, it would have been replaced by $\omega J_0/k$ in those equations. This choice is usually done in nuclear physics. Since the current (Eqs. 12 to 15) is conserved (up to order $1/m^2$), both prescriptions lead to the same result (at this order).

2.4 Comparison with Experiments

The measurement of the azimuthal, ϕ, and polar, θ, angular dependence of the cross section makes it possible to disentangle the four cross sections in Eq. 23. This has been achieved in a study of the p(e,e′π$^+$)n reaction performed at Bonn[39]. As can be seen in Fig. 7, the model leads to a fair agreement with the data. Experimental studies of the p(e,e′π^0)p reaction are fewer[40], and also well reproduced by the model (Fig. 8).

At $\theta_\pi = 0°$ (and also at $\theta_\pi = 180°$), with respect to the direction of the virtual photon, the two interference cross sections vanish, and the variation of the cross section with the electron polar angle only makes it possible to separate the transverse and the longitudinal cross sections, the values of which were determined at Saclay[41] and are plotted in Fig. 9. Here the pion pole dominates the longitudinal part of the cross section, since the other Born diagrams are forbidden by helicity conservation. This is not the case at larger angles, and the integrated longitudinal cross section (Eq. 22) is much smaller than the transverse one (see Fig. 10).

It is worthwhile to note that the extrapolation of the electroproduction cross section to the photon point is not linear, and is well accounted for by the model. The reason is that the $\gamma N\Delta$ operator is proportional to the momentum $\vec{k}$ of the virtual photon. When the four momentum squared q^2 increases (the mass Q being fixed), the increase of $\vec{k}$ is faster than the decrease of the form factor $F_\Delta(q^2)$ near the photon point, whereas the form factor decreases more quickly at higher momentum.

Therefore, the extension of the B-L model to the virtual photon case is good enough to be safely used when dealing with pion electroproduction reactions in nuclei in the kinematical domain defined by $0 < Q - M < 500$ MeV and $0 < -q^2 < 0.5 (\text{GeV/c})^2$, Q being

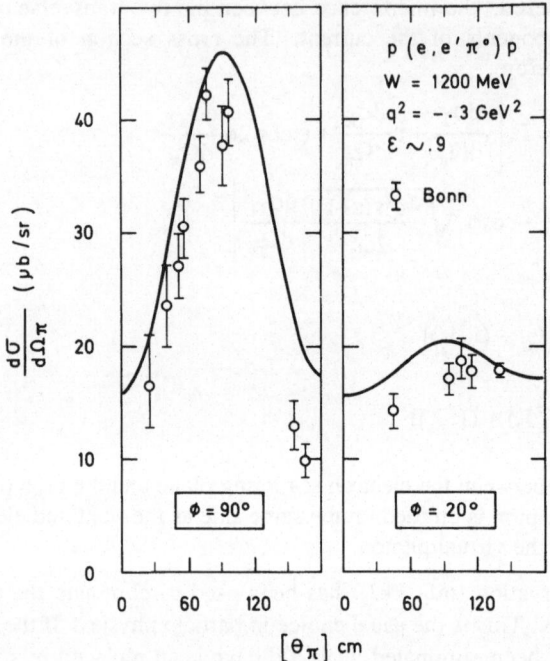

Fig. 8   The angular distribution[40] of the p(e,e′π⁰)p reaction cross section for Q = 1200 MeV and q² = −.3(GeV/c)², for two values of the azimuthal angle φ.[105]

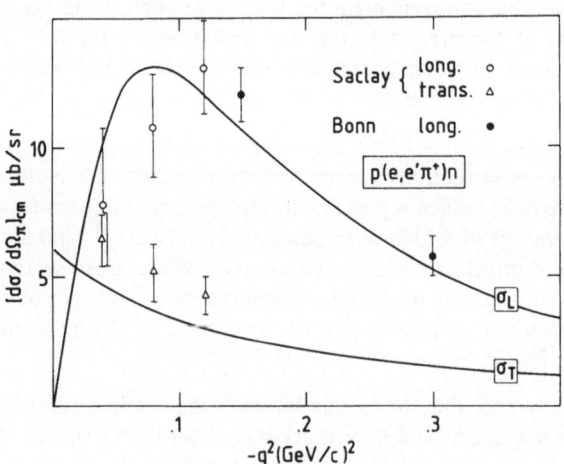

Fig. 9   The transverse and the longitudinal differential cross sections of the p(e,e′π⁺)n reaction, when the pion is emitted at θ_π = 0°, in the direction of the virtual photon. The experiments have been performed at Saclay[41] and Bonn[39]. The invariant mass of the πN pair is Q = 1175MeV, and the square of the mass of the virtual photon, q², is plotted on the abscissa.[105]

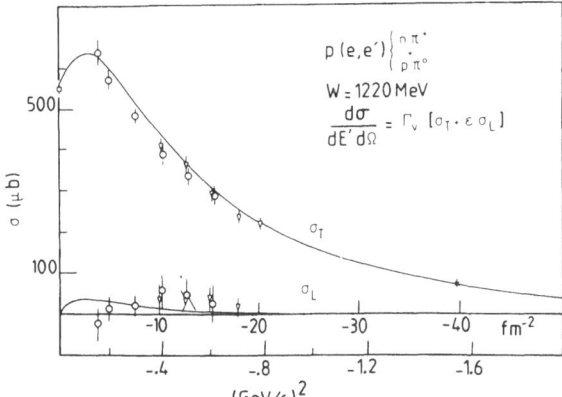

Fig. 10   The same as in Fig. 9, but for the integrated transverse and longitudinal cross section at $Q = 1220$ MeV. The experimental points have been obtained at Deutsches Elektronen-Synchroton (DESY)[42] and Bonn[43].

the total center of mass (c.m.) energy of the πN pair. It works also at momentum transfer as high as $1(\text{GeV}/c)^2$ (Fig. 10), but care must be taken here of the effects of the difference between the Sachs and the Dirac nucleon electromagnetic form factors.

On the experimental side, while the single pion photoproduction channels are fairly well known in the Δ–resonance region, they are less known above, and only a few studies of pion electroproduction reactions have been performed so far. Therefore one of the major tasks of future electron machines will be to allow for an extensive study of those electroproduction channels in the energy range $0 < Q - M < 1$ GeV and for momentum transfer up to $q^2 \geq 1(\text{GeV}/c)^2$.

2.5 Two-Pion Production

Double pion photoproduction reactions on free nucleons have been studied in several bubble chamber experiments[26,44-46]. Lüke and Söding [27,47] have constructed an isobar model which reproduces the shape, the magnitude of the cross section, and also the joint decay distributions. The two most important contributions come from the contact term (Fig. 11a) and the pion photoelectric term (Fig. 11b), which account respectively for 75% and 25% of the cross section. Other diagrams (Figs. 11c, d) are necessary for insuring gauge invariance, but do not play a significant role in the cross sections. Above $E_\gamma \sim 700$ MeV the resonant production (Fig. 11e) of a π–Δ pair begins to appear and absorption corrections must be considered.

An effective operator easy to use in a nucleus is obtained, starting from the effective Lagrangians $L_{\gamma\pi\pi}$, $L_{\gamma NN}$, $L_{\pi NN}$ and $L_{\pi\Delta N}$ and computing the non relativistic limit of each matrix element up to and including terms of order $(p/m)^2$. I deduce the new contact Lagrangian $L_{\gamma\pi N\Delta}$ by making the minimal substitution in the Lagrangian $L_{\pi N\Delta}$.

The space part of the corresponding current is[48]:

$$\vec{J} = -\frac{CG_3^2}{R^0 - E_R + i\Gamma/2}\ \vec{S}\cdot[\vec{q}_\pi - \frac{E_\pi}{M_\Delta}\ \vec{R}]$$

$$\cdot \{F_A(q^2)\vec{S}^\dagger + F_\pi(q^2)\vec{S}^\dagger\cdot(\vec{\mu} - \vec{k})\ \frac{2\vec{\mu} - \vec{k}}{(\mu - k)^2 - m_\pi^2}\} \tag{25}$$

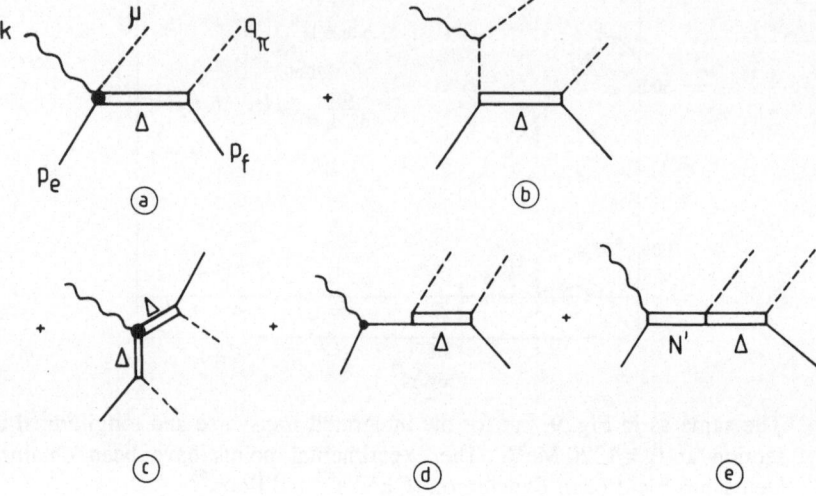

Fig. 11   The relevant diagrams for describing the double pion photoproduction reactions on a nucleon.

where $R^0 = \omega + p_i^0 - \mu^0$ is the actual energy of the intermediate Δ and

$$E_R = [M_\Delta^2 + (\vec{k}+\vec{p}_i-\vec{\mu})^2]^{1/2}$$

its on-shell energy. The momenta and the energies of each particle are labeled in Fig. 11. The value of the coupling constant G_3, the mass M_Δ and the width Γ are the same as before (section 2.1.4 and Ref. 25). The isospin coefficient is $C = 1$ for the reaction $\gamma p \to \pi^- \Delta^{++}[p\pi^+]$, $C = -\sqrt{2}/3$ for the reaction $\gamma n \to \pi^- \Delta^+[p\pi^0]$, $C = 1/3$ for the reaction $\gamma p \to \pi^+\Delta^0[p\pi^-]$, etc. (the particles inside the bracket indicate the Δ decay mode). By construction the model leads to a vanishing cross section when a $\pi^0-\Delta$ pair is emitted whatever the Δ charge is, in agreement with experiment. It also predicts a ratio between the cross sections $\sigma(\gamma p \to \pi^+\Delta^0[p\pi^-])/\sigma(\gamma p \to \Delta\pi^-\Delta^{++}[p\pi^+]) = 1/9$ (the ratio of the isospin coefficients) in agreement with experiments.

The p(e,e'$\pi\Delta$)N reactions have been studied in Ref. 49 where the dependence of the cross section on the momentum transfer q^2 has been checked (see Ref. 53).

2.6 Vector Meson Production

These channels have been studied in bubble chamber and counter experiments[24,50]. Three basic mechanisms (Fig. 12) are necessary to account for the cross section[27,51]. The vector dominance model assumes that the photon is converted into a ρ or a ω which subsequently scatters on the nucleon. The diffractive dissociation model assumes that the photon is coupled to two or three pions, of which one scatters on the nucleon and eventually recombines with the others to lead to a vector meson in the final state. The resonance production mechanisms assume that the photon creates a baryonic resonance which decays by emitting a vector meson.

Contrary to single pion photoproduction reactions, no effective operator, convenient to use in a nucleus, has been systematically worked out, although some pieces of the amplitude have been used in the calculation of the contribution of the vector meson

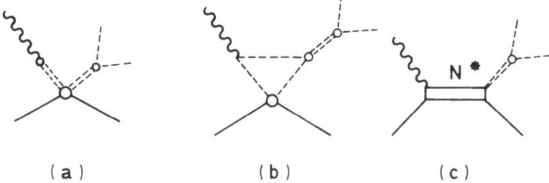

(a)   (b)   (c)

Fig. 12   The relevant diagrams for describing the photoproduction of vector meson.

exchange currents in nuclei. This work remains to be done.

On the experimental side, more data are needed to put the corresponding amplitudes on a firmer basis.

2.7 Strange Meson Production

Few measurements of the $\gamma N \to K\Lambda$ or $\gamma N \to K\Sigma$ cross sections have been performed so far[52,53]. Some angular distributions exist and even the total cross sections are known with a poor accuracy.

The structure of the amplitude is very close to the structure of the amplitude of the $\gamma N \to N\pi$ reactions[53]. It has been expanded into multipoles, and effective operators[54], based on a diagrammatic expansion, have been proposed.

The $\gamma p \to K^+\Lambda^0$ channel (or the $\gamma p \to K^+\Sigma^0$ channel) is particularly interesting. Since no resonances are coupled to the $K^+\Lambda^0$ state, the amplitude is entirely given by the Born terms, where mesons, nucleons or Δ's and Σ's are exchanged. Furthermore, the K^+–nucleus interaction is weak and is not dominated by resonant scattering as the π–nucleus interaction is, for instance. Therefore, this reaction provides us with a powerful tool, of which the weakness does not disturb the nucleus and makes possible the systematic study of hypernuclei.

However, the study of the electromagnetic properties of hypernuclei has not been performed so far. More data are also needed to check the model of the elementary interaction, since some coupling constants are not known precisely (especially those concerning the excited states of the K and the Λ or Σ).

This is a completely open field for the next generation of electron accelerators.

3. MESON ELECTROPRODUCTION ON FEW-BODY SYSTEMS

In a nucleus, the pion, which is photoproduced or which scatters on a bound nucleon, may escape freely or may undergo one or several rescatterings.

A powerful way to deal with these mechanisms is to expand the amplitude in a series of few relevant diagrams (Fig. 13). This diagrammatic approach is very well suited to the analysis of reactions induced in the few-body systems, and has two main advantages. On the one hand, it allows to deal easily with mechanisms which involve many partial waves. This is particularly convenient when dealing with processes occurring at high energy and/or involving few emitted particles. On the other hand, each diagram can be singled out by looking at its singularity and choosing the kinematics in such a way to enhance it. Indeed this possibility has been fully exploited at Saclay in the study of the

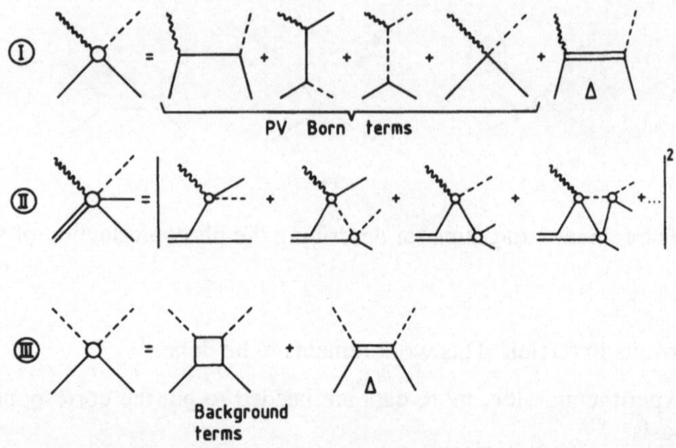

Fig. 13   The relevant diagrams in the description of the elementary operators $\gamma N \to \pi N(I)$, $\pi N \to \pi N(III)$. The multiple scattering series of the pion created at one bound nucleon is schematically depicted in (II). Contrary to the $\gamma N \to N\pi$ reactions, the $\pi N \to \pi N$ reactions are dominated by the Δ-isobar, and the background is very weak.

$d(\gamma,p\pi^-)p$, $d(\gamma,pp)\pi^-$ or the ^{4}He$(\gamma,p\pi^{\pm})$ reactions[9] and more recently of the $d(\gamma,\pi^{\pm})NN$ reaction[55]. The analysis of reactions induced in ^{3}He is in progress.

All the details of the calculation of each diagram are given in Ref. 9, where a general review of the experimental data can also be found. Here I will only update it and discuss two recent developments, the measurement of the inclusive spectra of pion photoproduced on deuterium by a monochromatic photon beam, and the effects of the ΔN interaction on the cross section of the $d(\gamma,p\pi^-)p$ reaction cross section.

3.1 The $d(\gamma,\pi^{\pm})NN$ Reaction[55]

This experiment has been made possible by the quality of the new Saclay "monochromatic" photon beam. It is obtained by positron in-flight annihilation in an hydrogen radiator, and the contribution of the bremsstrahlung tail is subtracted by repeating the measurement with a copper radiator. Since the shape of the bremsstrahlung spectra is not the same for the two radiators, a small undershoot appears on the low energy side of the annihilation peak, which is obtained by the subtraction procedure. This is clearly apparent in the lower part of Fig. 14, where its shape has been determined by the measurement of the $p(\gamma,\pi^+)$ reaction cross section.

This "monochromatic" photon beam has been used to determine the spectra of the pions emitted at a given angle in the $d(\gamma,\pi^+)nn$ reaction. A typical spectrum is shown in the upper part of Fig. 14. The broad peak is due to the pion photoproduction on a quasi-free nucleon. Its top corresponds precisely to the kinematics of the free nucleon reaction $p(\gamma,\pi^+)$, which is depicted in the bottom of the Fig. 14. Its width is due to the Fermi motion of the target nucleon. The peak, which appears at the high momentum end-point of the pion spectrum, is due to the strong s-wave interaction between the two nucleons which recoil here with a vanishing relative kinematical energy. When the theoretical cross section, which is drawn in the inset, is folded with the shape of the

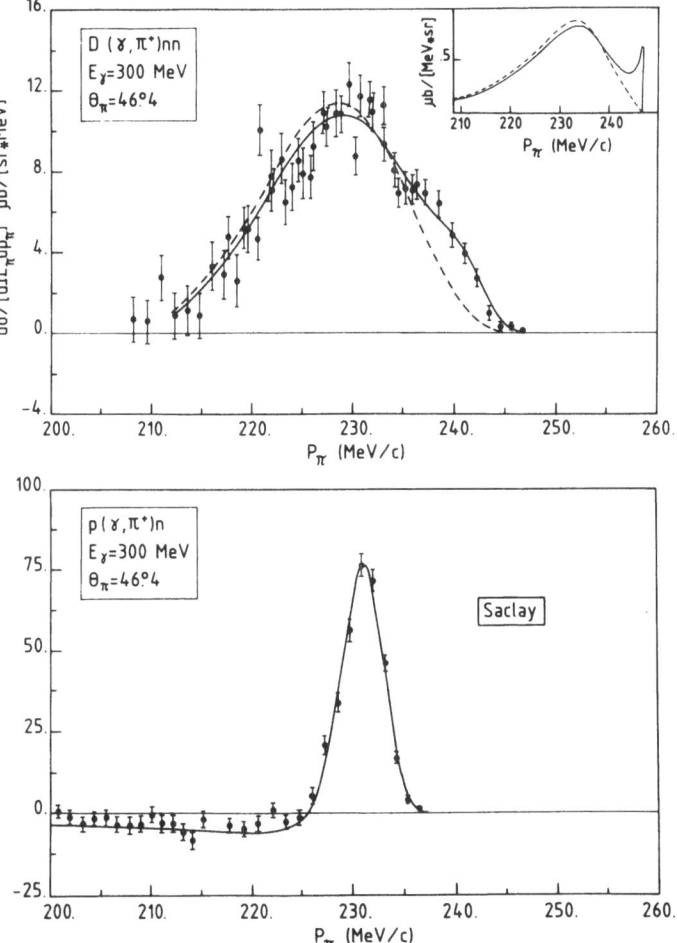

Fig. 14   The spectra of the pions emitted at $\theta_\pi = 46.6°$, when $E_\gamma = 300$ MeV, in the
$D(\gamma,\pi^+)nn$ and $p(\gamma,\pi^+)n$ reactions are plotted against their momentum. They
have been recently measured at Saclay with the monochromatic photon beam.
The theoretical cross section is drawn in the inset. It is folded with the experimental photon line shape, when it is compared to the data. The final state
interaction effects are (are not) included in the full (dashed) curves. The comparison is absolute: the experiment and the theory have been normalized to the
same number of incoming photons and to the same target thickness.

"monochromatic" photon spectrum, it reproduces fairly well the deuterium data. The
good agreement between the theory and the experiment, e.g. Fig. 14, is not only a measure of the accuracy of the measurements which can be performed with the modern intermediate energy photon beams, but also a check of the ability of this diagrammatic
method to accurately reproduce a wide bulk of experimental data[9]. It is worthwhile to
point out that there are no free parameters. The calculation relies very heavily upon the
free nucleon cross section and the deuteron wave function which are independently determined by the analysis of other reactions.

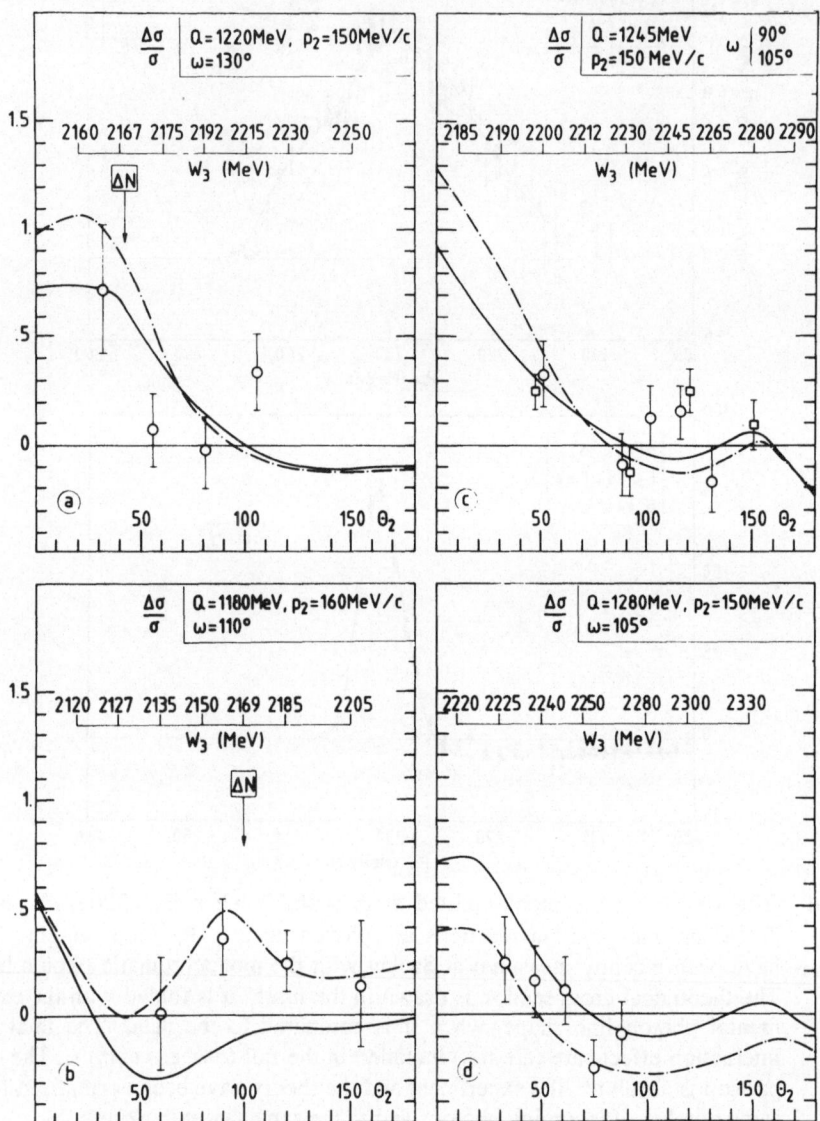

Fig. 15   The relative excess, with respect to the predictions of the quasi free mechanism, of the cross section of the reaction d(γ,pπ⁻)p as measured at Saclay[9,56].
The pion and one of the two protons were detected in coincidence, in such a
way that the mass Q and the momentum of the undetected proton are kept constant. The abcissa is the angle θ_2 of the undetected proton. The full line curves
include all the leading pion and nucleon rescattering graphs. The dashed line
curves show the effects of the part of the NΔ interaction which does not reduce
to the multiple scattering of the Δ constituents (see Refs. 9, 10 for a detailed
discussion).

Moreover, such a good agreement in the case of the d(γ,π^-)pp reaction tells us that the elementary $\gamma n \to p\pi^-$ reaction, which cannot be studied on a free neutron, is well under control. A detailed analysis of this experiment is given in Ref. 55. These measurements of inclusive pion spectra are a useful complement to the studies of more exclusive reactions (coincidence experiments), which I extensively discussed in Ref. 9.

This is the starting point of the analysis of inclusive pion spectra emitted on heavier nuclei which I will discuss later on (section 5). The study of the ^{3}He($\gamma,\pi^{\pm}$) reaction has been recently completed at Saclay, and the data are under analysis.

3.2 The N–Δ Interaction

The multiple scattering expansion reproduces a wide bulk of experimental data which have been obtained at Saclay, but a significant deviation remains near the NΔ threshold (when the mass of the πNN system is $W_3 = 2170$ MeV). It appears clearly in Fig. 15, where I show the Saclay data which have just been reanalysed[56]. It might be due to the part of the NΔ interaction (Fig. 16) which does not reduce to a sequential two-body scattering of the constituents of the Δ, and which looks like the nucleon-nucleon interaction (and is due to the exchange of virtual mesons).

As shown in Refs. 9, 10, it is possible to define an NΔ interaction and to add it to the multiple scattering series in a way which preserves unitarity and prevents double counting.

A coupled treatment of the NN and NΔ channels, in the K-matrix approximation, leads to a good account for the structure which exists near the NΔ threshold, as well as

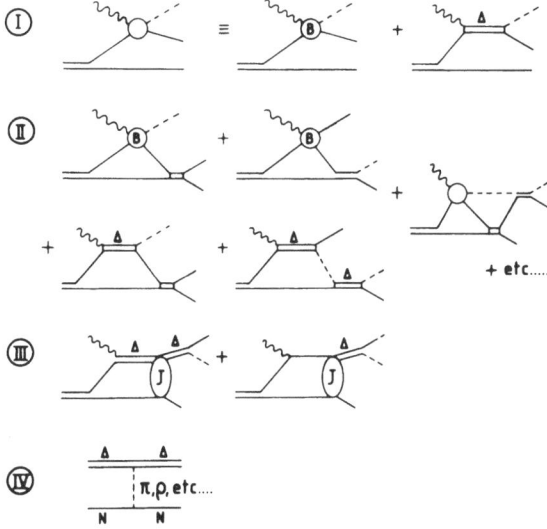

Fig. 16   The relevant diagrams in the analysis of the D($\gamma,p\pi^-$)p reaction. I: The quasi-free process where the elementary $\gamma n \to p\pi^-$ reaction amplitude has been split into the non-resonant Born terms and the Δ-resonance production amplitude. II: The dominant final state interaction diagrams which involve the rescattering of the Δ constituents. III: The diagrams which involve the part of the NΔ interaction which does not reduce to the rescattering of one of the Δ constituents. IV: A possible example of such a part of the NΔ interaction.

the resonant behaviour of the nucleon-nucleon $^1D_2$ phase shift[9,10]. It is worthwhile to point out that this part of the NΔ interaction interferes with the multiple scattering background. The effect is larger in Fig. 15b than in Fig. 15a.

It is also possible to define and to extract the NΔ phase shift near threshold[57]. Assuming that only the $^5S_2(N\Delta)$ and $^1D_2(NN)$ waves are coupled, I define the S-matrix as follows:

$$S = \begin{bmatrix} \eta e^{i\delta_1} & \sqrt{1-\eta^2}\, e^{i(\delta_1+\delta_2)} \\ \sqrt{1-\eta^2}\, e^{i(\delta_1+\delta_2)} & \eta e^{i\delta_2} \end{bmatrix}$$

where δ_1 and η are the known experimental NN phase shift and inelasticity, and δ_2 the ΔN shift which is to be determined. The width of the Δ is treated as explained in Refs. 9, 10, in such a way that the representation is unitary. This S-matrix parametrization is used in diagrams III of Fig. 16, and the values of δ_2 are extracted from the data. The results are summarized in Fig. 17, where each symbol corresponds to each different kinematics selected in Fig. 15. The overlap between each set of phase shifts is good and makes it possible to choose between K-matrix solutions[58] which fit equally well the NN phase shifts, but which predict different NΔ phase shifts. The determination of the NΔ amplitude is, therefore, a very strong constraint in the analysis of the coupled NN→NΔ systems, and a crucial observable in the search of dibaryonic resonances and the study of their nature. However, I would like to emphasize the two following points. On the one hand, only the lowest NΔ partial wave has been considered. The effects of higher partial waves (P-waves) is under study and may affect the behaviour of δ_2 at high energy. On the other hand, the large error bars are due to the fact that this experiment is at the limit of the capability of the present Saclay machine. This experiment is difficult to perform with a 1% duty factor beam. A CW machine will allow its systematic study and will make it possible to perform a full partial wave analysis of the NΔ final state. It will also allow its extension to the (e,e'Δ) experiment.

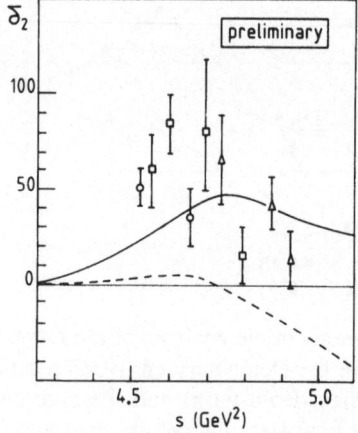

Fig. 17   The phase shift δ_2 of the NΔ S-wave is plotted against the c.m. squared total energy s. Each symbol corresponds to a given kinematical setting in Fig. 15. The two curves correspond to two K-matrix fits of the $^1D_2$ (NN) phase shift[58].

3.3 Prospects

All these studies have allowed us to get a fair understanding of the basic mechanisms which govern the Δ creation and propagation in the few-body systems.

However the limits of the capabilities of the present generation of electron accelerators have been reached. Future machines will allow for more systematic studies of these meson electroproduction reaction induced on the few-body systems. On the one hand, the increase of the duty factor will make it possible to perform more easily coincidence experiments, which are not possible or difficult to perform today. On the other hand, the increase of the available energy will make it possible not only to extend the present measurements to the study of reactions induced by virtual photons, but also to study the creation and the propagation the in nucleus of baryonic resonances with higher mass than the Δ, vector mesons, and strange baryons (Λ and Σ) .

This is an open field, the study of which should be one of the axes of the research programs initiated at the new electron facilities. On the one hand, the simultaneous study of the NN, the NΔ, the NN', and the ΛN or ΣN channels is the necessary condition for a good understanding of the structure and the interaction of baryons. On the other hand, the study of the creation and the propagation of vector mesons in nuclei is complementary to the study of the creation and the propagation of pions and leads to strong constraints on the interaction of vector mesons with nucleons.

4. VIRTUAL MESONS IN THE FEW-BODY SYSTEMS

Contrary to real photoproduction reactions, the analysis of virtual pion photoproduction reactions is not parameter free.

The pion which is photoproduced on a bound nucleon may also be kept inside the nucleus and be reabsorbed by another nucleon. Although the method of calculation of this pion reabsorption diagram is the same as the pion rescattering one (see Fig. 18), the

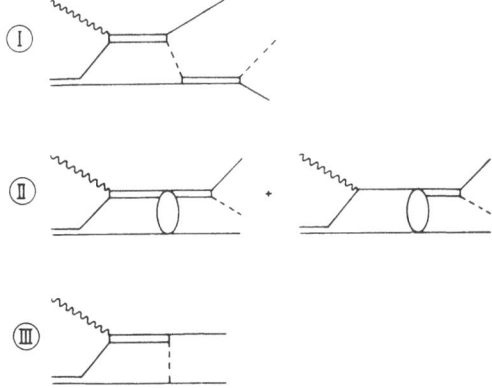

Fig. 18   The different pieces of the ΔN interaction. I: The exchange part is mediated by a pion which can propagate on shell. It is very well described by the multiple scattering of the Δ constituents (pion or nucleon). II: The direct part of the NΔ interaction is mediated by the exchange of virtual mesons, and looks like the interaction between two nucleons. III: The NΔ and NN systems are strongly coupled through the NN → NΔ transition amplitude which dominates many break-up channels in reactions induced on the few-body systems.

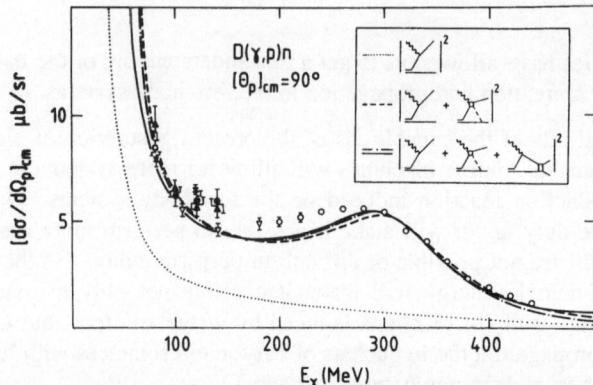

Fig. 19   The excitation function at $\theta_p = 90°$, of the d(γ,p)n reaction is plotted against the
incoming photon energy E_γ. The high energy experimental points have been
obtained at Bonn[60]. At low energy the references can be found in Refs. 28, 59.
The dotted and dashed curves correspond to the plane wave calculation without
and with the exchange current contribution respectively. The dash-dot and the
full curves include also the neutron-proton final state interaction, in the S and
S+P states respectively.

virtual pion which is reabsorbed is highly off its mass shell, and the free pion photoproduction operator, as well as the pion absorption operator, should be corrected. Two ways
are usually followed to overcome this difficulty.

On the one hand, since it is far off-shell, the exchanged pion is sensitive to the finite
size of the nucleon, and I use at each pion-baryon vertex a monopole form factor

$$F_\pi(q^2) = \frac{\Lambda_\pi^2 - m_\pi^2}{\Lambda_\pi^2 - q_\pi^2}$$

where q_π^2 is the squared mass of the virtual pion.

On the other hand, other virtual mesons can also be emitted and reabsorbed. Among
them the ρ–meson exchange diagram plays an important role (in which case I use a
dipole ρ–baryon form factor with a cut-off mass equal to two times the nucleon mass).

I have determined the values of the cut-off mass Λ_π and the ratio G_ρ^2/G_π^2 between the
square of the rho- and the pion-baryon coupling constants by fitting[28,59] the 90° excitation function of the d(γ,p) reaction cross section (Fig. 19). It turns out that, in this reaction, the ρ–exchange mechanism is negligible below the pion production threshold, and
only affects significantly the Δ–N $\rightarrow$ N–N transition in the resonance region. It is therefore possible to separately determine the cut-off mass $\Lambda_\pi = 1.2$ GeV at low energy and
the ratio $G_\rho^2/G_\pi^2 = 1.6$ in the Δ region. They lie in the range of the uncertainties of the
currently accepted values[28].

Once the choice is made there are no other free parameters, and the model reproduces
decently the available experimental data.

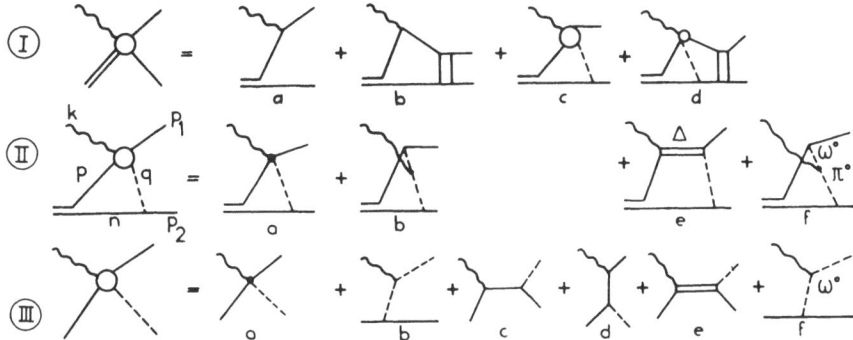

Fig. 20   The deuteron photodisintegration amplitude is expanded in terms of leading
diagrams I. The pion reabsorption amplitude Ic is expanded into the relevant
diagrams II, which come from the expansion III of the elementary $\gamma N \to N\pi$
amplitude in terms of Born terms (IIIa, b) the nucleon exchange terms (IIIc, d)
the Δ-formation terms (IIIe) and the ω^0 exchange terms (IIIf).

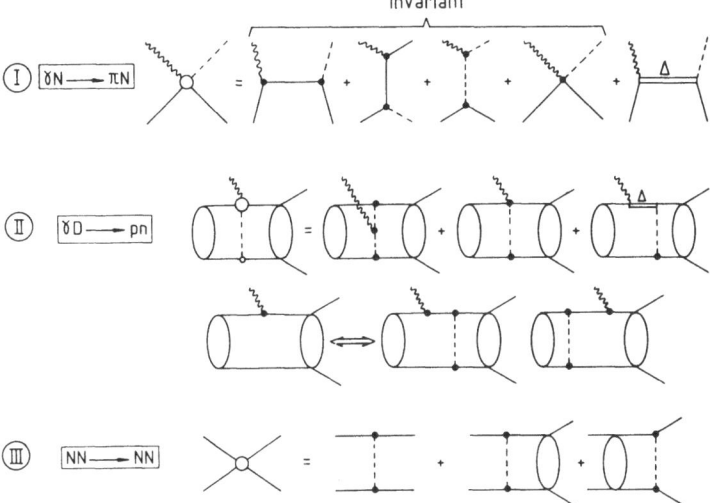

Fig. 21   (a) In the elementary $\gamma N \to \pi N$ reaction amplitude, only the sum of all the
Born terms is gauge invariant. (b) When the pion is created on a nucleon
bound in a nucleus, each diagram contributes to the meson exchange amplitude. However, the diagrams that correspond to the nucleon Born term are
already included in the wave function of the initial and the final states, and
should be disregarded. (c) These wave functions are obtained by iterating the
driving terms in the nucleon-nucleon interaction, and gauge invariance requires
that if the photon interacts with the nucleons, it also interacts with the
exchanged mesons that bind them together.

4.1 Gauge Invariance

The relevant diagrams in the analysis of the d(γ,p)n or the d(e,e'p)n reactions are depicted in Fig. 20. To each graph, in the elementary photoproduction amplitude, corresponds a graph in the pion reabsorption amplitude in deuterium. However the nucleon Born terms (IIIc, d) must be disregarded, since they are already included in graphs Ia, b, when the nuclear wave functions are generated by OBE potential. This important point is illustrated in Fig. 21, where the nucleon-nucleon scattering amplitude (or the wave function) is expanded into its driving terms and the multiple scattering terms. In other words, since these nucleon Born terms are already included in the normal part of the photodisintegration amplitude, and since only the sum of all the Born terms are gauge invariant, the gauge invariance of the full photodisintegration amplitude requires that, besides nucleons, the electromagnetic field couples also to pions which bind the nucleons together. A schematic explanation is given in Fig. 21, and a more formal demonstration is given in the Appendix.

This statement can be extended to the Born part of the vector meson exchange currents. The Δ-resonance part of the exchange currents, and the ω^0-π^0 exchange currents are gauge invariant by themselves, and are not linked to the nuclear wave function, unless those isobars are explicitly included in the wave functions.

4.2 The Two-Body Break-Up Channels

This model provides us with a good representation of the angular distributions of the d(γ,p)n reaction. As an example, the proton angular distributions at $E_\gamma = 260$ MeV and 95 MeV are plotted in Figs. 22, and 23. While final state interactions do not affect very much the unpolarized cross sections, they are essential to reproduce the proton or the neutron polarization observables (Fig. 23). For instance, the neutron polarization at $E_\gamma = 95$ MeV is different from zero only if neutron-proton rescattering in the P-wave is taken into account.

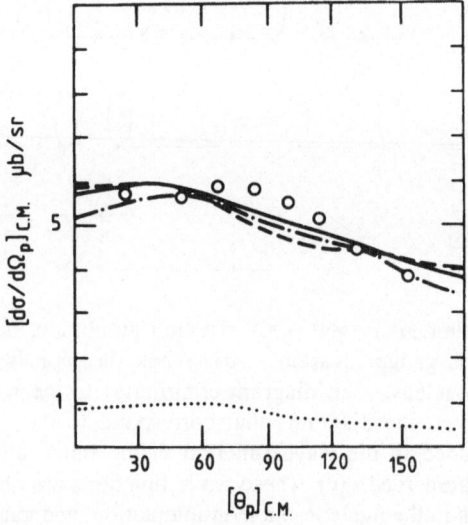

Fig. 22   The angular distributions of the protons emitted in the D(γ,p)n reaction when $E_\gamma = 260$ MeV. The experimental points, as well as the meaning of the curves, are the same as in Fig. 19.

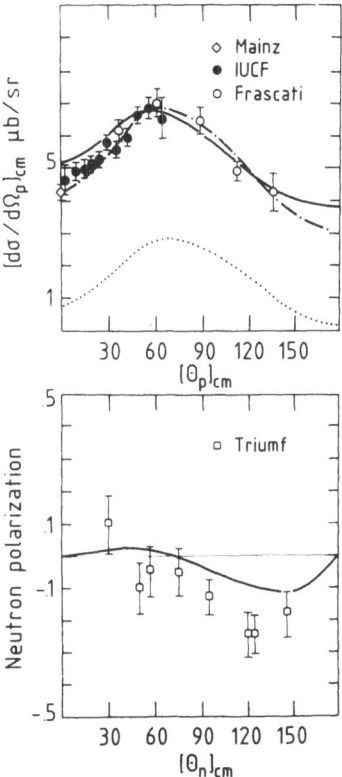

Fig. 23   In the upper part, the angular distribution of the unpolarized differential cross section of the D(γ,p)n reaction at $E_\gamma = 95$ MeV is plotted. The experiments have been performed at Mainz[61], IUCF[62], and Frascati[63]. The meaning of the curves is the same as in Fig. 19. In the lower part, the polarization of the neutron emitted in the D(γ,n)p reaction, when $E_\gamma = 95$ MeV, is plotted against the neutron angle. It is measured along the k×n axis. The experimental points come from the p(n,d)γ reaction recently performed at TRIUMF[64].

Final state interactions help also to reproduce the cross section of the d(e,e'p)n reaction[65]. As an example, Fig. 24 shows the angular distribution of the proton detected in coincidence with the electron at the top of the quasi-elastic peak. This experiment probes the nucleon momentum in deuterium between 0 and 150 MeV/c. Final state interactions decrease the plane wave cross section by 10 to 20%. I refer the reader to Ref. 28 for a detailed analysis of these data and for a discussion of the limits of the model when high momentum components are probed in the nuclear wave function.

The most striking feature of the model is that it leads also to a good accounting for the forward angle ($\theta_p = 0°$) photodisintegration cross section (Fig. 25), which has always been a puzzle. Details are given in Ref. 28. I wish only to point out here that the agreement is due to the fact that I compute directly the photodisintegration amplitude from the space part of the current (I do not use the Siegert hypothesis) and to the fact that all terms of order $1/m^2$ are taken into account from the beginning of the calculation.

The extension of this formalism to the photo- and the electrodisintegration of the three body systems is straightforward. The only change is the nuclear wave function[20]

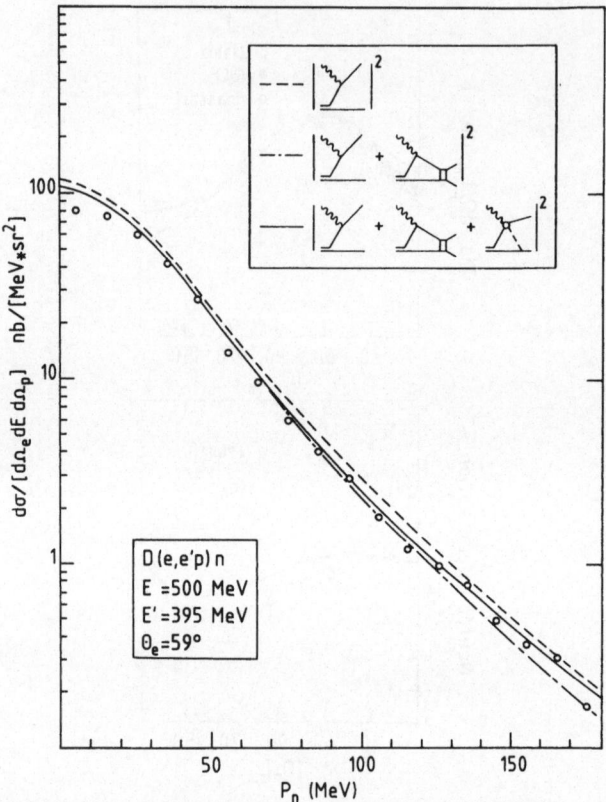

Fig. 24   The angular distribution of the protons emitted in the D(e,e′p)n reaction, when
E = 500 MeV, E′ = 395 MeV and θ_e = 59°, is plotted against the momentum p_n
of the recoiling neutron. The dashed curve corresponds to the Born approximation. The neutron-proton final state interactions are included in the dash-dot
curve. The experiment has been recently performed at Saclay[65]

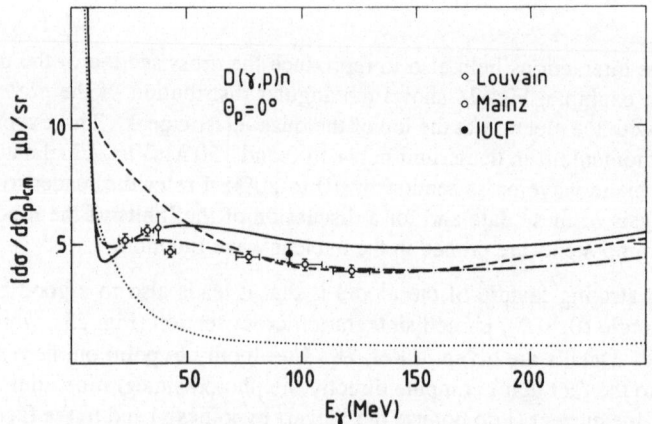

Fig. 25   The excitation function at θ_p = 0°, of the d(γ,p)n reaction is plotted against the
incoming photon energy. The meaning of the curves is the same as in Fig. 19.
The references for the experimental points can be found in Ref. 28.

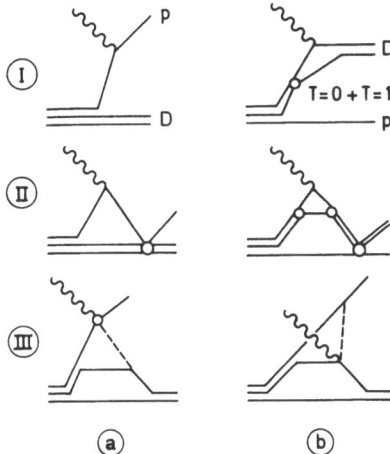

Fig. 26   The relevant diagrams in the analysis of the ^{3}He(γ,p)D reaction.

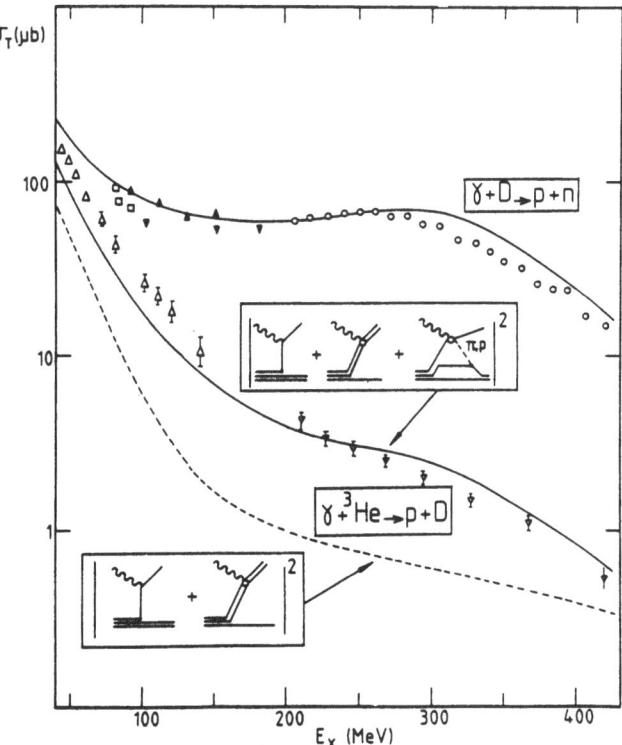

Fig. 27   The integrated cross section of the two-body photodisintegration of the deuteron and the ^{3}He nucleus are plotted against the energy of the incoming photon. The high energy data points have been recently measured at Bonn[60,66]. The meaning of each curve is explained in the insets.

and there are no free parameters. The relevant diagrams of the two-body break-up are shown in Fig. 26, and the calculation is schematically described in Ref. 22.

The total two-body photodisintegration cross section of ^{3}He is plotted in Fig. 27 together with the total photodisintegration cross section of deuteron. In both cases the contribution of the exchange current is essential to reproduce the data[60,66]. The angular distribution of the proton measured at $E_\gamma = 240$ MeV is plotted in Fig. 28. The meson exchange contribution is very important and the cross section is very sensitive to the details of the model. For instance, the fit to the experimental data[66,67,68] is significantly improved when the d-wave parts of the spectator nucleon and the deuteron, which is emitted in the meson exchange diagram, are also taken into account (The d-wave part of the active deuteron is also important[59], but it is always taken into account in the two-body operator).

It is only the use of a good, and realistic, three-body wave function and of good elementary operators which allows such a good fit. Had I used a more phenomenological three-body wave function (as a cluster representation which fits the three-body form factors[69] for instance) the disagreement between the theory and the experiment would have been catastrophic.

The proton polarization (Fig. 29) in the ^{3}H(γ,p)d reaction is even more sensitive to the details of the model. These final state interactions play a significant role and help to

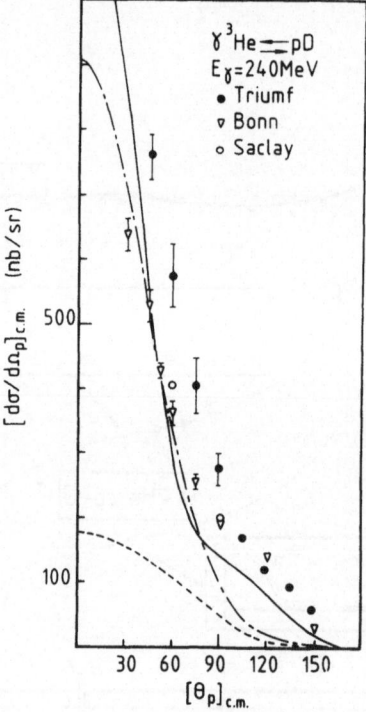

Fig. 28   The angular distribution of the protons emitted in the ^{3}He(γ,p)d reaction when $E\gamma = 240$ MeV. The experiments have been performed at Bonn[66], Saclay[67], and TRIUMF[68]. The dashed curve corresponds to the Born approximation. The dash-dot curve includes the meson exchange contributions, where only the S-wave parts of the final deuteron and the three-body system are retained. Their D-wave parts are included in the full line curve.

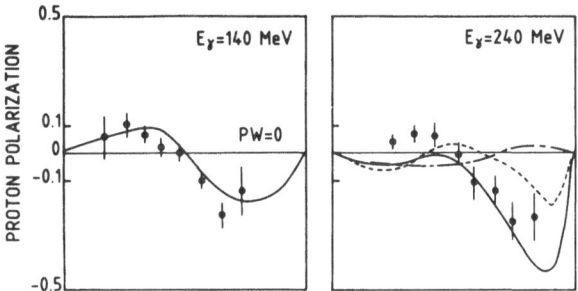

Fig. 29   The proton polarization in the ^{3}He(γ,p)d reaction[68]. Only the S wave parts of
the three-body and the deuteron wave function are retained in the dot-dash
curve (plane wave). The D-wave parts are included in the broken line curve.
The final state rescattering of the proton and deuteron in the S, P and D waves
is included in the full line curve.

bring the theory close to the data. For instance, the proton polarization at $E_\gamma = 140$ MeV
vanishes in a plane wave treatment. It is worthwhile to point out that these data have
been obtained in the inverse reaction: the radiative capture of polarized protons on deuterium[68]. This is a good example of the complementarity of high intensity proton and
electron machines.

The model provides us also with a good understanding of the two-body electrodisintegration of ^{3}He. As in the deuteron case, the final state interaction in the emitted p-d
pair and the meson exchange currents help to satisfactorily reproduce the data recently

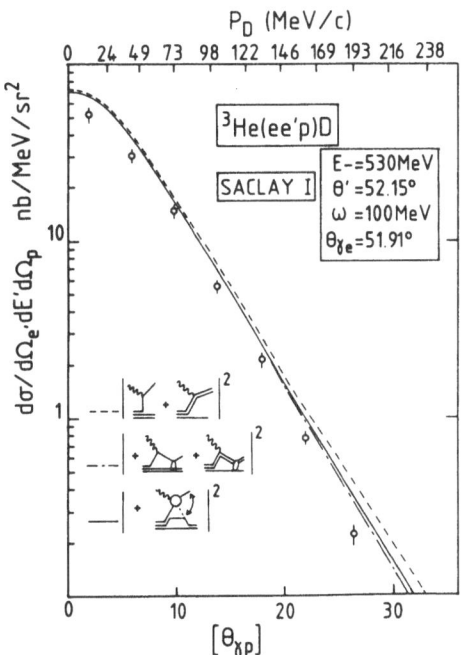

Fig. 30   The angular distribution of the proton emitted in coincidence with the scattered
electron in the ^{3}He(e,e'p)D reaction[70] is also plotted against the momentum of
the recoiling deuteron (upper scale).

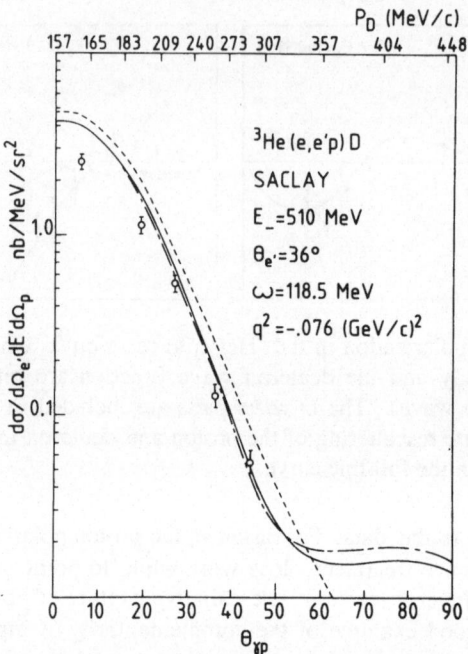

Fig. 31   The same as in Fig. 30, but for higher momentum of the recoiling deuteron.

obtained at Saclay[70], Figs. 30, 31.

Recently the ^{3}He(e,e'd)p reactions have been studied when the undetected proton is almost at rest in the laboratory[71]. This kinematics enhances the contribution of diagram Ib in Fig. 26, where the incoming photon interacts with a correlated p-n pair (either in the T = 0 or the T = 1 state) and ejects a deuteron in the final state. The data are shown in Fig. 32. This mechanism can be viewed either as a consequence of the antisymmetry of the two protons in the p-d final state, or as a final state interaction mechanism when the proton interacts with one nucleon to form the emitted deuteron. Both descriptions are equivalent and lead to the same matrix element[22].

This brief review shows that the two-body break-up channels are well understood. However, some discrepancies still remain. On the one had, the nucleon momentum distribution of ^{3}He seems to be slightly overestimated for small values of the nuclear momentum (Figs. 4, 30). On the other hand, the unpolarized angular distribution of the ^{3}He(γ,p)d reaction (Fig. 28) is underestimated by the model at backward angles, while it is well reproduced at forward angles.

The source of these discrepancies may be the three-body wave function itself. Although a general consensus is now reached on the stability and the reliability of the numerical solutions of the Faddeev equations, all the existing wave functions underbind the three-body system. This shortcoming has sizeable consequence when the low momentum components of the wave function are probed[22], and it is very likely that the use of a wave function, which produces the good binding energy of the three-nucleon system, will help to bring the theory close to the data in Fig. 30. Such a wave function might also help to reproduce the cross section of the ^{3}He(γ,p)d reaction at backward angles, which involves high momentum components.

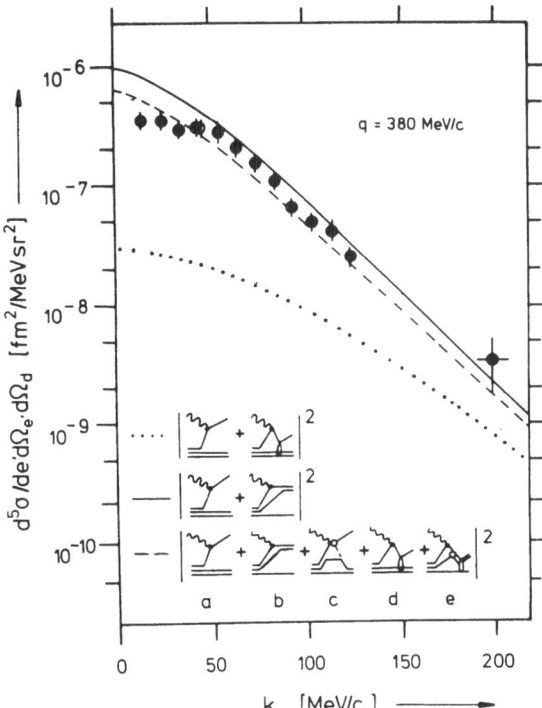

Fig. 32   Experimental coincidence cross sections[71] of the ^{3}He(e,e'd)p reaction as function of recoil momentum (k) at fixed momentum transfer (q). The curves correspond to photon-proton coupling including FSI (dotted), photon-proton and photon-pn coupling in PWIA (solid) and a full calculation including FSI (dashed). The relevant diagrams are shown in the inset.

However, the cross section of these two-body break-up channels is very sensitive to the small components of the three-body wave function (see Fig. 28). So far I have retained only the dominant S- and D-waves. The small P-wave components may also play a role, and they must be included before making a definite statement on the source of the disagreement.

These discrepancies may also be the signature of mechanisms which involve the three nucleons of the target. Some examples are given in Fig. 33. The photoproduced pion may scatter twice, or two pions may be created on one nucleon and subsequently reabsorbed. Those diagrams are expected to contribute at backward angles, or at high energy, when the diagrams involving two nucleons are suppressed by the nuclear form factors.

Although all these possible explanations must be investigated, and open new perspectives for looking for non trivial mechanisms, it is important to begin to firmly establish the basis of our knowledge of the two-body mechanisms.

4.3 The Three-Body Break-Up Channels

Since the neutron and one of the protons must recombine into the deuteron, the corresponding form factors strongly suppress the two-body mechanisms in the two-body

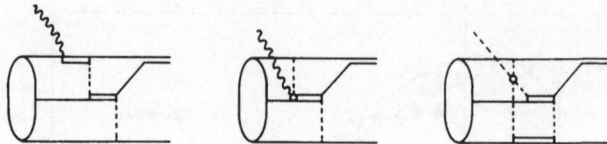

Fig. 33   Possible three-nucleon mechanisms in pion or photon induced reactions at
intermediate energy. They involve the double scattering of the pion, the creation of two pions at one nucleon and/or the creation of two Δ's.

break-up channels.

On the contrary, in the three-body break-up channels, it is possible to select kinematical conditions where these two-body mechanisms are not suppressed.

The relevant diagrams are shown in Fig. 34, and the study of the ^{3}He(e,e'p)np (Refs.
70, 72) and ^{3}He(γ,p)np (Ref. 73) are in progress at Saclay. Fig. 35 shows the preliminary
results of the study of the ^{3}He(γ,p)X reaction. This is the spectrum of the proton emitted
at a fixed angle. While the peak, which appears at the highest momentum, corresponds to
the two-body break-up channel, the dominant effect comes from the disintegration of a
nucleon pair almost at rest. The top of the corresponding peak corresponds precisely to
such a kinematics, and its width is due to the Fermi motion of the pair. The curve is the
convolution of the model[22] and of the monochromatic photon line shape. It is obtained
by in flight annihilation of positrons, and the negative tail comes from the subtraction of
the bremsstrahlung background (I refer to Ref. 55 for a detailed discussion of the experimental method). The parameters are the same as in the treatment of the two-body
break-up channels, but although the cross section is still sensitive to the correlated two-
nucleon wave function, it is only sensitive to the long range part of the spectator nucleon
wave function, which is basically given by its static properties, and does not involve the
small components of the three-nucleon wave function. In that particular kinematics, the

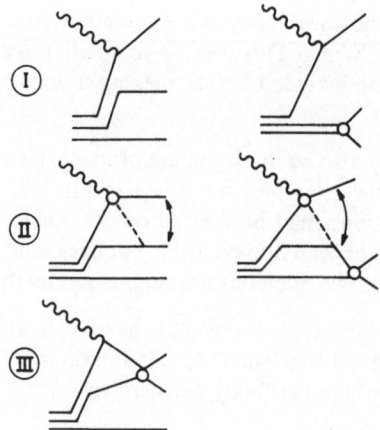

Fig. 34   The relevant diagrams in the analysis of the ^{3}He(γ,p)np reaction.

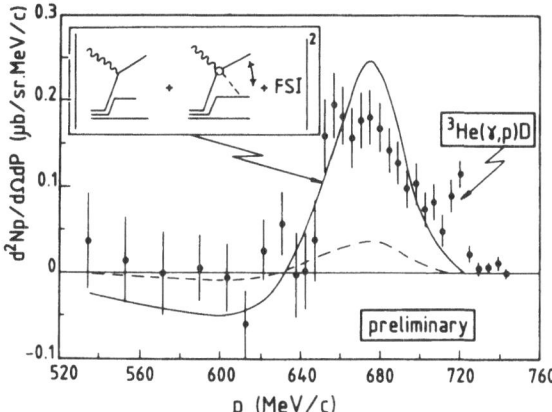

Fig. 35   The spectrum[73] of the proton emitted at $\theta_p = 23°$ in the reactions ^{3}He(γ,p)X induced by monochromatic photons of $E_\gamma = 310$ MeV. The curve is the result of the folding of the model[22] and the measured beam line shape (see Ref. 55 for the discussion of the experimental method). The dashed curve does not take into account the exchange current graph (on the right of the inset).

three-body break-up channel is dominated by the resonant exchange current contribution (where the Δ is created in one nucleon and its decaying pion if reabsorbed by another). The contribution due to the disintegration of a correlated pair (diagram I in Fig. 34) is suppressed here.

This is not the case in the spectrum of the proton emitted in coincidence with the scattered electron in the ^{3}He$(e,e'p)$np reactions, which is shown in Fig. 36, where the exchange current contribution is small. On the one hand, the energy transferred to the nucleus is smaller ($\omega = 200$ MeV instead of 310 MeV). On the other hand, exchange currents do not contribute to the longitudinal component which is a sizeable contribution of the cross section, since the electron is scattered at a forward angle ($\theta_{e'} = 25°$).

The two measurements are complementary, and illustrate how the mechanisms involving two nucleons can be singled out and studied in the three-body break-up channels. Their systematic extension to various kinematics is under way at Saclay. More specifically, the separation of the longitudinal and transverse cross sections.

4.4 The Electromagnetic Form Factors

Those two-body mechanisms are also very important in the analysis of the electromagnetic form factors of the few-body systems (Fig. 37). They correspond to the popular meson exchange mechanisms. They have been more systematically studied than in the break-up channel, although their understanding relies upon a good control of the nuclear wave function of which they are sensitive to the small components. I refer the reader to Ref. 74, and I wish only to emphasize the following points: While gauge invariance constrains the leading terms (order $1/m$), low energy theorems and PCAC constrain also the terms of order $1/m^2$ (see for instance Ref. 9). An elegant way to satisfy

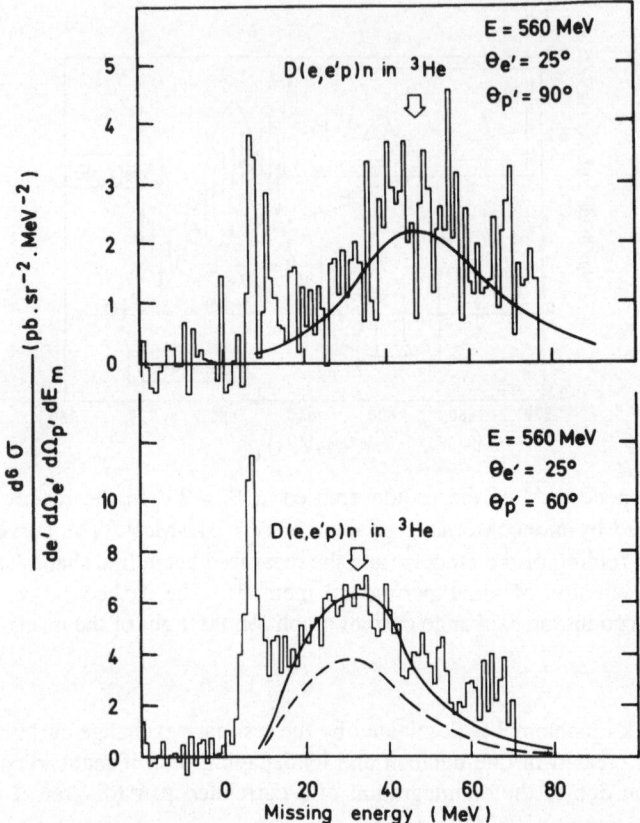

Fig. 36   The same as in Fig. 35, but for the ^{3}He(e,e'p)np reaction when E = 560 MeV,
ω = 200 MeV, $\theta_{e'}$ = 25° and θ_p = 60° or θ_p = 9°. The abscissa is the missing
mass of the undetected pn pair. The arrows indicate the momentum of the pro-
ton emitted in the D(e,e'p) reaction for the same kinematics.

these two requirements is to deduce the exchange currents from the pseudo-vector πNN
Lagrangian[9,28,59]. It is remarkable that such a procedure allows us to reproduce not only
the magnetic form factor but also the charge form factors of ^{3}He and ^{3}H, see (Ref. 74),
for momentum transfer as high as $q^2 \sim 1$ (GeV/c)^2.

As in the case of the break-up reactions, this good agreement, at such high momentum transfer, is achieved by introducing two extra parameters, which are not constrained
by gauge invariance and PCAC: the cut-off mass of the πNN form factor and the
ρ–nucleon coupling constant. I have summarized their values in Table 1, and it is
remarkable that the analysis of different channels leads to the same set of parameters.

We are naturally led to ask ourselves the following questions: What is the physical
meaning of this πNN form factor? Is the ρ–exchange concept really relevant?

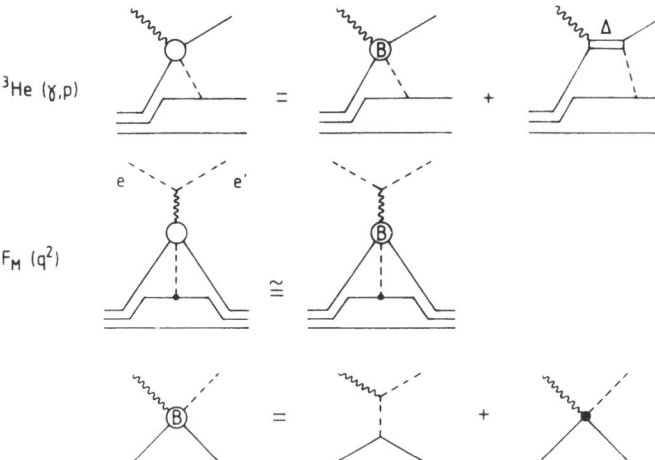

^{3}He (γ,p)

F_M (q^2)

Fig. 37   The relevant meson exchange diagrams in the ^{3}He(γ,p) reaction and the magnetic form factor of ^{3}He. The symbol B stands for the non-resonant Born terms
of the elementary operator, which are not implicitly included in the nuclear
wave functions.

Table 1   The values of the cut-off mass Λ_π of the πNN monopole form factor and of
the ρ–nucleon coupling constant $G_\rho = g_\rho(1+\kappa_v)$

		Break-up Channels		Electromagnetic form factors	
Experiment	NN ---- NN	Photo-disintegration	Pion-disintegration	Mathiot	Sauer
	Bonn[75]	[28]	[76]	[77]	[74,78]
Λ_π(GeV)	1.3	1.2	1.2	1.25	1.2
G_ρ^2/G_π^2	2.6	1.6	2.4	2.2	2.26
ρNN form factor (GeV)	monopole $\Lambda_\rho = 1.5$	dipole $\Lambda_\rho = 2m_N$	dipole $\Lambda_\rho = 2m_N$	monopole $\Lambda_\rho = 1.25$	monopole $\Lambda_\rho = 1.2$

4.5 The Limit of the Classical Description

The ρ–exchange mechanisms have been criticized[16], since it occurs at very short distance (~ .3 fm) and may simulate more subtle mechanisms which involve the quark degree of freedom of the nucleons.

The use of πNN form factor is a way to go beyond the description of a nucleus in terms of point-like structure nucleons, and to take into account their finite size. The cut-off mass of $\Lambda_\pi = 1.2$ GeV, which is needed to reproduce the data, corresponds to a core radius of the nucleon of approximately .5 fm, very close to the "little bag" radius[79] . If this were the reality, the problem of quarks in nuclei would reduce to the understanding of the various nucleon form factors. Indeed it is possible[80] to reexpress the pion exchange amplitude in terms of the direct coupling of the pions to the quarks of a bound nucleon (Fig. 38). This approach provides us with a dynamical model for the πNN form factors. At leading order (1/m), the structure and the strength of the meson exchange operators are the same as in the case of the coupling of pion to the nucleon, and the agreement with the magnetic form factor is the same. For the charge form factor, where the exchange current contributes at order $1/m^2$, this quark approach leads to a better agreement than the traditional treatment of meson exchange currents, in which the pseudo-scalar coupling is used at the πNN vertex. However, it is known for a while that the pseudo-vector coupling should be preferred (see Section 2.2 and Ref. 9), and a recent calculation[74] of the three-nucleon system charge form factor, which uses it, leads to a fair agreement with the experimental data[81]. Therefore, both approaches, pion-nucleon and pion-quark coupling, lead to the same agreement with the experimental data, provided the same nucleon size is chosen.

But if the nucleon radius is of the order of 1 fm, as in other current models[82], the cut-off mass is only $\Lambda_\pi \sim 600$ MeV, and the pion exchange mechanisms are strongly suppressed. Room is left, even at low momentum transfer for more complex mechanisms where the quarks of two distinct nucleons are mixed together[16,83–85] and which give rise

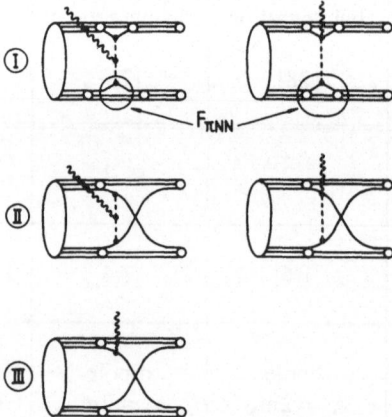

Fig. 38   I: A possible description of the exchange current, in terms of the direct coupling of the pion to the constituent quarks. II: The graphs which must be considered when the final state quark wave function is antisymmetrized. Their importance may be hidden by the phenomenological determination of the πNN form factor. III: The quark interchange mechanism which is expected to be very important at very high momentum transfer.

to six-quark clusters in the ground state wave function. Some examples are shown in Fig. 38. Of special interest are the diagrams II, in which two quarks are interchanged at the very time when the pion absorbs the photons. Due to the antisymmetry of the quark in the final state wave function, these mechanisms must be treated on the same footing as the direct mechanisms (Fig. 38-I) which have been only considered in Ref. 80. While these effects of the Pauli principle are expected to be small if the nucleon size is small enough, they might be significant when their size allows the nucleon to overlap in a nucleus. It is very likely that the complete treatment of these quark interchange mechanisms is the way to reconcile the apparent contradiction between the large nucleon size predicted by current nucleon models and the small size required by the conventional analysis of the electromagnetic properties of the few-body systems. Works in that direction are already underway[86,87].

For sake of completeness, the pure quark interchange diagram (Fig. 38-III) must be considered[16], but it is expected to contribute at very high momentum transfer and to govern the asymptotic behaviour of the amplitudes.

Let me summarize. We have now a consistent and successful framework to analyse the reactions induced by photons and pions in the few-body systems. Their extensive studies, performed during the last twenty years, have allowed us to understand the basic mechanisms which involve the pion and Δ's degrees of freedom in the nucleus. However, we have reached the limits of this standard model, and it is likely that the current phenomenology hides some more fundamental processes.

4.6 Prospects

Obviously, we have to go beyond the correlations which are due to the exchange of pions or the creation of Δ's. On the one hand, the short range correlations are still badly known and must be studied extensively. On the other hand, we must look for experiments where the effects of pions and Δ's degrees of freedom are strongly suppressed, and where no free parameters are left to play with. To close this chapter, I will discuss now two such experiments: the low energy part of the quasi-elastic peak, and the ^{3}He(e,e′NN) experiment.

4.6.1 The Low Energy Side of the Quasi-Elastic Peak. Here the electron scatters on nucleons moving inside ^{3}He with high momentum and far from their mass shell: this kinematical region is forbidden in the scattering of electrons on free nucleons. This part of the spectrum of the electrons inelastically scattered at $\theta_{e'} = 8°$ on ^{3}He, is plotted on a logarithmic scale in Fig. 39, for the two values of the energy $E_- = 7.26$ GeV and $E_- = 3.26$ GeV (which has already been shown in Fig. 4). The pure quasi-elastic mechanism, where the virtual photon interacts with one nucleon and where the two others are spectators, is unable to reproduce its low energy part. Although a factor two is still missing when $q^2 \sim 1.(GeV/c)^2$, the final state interaction effects, which corresponds to the interaction of the virtual photon with a correlated nucleon pair, improves dramatically the agreement between the model and the data. The reason is simple: this active pair is almost at rest in ^{3}He and the corresponding amplitude is sensitive to the low momentum components of the wave function. Just near threshold the strong low energy p-d scattering dominates the spectrum.

One of these two-nucleon mechanisms has been beautifully singled out in the study of the ^{3}He(e,e′d)p reaction[71] recently performed at Amsterdam. The experiment has been discussed in section 4.2, and it can be seen in Fig. 32 that the model leads to a good agreement with the data, provided that final state interactions are taken into account. Although the squared mass of the virtual photon is smaller than in the inclusive electron

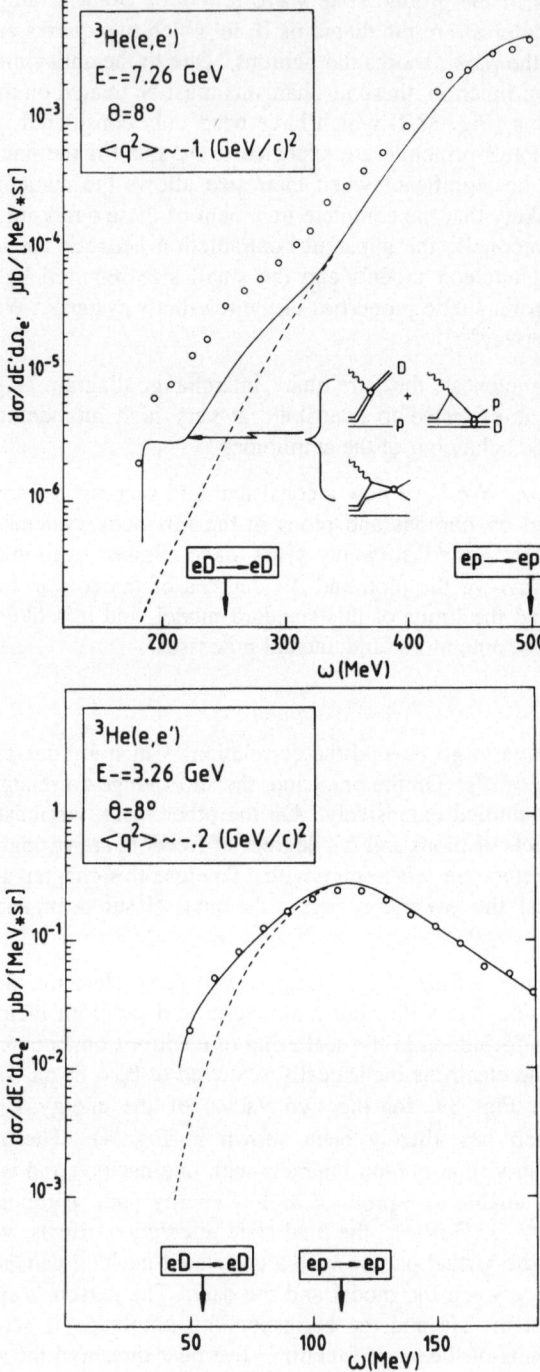

Fig. 39   The low energy part of the spectra of electrons inelastically scattered by ³He
are plotted on a logarithmic scale. The arrows correspond to elastic scattering
of electrons on free protons and deuterons at rest. The broken line curves take
into account only the quasi-elastic scattering of an electron on a nucleon pair,
the p-D rescattering for the two-body break-up channel, and the nucleon-
nucleon rescattering in the active pair for the three-body channel.

spectrum, this coincidence experiment tells us that the corresponding two-nucleon exchange mechanism is relevant.

These final state interaction effects prevent us here from extracting the high momentum part of the three-body wave function as was done in Ref. 88. As can be seen in Fig. 40, the same disagreement occurs also in the d(e,e') reaction near $q^2 = -1.(GeV/c)^2$, and it is remarkable that the theoretical and the experimental ratio between the ^{3}He(e,e') and d(e,e') cross sections are the same. Two consequences immediately follow: in that energy and momentum range, the ^{3}He(e,e') cross section is dominated by the two-nucleon mechanisms, and the source of disagreement between the deuterium data or the ^{3}He data and the model is the same.

It does not come from a lack of knowledge of the high momentum components in the wave function, since the same momentum range has also been probed in coincidence experiments which have been performed recently at Saclay[65,70]. As can be seen in Fig. 31, the same model leads to a good agreement, provided that the final state effects are also taken into account. Had I used the modified momentum distribution of Ref. 88, which fits the ^{3}He(e,e') reaction cross section at low energy and high momentum transfer, the disagreement would have been catastrophic.

The main difference between the two experiments is the low value, $q^2 = -.078 (GeV/c)^2$ (as compared to $1.(GeV/c)^2$) of the mass of the virtual photon in the Saclay experiment. At the top of the quasi-free peak the nucleons are close to their mass shell and their on-shell electromagnetic form factor gives a good account of their internal structure. Far away from the quasi-free peak, they are highly off their mass shell, each of them being deformed and polarized by the proximity of the others. The free nucleon electromagnetic form factors are not a good description of their structure and a full description of the two-nucleon system in terms of its quark constituents must be used. However, those effects occur only at short distance and only appear when the wave length of the virtual photon is small enough to resolve them: this is the case around $|q^2| = 1.(GeV/c)^2$ but not below $|q^2| \sim .2 (GeV/c)^2$.

Several attempts have been made to deal with these quark degrees of freedom in nuclei. I have reviewed elsewhere (Refs. 3, 11, 89) and more particularly Ref. 15 the corresponding analyses of the low energy side of the quasi-elastic peak, which I will not discuss further here.

The relevance of such a description is one of the most fascinating issues in nuclear physics, and the study of the low energy side of the quasi-elastic peak, at large momentum transfer, is one of the few places where it could be checked. What experiments should be done now?

First of all, the separation between the corresponding transverse and longitudinal response functions should be carried out in the momentum range of a few (GeV/c). The two-body mechanisms, which dominate here, contribute differently to each response function[22]. Secondly, the two-body and the three-body break-up channels must also be disentangled. On the one hand, the study of the ^{3}He(e,e'd)p reaction has already allowed to pin down those two-body mechanisms of low momentum transfer. Such an experiment should now be performed in the momentum range of a few (GeV/c)^2. On the other hand, the only way to single out these two-body mechanisms in the three-body break-up channels is to perform a triple coincidence experiment of the type ^{3}He(e,e'NN)X. Such an experiment will be possible with the new generation of high energy and high duty factor electron machines. However, its interest is not restricted to the low energy side of the quasi elastic peak, but lies also in a broader range of kinematics.

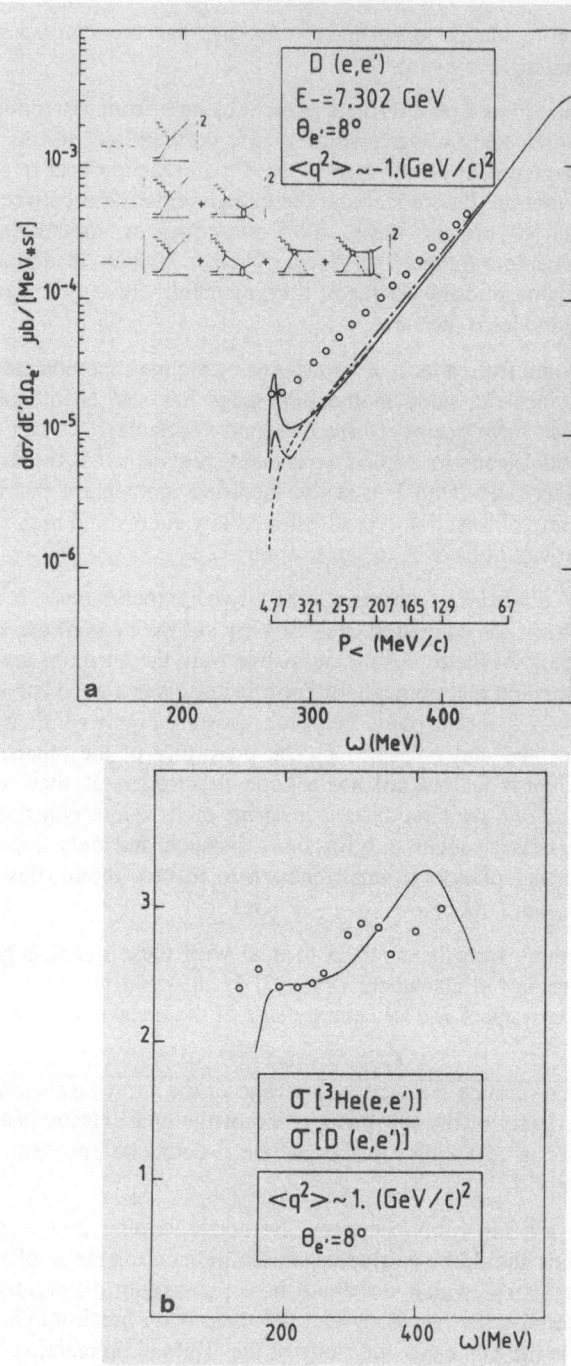

Fig. 40   a) The low energy part of the spectrum of electrons inelastically scattered at 8°
from deuterons, when $q^2 \sim -1.(GeV/c)^2$. The values of the nucleon momenta
$p_<$, which is involved in the reaction, are also plotted on the abscissa. b) The
ratio of the cross section of the $^3He(e,e')$ and $D(e,e')$ reactions near
$q^2 = -1.(GeV/c)^2$.

4.6.2 The ^{3}He(e,e'NN)X Reaction and the Two or Three-Body Correlations. Let me start with the (e,e'pp) reaction. The transverse part of its cross section (and also the cross section of the (γ,pp) reaction) is strongly suppressed. The coupling of a transverse photon to a pair of protons, which has no dipole moments, is very weak: the photon must be absorbed by higher multipoles. Moreover, charged mesons cannot be exchanged between two protons, and, therefore, the contribution of exchange currents is vanishing.

The only possible contribution is the two-body mechanism, where a Δ^+ is created on one of the protons and decays by emitting a neutral pion which is reabsorbed by the other proton. However, a selection rule strongly suppresses also this mechanism, starting from a $^1S_0$(pp) state, a magnetic dipole induces a transition to the $^3S_1(\Delta N)$ state, which cannot decay in the pp channel, since its spin and parity are $J^\pi = 1^+$ and since the only pp state with spin J = 1 has a negative parity. The transition involves either small P wave components of the nuclear wave function or higher multipoles.

For instance, in ^{3}He the (γ,pp) reaction cross section is suppressed by two orders of magnitude with respect to the (γ,pn) reaction cross sections. This is depicted in Fig. 41, where the cross section of the disintegration of a pn or a pp pair, at rest in ^{3}He , is plotted against the energy of the incoming photon. The calculation follows the lines of Ref. 22. In the pp channel, the background, due to the final state interactions and mechanisms where the photon disintegrates a pn pair but where the active and the spectator protons are detected, is one order of magnitude smaller in the Δ region but dominates below the

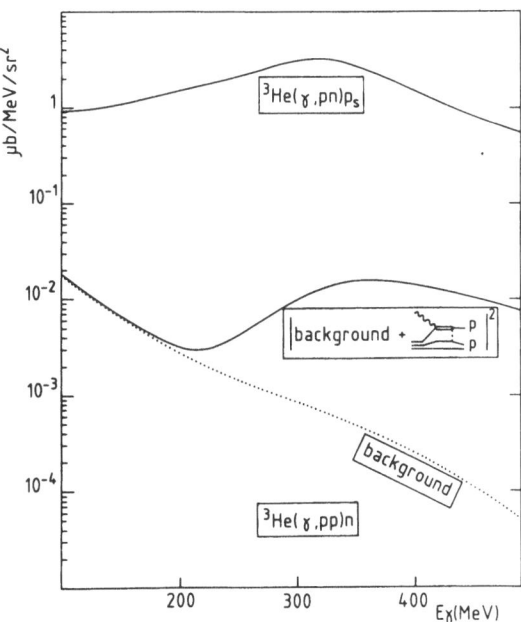

Fig. 41   The photodisintegration cross section of the pn pair (upper part) and a pp pair (lower part) at rest in ^{3}He. One of the detected protons is assumed to be emitted at $\theta_p = 90°$, with respect to the incoming photon, in the center of mass frame of the active pair. In the pp channel the background is due to all the graphs arising from final state interactions and corresponding to pion reabsorption in a pn active pair. It does not include the pion reabsorption graph in the pp active pair, which dominates the (γ,pp) cross section. See Ref. 22 for details of the model.

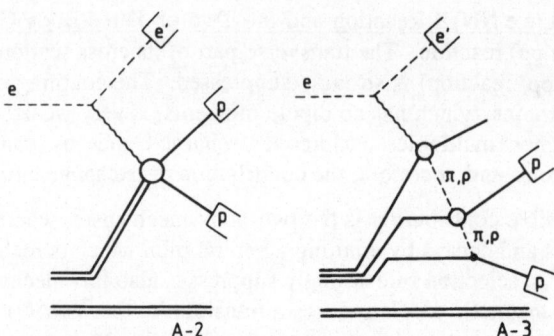

Fig. 42   The two relevant diagrams in the analysis of the (e,e′pp) reaction. The diagram
at the left does not contribute to the transverse parts, but only to the longitudinal cross sections (at lowest orders).

pion production threshold.

However, the ratio of the (γ,pp) reaction cross section and of the (γ,pn) reaction cross section has been recently measured at Bonn[90] and Tokyo[91], and is of the order of 6%. This value is significantly larger than the value which is computed when the photon interacts with pp or pn pairs.

Although in such heavy nuclei charge exchange nucleon rescattering may lead to such a figure, in the few body systems a very likely mechanism is the absorption of the photon by the three nucleons (Fig. 42). The pion which is created on one nucleon is reabsorbed by a correlated nucleon pair, and two fast nucleons are emitted almost back to back. Although no calculation of the corresponding cross section has already been done, let me discuss the main aspects of this new mechanism. Above the pion production threshold, the photoproduced pion can propagate on its mass shell, and the effect of this diagram can be considerably enhanced by judiciously choosing the kinematics in order to maximise the influence of the corresponding triangular singularity[9]. This diagram is to be related to the pion true absorption mechanism, which contributes to the imaginary part of the optical potential describing the motion of the pion in the nucleus after its creation on one of the nucleons. It is known[3,92] that this mechanism decreases the pion photopro- duction total cross section in a nucleus by roughly 10%. It is, therefore, reasonable to consider this three nucleon mechanism as the explanation of the large number of detected pp pairs.

Below the pion production threshold, both pions are off their mass shell and are vir- tual. This two pion exchange diagram is therefore a typical example of a three-body exchange mechanism. Because of gauge invariance its study will lead to strong con- straints on the three-body forces in a nucleus, in the same way as the study of the two- body exchange currents has already constrained the one boson part of the two nucleon potential.

The mechanism which is responsible for the longitudinal part of the (e,e′pp) reaction cross section is quite different since the virtual photon couples directly to the charge of both protons. Since the two body exchange currents do not contribute here, the study of the longitudinal part of the (e,e′pp) reaction cross section is the best way to determine the two nucleon correlation function in a nucleus.

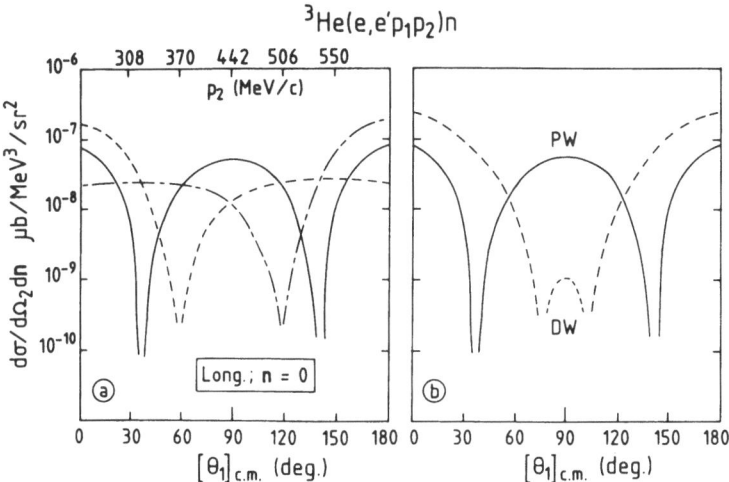

Fig. 43 The longitudinal part of the ³He(e,e′pp)n reaction cross section, when $E_- = 560$ MeV, E′ = 360 MeV and $\theta_e = 25°$. a) Plane-Wave (PW) result. The spectator neutron is assumed to be at rest, and the cross section (full line) is plotted against the angle θ_1 of one of the protons, measured in the c.m. frame of the outgoing proton pair. The momentum p_2 of the other proton is also plotted on abcissa. The dotted line and the dash-dotted line represent the contribution of each term in Eq. 26, where the virtual photon couples respectively to the proton 1 or the proton 2. b) The Distorted Wave (DW) result is compared to the Plane Wave (PW) result.

For instance, starting from the matrix elements of Ref. 22, and neglecting final state interactions, the longitudinal part of the ³He(e,e′pp) reaction cross section can be reduced to the form:

$$\frac{d\sigma}{dE'd\Omega_e \cdot dp_1 d\Omega_1 d\Omega_2} = A\Gamma_v F^2(q^2) \mid \phi(\vec{p}_1 + \frac{\vec{k}}{2}, \vec{k}) + \phi(\vec{p}_2 + \frac{\vec{k}}{2}, \vec{k}) \mid^2 \tag{26}$$

where $\phi(\vec{p},\vec{q})$ is the three-body wave function, in the momentum space, for the L=0, S=0, T=1 states, where Γ_v is the virtual photon flux factor (Eq. 19) and A the phase space factor. Each term corresponds to the absorption of the virtual photon by one of the protons. Due to their indiscernability, the angular distribution, against the angle of one of the protons in the center of mass of the pair, is symmetric around $\theta_p = 90°$, where the interference is purely constructive. This longitudinal cross section is plotted in Fig. 43a, for the same kinematical conditions already achieved in the more inclusive ³He(e,e′p)np reactions, Fig. 36. The range of momentum probed (300 < p < 700 MeV/c) contains a node of the $S_0(pp)$ wave function. It appears clearly in each term of Eq. 26, which are plotted separately in Fig. 43a. Due to their interference, two nodes appear in the full plane wave cross section. In that particular kinematics the large momentum components of the two nucleon wave functions in ³He are probed, and consequently the plane-wave cross section is small. Therefore, the corrections due to the final state interactions are not negligible and significantly modify this simple picture (Fig. 43b). But they can be computed with well known methods[22,28], and they are under control.

These two examples (Figs. 41 and 43) illustrate how the study of the (e,e′NN) reactions will allow us to disentangle the two and three body correlations in nuclei. Few experiments have been already done or are in progress. But the new generation of electron accelerators will allow us to undertake a systematic experimental program. Up to now, all the two body mechanisms[22] have been computed for reaction induced in ^{3}He . The extension to heavier nuclei and more particularly to ^{4}He, will be discussed in the next section. The calculation of the three body mechanisms (Fig. 42) is in progress.

I believe that all those experimental and theoretical efforts will be rewarding, since I really think that three arm coincidence experiments are the most promising future of the study of nuclear matter with the electromagnetic probe.

5. VERY INELASTIC ELECTRONUCLEAR REACTIONS IN HEAVY NUCLEI

All the mechanisms which dominate the electronuclear reactions induced in the few body systems occur also in heavier nuclei. Although more global methods (like the Δ−hole formalism[93] for instance) exist, the diagrammatic method, which has been tailored to, and checked by, the analysis of reactions induced in the few body systems can also be used. This is a good way to get a first order understanding of the salient features of the absorption of photons by nuclei, and I reproduce and update here the discussion of Refs. 3, 94.

5.1 One Arm Experiments

Fig. 44 shows schematically what happens to a pion which is created inside a nucleus. When it escapes the nucleus and is emitted at a given angle (part a), its momentum spectrum is very different from the spectrum of the pions emitted on a quasi-free nucleon. Since the pion energy is close to the Δ(1236) resonance energy, it interacts strongly with the A-1 nucleons: when the pion distorted wave function is used instead of a plane-wave, the quasi-free contribution is reduced by a factor four and comes close to the experimental data[94]. This is the most economical way to couple the elastic pion scattering channel to the channels which break up the A-1 residual nucleus: the quasi elastic pion scattering on the A-1 nucleons (which leads to the optical potential $V_\pi + iW_\pi$ (scatt.)) and the pion absorption by the A-1 nucleons (which leads to the potential iW_π (abs.)).

For the ^{12}C(γ,π^-) reaction the differential cross section is:

$$\frac{d\sigma}{d\Omega_\pi dT_\pi} = f_\pi \sum N_l \int K \phi_l(\vec{p}_{A-1}) \frac{d\sigma}{d\Omega_\pi}(Q,\theta_\pi) d\Omega_n \qquad (27)$$

where K is a purely kinematical factor, and where the integral runs over the available phase space of the two body undetected system made of the struck nucleon and the recoiling nucleus. I assume that it is left in a sharp hole state, of which the excitation energy ε_l (for ^{12}C, $\varepsilon_s = 38$ MeV and $\varepsilon_p = 17.5$ MeV) is determined by the analysis[2] of the (e,e′p), reaction. The momentum distribution $\phi_l(\vec{p}_{A-1})$ is the square of the nucleon orbital wave function, and is also chosen in a way which leads to a good agreement with the corresponding experimental momentum distribution. The number of neutrons and protons in each orbital are respectively N_l and Z_l (for ^{12}C, $Z_s = N_s = 2$ and $Z_p = N_p = 4$) . The energy and momentum are conserved at each vertex and relativistic kinematics is used. The B-L model[25] is used to compute the elementary cross section: $\frac{d\sigma}{d\Omega}(Q,\theta_\pi)$.

When plane waves are used the model is a straightforward extension of the model which leads to a very good agreement with the d$(\gamma,\pi^\pm)$ reaction (see Fig. 14 and section 3.1).

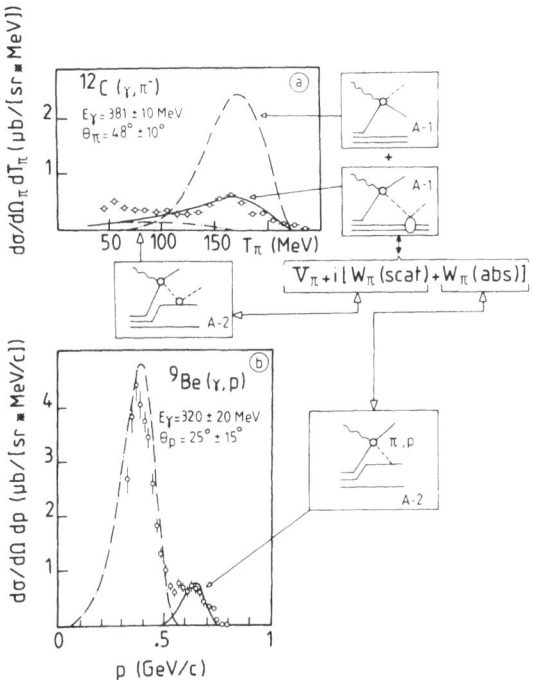

Fig. 44 The spectra of the pions emitted in the ^{12}C(γ,π^-) reaction[95], and the proton emitted in the ^{9}Be(γ,p) reaction[96]. The meaning of the curves is explained in the insets, as well as the connection between the different parts of the spectra and the pion optical potential.

The distortion of the pion wave, due to the final state interaction, is computed according to Ref. 92 in the semi-classical approximation. The net effect is an overall attenuation factor f_π which is related to the depth W_π of the imaginary part of the optical potential, which induces a loss in the flux of the pion which elastically scatters on the A-1 hole state.

Of course these pions, which have disappeared from the pion elastic scattering channel, appear elsewhere in the phase-space. The inelastically scattered pions have lost energy and fill in the low energy part or the measured pion spectrum which exhibits a significant excess of cross section. I have computed the corresponding diagrams using the two-body matrix elements of the $\gamma d \rightarrow pp\pi^-$ reaction cross section[9]. I have only changed the two-nucleon wave function and used an harmonic oscillator wave-function, which reproduces the single particle properties of ^{12}C (binding energies, radius and spectral functions). The Fermi motion and the binding energy of the active two-nucleon pair have also been taken into account. For the ^{12}C(γ,π^-) reactions again the cross section takes the form:

$$\frac{d\sigma}{d\Omega_\pi dT_\pi} = f_\pi NZ \int K\phi(\vec{p}_{A-2}) \frac{d\sigma}{d\Omega_\pi} (\gamma np \rightarrow \pi^- p_1 p_2)\, d\vec{p}_{A-2}\, d\Omega_2 \qquad (28)$$

where K is a purely kinematical factor and where the integral runs over the available

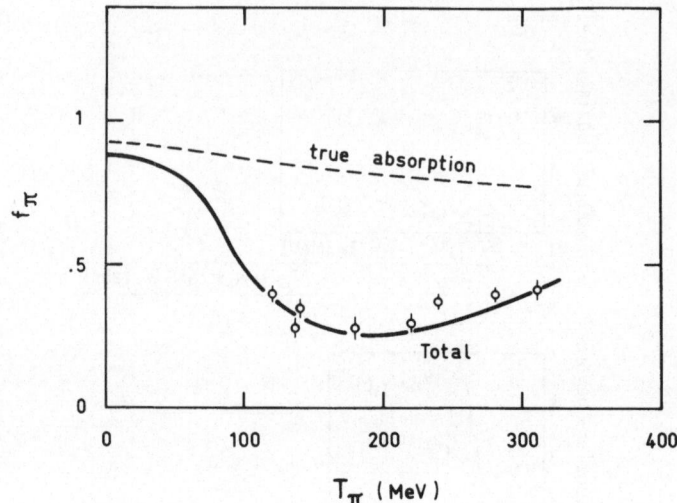

Fig. 45   The pion attenuation factor f_π. The full curve is computed as explained in Ref.
92. The experimental points have been obtained as the ratio of the area of the
experimental to the plane wave quasi-free peaks[95,97]. The dashed curve takes
only into account the time absorption mechanism, and is computed by retaining
only the corresponding imaginary part W_π (abs) of the optical potential.

phase space of the three-body system made of the two active nucleons and the recoiling
nucleus which are not detected. The cross section is proportional to the number NZ of pn
pairs.

Here again, I have taken into account the rescattering of the emitted pion by the A-2
recoiling nucleons, by multiplying the plane wave cross section by the attenuation factor
f_π . It reduces the two-nucleon cross section by only a factor two (instead of four at the
quasi-free peak because a 100 MeV pion suffers less scatterings than a 200 MeV pion
(see Fig. 45). Again those scattered pions loose energy and, as above, should fill in the
lower energy part of the pion spectra. Although I have not yet computed these diagrams
which involve three active nucleons, I think that the pion spectra can be understood as
the incoherent sum of mechanisms which involve few nucleons in the nucleus. After the
second scattering the pion looses enough energy to escape the nucleus without suffering
any interaction (no more than 20% of the pions are lost below $T_\pi = 50$ MeV). These rescattering mechanisms are also the bases of the cascade models which treat these pion
spectra at a more macroscopic level, but which also fairly reproduce the data[95].

The pions which have been absorbed in the nucleus do not escape but are responsible
for the peak which appears in Fig. 44b at the high energy part of the spectrum of the protons emitted at a given angle[90,96]. Its maximum corresponds to the two-body photodisintegration of a proton-neutron pair at rest inside the nucleus, and its width is due to the
Fermi-motion of the center of mass of the pair. It follows the two-body kinematics of the
$\gamma d \to pn$ reaction, when the proton angle and the incoming photon energy are varied.

The peak which appears at lower energies corresponds to the recoil proton associated
with the quasi-free pions.

5.2 The Quasi-Deuteron Model

In fact the effective phase space for creating a pion at a nucleon is increased, since, besides real pions, virtual pions, which remain inside the nucleus, can also be created. This is really a meson exchange contribution, which has been beautifully singled out and checked in the few body systems (Section 4.3 and Figs. 35-36). However, such a quantitative treatment cannot be performed in heavy nuclei, since we do not know how to deal with the short range behaviour of the two nucleon wave function. This is the reason why the quasi deuteron model[98] is used and is so popular. The corresponding cross section for the $^{12}C(\gamma,p)nB$ matrix element takes the form:

$$\frac{d\sigma}{d\Omega_p dT_p} = f_p \frac{L}{A} NZ \int K\phi \, (\vec{p}_{A-2}) \frac{d\sigma}{d\Omega_p} (\gamma d \to np) \, d\Omega_n \qquad (29)$$

where again K is a kinematical factor, and where the integral runs over the phase space available for the two body system made of the struck neutron and the recoiling nucleus B. I assume that it is left in a sharp two-particle two-hole state (with a mean binding energy $<\varepsilon> \sim 40$ MeV), although it is also possible to sum over all the possible two-particles two-hole states built on the ^{12}C orbitals. The momentum distribution of the center of mass of the pn pair is $\phi(\vec{p}_{A-2})$. I deduce it from the harmonic oscillator representation of the single particle orbitals. The distortion of the out going proton is taken into account by the attenuation factor f_p (Fig. 46) which is related to the imaginary part of the proton optical potential according to Ref. 92. As can be seen in Fig. 46, it is roughly constant around $f_p \sim .75$, in the range of proton momentum considered. This model tells us that the cross section of the $A(\gamma,pn)B$ reaction is proportional to the number NZ of neutron-proton pairs, and to the cross section of the $\gamma d \to pn$ reaction. It is assumed that the shape of the two-nucleon wave function is the same as the real deuteron wave function, and the scaling coefficient L/A is nothing but the ratio of their norms. The density of a nucleus is higher than the density of the deuterium. Fig. 47 shows the excitation function of the area under the quasi-deuteron peak which appears in

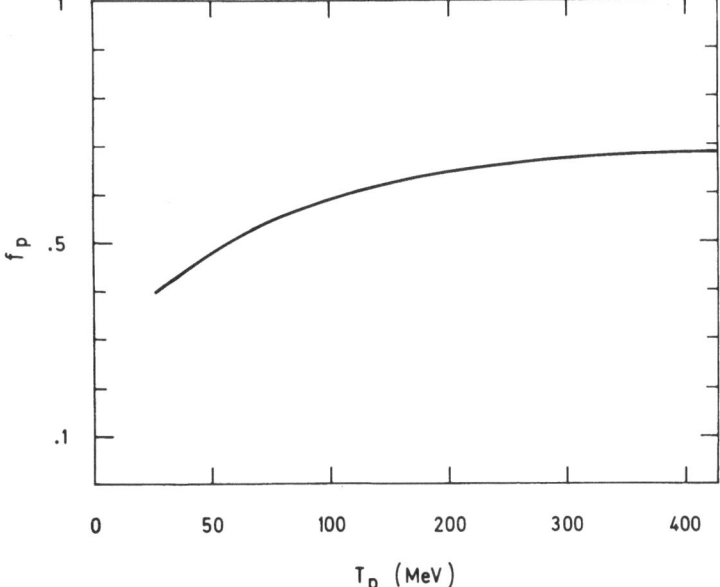

Fig. 46   The proton attenuation factor f_p.

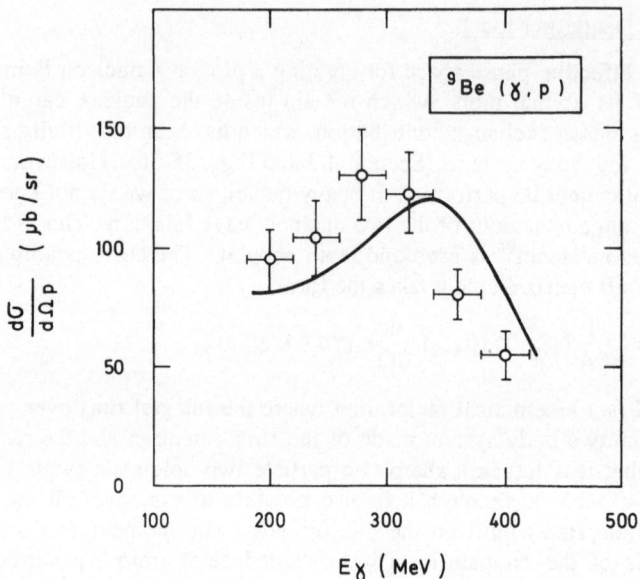

Fig. 47   The area integrated under the quasi-deuteron peak observed at $\theta_p = 25 \pm 5°$ in
the ^{9}Be(γ,p) reaction[96]. The curve is the meson exchange contribution when
L=7.

Fig. 44 at the high energy part of the spectra of the protons emitted at $\theta_p = 25°$ in the
^{9}Be(γ,p) reactions. The curve is obtained assuming that the Levinger factor remains constant between $E_\gamma = 50$ MeV and 400 MeV at the value L = 7. The model overestimates
the ^{9}Be cross section above $E_\gamma = 300$ MeV, but this is presumably due to the fact that the
elementary γd $\to$ pn reaction cross section which I use overestimates also the experimental values at forward angles in that energy range (see Ref. 28). It is also possible to determine the experimental value of the Levinger factor by fitting the largest set of experimental data. In Fig. 48, I have plotted the ratio of the area under the quasi-deuteron
peak, which has been measured in the study of the ^{9}Be(γ,p) reaction[96] and the ^{12}C(γ,p)
reaction (Ref. 90), and the quantity σ_D NZ/A. It is remarkable that these two experiments, which have been performed by two independent groups using different methods
(tagged photons[90], and photon difference[96] lead to the same energy dependence of L_{exp}.
I have also plotted the values of L_{exp} which come out the analysis of the total absorption
cross section on light nuclei[99] and heavy nuclei[100] below the pion threshold. For those
integrated cross sections, Eq. 29 becomes:

$$\sigma = \frac{L}{A} NZ \int \phi(\vec{P}_{A-2})\, \sigma_D(W)\, d\vec{P}_{A-2} \tag{30}$$

and the integral over the three body phase space guarantees its good threshold behaviour.
At high energy, when no kinematical cuts affect it, the cross section reduces to:

$$\sigma = \frac{L}{A} NZ < \sigma_D > \tag{31}$$

where $< \sigma_D >$ is a mean value of the deuteron photodisintegration. This is the original
Levinger formula[98].

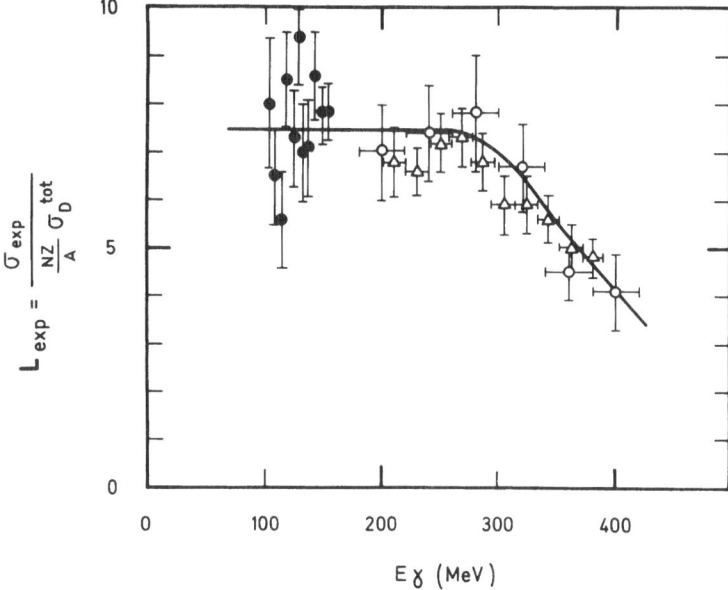

$$L_{exp} = \frac{\sigma_{exp}}{\frac{NZ}{A}\sigma_D^{tot}}$$

E_γ (MeV)

Fig. 48   The experimental value of the Levinger factor deduced from total photoabsorption measurements performed at Mainz (solid circles[99]) and Saclay (hatched area[109]) and from the analysis of the ^{9}Be(γ,p) reaction (open circles[96]) and the ^{12}C(γ,p) reactions (open triangles[90]). The curve is a mean value of L_{exp} which is used to compute the quasi-deuteron cross section in integrated cross sections.

It is very satisfactory that the Levinger factor, L, is constant when the complete theoretical deuteron cross section is used. The phase space integration and the good kinematical treatment of the binding effects in Eq. 30 allow to reproduce the total photoabsorption cross section below the pion production threshold, contrary to the analyses[101,102] which are based on the approximate Eq. 31. Moreover, its value is L = 7±1, in good agreement with the theoretical value which was estimated thirty years ago by Levinger[98], and which is based on the effective range theory of the nuclear forces.

5.3 The Integrated Cross-Sections

Fig. 49 shows the integrated cross section of the ^{12}C(γ,π^+) reaction measured at Bonn[95]. It exhibits a different shape than the free nucleon cross section. This is due to two trivial effects. On the one hand, the Fermi motion of the nucleon, which absorbs the incoming photon, reduces the height and broadens the free nucleon cross section. On the other hand, the binding of the target nucleon shifts the cross section towards higher energies: besides the emitted pion and proton energies, the incoming photon has to provide the target nucleon with its binding energy ($\sim$ 17.5 MeV in the p-shell and $\sim$ 38.5 MeV in the s-shell) to make it free. The losses of the pion flux, due to true absorption (which I have computed, in a semi classical way[92], assuming that a part of the pions emitted at a nucleon are absorbed by a correlated pair among the A-1 nucleons, and retaining only the true absorptive part iW_π (abs.) of the optical potential), reduces the quasi-free cross section by less than 20% at the Δ (1236) peak, but does not affect very much the cross

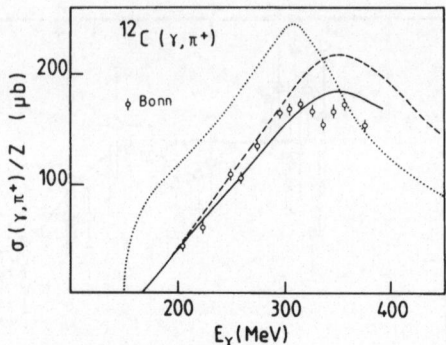

Fig. 49   The integrated cross section of $^{12}C(\gamma,\pi^+)$ reaction as measured at Bonn[90]. Dotted line curve: free nucleon cross section. Broken line curve: quasi-free pion production (including Fermi motion and binding effects). The full line curve includes also the correction for "true absorption" of the pion (see text).

section at lower energies.

This result may appear surprising, since the outgoing pion is known to suffer strong final state interactions (Fig. 44a). The explanation is given in Fig. 50. When the optical wave-functions of the pion and the proton, emitted in the quasi-free mechanism, are used instead of plane-waves, the quasi-free contribution is strongly reduced. This DWIA treatment is fully described in Ref. 92. However, I am dealing with an inclusive cross section, and I have to add the contribution of the mechanisms which contribute to the loses of the flux of the particles which are created at one nucleon: the inelastic scatterings of the proton or the pion which break up the A-1 nucleons. As above (Fig. 44a), I have computed their contribution using the two-body matrix elements which are given in Ref. 9, an harmonic oscillator wave function and counting the number of active nucleon-nucleon pairs in ^{12}C. The sum of the contributions of these two channels is very close to the experimental data and to the quasi-free contribution, when it is corrected for true pion absorption effects (Fig. 49).

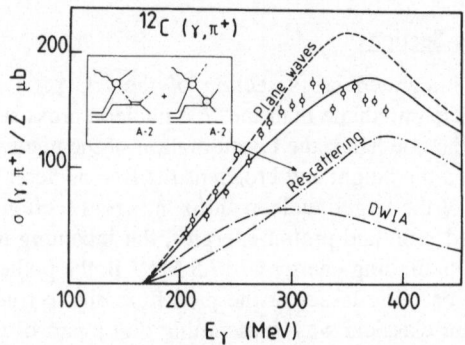

Fig. 50   The contribution to the $^{12}C(\gamma,\pi^+)$ integrated cross section of the one-body (dot-dash curve) and the two-body mechanisms (double dot-dash curve). Their sum is the full line curve. The quasi-free cross section is also shown (broken line curve).

The lesson is that the incoming photon sees only the one nucleon current, and does not know what happens to the pions which escape the nucleus. This is a consequence of unitarity. In other words, the pion or the photon propagate nearly on shell (see Ref. 9) far away from the target nucleon and escape the interaction volume of the incoming photon, before suffering a scattering.

The quasi-free pion photoproduction contribution alone does not reproduce the integrated cross section of the reaction $^{12}C(\gamma,p)$ measured at Bonn[90] which is depicted in Fig. 51. But the meson exchange current contribution (computed in the quasi-deuteron model) accounts fairly well for the excess of the measured cross section. I have used the experimental values of the Levinger factor shown in Fig. 48.

Here the pion is virtual and must be reabsorbed by another nucleon within the interaction volume of the incoming photon. Besides the one nucleon current the photon is also sensitive to the two-nucleon current.

These two dominant mechanisms allow one to reproduce the main tendencies of the total photoabsorption cross sections (Fig. 2). Here the small additional contribution due to the coherent π^0 production ($^{12}C(\gamma,\pi^0)^{12}C$) helps to improve the agreement just above the pion threshold, but is negligible in the Δ energy range and above.

The two dominant contributions are proportional to the number A of nucleons. This is obvious for the one body part (the quasi free photoproduction) and the two body part (the quasi deuteron) is proportional to NZ/A, which in turn behaves as A. Therefore, this simple model provides us with a good understanding of the "universal curve" (Fig. 2) which is obtained when the total photoabsorption cross sections of various nuclei are divided by their mass number. The difference of its shape, from the shape of the free nucleon cross section, comes not only from trivial effects, which have to do with the bulk properties of nuclear matter (Fermi motion and binding energies), but also from the nature of the behaviour of the Δ in nuclear matter (through the coupling between the $N\Delta \rightarrow NN$ channels).

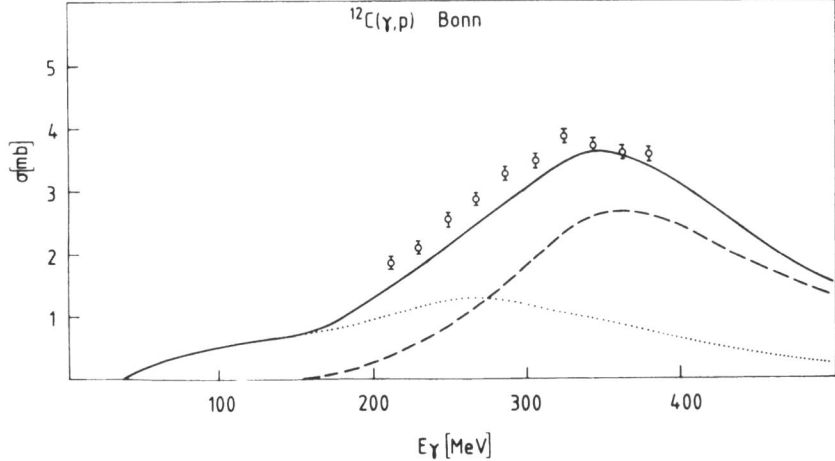

Fig. 51   The integrated cross section of the $^{12}C(\gamma,p)$ reaction[90] is plotted against the energy of the incoming photon. Dashed line: recoiling proton associated with the photoproduced pions. Dotted line: meson exchange contribution computed in the quasi-deuteron model.

In fact this close similarity of the "universal" photoabsorption cross section per nucleon with the free nucleon cross section is a direct consequence of the nature of the photon which sees the entire nuclear volume. This is not the case of the pion which is strongly absorbed at the surface of the nucleus. The pion-nucleus total cross section is quite different from the $\pi - N$ cross section.

This is exactly the result which is found in a more elaborate treatment, the Δ - hole model[93]. This is an elegant way to take unitarity into account. The incoming photon is assumed to change a nucleon into its first excited state, the Δ resonance, which subsequently propagates in the nucleus. The binding of the nucleon and the Δ are taken into account by assuming that they are moving in a mean nuclear potential, and unitarity demands that the Δ - hole state is strongly coupled to all the possible other channels. This coupling strongly affects the width and the position of the resonance (as compared to the free nucleon case) through the spreading potential, a sizeable part of which is due to the $\Delta- N \rightarrow N - N$ transition. This is also a sizeable part of the exchange current contribution in my diagrammatic approach. It is here that the phenomenology enters both

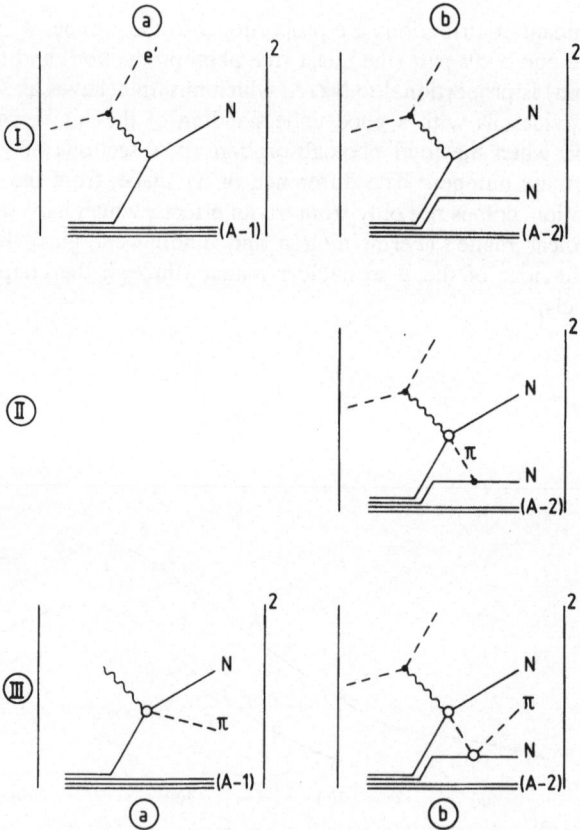

Fig. 52   The relevant diagrams in the analysis of electronuclear reactions on light nuclei at intermediate energy. I: the one-body current contribution (quasi-elastic scattering) is split into its different parts according to the number of particles in the final state. II: the meson exchange diagram. III: the quasi-free pion electroproduction diagram and the pion rescattering diagram.

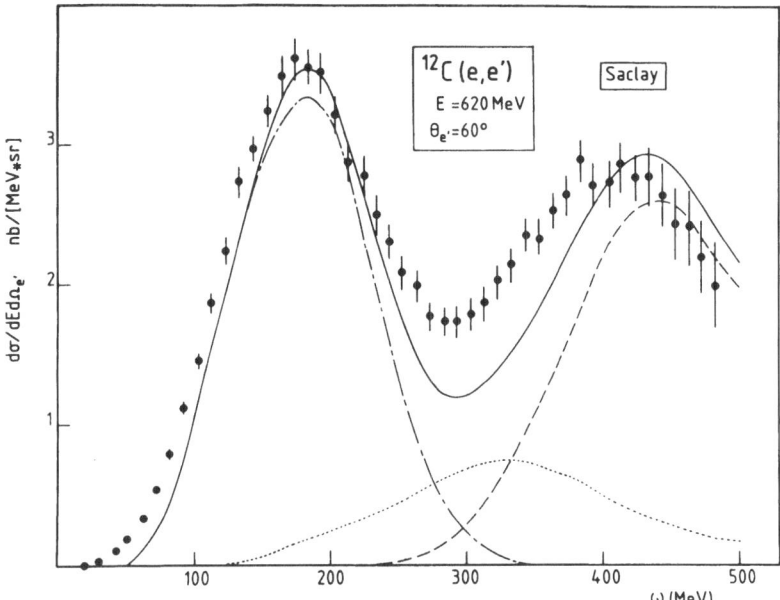

Fig. 53   The spectrum of the electrons inelastically scattered by ^{12}C at $\theta = 60°$ when E=620MeV, is plotted against the energy ω of the virtual photon. The dashed line corresponds to the pion electroproduction channel. The dash-dot line corresponds to quasi-elastic scattering of the electron. The dotted line corresponds to the meson exchange current contribution, estimated in the framework of the quasi-deuteron model.

models. In the Δ- hole model, the spreading potential is usually empirically determined by fitting the cross section of pion elastic scattering by selected nuclei. In my calculation, I parametrize the exchange current contribution by the quasideuteron model in which the Levinger factor is also determined in an empirical way.

However, the Δ- hole model does not treat consistently the non resonant Born terms in the elementary operator[18]. They dominate near threshold and they represent half of the charged pion photoproduction cross section[25] at the resonance, with which they strongly interfere (see Fig. 12 in Ref. 18). In the diagrammatic method they are automatically taken into account in the elementary operator. It is important to realize that these non resonant Born terms are responsible for the non resonant meson exchange currents which dominate the total photoabsorption cross section below the pion production threshold. Nevertheless the physics is the same and both approaches lead to roughly the same results, although the phenomenology is not introduced in the same way. The extension to heavy nuclei of the diagrammatic approach, which has been checked in the few body systems, in an economical and pedagogical way to understand the basic mechanisms which govern the absorption of photons by nuclei.

5.4 Deep Inelastic Electron Scattering

The same mechanisms are also responsible for the total absorption of virtual photons by a nucleus, but a new mechanism becomes important, the quasi- free electron

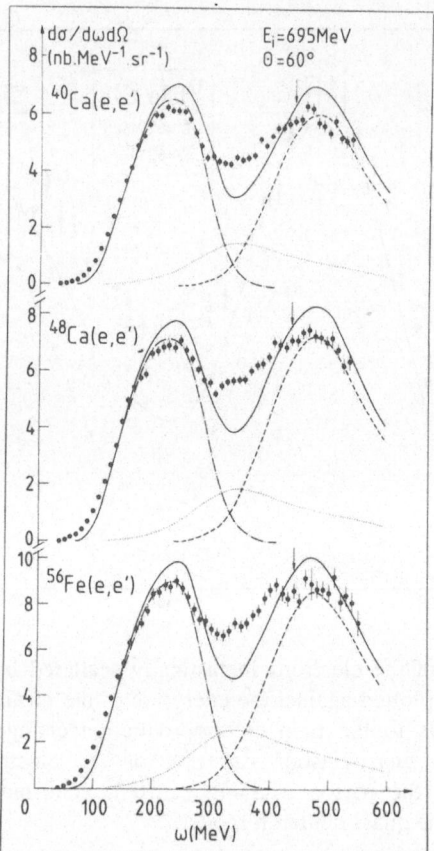

Fig. 54   The same as Fig. 53, but for ^{40}Ca, ^{48}Ca, and ^{56}Fe when E = 695 MeV and
θ = 60°.

scattering (Fig. 52). As an example, the cross section of electron deep inelastic scattering
on ^{12}C, ^{40}Ca, ^{48}Ca and ^{56}Fe are plotted in Figs. 53 and 54. They have been recently measured at Saclay[5-6]. Besides the pion electroproduction peak, which appears on the right of
the figure for high values of the energy ω of the virtual photon, the peak which appears at
lower energy, ω, of the virtual photon, the peak which appears at lower energy
corresponds to the absorption of the virtual photons by a quasi-free nucleon nearly at rest
in the nucleus (this is kinematically forbidden to a real photon). The top of the peak
corresponds precisely to the energy of the elastic scattering of the incoming electron on a
nucleon at rest, and the width is due to its Fermi motion in the nucleus. Between these
two peaks, the contribution of the exchange currents (diagram II in Fig. 52) helps to
reproduce the experiment. However, the situation is not as simple as in the case of real
photon absorption. When the contribution of the quasi-elastic peak is computed, the closure approximation is usually made. The one-particle one-hole diagram I.a, the two-
particle two-hole diagram I.b and all the possible other diagrams are added incoherently.
If the available energy is high enough, this sum is equivalent to the one-nucleon contribution, since all those different final states form a complete basis.

When the complete deuteron electrodisintegration cross section is used in the quasi-deuteron model, diagram I.b is counted twice, since the coherent sum of diagram I.b and II is used, and since diagram I.b is already contained in the quasi-elastic contribution. This is not the case for the real photon absorption, since the kinematics of diagrams I.a and I.b require a very large nucleon momentum in the nucleus: the one-body current contributes only through the interference between diagrams I.b and II.

There are two ways to overcome this double counting problem. The first one is to start with a cluster expansion of the target nucleus ground state wave function and to compute its overlap with all the possible one-particle one-hole and multi-particle multi-hole states (not only those where the particles are free, but also those where they interact in the final state). This is a very ambitious program, which has only been possible in the few-body systems (section 4), and I am not sure that it could be easily achieved in the case of a nucleus as complex as ^{12}C.

The second way is to retain only the exchange part (diagram II) in the elementary deuteron electrodisintegration cross section, but the price to pay is the need of an energy dependent Levinger factor (the same as in the real photon case) to simulate the interference between diagrams I.b and II: the corresponding values, extracted from the analysis of the (γ,p) reaction cross sections, as given in Fig. 55, and are used to compute the quasi-deuteron contribution in Figs. 53 and 54.

Like the total photonuclear absorption cross sections, the deep inelastic electron scattering cross sections exhibit a remarkable scaling behaviour. This is clearly apparent in Fig. 56 where the spectra of the electrons inelastically scattered at 60°, when $E \sim 680$ MeV, by various nuclei[5,6] are divided by the corresponding mass number.

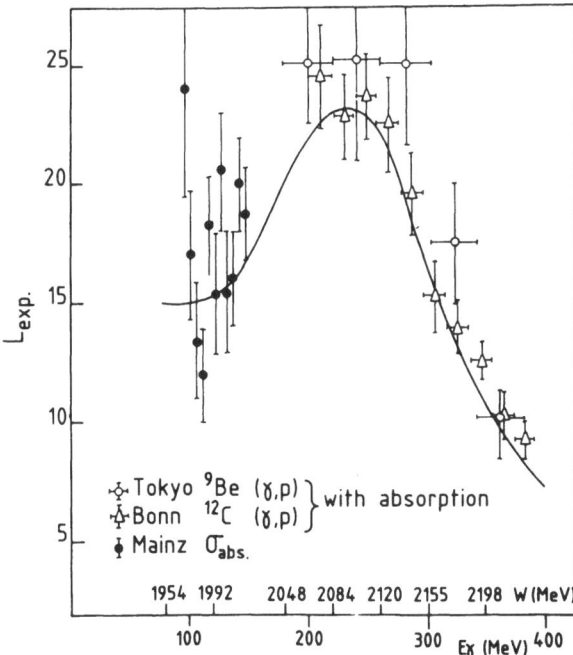

Fig. 55   The experimental values of the Levinger factor L, when only the exchange part is retained in the amplitude of the $\gamma D \to pn$ reaction. The meaning of the symbols is the same as in Fig. 48.

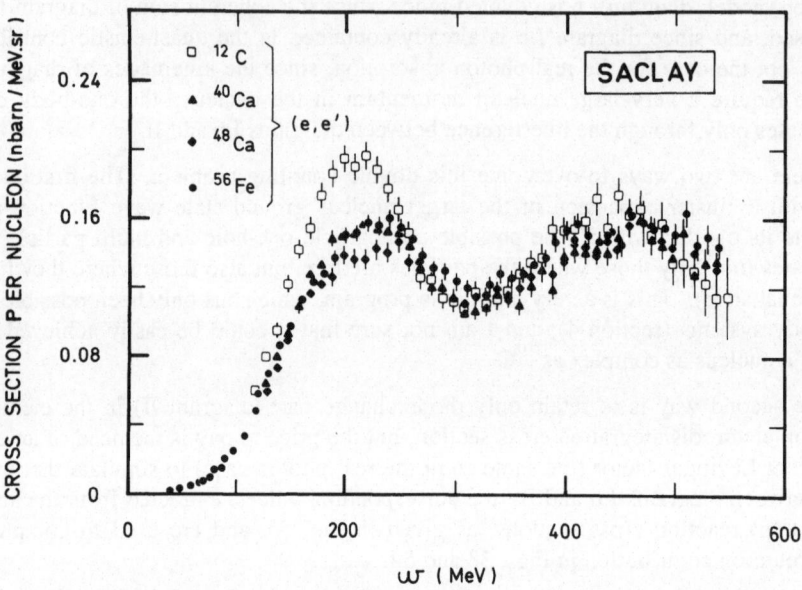

Fig. 56   The "universal curve" obtained by dividing the cross section of electrons inelastically scattered at $\theta = 60°$ by the corresponding mass number, when $E \sim 680 \mathrm{MeV}$, on various nuclei.

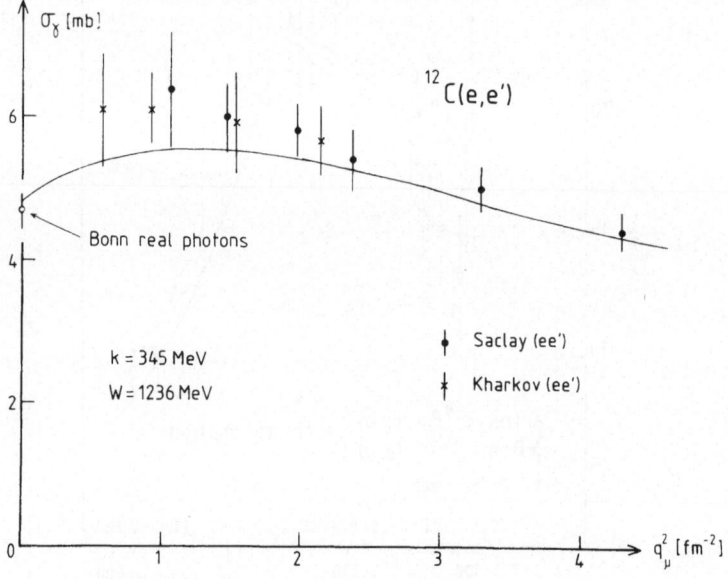

Fig. 57   The extrapolation of the ^{12}C(e,e′)X reaction cross section to the photon point.

In this energy range, the cross sections are the same for a wide range of nuclei. Here the absorption of a real or a virtual photon is mainly sensitive to the bulk properties of nuclear matter. This similarity of the basic mechanisms involved in total photoabsorption, and deep inelastic scattering, reaction cross sections is even more apparent in Fig. 57, where the $^{12}C(e,e')x$ reaction cross section extrapolates fairly well to the total photoabsorption cross section when the four momentum q^2 of the virtual photon vanishes. It is worthwhile to point out that the extrapolation is not linear as in the case of a free nucleon (Fig. 10).

In the quasi-elastic scattering region, the shape of the cross section changes with the mass number. Here the photo-absorption process is mainly sensitive to the size of the nucleus and probes the nucleon momentum distribution. The transverse and longitudinal response functions have been singled out in that region, but not under the Δ peak. This interesting topic is outside the scope of these lectures, and I refer to Refs. 5-6 for a discussion.

5.5 Prospects

So far only integrated quantities have been measured and the separation between the transverse and the longitudinal response functions under the Δ peak remains to be done. The limitation of the energy (<700 MeV) of the present generation of high intensity electron accelerators has prevented to perform it. However, it is very likely that the new injector at SLAC will soon allow it[103].

The next problem, which is left open, is the structure of the continuum, which can be studied by means of coincidence experiments of the type $(e,e'p)$ or $(e,e'NN)$.

When analysing coincidence experiments of the type $(e,e'p)$, one has always been faced with two problems[2]: the occupation number of a given shell is always smaller than expected in a pure shell model, and the structure of deep lying hole states is still not well understood.

To get a feeling of what is happening, let me come back to the basic mechanisms, which happen in the simplest nucleus, the ^3He. All of the three nucleons are basically in a relative S state and the details of its wave function, especially the correlation functions, are known with enough accuracy to allow for precise calculations. Moreover, there are no deep lying hole states.

A typical missing mass spectrum, measured at Saclay[70], is presented in Fig. 58. Besides the deuteron peak, a significant contribution is found in the continuum. The cross sections of the $^3He(e,e'p)$ reaction, corresponding to these two parts of the spectrum, are very well reproduced, provided a good correlated wave function[20] is used, and provided that final state interactions and meson exchange current effects are taken into account (see for instance Sections 4.2 and 4.3, and Figs. 30-31 and 36). However, the two body break-up channel represents only two thirds of the total inelastic electron scattering cross section at the quasi-elastic peak (see Fig. 4). The rest is due to the three body break-up channel. The reason is that only two thirds of the protons are coupled to a proton-neutron pair of which the isospin is zero, and the overlap of the ^3He and the deuteron wave functions is close to 1.3 instead of 2. The remaining one third of the protons are coupled to $T = 1$ proton-neutron pairs which overlap only with the continuum. This extreme example shows that the effects of the ground state correlation is to shift a sizeable part of the one-particle one-hole strength toward the continuum.

It is very likely that such a mechanism occurs also in heavier nuclei, and it must be perfectly understood before any sensible statement is made on the nature of the deep

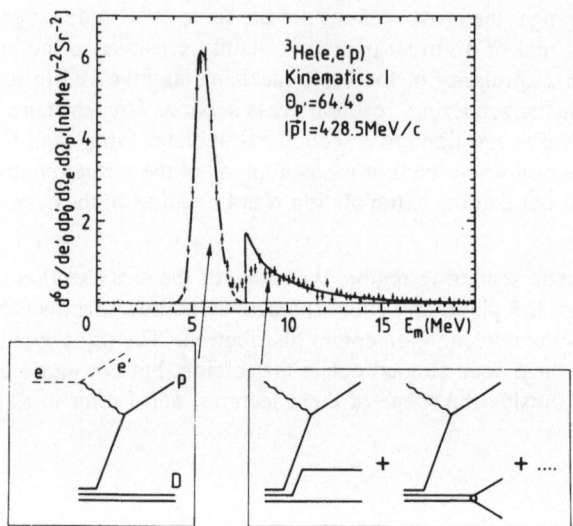

Fig. 58   A typical missing mass spectrum, measured at Saclay[70] in the study of the
          ^{3}He(e,e′p) reaction. The peak corresponds to the two body break-up. The
          correlations in the ^{3}He ground state shift a sizeable part of the one-particle
          one-hole strength toward the continuum.

lying hole states. It is in the a background above in which they show up. But it offers us
also a unique way to study the correlation functions in the nucleus.

The first priority is to perform coincidence experiments of the type (e,e′p) or (e,e′pp).
In the preceding section I have put the emphasis on their study in the few body systems,
where the basic mechanisms have been checked. They are well under control. They
must now be performed in heavier nuclear systems, when the missing energy is well
above the removal energy of the deep lying hole states.

It is in that part of the missing mass spectrum that the exchange currents and the
correlations can be studied in a nucleus. Let me come back again to the ^{3}He(e,e′p)np
reaction of which a typical set of cross sections is plotted in Fig. 59. Contrary to a single
arm experiment, a two arm experiment allows for the measurement of four response
functions (see also Eq. 23). Besides the transverse and the longitudinal one, it is necessary to consider also the transverse-transverse and the transverse-longitudinal interference response functions. The top of the peak, which appears around 30 MeV in these
missing mass spectra, corresponds to the electrodisintegration of a correlated neutron-
proton pair at rest in ^{3}He. Its width is due to the Fermi motion of this pair. The
exchange currents do not affect at all the longitudinal response function, but dominate
the transverse one and affect a little the transverse- longitudinal interference term. The
separation of these four response functions, in a two arm experiment, allows us to study
the structure of the continuum, to disentangle the dominant mechanisms and to study the
two body correlation functions. But non coplanar experiments are required, since it is
necessary not only to vary the electron scattering angle θ (and therefore the virtual photon polarization ε), but also the angle φ between the electron scattering plane and the

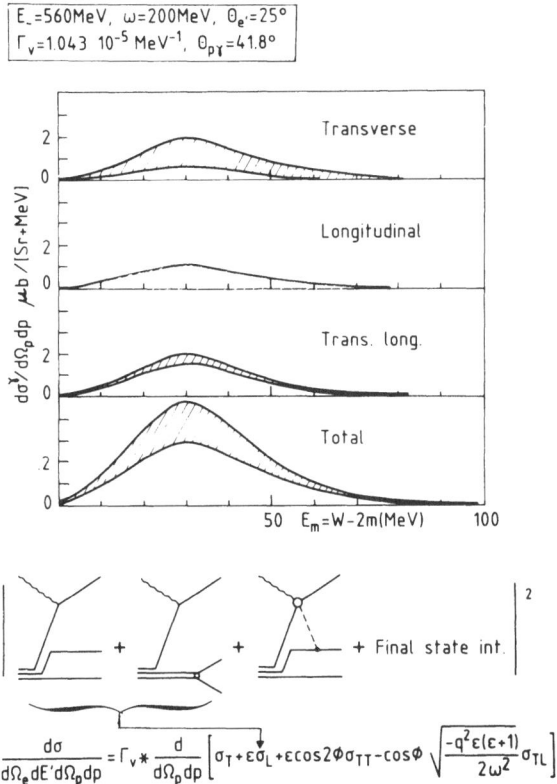

Fig. 59   The response functions in the ^{3}He(e,e'p)pn reaction are plotted against the missing mass of the two undetected nucleons. They are defined in the lower part of the figure. The transverse-transverse part of the cross section is very small and has not been drawn. The longitudinal part is free of meson exchange currents, which dominate the transverse cross section (hatched area) and represent half of the total cross section.

hadron production plane. Those are very difficult experiments which require very large solid angle detectors.

However, the interference response functions vanish when the proton is emitted in the direction of the virtual photon (forward or backward). It is therefore possible, in this particular case, to determine the longitudinal and the transverse functions without performing a non coplanar experiment, but only a simple Rosenbluth plot. This possibility is used at Bates[104] and Saclay[72]. But the maximum energy of the present generation of electron Linacs (about 700 MeV) does not allow us to probe the very high momentum components of the nuclear wave function. When the proton is emitted in the forward direction, the neutron of the active neutron-proton pair recoils with the smallest momentum which is allowed by the kinematics.

Although the maximum energy of the electron beams should be significantly increased, this kind of experiment should be systematically undertaken with the present generation of electron Linacs on few body systems and then on heavier nuclei.

6. CONCLUSION

In these lectures I have reviewed the basic features of meson photo- and electroproduction reactions on nuclei. They are a good example of very inelastic electronuclear reactions in which both high energy and high momentum are transferred to the nucleus.

I have adapted, to the reactions induced in nuclei, a diagrammatic method which was popular ten or twenty years ago for analyzing the reactions occurring between elementary particles. This is an expansion of the S matrix around its singularities, and I have shown how the systematic study of exclusive experiments has allowed us to single out each dominant mechanism. There are no conceptual difficulties to extend this method to relativistic nuclear physics, and it provides us with a link between conventional nuclear physics and high energy particle physics. This is a good example of the overlap of two different fields, and I really believe that it is rewarding that nuclear physicists learn and adopt these methods which have been developed elsewhere.

I have deliberately not dealt with "coherent" meson photo- and electroproduction reactions, in order to put the emphasis on the use of the nucleus as a laboratory, where mechanisms, which cannot occur on a free nucleon, can be singled out and studied. The creation of virtual mesons in nuclei, or the additional decay channels of a Δ created in a nucleus, are two typical examples. They are related to the fundamental mechanisms which govern the long range and intermediate range part of the nucleon-nucleon interaction.

So far the meson and Δ degrees of freedom have been systematically studied in nuclei. Although the use of hadronic (as pions for instance) and electromagnetic probes have been complementary, it is only the study of the electromagnetic properties of nuclei which has allowed us to put on firm ground the treatment of these non nucleonic degrees of freedom. The electromagnetic probe is weak enough to allow for a safe treatment of the absorption processes, but its most important quality is to satisfy gauge invariance. This principle leads to strong constraints on the electromagnetic transition matrix elements.

What should we do now?

Obviously we have reached a limit. We have to go beyond those meson and Δ degrees of freedom in nuclei, and to look for quark and gluon effects which are not occurring in a free nucleon, but which occur at short distance in nuclei. Although weak interactions are sensitive both to quarks and gluons, the electromagnetic probe is only sensitive to quarks (since the gluons are neutral), and provides us again with a filter. Moreover, it provides us with a way to get rid of the meson and Δ degrees of freedom, which dominate at long range, but still contribute at short range. They do not affect the longitudinal response functions, but only the transverse ones.

Future machines should allow us to achieve this goal, and, as in the past, the complementarity of different kind of beams and probes will be necessary. While the so called "kaon factories" will allow for the study of both quark and gluon degrees of freedom, electron accelerators will us allow to single out the quark degrees of freedom in nuclei. Their characteristics are now well defined (see for instance Refs. 11-12): a nominal energy between 2 and 4 GeV, a duty factor close to unity and a high intensity of about $100 \, \mu A$.

The main axes of research are also well defined, and have naturally emerged in the different parts of these lectures. Firstly, the elementary operators must be perfectly understood. The study of the meson photo- and electroproduction reactions induced on free nucleons is not only the necessary step before the study of the same reactions in nuclei, but also provides us with strong constraints on the internal structure of the baryons. Secondly, the study of the few body systems at short distance is the way to look for the quark effects which do not occur with a free nucleon and which cannot be studied during the scattering between free baryons. Thirdly, in more heavy nuclei, it is possible that multi-quark clusters exist. Their discovery and their study would be a strong constraint on the dynamics of quarks in baryonic systems.

All this study has to do with the modification of the confinement mechanism in the nucleus, and I really believe that particle physicists will gain a lot by learning and using the methods which have been successful to analyse the many body nuclear problem.

These issues are fascinating. But it is my feeling that no single experiment (smoking gun!) will alone solve these problems. A continuous effort, both on the theoretical and the experimental side, is the only way which is left open.

ACKNOWLEDGEMENTS

This series of lectures was a good opportunity to summarize various works and to make a link between them. I would like to thank the organizers, H. Caplan, M. Preston and E. Tomusiak, of this NATO Advanced Study Institute for their invitation and their warm hospitality. I also thank F. Lepage and D. Bouziat for having diligently prepared the various drafts and figures.

APPENDIX: Gauge Invariance and Exchange Currents[28]

Although gauge invariance is not satisfied by the elementary nucleon current, Eq. 1-2, when the nucleons are bound in a nucleus and off-mass shell, it must be satisfied by the full nuclear current.

The nucleons are off the mass shell because they interact by exchanging pions (and mesons). In the diagram where the photon couples directly to the nucleon (Figs. 20 Ia, b and 21II), the nucleon Born terms appear explicitly when the nuclear wave functions are expanded in such a way as to exhibit the pion exchange part of the NN interaction (Fig. 21 III). The nuclear current can be put in the form:

$$\vec{k} \cdot \vec{J}^N = \langle f | \frac{V_{OPE}^{\pi^+}}{p_a^0 - E_a} \frac{\vec{k} \cdot (2\vec{p}_i + \vec{k})}{2m} + \frac{\vec{k} \cdot (2\vec{p}_f - \vec{k})}{2m} \frac{V_{OPE}^{\pi^-}}{p_b^0 - E_b} | i \rangle \qquad [A1]$$

$$\omega J_0^N = \langle f | V_{OPE}^{\pi^+} \frac{\omega}{p_a^0 - E_a} + \frac{\omega}{p_b^0 - E_b} V_{OPE}^{\pi^-} | i \rangle \qquad [A2]$$

Using Eq. 16 or equivalently.

$$\frac{\vec{k} \cdot (2\vec{p}_f - \vec{k})}{2m} = -(p_b^0 - E_b) - \omega$$

$$\frac{\vec{k} \cdot (2\vec{p}_i + \vec{k})}{2m} = -(p_a^0 - E_a) + \omega . \qquad [A3]$$

[A1] can be reexpressed as:

$$\vec{k} \cdot \vec{J}^N = <f| \ V_{OPE}^{\pi^+}(-1 + \frac{\omega}{p_a^0 - E_a}) + (1 + \frac{\omega}{p_b^0 - E_b}) \ V_{OPE}^{\pi}|i> \tag{A4}$$

and the current conservation law takes the form:

$$\omega J_0^N - \vec{k} \cdot \vec{J}^N = <f| \ V_{OPE}^{\pi^+} - V_{OPE}^{\pi}|i> . \tag{A5}$$

The pion exchange (ex) contribution, which we have defined in Ref. 29 and corrected in Ref. 28, in order not to double count the pion photoelectric term, is precisely such that:

$$\omega J_0^{ex} - \vec{k} \cdot \vec{J}^{ex} = <f| -V_{OPE}^{\pi^+} + V_{OPE}^{\pi}|i> . \tag{A6}$$

The demonstration follows the same lines as the demonstration of the gauge invariance of the elementary currents, Eqs. 12-15. Therefore,

$$\omega(J_0^N + J_0^{ex}) - \vec{k} \cdot (\vec{J}^N + \vec{J}^{ex}) = 0 . \tag{A7}$$

REFERENCES

[1] B. Frois, Lecture Notes in Physics **137,** 55 (1981); Proc. Int. Conf. on Nucl. Phys., Florence, Aug. 29 - Sept. 3 (1983); P. Blazi and R. Ricci, Eds. Tipographia Compositori, Bologna, 221 (1983); B. Frois et al., Phys. Rev. Lett. **38,** 152 (1977).

[2] J. Mougey, Nucl. Phys. **A335,** 35 (1980); S. Frullani and J. Mougey, Advances in Nucl. Phys. **14,** 1 (Plenum Press 1984); P. de Witt Huberts, "Second Workshop on Perspectives in Nuclear Physics at Intermediate Energies", Trieste, March 25-29, 1985; S. Boffi et al., eds. to be published by World Scientific Pub. Co., Singapore.

[3] J. M. Laget, Lecture Notes in Physics, **137,** 148 (1981); 4th Course of the International School of Intermediate Energy Physics; R. Bergere et al., eds. World Scientific, 272 (1984).

[4] J. Arends et al., Phys. Lett. **98B,** 423 (1981); C. Chollet et al., Phys. Lett. **127B,** 331 (1983); P. Carlos et al., Nucl. Phys. **A431,** 573 (1984); L. Ghedira et al., Contributed paper A12, PANIC, Heidelberg, August 1984; 8eme Session d'Etude Biennale de Phys. Nucl., AUSSOIS, 4-8 February 1985. LYCEN 8502, Université de Lyon, p S.9.1.

[5] Z. Meziani et al., Phys. Rev. Lett. **52,** 2130 (1984); Phys Rev. Lett. **54,** 1233 (1985).

[6] P. Barreau et al., Nucl. Phys. **A402,** 515 (1983).

[7] J. Dechargé and D. Gogny, Phys. Rev. **C21,** 1568 (1980).

[8] J. M. Cavedon et al., Phys. Rev. Lett. **49,** 978 (1982).

[9] J. M. Laget, Phys. Report **69,** 1 (1981).

[10] J. M. Laget, Nucl. Phys. **A358,** 329C (1981).

[11] A. Gerard and J. M. Laget, Rapport Prospectif sur la Physique Photo et Electronucleaire, Note CEA-N-2360 (1983).

[12] Rénovation de l'Accélérateur Linéaire d'Electrons de Saclay (ALS), Note CEA-N-2359 (1983).

[13] R. Vinh Mau, Mesons in Nuclei, eds. M. Rho and Wilkinson, 151 (North Holland 1979).

[14] K. Holinde and R. Machleidt, Nucl. Phys. **A256,** 479 (1976).

[15]　J. M. Laget, International Conference on Clustering Aspects of Nuclear Structure and Nuclear Reactions, J. S. Lilley and M. Nagarajan eds, D. Reidel Pub. Co. (Dordrecht, Holland) 295 (1985).

[16]　S. J. Brodsky, Comments Nucl. Part. Phys. **12**, 213 (1984).

[17]　D. Day *et al.*, Phys. Rev. Lett. **43**, 1143 (1979).

[18]　J. M. Laget, Nucl. Phys. **A358**, 275c (1981).

[19]　C. Marchand *et al.*, Phys. Lett. **153B**, 29 (1985).

[20]　C. Hajduk *et al.*, Nucl. Phys. **A322**, 329 (1979); R. A. Branderburg *et al.*, Phys. Rev. **C12**, 1368 (1975).

[21]　R. V. Reid, Ann. Phys. (N.Y.) **50**, 411 (1968).

[22]　J. M. Laget, Phys. Lett. **151B**, 325 (1985).

[23]　A. M. Bernstein, "3rd course of the International School of Intermediate Energy Nuclear Physics", (Verona 1981); R. Bergere *et al.*, eds. World Scientific Pub. (1982); N. de Botton, Lecture Notes in Physics **108**, 339 (1979).

[24]　H. Genzel *et al.*, Landölt-Börnstein, Numerical date and relationship in science and technology, Vol. **8**, (Springer Verlag, Berlin, 1973); T. Fujii *et al.*, Nucl. Phys. **310**, 395 (1971); G. Von Holtey, Springer Tracts in Modern Physics **59**, 3 (1971).

[25]　J. Blomqvist and J. M. Laget, Nucl. Phys. **A280**, 405 (1977).

[26]　A.B.B.H.H.M. collaboration, Phys. Rev. **175**, 1669 (1968).

[27]　D. Lüke and P. Söding, Springer Tracts in Modern Physics **59**, 39 (1971).

[28]　J. M. Laget, Can. J. of Phys. **62**, 1046 (1984).

[29]　J. M. Laget, Nucl. Phys. **A312**, 265 (1978).

[30]　S. Boffi *et al.*, Nucl. Phys. **A386**, 599 (1982).

[31]　J. D. Bjorken and S. Drell, Relativistic Quantum Mechanics (McGraw-Hill, New York, 1964).

[32]　R. Wittman *et al.*, Phys. Lett. **142B**, 336 (1984).

[33]　J. Sabutis, Phys. Rev. **C27**, 778 (1983).

[34]　M. K. Singham and F. Tabakin, Ann. Phys. **135**, 71 (1981).

[35]　L. Tiator and L. E. Wright, Phys. Rev. **C30**, 989 (1984).

[36]　L. Tiator and D. Dreschsel, Nucl. Phys. **A360**, 208 (1981).

[37]　P. Bosted and J. M. Laget, Nucl. Phys. **A296**, 413 (1978).

[38]　N. Dombey, Hadronic interaction of electrons and photons, Eds. J. Cuming and H. Osborn, Academic Press Inc., New York, 17 (1971).

[39]　H. Breuker *et al.*, Nucl. Phys. **B146**, 285 (1978).

[40]　K. Batzner *et al.*, Nucl. Phys. **B76**, 1 (1974); J. C. Alder *et al.*, Nucl. Phys. **B40**, 573 (1972); P. Sidelle *et al.*, Nucl. Phys. **B35**, 93 (1971).

[41]　G. Bardin *et al.*, Nucl. Phys. **B120**, 45 (1977).

[42]　W. Bartel *et al.*, Phys. Letter **35B**, 181 (1971).

[43]　U. Batzner *et al.*, Phys. Letter **39B**, 575 (1972).

[44]　G. Gianella *et al.*, Nuovo Cim. **63A**, 892 (1969).

[45]　P. Benz *et al.*, Nucl. Phys. **B79**, 10 (1974).

[46]　F. Carbonara *et al.*, Nuovo Cim. **36A**, 219 (1978).

[47]　D. Lüke *et al.*, Nuovo Cim. **53B**, 235 (1968).

[48]　J. M. Laget, Phys. Rev. Lett. **41**, 89 (1978).

[49]　P. Joos *et al.*, Phys. Lett. **52B**, 481 (1974).

[50]　T. H. Bauer *et al.*, Rev. Mod. Phys. **50**, 261 (1978).

[51]　G. Wolf, Springer Tracts in Modern Physics **59**, 77 (1971).

[52]　See Ref. 24, for a compilation.

[53]　A. Donnachie, in "High Energy Physics", Vol. V, p. 1, E.H.S. Burhop ed. (Academic Press, 1972).

[54] H. Thom, Phys. Rev. **107,** 1322 (1966); A. Bernstein *et al.,* Nucl. Phys. **A358,** 1916 (1981); T. W. Donnelly, "Workshop on Electron and Photon Interaction at Medium Energies" (Bad-Honnef, F. R. Germany, Oct. 29-31, 1984) - Lecture Notes in Physics **234,** 309 (1985); J. Cohen, Phys. Lett. **153B,** 367 (1985); S. S. Hsiao and S. R. Cotanch, Phys. Rev. **C28,** 1668 (1983).

[55] J. L. Faure *et al.,* Nucl. Phys. **A424,** 383 (1984).

[56] G. Audit *et al.,* private communication.

[57] G. Audit and J. M. Laget, in preparation.

[58] B. J. Edwards *et al.,* Phys. Rev. **D22,** 2772 (1980).

[59] J. M. Laget, Nucl. Phys. **A312,** 265 (1978).

[60] J. Arends *et al.,* Nucl. Phys. **A412,** 509 (1984).

[61] R. J. Hughes *et al.,* Nucl. Phys. **A267,** 329 (1976).

[62] H. O. Meyer *et al.,* Phys. Rev. Lett. **52,** 1759 (1984); Phys. Rev. **C31,** 309 (1985).

[63] E. de Sanctis *et al.,* Phys. Rev. Lett. **54,** 1639 (1985).

[64] J. M. Cameron *et al.,* Phys. Lett. **137B,** 315 (1984).

[65] M. Bernheim *et al.,* Nucl. Phys. **A365,** 349 (1981); S. Turck-Chieze *et al.,* Phys. Lett. **142B,** 145 (1984).

[66] H. J. Gassen *et al.,* Z. Phys. **303,** 35 (1981).

[67] P. E. Argan *et al.,* Nucl. Phys. **A237,** 447 (1975).

[68] R. Abegg *et al.,* Phys. Lett. **118B,** 55 (1982); J. M. Cameron *et al.,* Nucl. Phys. **A424,** 549 (1984).

[69] T. Lim, Phys. Lett. **43B,** 349 (1973).

[70] E. Jans *et al.,* Phys. Rev. Lett. **49,** 974 (1982).

[71] P. H. M. Keizer *et al.,* Phys. Lett. **157B,** 255 (1985).

[72] J. Morgenstern, "Second Workshop on Perspectives in Nuclear Physics at Intermediate Energies", Trieste, March 25-29, 1985, S. Boffi *et al.,* eds. to be published by World Scientific Pub. Co., Singapore.

[73] N. D'Hose *et al.,* "8 eme Session d'Etude Biennale de Physique Nucleaire". Aussois, 4-8 Fevrier 1985. LYCEN 8502 - Universite de Lyon 1 - p. S10.1- .

[74] P. Sauer, same as Ref. 72

[75] R. Machleidt, Lecture Notes in Physics **197,** 352 (1983).

[76] J. M. Laget, same as Ref. 72.

[77] J. F. Mathiot, Nucl. Phys. **A412,** 201 (1984).

[78] C. Hajduk *et al.,* Nucl. Phys. **A405,** 581 (1983).

[79] G. E. Brown and M. Rho, Phys. Today **36,** 24 (1983), and references therein.

[80] M. Beyer *et al.,* Phys. Lett. **12B,** 1 (1983).

[81] J. M. Cavedon *et al.,* Phys. Rev. Lett. **49,** 986 (1982); F. P. Juster *et al.,* Phys. Rev. Lett. **55,** 2261 (1985).

[82] A. W. Thomas, Adv. Nucl. Phys. **13,** 1 and references therein (Plenum Press, 1983).

[83] L. S. Kisslinger, Phys. Lett. **112B,** 307 (1982); (see also Ref. 72).

[84] G. A. Miller *et al.,* Phys. Rev. **C27,** 1669 (1983).

[85] H. J. Pirner *et al.,* Phys. Rev. Lett. **46,** 1376 (1981).

[86] M. Chemtob and S. Furui, Nucl. Phys. A. (in press).

[87] P. Guichon and G. A. Miller, private communication.

[88] J. Sick *et al.,* Phys. Rev. Lett. **45,** 71 (1980).

[89] J. M. Laget, 7eme Session d'Etudes Biennales de Physique Nucleaire, Aussois, March 14-18, 1983. J. Meyer ed., LYCEN 8203 (IPN-Lyon), p. C12.1; J. M. Laget, "Workshop on the use of Electron Rings for Nuclear Research in the Intermediate Energy Region" - Lund, Oct. 5-7, 1982. J. O. Adler and B. Schroder eds. Lund, (1983), p.4.

[90]   J. Arends *et al.*, Z. Phys. **A298,** 103 (1980).

[91]   S. Homma *et al.*, Phys. Rev. Lett. **52,** 2026 (1984).

[92]   J. M. Laget, Nucl. Phys. **A194,** 81 (1972).

[93]   J. H. Koch *et al.*, Ann. of Phys. (N.Y.) **154,** 99 and references therein (1984),

[94]   J. M. Laget, Symposium on Δ -nuclear dynamics. Argonne, May 1983 - T. H. S. Lee ed., ANL-PHY-81-1, CONF-830588, p. 329.

[95]   J. Arends *et al.*, Z. Phys. **305,** 205 (1982); B. A. Mecking, Lecture Notes in Physics **108,** 382 (1979).

[96]   S. Homma *et al.*, Phys. Rev. Lett. **45,** 706 (1980); Phys. Rev. Lett. **53,** 2536 (1984).

[97]   K. Baba *et al.*, Nucl. Phys. **A306,** 292 (1978).

[98]   J. S. Levinger, Phys. Rev. **84,** 43 (1951).

[99]   J. Arhens *et al.*, Nucl. Phys. **A335,** 67 (1980).

[100]   P. Carlos *et al.*, Nucl. Phys. **A431,** 573 (1984).

[101]   A. Leprêtre *et al.*, Nucl. Phys. **A367,** 237 (1981).

[102]   J. S. Levinger, Phys. Lett. **82B,** 181 (1979).

[103]   R. Arnold, these proceedings.

[104]   W. Bertozzi *et al.*, private communication.

[105]   In the original paper, the experimental values have been obtained with a slightly different definition of the flux factor Γ_v (Eq. 19). I have multiplied the theoretical curves by $k/k\gamma$, where $k\gamma = (Q^2 - m^2)/2m$ is the energy of the real photon which leads to the same πN invariant mass Q.

DYNAMICAL EQUATIONS FOR BOUND STATES AND SCATTERING PROBLEMS

J. A. Tjon

Institute for Theoretical Physics
Princetonplein 5, P.O. Box 80.006
3508 TA Utrecht, The Netherlands

INTRODUCTION

One of the main objectives in nuclear physics has been to explain the properties of nuclei in terms of the basic underlying force between nucleons. In such studies one in general has to deal with the dynamics of strongly interacting particles. Except for the few nucleon system and infinite nuclear matter, there is not much hope that we will be able to solve the dynamical equations for an N-particle system in its full complexity. In the past two decades considerable technical progress has been made for the three- and four-nucleon system[1-3]. In particular, the state of art is such that with present-day computers bound state and scattering solutions of the nonrelativistic Schrödinger equations for the tri-nucleon system can be obtained for realistic nucleon-nucleon potentials, neglecting the Coulomb interaction. Extensive use is thereby made of the integral equation formulations for the T-matrix such as those of the Faddeev-Yakubovsky type in which boundary conditions are automatically built in. In particular, for N = 3 we have for 3 → 3 scattering the well known Faddeev equations for the three-body T-matrix

$$T = \sum_{n=1}^{3} T^n , \quad T^n = t_n - \sum_{m \neq n} t_n G_0(z) T^m \tag{1.1}$$

where t_n is the T-matrix of the subsystem of the two particles $k \neq n$, $l \neq n$ and G_0 is the free Green's function

$$G_0(z) = \left[\sum_{n=1}^{3} p_n^2/2m_n - z \right]^{-1}$$

with $z = E + i\varepsilon$, E being the total energy of the three-body system. Boundstates corresponding to the poles of the T-matrix, are determined as solutions to the corresponding homogeneous integral equation for negative energy z.

In the study of the few-body systems one often uses a quasi particle approach[4,5] to reduce the complexity of the problem. For the case of the three-body problem it essentially consists of approximating the two-body T-matrix by a separable form

$$t(p,p',z) = g(p)g(p')\tau(z) \tag{1.2}$$

with p the relative momentum between the two particles. The function $\tau(z)$ may be interpreted as the propagator of a quasi particle with a wave function essentially being given by the form factor g. For the propagator τ we find

$$\tau(z)^{-1} = 1 + 4\pi \int_0^\infty p^2 dp \ g(p)^2 \left[p^2/2\mu - z \right]^{-1} \tag{1.3}$$

with μ the reduced mass. Using Eq. 1.2 the Faddeev equations reduce to effective Lippmann-Schwinger (LS) equations describing the scattering of the quasi particle on the third particle. The potentials are energy dependent and the free Green's function is given by the propagator τ.

The investigations in few nucleon systems using potentials which describe accurately the two-nucleon data up to a few hundred MeV have led in recent years to the insight that a non-relativistic description in terms of pair-wise forces is not in accordance with the experimental findings. Possible reasons for this are the presence of non-nucleonic degrees of freedom like mesons and quarks and effects of special relativity. In the actual physical situation we are interested in, like scattering of nucleons on nuclei at intermediate energies and electron scattering at moderate momentum transfers, relativistic effects are expected to play an important role. As a result, relativistic dynamical models are of interest to study. In many treatments, effects of relativity and non-nucleonic degrees of freedom on dynamical and electromagnetic (em) properties of composite systems are considered to be small and, as a consequence, are treated perturbatively. However care should be exercised that it is done in a consistent manner, since internal dynamics and current conservation are closely related. Also, in such cases a careful study of proper dynamical models can shed light on how to do it appropriately.

The attempts to extend the non-relativistic theory to accommodate for Lorentz invariance can roughly be divided in approaches using a relativistic Hamiltonian description, sometimes called relativistic constrained dynamics, and the quantum field theoretical framework. In the relativistic Hamiltonian approach which has its origin in the classical papers of Dirac[6], the particles are implicitly assumed to be on mass-shell, and the various possible dynamical forms are studied constrained by Poincare invariance. Allowed interactions are then constructed as generators of time translations. A major difficulty thereby has been to incorporate the requirement of a proper cluster decomposition, i.e., when a subcluster of particles is well separated in coordinate space from the rest, the S-matrix of the whole system should factor out as a product of S-matrices of the two sub-systems[7]. For an extensive review see Todorov[8].

In these lectures we describe some studies based on field theoretical approaches. A canonical procedure for the study of the scattering amplitude is to carry out a relativistic covariant perturbation analysis in terms of Feynman diagrams. The main virtue of this is that it is manifestly covariant. However in the intermediate states the particles are allowed to be both off the energy and off the mass shell, while the total four momentum is conserved. As a result we are dealing with four dimensional integrations in the loops of the diagrams in contrast to three-dimensional ones in the non relativistic case. In the next section we will be concerned with relativistic quasi potential equations where the fourth variable has been eliminated in some way. In section 3 we consider the Bethe-Salpeter equation and show some of the uncommon features of this framework. Section 4 deals with the nucleon-nucleon interaction at intermediate energies employing various dynamical equations. As an example of Faddeev-like models, where strong emphasis is put on three-particle unitarity the Kloet-Silbar calculation is considered. Next, the Bethe-Salpeter approach for a relativistic one boson exchange model with isobaric degrees of freedom is discussed, and it is shown how to modify the model to also satisfy three-particle unitarity. As a last topic, certain aspects of the Dirac approach of elastic nucleon nucleus scattering is treated. In particular, the optical Dirac potential is determined starting from the relativistic meson theoretical model discussed in section 4.

QUASI POTENTIAL EQUATIONS

To discuss the basic features of relativistic dynamical equations, let us for simplicity consider a field theory of two types of scalar particles with mass m and μ, interacting through a ϕ^3 theory. One procedure for the study of the scattering amplitude is to carry out a relativistic covariant perturbation analysis in terms of Feynman graphs. It provides a manifestly covariant scheme and satisfies automatically the proper clustering property. However, the particles corresponding to the internal lines are allowed to be off the mass shell. Let us consider the two-particle system. Our starting point for the description of elastic scattering of two particles with mass m is the Bethe-Salpeter (BS) equation for the four point function[9]. Formally it has the form

$$T(s) = V - V\, G_0(s)T(s) \tag{2.1}$$

where s is the total invariant mass squared of the two-particle system and G_0 is the two-particle Green's function

$$G_0(s) = -i\left[p_1^2 - m^2 + i\varepsilon\right]^{-1}\left[p_2^2 - m^2 + i\varepsilon\right]^{-1} \tag{2.2}$$

with p_i the four-momentum of the i-th particle in the intermediate state. The interaction V consists of all irreducible diagrams defined as the set of Feynman graphs, which when they are cut internally to become disconnected, more than two lines are encountered. The diagrammatic representation of Eq. 2.1 is shown in Fig. 1. Except for the renormalization of the lines in the intermediate states, Eq. 2.1 is an exact equation. Using the cutting rules of Cutkowsky[10] one can easily verify that the BS equation satisfies two-particle unitarity as long as no additional particles μ can be produced energetically; that is, for $2m < \sqrt{s} < 2m + \mu$. For physical scattering we have that the external particles are on the mass shell i.e. $p_i^2 - m^2 = p'^2_i - m^2 = 0$ and on the energy shell. The latter implies that we have $E = \sqrt{\vec{p}^2 + m^2}$, $|\vec{p}| = |\vec{q}|$ and $p_0 = q_0 = 0$ in the cm system $P \equiv (\sqrt{s}, \vec{0})$.

Although Eq. 2.1 looks similar to the well known Lippmann- Schwinger (LS) equation from nonrelativistic scattering theory, the structure is different because of the presence of the additional integration variable in the intermediate state, being the relative energy variable k_0. This variable represents retardation effects of the propagating

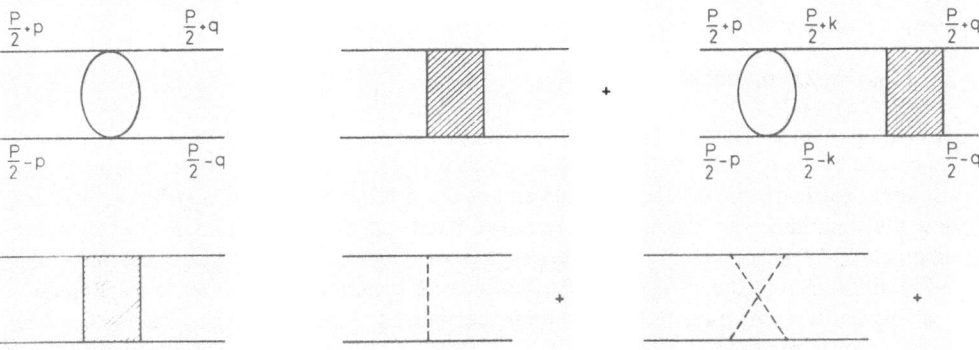

Fig. 1   Diagrammatic form of the Bethe-Salpeter equation.

particles due to special relativity. It leads to some uncommon features such as the presence of Regge daughters and so-called abnormal solutions[11]. Similarly as in the LS case we may obtain bound states with mass $M_B^2 = s$ as solutions to the homogeneous BS equation. They correspond to poles of the T-matrix, while the bound state wave functions $|\psi>$ are related to the residues. Near the pole we have

$$T(s) \approx \frac{|\bar\psi><\bar\psi|}{s-M_B^2} \tag{2.3}$$

with

$$|\psi> = G_0|\bar\psi>. \tag{2.4}$$

To formulate relativistic equations closer to the structure of the LS equation Eq. 2.1 can be recast into a three-dimensional integral equation by constraining the relative energy variable k_0 to be some function of the three momentum $\vec{k}$. For any given propagator $G_{QP}(s)$, Eq. (2.1) can be rewritten as

$$T(s) = W - W\, G_{QP}(s)T(s) \tag{2.5}$$

where the new interaction W is defined as

$$W = V - V[G_o(s)-G_{QP}(s)]W . \tag{2.6}$$

The modified propagator G_{QP} contains a δ-function in k_0. Since the three momentum in Eq. 2.5 is in general off the energy-shell a simple exercise shows that we cannot choose k_0 such that both particles are simultaneously on the mass shell. To limit the possible choices for G_{QP} we like to impose certain conditions such as that the relativistic covariance should be manifest and that it reduces to the LS propagator in the nonrelativistic limit. Furthermore, the imaginary part of G_{QP} should lead to two-particle unitarity. Even with the above conditions, G_{QP} is not uniquely determined. There are two choices which are widely used in the literature in the study of nucleon-nucleon scattering with quasi potential (QP) equations. One choice[12,13] is to treat the nucleons in a symmetrical way by writing down a dispersion relation for G_{QP}, which has the same discontinuity as the Green's function (2.2)

$$G_{QP1} = 2\pi \int_{4m^2}^{\infty} \frac{ds'}{s'-s-i\varepsilon}\, \delta^+[(\frac{P'}{2}+k)^2-m^2]\delta^+[(\frac{P'}{2}-k)^2-m^2] . \tag{2.7}$$

In the cm system we get

$$G_{QP1} = 2\pi\frac{1}{E_{\vec{k}}}\left[4E_{\vec{k}}^2-s-i\varepsilon\right]^{-1}\delta(k_0) \tag{2.8}$$

with $E_{\vec{k}} = \sqrt{\vec{k}^2+m^2}$. The other choice[12] is to put one of the particles, for example particle 2, on the mass shell i.e.

$$G_{QP2} = -2\pi\left[p_1^2-m^2+i\varepsilon\right]^{-1}\delta^+\left(p_2^2-m^2\right) . \tag{2.9}$$

It has the main advantage that em properties can be discussed in a nice way. However, since one of the particles is treated in a special way, it violates the symmetry principle of identical particles. Of course, this can be corrected for in an ad hoc way by symmetrizing the amplitudes or the integral equation itself. If we would like to calculate the interaction W successively by including the one meson exchange and the one loop corrections such as the cross box contribution with the choice (2.9) it leads to nonphysical singularities. As a result, the potentials become complex even in the bound state and elastic energy region[15].

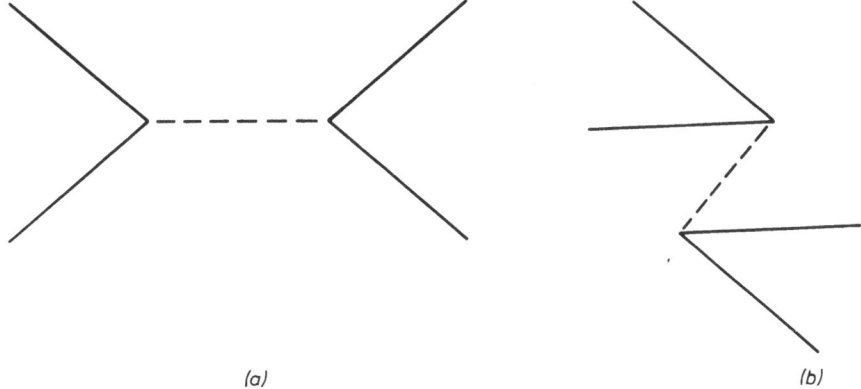

(a)   (b)

Fig. 2   Examples of some old-fashioned time ordered graphs.

As discussed above this way of deriving QP equations does not lead to a formalism where both particles are on the mass shell simultaneously. One way to get that is to use the Hamiltonian formalism. The interaction Hamiltonian for a ϕ^3 theory is given by

$$H_{int} = g \int d\vec{x} \, \phi(x)^3 . \tag{2.10}$$

We may apply the old-fashioned perturbation theory to determine the resolvent $[H-z]^{-1}$ and the corresponding T matrix by expanding it in terms of time ordered graphs[16] . The internal lines correspond to on the mass shell particles and carry only three momentum. Examples are shown in Fig. 2. Any ordering of the vertices can occur. The diagram rules for a given graph are simply given by

1.   each vertex except the last one represents

$$g(2\pi)^3\delta(\textstyle\sum\vec{p}_i)$$

2.   each internal line i

$$\frac{d\vec{p}_i}{(2\pi)^3 2\omega_i} \quad \text{with } \omega_i = \sqrt{\vec{p}_i^2+m^2}$$

3.   each intermediate state

$$[z-E_\gamma]^{-1}$$

where $E_\gamma = \sum_i \omega_i$ and $Z=E_{ini}+i\epsilon$, E_{ini} and E_γ being the total energy of the initial and inter-mediate states, respectively. As compared to a Feynman graph analysis, this expansion has the drawback that it is not manifestly covariant. Although each graph is not Lorentz covariant, the sum of graphs is when the external particles are on the energy shell. Another disadvantage is that one gets many more graphs to·consider. However, by chos-ing a special frame, the so-called infinite momentum frame (IMF), many graphs can be eliminated. Let us assume that the total three momentum $\vec{P}$ is along the positive z-axis

$\vec{P} = (0,0,P)$ with $P > 0$. Then by going to a frame, which moves in the -z direction infinitely fast i.e. $P \to \infty$, all graphs where particles are created or annihilated at a vertex out of the vacuum will vanish while all the other graphs have a finite limit. As an example, in Fig. 2 only graph (a) will survive. The momentum of the i-th particle in a given state can be written as

$$\vec{p}_i = x_i \vec{P} + \vec{q}_i \tag{2.11}$$

where $\vec{q}_i$ is transverse

$$\vec{P} \cdot \vec{q}_i = 0 .$$

Then

$$\sum_i x_i = 1$$

$$\sum_i \vec{q}_i = 0 . \tag{2.12}$$

In general x_i can also be negative, with the exception of the initial and final state where for large enough P all $x_i > 0$. For $P \to \infty$ we have

$$E_i = |x_i| P + \frac{s_i}{P} + O\left[\frac{1}{P^3}\right] \tag{2.13}$$

with

$$s_i = \left[\vec{q}_i^2 + m^2\right] / x_i .$$

Hence for the initial state we have in view of Eq. 2.12

$$E_{ini} = P + \frac{1}{P} \sum_1 s_i + O\left[\frac{1}{P^3}\right] . \tag{2.14}$$

As a result all graphs where one of the x_i's in the intermediate state is negative for $P \to \infty$ leads to vanishing small contributions because of the vanishing energy denominators. In particular this happens when we have Z-type graphs. For the two particle system a three-dimensional integral equation of the QP type can be derived in the IMF frame. It has the form

$$<\vec{q}'x'|T|\vec{q}x> = <\vec{q}'x'|V|\vec{q}x> + \int d_2\vec{q}'' \int_0^1 dx'' \frac{<\vec{q}'x'|V|\vec{q}''x''>}{x''(1-x'')}$$

$$\cdot \left[s - \frac{q''^2 + m^2}{x''(1-x'')} + i\varepsilon \right]^{-1} <\vec{q}''x''|T|\vec{q}x> \tag{2.15}$$

with

$$s = \frac{\vec{q}^2 + m^2}{x(1-x)}$$

and where $\vec{q}$ is the relative transverse momentum between the two particles. For the N = 3 system integral equations can also be written down which satisfy the cluster decomposition property[17]. Since x_i does not transform like a scalar, Eq. 2.15 is not covariant. The equation can be made manifestly covariant by introducing the light cone variables $x^{\pm} \equiv x_0 \pm x_3$ and letting x^- play the role of time, which is an allowed form of constraint dynamics as has been shown by Dirac, the so-called light front dynamics. The above

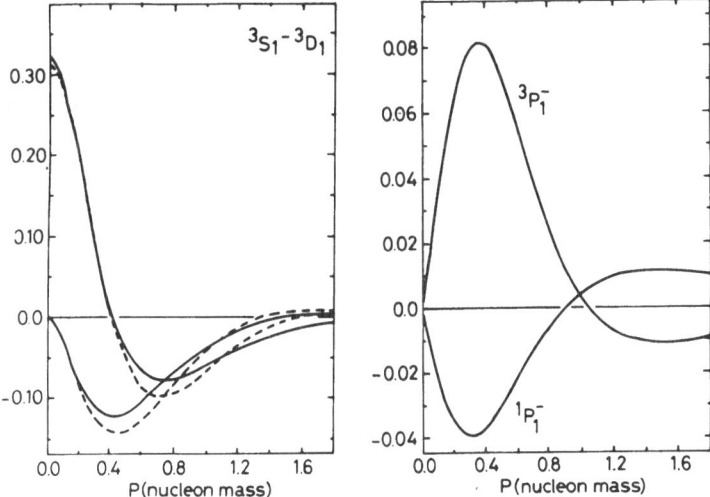

Fig. 3   The positive and negative energy state components of the deuteron vertex function for the quasipotential equations with prescription QP2. The dashed line is
the vertex function using the Reid soft core interaction.

expressions in the old-fashioned perturbation theory are simply modified by replacing
everywhere p_{i3} by $p_i^+ = p_{i0} + p_{i3}$. Under a boost transformation in the z-direction with
velocity beta, p_i^+ transforms like

$$(p_i')^+ = e^\beta p_i^+ .$$

As a consequence, the new $x_i \equiv \dfrac{P_i^+}{P^+}$ is invariant under these transformations. Unfortunately, the resulting equations break manifest rotational symmetry because of the special treatment of the longitudinal direction. In the presence of spin the situation is more
complicated because the spinors carry additional factors of P. The approach of light
front dynamics has been used to study, in particular, the em form factors at high momentum transfers[18–21].

   We now discuss the problem of consistent treatment of the elastic charge form factors
of the deuteron at moderate momentum transfers. Due to the presence of virtual mesonic
currents, the em one-body operators are not sufficient any more to describe the form factors. Estimates have been made within the framework of perturbation theory[20]. Assuming that the dynamics is not changed, effective two-body em operators are constructed.
In particular, the pair excitation graph turns out to give a sizable contribution. However,
since the Ward identity has to hold for any consistent calculation, the inclusion of the
negative energy spinors state of the nucleon in the pair graph necessarily leads to additional graphs where the negative energy spinor states propagate between the exchange of
pions, thereby modifying the dynamics.

   To do a consistent analysis of the deuteron, we may adopt the GP equations where
the spectator particles are put on the mass shell i.e. propagator choice (2.9)[23,24]. In the
actual calculations, the spin complication can most conveniently be accounted for by
using the helicity formalism[25,26]. The two-nucleon system has four spin states to be
characterized by the helicities λ_1, λ_2 and four energy states to be characterized by the

ρ–spinor ρ_1, ρ_2^{27}. $\rho_i = 1/2$ corresponds to the positive energy spinor state of particle i and $\rho_i = -1/2$ to the negative energy spinor state. Since particle 2 has been put on the mass shell, we may have 8 components for the deuteron, which is further reduced to 4 components because of parity conservation. Besides the known positive energy states $^3S_1^+$, $^3D_1^+$ there are two negative energy P-wave components $^3P_1^-$, $^1P_1^-$. Assuming some specific model for the interaction between the nucleons one can calculate both the bound state dynamics and em properties for such a model. This has been studied for the one boson exchange (OBE) model where a pseudo vector coupling has been used for the pion[23]. Near the deuteron pole the T-matrix behaves like Eq. 2.3. From the vertex function $\underline{\psi}$ the deuteron wave function can be computed as

$$\psi_D(p) = (\not{p}_1-m)^{-1}(\not{p}_2-m)^{-1}\, \underline{\psi}(p_1) . \tag{2.16}$$

In Fig. 3 are shown the various components of $\underline{\psi}$ as a function of the relative three-momentum p. From this figure we see that the P-waves are only a fraction smaller than the S, D components, but one should keep in mind that they are significantly smaller if we had plotted the deuteron wave function itself. Indeed, the probability of these P-wave components is a factor of 100 smaller than the D-state probability. The em form factors can be calculated from the deuteron current in the Breit frame. Since the deuteron has been calculated in the cm system, in order to calculate the em vertex functions, boost transformations have to be applied to go to the Breit frame. In the relativistic impulse approximation these can all be done. Gauge invariance is satisfied due to the normalisation condition. In Fig. 4 are shown the results for the various em form factors. The non-relativistic result is modified much less than expected from the perturbational estimate. This can be traced back to a curious cancellation between the dynamical correction to the deuteron wave function and the two-body charge operators. Effective two-body current operators have recently been used and constructed subject to the condition that they should satisfy the continuity equation for a given nonrelativistic potential V^{28-31}

$$i[V,\rho_{em}] + \vec{\nabla}\cdot\vec{j}_{em} = 0 . \tag{2.17}$$

Such an approach strictly makes sense only if a microscopic picture like meson theory is available for the given interaction. At some stage the precise nature of the dynamics

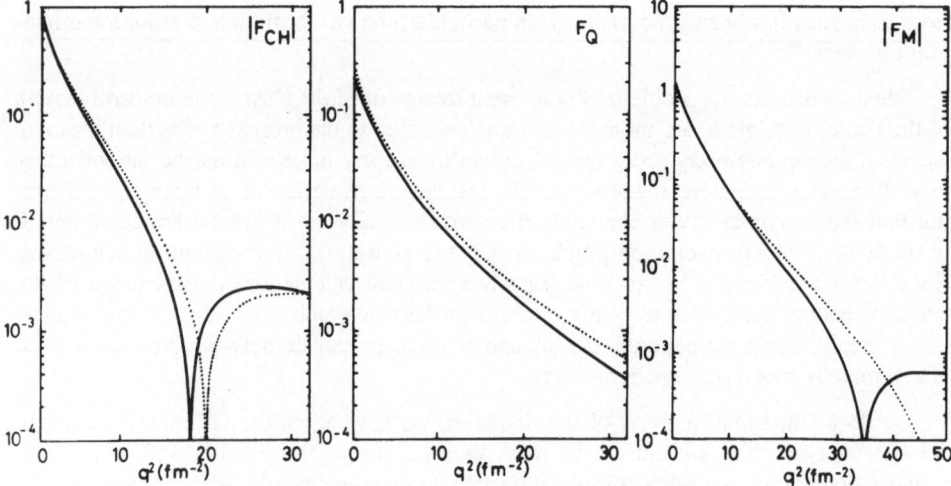

Fig. 4   The charge, quadrupole and magnetic form factors of the deuteron for the wave function shown in Fig. 3. For reference the Reid result (⋯) is also shown.

such as the relativistic aspects may be important, and, as a result, one has then to go beyond a dynamical description in terms of a nonrelativistic potential.

BETHE-SALPETER EQUATION

To illustrate some of the unusual features of the BS equation, let us again for simplicity consider the case of a field theory of scalar particles. The BS equation is an integral equation in four variables. In the two-particle cm system $P = (2E,\vec{0})$ it simplifies considerably by a partial wave decomposition. The two-particle propagator has the form

$$S(k_0,|\vec{k}|) = \left[(E+k_0^2)-E_k^2+i\epsilon\right]^{-1}\left[(E-k_0)^2-E_k^2+i\epsilon\right]^{-1} \tag{3.1}$$

where k is the relative four momentum (see Fig. 1). For the two-body T-matrix a partial wave representation can be written down

$$T(p_0,\vec{p};q_0,\vec{q}) = \sum_{l=0}^{\infty} (2l+1)P_l(\cos\theta)T_l(p_0,|\vec{p}|;q_0,|\vec{q}|) \tag{3.2}$$

with θ the scattering angle between $\vec{p}$ and $\vec{q}$. Due to rotational symmetry the BS equation reduces to a two-dimensional integral equation

$$T_l(p_0,|\vec{p}|;q_0,|\vec{q}|) = V_l(p_0,|\vec{p}|;q_0,|\vec{q}|) - \frac{i}{\pi^2}\int_0^\infty \vec{k}^2\,d|\vec{k}|\int_{-\infty}^\infty dk_0$$

$$\cdot V_l(p_0,|\vec{p}|;k_0,|\vec{k}|)S(k_0,|\vec{k}|)T_l(k_0,|\vec{k}|;q_0,|\vec{q}|). \tag{3.3}$$

For physical scattering, the external particles are on the energy and the mass shell i.e. $p_0 = q_0 = 0$ and $|\vec{p}| = |\vec{q}| = \hat{p}$ with $E = \sqrt{\hat{p}^2+m^2}$. The corresponding T_l operator can be related to the phase shift δ_l and inelastic parameter η_l as follows

$$T_l(0,\hat{p};0,\hat{p}) = \frac{2E}{\hat{p}}\left[\eta_l e^{2i\delta_l}-1\right]/2i. \tag{3.4}$$

Eq. 3.3 is in general a singular equation because of the occurrence of the singularities of the various propagators. In particular, the Green's function (3.1) has four poles. In the k_0 plane they are located at

$$k_0 = E \pm (E_k-i\epsilon)$$

$$k_0 = -E \pm (E_k-i\epsilon) \tag{3.5}$$

When $|\vec{k}|$ varies two of the poles can coincide, so that the integration path is pinched. This happens when $k_0 = 0$ and $E = E_k$, i.e. for the on shell situation. It is precisely due to this pinching that the T-matrix develops a two-particle cut in the energy plane. There is an obvious example where Eq. 3.3 can be solved in closed form. Consider the Zachariasen model[32] in a ϕ^4 theory in which only the set of graphs shown in Fig. 5 are taken into account. Let us assume that the vertex function is given by a separable form

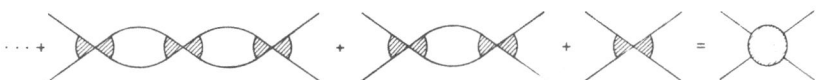

Fig. 5   The series of diagrams considered in the ϕ^4-theory.

(denoting $p \equiv |\vec{p}|, q \equiv |\vec{q}|$)

$$V_l(p_0,p;k_0,k) = g_l(p_0,p)g_l(k_0,k) . \tag{3.6}$$

Then Eq. 3.3 can immediately be solved to give

$$T_l(p_0,p;q_0,q) = g_l(p_0,p)g_l(q_0,q)\tau_l(E) \tag{3.7}$$

with

$$\tau_l(E) = \left[1 + \frac{i}{\pi^2} \int_0^\infty k^2 dk \int_{-\infty}^\infty dk_0 \, S(k_0,k)g_l(k_0,k)^2 \right]^{-1} . \tag{3.8}$$

The whole singularity structure in the E-plane is determined by τ_l. From Eq. 3.8 we see that T_l is analytic in E with a cut starting from $E = m$ to ∞. Its discontinuity is given by

$$\text{Im } \tau_l^{-1}(E) = -2 \int_0^\infty k^2 dk \int_{-\infty}^\infty dk_0 \, g_l(k_0,k)^2 \, \delta^+\!\left[(E+k_0)^2 - E_k^2 \right] \delta^+\!\left[(E-k_0)^2 - E_k^2 \right] . \tag{3.9}$$

The two delta–functions correspond precisely to those used in dispersion relation (Eq. 2.7). As a result, T_l satisfies two-particle unitarity for $E \geq m$. On shell scattering corresponds for all energies to pure elastic scattering, i.e., $\eta_l = 1$. Eq. 3.7 can also support a bound state with mass $M_B = 2E$ if τ_l has a pole

$$\tau_l^{-1}(E) = 0 \tag{3.10}$$

with $E < m$. The corresponding bound state wave function is then given by

$$\psi(k_0,k) = S(k_0,k) \, g_l(k_0,k) . \tag{3.11}$$

Similarly, as in the nonrelativistic case, interactions of a separable form can be a practical tool to study the properties of few body systems using in this case field theoretical Fadeev type of equations. In particular, relativistic separable interactions may be constructed to parameterize the two-nucleon interactions by fitting to the experimental phase shifts[33] and subsequently using this to study relativistic effects in the few nucleon system. Currently this is being done for the tri-nucleon properties like the binding energy and the em form factors of the triton[34].

For a ϕ^3–theory the situation is more complicated due to the presence of singularities in the interaction V_l. To see what happens in this case let us for convenience approximate V_l by the one meson exchange diagram. In this approximation we get[35]

$$V_l(p_0,p;k_0,k) = \frac{g^2}{2pk} \, Q_l\!\left[\frac{p^2+k^2-(p_0-k_0)^2+\mu^2-i\varepsilon}{2pk} \right] . \tag{3.12}$$

In the k_0 plane we have logarithmic cuts arising from the Q_l–function. There are four logarithmic branch points located at

$$k_0 = p_0 \pm \sqrt{(p \pm k)^2 + \mu^2 - i\varepsilon} . \tag{3.13}$$

In contrast to the Green's function singularities, those from V_l depend on the initial momenta. Eq. 3.3 cannot be solved any more in a closed form. Also Eq. 3.3 is not immediately tractable for numerical analysis due to the presence of the various singularities occurring in the integration range of k and k_0 even in the bound state region $E < m$.

Let us first consider the bound state problem. To reduce the BS equation to a non-singular integral equation we can apply a Wick rotation by introducing $p_0 \rightarrow p_0 e^{i\alpha}$, $q_0 \rightarrow q_0 e^{i\alpha}$ and $k_0 \rightarrow k_0 e^{i\alpha}$. From the Eq. 3.5, we see that for $E < m$ the poles of the Green's function never cross the imaginary axis in the complex k_0 plane when we vary the three momentum k. In the process of changing α from 0 to $\pi/2$, the branch points of V_I move into the complex plane without intersecting the integration path. Hence, the Wick rotation[36] can be performed and we end up with an equation along the imaginary k_0 axis. The resulting integral equation becomes

$$T_I(ip_4,p;iq_4,q) = V_I(ip_4,p;iq_4,q) + \frac{1}{\pi^2} \int_0^\infty k^2 \, dk \int_{-\infty}^\infty dk_4$$

$$\cdot V_I(ip_4,p;ik_4,k)S(ik_4,k)T_I(ik_4,k;iq_4,q) \ . \tag{3.14}$$

Note that the matrix elements V_I and S are real. Hence, if there is a bound state, its corresponding wave function is real along the imaginary k_0 axis. Furthermore, Eq. 3.14 has become a nonsingular integral equation, and consequently it can be solved by standard methods like matrix inversion after discretization. In the scattering region $E < m$, two of the poles of S can pinch as discussed earlier. It happens at $k_0 = 0$. As a result, on performing the Wick rotation the contour has to be deformed so that we get an additional contribution from these poles when they have crossed the imaginary axis (see Fig. 6). The Wick rotated equations become in the case of $E > m$

$$T_I(ip_4,p;0,\hat{p}) = V_I(ip_4,p;0,\hat{p}) + \frac{1}{\pi^2} \int_0^\infty k^2 \, dk \int_{-\infty}^\infty dk_4$$

$$\cdot V_I(ip_4,p;ik_4,k)S(ik_4,k)T_I(ik_4,k;0,\hat{p})$$

$$- \frac{1}{8\pi E} \int_0^{\hat{p}} k^2 dk \Bigg\{ V_I(ip_4,p;E-E_k,k)S_R(k) \, T_I(E-E_k,k;0,\hat{p})$$

$$+ V_I(ip_4,p;-E+E_k,k)S_R(k) \, T_I(-E+E_k,k;0,\hat{p}) \Bigg\} \tag{3.15}$$

with

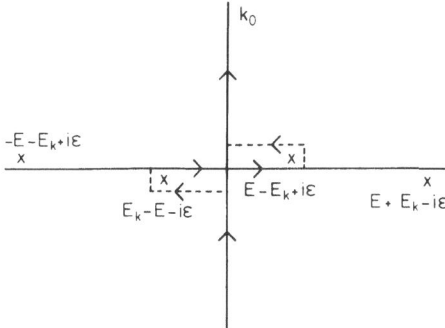

$$k_0$$

$$-E-E_k+i\varepsilon$$
$$\times$$

$$E-E_k+i\varepsilon$$

$$E_k-E-i\varepsilon$$

$$E + E_k - i\varepsilon$$

Fig. 6   Deformation of the integration path in k_0 plane when the two poles have passed each other. The location of the poles of the Green's function are given by crosses.

$$S_R(k) = \frac{1}{E_k} \frac{1}{E-E_k+i\varepsilon} .$$ (3.16)

The integration range of k in the single integral of Eq. 3.15 is finite, since the additional pole contributions are present only when the poles in S have crossed each other. The propagator S_R is essentially the same one as is used in the QP2 prescription of the quasi potential approach in section 2. There, however, the integration range extends to ∞. To derive Eq. 3.15 we have set the initial state to the on shell value. From Eq. 3.15 we see the appearance of the unknown T_l operator for real k_0 arguments. Hence, we have to supplement Eq. 3.15 by equations for T_l with $k_0 = \pm (E-E_k)$. They can simply be obtained from Eq. 3.15 by setting p_0 to these values. One can readily verify that the above procedure is allowed as long as $\sqrt{s} < 2m + 2\mu$, i.e., up to the threshold of two meson production. Above this energy the branch point of the logarithmic cut can cross the imaginary axis so that the Wick rotation cannot be performed any more. The additional contributions give rise to two meson production processes.

In the region where one meson production is possible, i.e., $\sqrt{s} > 2m + \mu$, the two-particle unitarity is not satisfied any longer, because in the single integral the Q_l-function can become complex. Physically this corresponds to the possibility that the meson which is exchanged can be energetically produced. The Wick rotated equation still contains singularities in the one-dimensional integral arising from the Green's function, but they are effectively of a LS type. They can be removed by subtraction procedures such as the Kowalski-Noyes technique[35,37,38] or using Fredholm theory, so that the resulting equations become numerically tractable.

Up to now we have considered the case of scalar particles. Since we will be interested in the two nucleon problem, I shall discuss the spin complication. A complete basis of two-nucleon helicity states can be constructed from the positive and negative solutions of the Dirac equation $u_\lambda^{(\rho)}(\vec{p})$ with $\rho = \pm 1/2$. Using the convention of Jacob and Wick[39] and Kubis[25] for the two-particle spinors, it is convenient to form states which are even or odd in ρ spin space

$$|\vec{p}\lambda_1\lambda_2 +> = U_{\lambda_1}^{(+)}(\vec{p}) U_{-\lambda_2}^{(+)}(-\vec{p})$$

$$|\vec{p}\lambda_1\lambda_2 -> = U_{\lambda_1}^{(-)}(\vec{p}) U_{-\lambda_2}^{(-)}(-\vec{p})$$

$$|\vec{p}\lambda_1\lambda_2\, e> = \frac{1}{\sqrt{2}}\left[U_{\lambda_1}^{(+)}(\vec{p}) U_{-\lambda_2}^{(-)}(-\vec{p}) + U_{\lambda_1}^{(-)}(\vec{p}) U_{-\lambda_2}^{(+)}(-\vec{p}) \right]$$

$$|\vec{p}\lambda_1\lambda_2\, 0> = \frac{1}{\sqrt{2}}\left[U_{\lambda_1}^{(+)}(\vec{p}) U_{-\lambda_2}^{(-)}(-\vec{p}) - U_{\lambda_1}^{(-)}(\vec{p}) U_{-\lambda_2}^{(+)}(-\vec{p}) \right] .$$

With this basis set the helicity amplitudes can be defined

$$\Phi = <\vec{p}'\lambda_1'\lambda_2'\rho' | T | \vec{p}\lambda_1\lambda_2\rho>$$

for which a partial wave expansion can be written

$$\Phi = \sum_J (2J+1)\Phi^J d_{\lambda\lambda'}^J(\theta)$$ (3.17)

with $\lambda = \lambda_1-\lambda_2$, $\lambda'=\lambda_1'-\lambda_2'$ and d^J the known rotation matrices. In the partial wave state representation parity conservation implies

$$<J\lambda_1'\lambda_2'\, \rho' | T | J\lambda_1\lambda_2\, \rho > = \eta_\rho\eta_{\rho'} <J-\lambda_1'-\lambda_2'\, \rho' | T | J-\lambda_1-\lambda_2\, \rho >$$ (3.18)

where $\eta_\rho, \eta_{\rho'}$ are the intrinsic parities of the initial and final states. Since the positive and negative energy spinor states have opposite intrinsic parity, it follows that the $\rho = +,-$ states have opposite intrinsic parity to the $\rho = e,0$ states. By forming eigenstates of overall parity

$$|Jr\lambda_1\lambda_2 \rho> = \frac{1}{\sqrt{2}} \left[|J\lambda_1\lambda_2 \rho > + r|J-\lambda_1-\lambda_2 \rho > \right]$$ (3.19)

with $r = \pm 1$, the BS equation can be partial wave decomposed to yield a coupled set of two-dimensional integral equations with 8 discrete channels. Knowing the helicity amplitudes, one then can determine the amplitudes in the LSJ basis. For details we refer to Kubis[25].

THE NUCLEON-NUCLEON INTERACTION AT INTERMEDIATE ENERGIES

As a direct application of the previously described dynamical equations, we may consider the two-nucleon system. It has received much interest in the past few years because of the experimental discovery of the resonant structure at around 600 MeV T_{lab} energy in polarized pp scattering[40]. Phase shift analysis[41,42] has indicated the presence of looping behaviour in the Argand plot of various I = 1 partial waves such as the $^1D_2, ^3F_3$ and $^1G_4$ channels. Possible explanations have been proposed in terms of exotic di-baryon resonances and pure kinematical threshold effects giving rise to pseudo resonance behaviour.

Here we describe in some detail the conventional description based on meson field theory. The theoretical approaches can roughly be divided into relativistic Faddeev-like models[44-49] and two-particle like models[50-54]. At energies above 300 MeV the two-nucleon amplitude becomes inelastic and in the region up to 1 GeV predominantly single pion production takes place. In view of the success in describing low energy NN scattering in terms of the OBE mechanism, it is natural to study relativistic extensions of such models, combined with Δ degrees of freedom which are known to be important to account for the inelastic processes. In so doing, one may hope that the pion production processes can be described in a reasonable way.

In the Faddeev-like approaches much emphasis is put on the three-particle aspects. As a representative of these models let me describe the Kloet-Silbar calculation[44], where the effects of pion deuteron channels are neglected. The starting point is a Lee type model[53] for NN scattering. The two nucleons are made distinguishable by assuming that one of them, called N', can emit a pion but not absorb it, whereas the other nucleon N can only absorb the pion but not emit it. As a result, not more than three particles can occur in an intermediate state of NN' scattering. To satisfy three particle unitarity, the N' particle is renormalized. Since in this model the N' is essentially a quasi particle of a N and π, the dressing of N' is equivalent to treating the N' as a πN scattering state for which the T-matrix is of a separable form. The corresponding form factor is assumed to be a function of the square of the relative four momentum. In describing the NN' state, it is assumed that the N particle is on the mass shell, whereas N' is not. To ensure relativistic covariance a Blankenbecler-Sugar type of Green's function is used in the description of πN scattering. In the rest system of the N' it has the form

$$G(S_{\pi N}) = \frac{\omega+E}{\omega E} \left[(\omega+E)^2-S_{\pi N} \right]^{-1} \delta\left[k_0-\frac{1}{2}E+\frac{1}{2}\omega \right]$$ (4.1)

where ω and E are the energies of the π and N in the intermediate state and k_0 is the fourth component of the relative πN momentum. It should be noted that it would have been more consistent with the way the Born term is computed in this model to have used the Gross type of prescription where the N is put on the mass shell. Similarly the Δ particle is considered as a p-wave quasi particle πN state by allowing the Δ to decay only into a N and π. Let p and q be the relative πN four momentum in the initial and final state.

Then a p-wave driving force in the πN cm system should have the form

$$V_{\pi N}(p,q) = g(p^2)g(q^2)\vec{p}\cdot\vec{q} . \tag{4.2}$$

To exhibit the explicit relativistic covariance use is made of the so-called three-dimensional magic vectors to rewrite the scalar product $\vec{p}\cdot\vec{q}$ so that it becomes frame independent. The diagrammatic representation of the resulting integral equations for NN′ scattering is shown in Fig. 7. It should be interpreted in the time ordered sense. The Born term is calculated in the overall cm system of NN′ with N put on the mass shell. Although relativistic covariance is satisfied in this model the clustering property does not hold[54]. In practice the breakdown of this property may not have a significant effect. More serious is that, due to the different treatment of the two nucleons, this approach suffers from the fact that the Pauli principle is violated. The model has been used to study elastic scattering and pion production in nucleon-nucleon scattering at intermediate energies. In particular, the resonant structures found experimentally can be accommodated for, although details of the lower partial waves, such as the state dependence of the P-waves are not well reproduced. This may be due to the neglect of ρ exchange in the driving force between NN′ and NΔ states.

We now turn to discuss an approach where emphasis is put on the two-particle aspects, although I will show that such a model can be extended to also satisfy three-particle unitarity. It has the advantage that the Pauli principle and the cluster property are manifestly satisfied. The model is essentially a relativistic extension of the OBE model including Δ degrees of freedom[50]. The starting point is the Bethe-Salpeter equation where in addition to the NN intermediate state NΔ and $\Delta\Delta$ states are also allowed. Hence we are dealing with coupled channel BS equations. In Fig. 8 is shown the graphical representation of the model. The matrix elements of the interaction between NN states are the same ones as already discussed in the deuteron problem. They are given by the OBE model, i.e. the nucleon-nucleon interaction is described by the exchange of $\pi,\sigma,\rho,\omega,\eta,$ and δ mesons. The transition interaction between the NN and NΔ , $\Delta\Delta$ states is given by π and ρ–exchange. No direct coupling is assumed between the pion and two $\Delta's$. To regularize the high momentum behaviour of the kernel of the integral equations, form factors are needed. These are assumed to be of the form

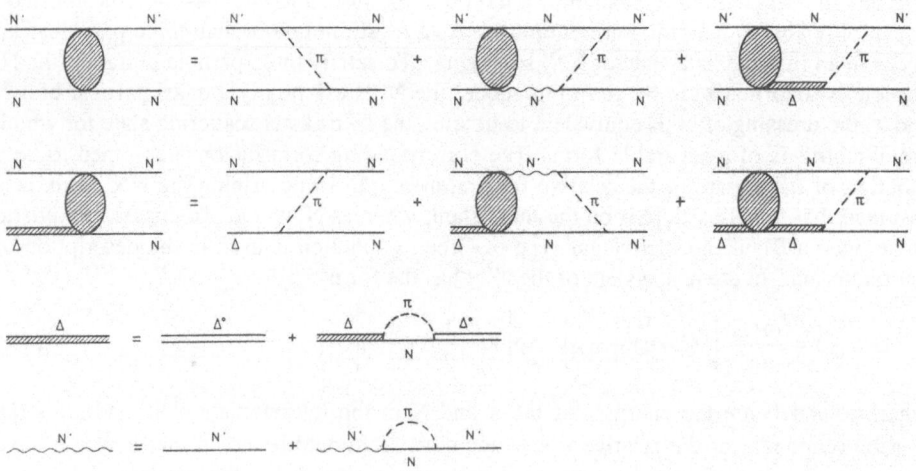

Fig. 7   Time ordered representation of the integral equations solved by Kloet and Silbar in a three-body model.

$$f_{OBE}(t) = \left[\frac{\Lambda^2}{\Lambda^2 - t} \right]^{\alpha} \tag{4.3}$$

where in most calculations $\alpha = 1$, and two cutoff masses are used; one for the meson NN vertex, Λ_N; and one for the meson NΔ vertex, $\Lambda_{N\Delta}$. These are taken typically to be $\Lambda_N^2 = 1.8 m_N^2$ and $\Lambda_{N\Delta}^2 = 1.3 m_N^2$.

Since both nucleons are off the mass shell for a given physical partial wave state, there are eight states coupled to it if we assume parity conserved forces. Attempts with γ_5 coupling for the pion to fit the various phase shifts were not successful due to the strong coupling to negative energy spinor states[26]. Clearly pair suppression is needed in the form of a σ term or the use of pseudovector coupling. In this latter case the negative energy NN states are of minor importance, and their neglect can be compensated for by slight changes in the coupling parameters. Because of practical reasons, in most calculations the negative energy spinor states of the nucleons and Δ's have been neglected.

The Δ is treated as a Rarita-Schwinger propagator with a complex mass. Neglecting the negative energy states, it is given by

$$P^{\mu\nu}(p_0, \vec{p}) = \left[p_0 - \sqrt{\vec{p}^2 + m_\Delta^2} \right]^{-1} \sum_\sigma \Delta^\mu(\vec{p}, \sigma) \Delta^\nu(\vec{p}, \sigma) \tag{4.4}$$

where $\sigma = \pm 1/2, \pm 3/2$ and Δ^μ are the positive energy Rarita-Schwinger spinors for spin 3/2 particles[55]. Furthermore, $\mu_\Delta = m_0 - i\Gamma(q)/2$ with $m_0 = 1236 \text{MeV}$ and the width Γ is parameterized by the Bransden-Moorhouse form[56]. The q represents the πN three momentum in the πN cm system. It is related to the invariant πN mass by

$$q^2 = \frac{[S_{\pi N} - (m_\pi - m_N)^2][S_{\pi N} - (m_\pi + m_N)^2]}{S_{\pi N}}. \tag{4.5}$$

The invariant mass $S_{\pi N}$ should now be expressed in terms of variables of the NΔ system. One simple approximation due to VerWest consists of assuming that the Δ particle picks up all the available energy in the NΔ system, i.e.

$$S_{\pi N} = (\sqrt{s} - m_N)^2. \tag{4.6}$$

The approximation essentially amounts to neglecting the motion of the nucleon. One can correct for it if we assume that we may take the nucleon four momentum on the mass shell. Instead of Eq. 4.6 we then have

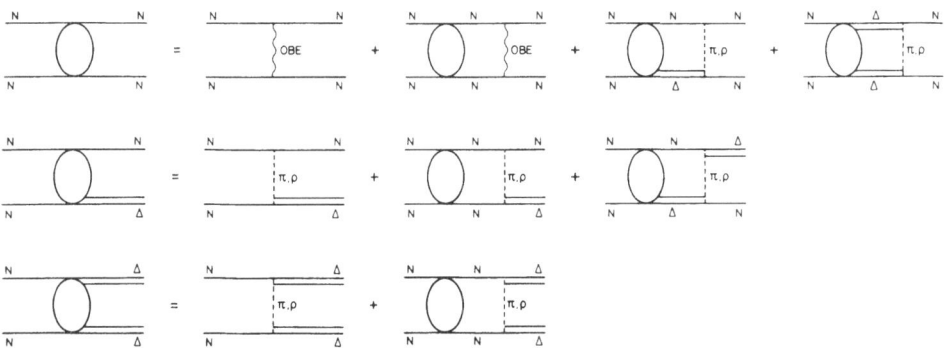

Fig. 8   Diagrammatic representation of the coupled Bethe-Salpeter equations with iso-baric degrees of freedom and the one boson exchange mechanism.

$$S_{\pi N} = \left[\sqrt{s} - \sqrt{p_1^2 + m_N^2} \right]^2 .$$ (4.7)

The effect of this modification is that the inelasticity of the NN amplitude at a given energy tends to get smaller. In the actual phase shift analysis of the experimental data, theoretical information is needed for the high partial waves. At low energy the one pion exchange provides a good description for the high partial waves. This is not sufficient any more at higher energies because of the inelasticities. As a result it is interesting to see what kind of model dependence one can expect. In particular, for the meson models as have been discussed here, the dominant contribution to the inelasticity comes from the NΔ box. Large differences have been found in the literature between the various models. Much of it can be ascribed to the treatment of the relation between the invariant πN and NN energy as is given by Eqs. 4.6 and 4.7, although some sensitivity is found for the choice of form factors[47] . Since there is no theoretical justification to prefer one above the others, model dependence exists and, as a consequence, it may affect the phase shift analysis.

A QP approximation can be made to the BS equations. To satisfy manifestly the Pauli principle, we have taken the Blankenbecler-Sugar prescription. The $q_0 = 0$ choice also has the virtue that no unphysical singularities appear in the transition interaction as we would have obtained if we, for example, took the nucleon in the NΔ state to be on the mass shell.

Extensive scattering calculations have been carried out both within the framework of BS and QP equations. Similarly, as in the Faddeev-like models, we find that the $I = 1$ resonant structures can be accommodated for within a meson theoretical approach. The structures can be ascribed in principle to a combination of kinematic threshold effects due to the looping of the NΔ-box, as happens in the $^1G_4$ and more complicated dynamical poles which are located near the NΔ branchpoints in the 2nd Riemann sheet of the energy variables. There is some evidence from phase shift analysis that there may be a resonant structure in the $I = 0$ $^1F_3$ channel. Since the NΔ channel is absent for $I = 0$ such a structure can unlikely be explained by coupled channel models as we have described here. Moreover, the state dependence in the P-waves can be improved by introducing the ρ-exchange in the transition interaction, thereby obtaining a reasonable overall fit to the various partial waves.

The basic mechanism of pseudo-resonance behaviour can be illustrated using an exactly soluble two-particle model[58] . It is a simple coupled channel model with separable potentials, neglecting spin. The channels correspond to the NN and NΔ states, while the potentials are assumed to be of the form

$$V_{ij}(p,q) = \lambda_{ij} g_i(p) g_j(q) .$$ (4.8)

The propagator of the NN channel can be written as

$$G_1(p,s) = \left[p^2 - \hat{p}^2 \right]^{-1}$$ (4.9)

where $\hat{p} = \frac{1}{2}\sqrt{s - m_N^2}$ is the on shell momentum. The Δ is treated as a fixed complex mass particle of mass $M_\Delta = M_0 - i\Gamma/2$ with $M_0 = 1.236$GeV and $\Gamma = 0.12$GeV, so that the NΔ propagator becomes

$$G_2(p,s) = \left[p^2 - p_R^2 \right]^{-1}$$

where p_R is the NΔ relative momentum, which is related to the NN energy through

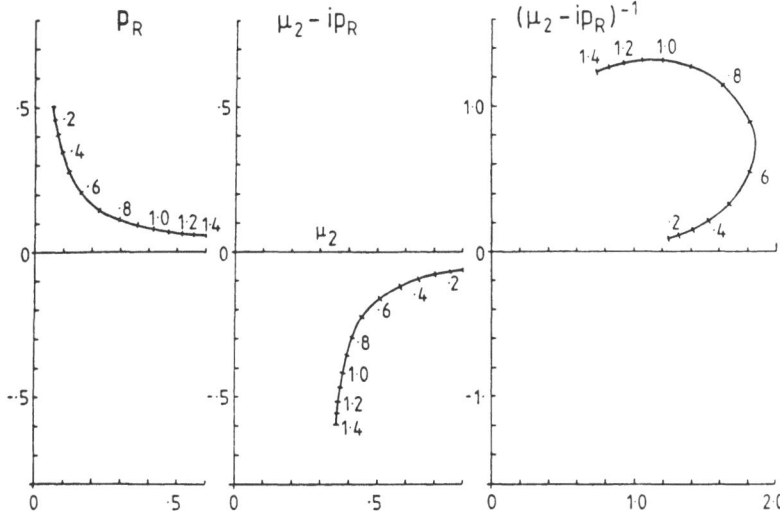

Fig. 9 Looping behaviour of simplified $N\Delta$ box. Behaviour in the complex plane of p_R, μ_2-ip_R and $[\mu_2-ip_R]^{-1}$ for $\mu_2 = 0.3$ GeV. The markings on the curves are T_{Lab} in 0.1 GeV intervals.

$$p_R = \left\{ \frac{[s-(m_N+M_\Delta)^2][s-(M_\Delta-m_N)^2]}{4s} \right\}^{\frac{1}{2}}. \tag{4.10}$$

In such an extremely simplified model of the isobar model, the looping behaviour of the $N\Delta$–box can be discussed. The $N\Delta$ box distribution is given by

$$B_{N\Delta}^{(2)}(s) = \lambda_{12}^2\, g_1(\hat{p})^2 \int_0^\infty p^2 dp\, g_2(p)^2 G_2(p,s). \tag{4.11}$$

Let us choose the form factor

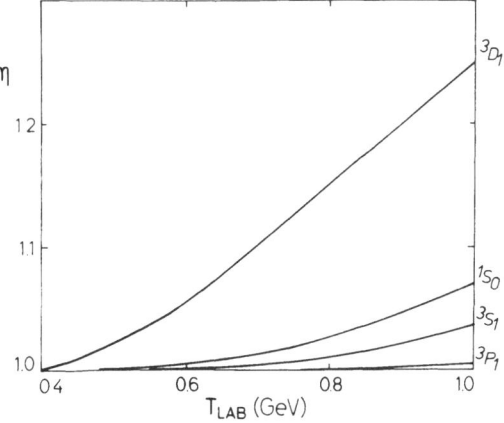

Fig. 10 The inelastic parameter for various low partial waves as a function of energy for the case that only NN intermediate states are present.

$$g_i(p) = \left[p^2 + \mu_i^2 \right]^{-\frac{1}{2}} \tag{4.12}$$

then

$$B_{N\Delta}^{(2)}(s) = \frac{\pi}{2} \lambda_{12}^2 \, g_1^2(\hat{p}) \left[\mu_2 - i p_R \right]^{-1}. \tag{4.13}$$

We may now discuss the energy dependence of Eq. 4.13 near the $N\Delta$ threshold $s_{thr} \approx (m_N + M_0)^2 - \Gamma^2/4$, which corresponds to $T_{Lab} \approx 0.64$ GeV. From Eq. 4.10 we see that for $s < s_{thr}$, p_r is almost purely imaginary and for $s > s_{thr}$, p_R is almost purely real. As a result, $B^{(2)}$ can change rapidly if we move past s_{thr}. It is precisely due to the square root singularity of Eq. 4.10 that the $N\Delta$–box has a looping behaviour in the argand plot. In Fig. 9 are shown as a function of energy p_R, $\mu_2 - i p_R$ and $[\mu_2 - i p_R]^{-1}$ for $\mu_2 = 0.3$ GeV, clearly exhibiting looping due to the $N\Delta$ branchpoint. Obviously no dynamical poles are present in $B^{(2)}$.

Since the mesons are allowed to propagate virtually, some inelastic effects are present in the BS equation. Consequently, two-particle unitarity is satisfied only up to one pion production threshold. As noted by Levine et al.[35], the BS equation fails to satisfy the unitarity bound in the inelastic region, i.e. the inelastic parameter η can become greater than one. This is illustrated in Fig. 10 for NN scattering, where we have taken the transition interaction to vanish in order not to mask the effect through the production of the $N\Delta$ states. Let us consider a diagram out of a given set of diagrams. Cut it so that it becomes disconnected and reconnect the lines in all possible ways to let it become connected again. From the Cutkowsky rules it follows that unitarity is only satisfied if all newly generated graphs belong to the given set. For the case where the ladder approximation has been made for the BS equation, it can easily be verified that in the region $2m_N < \sqrt{s} < 2m_N + m_\pi$, only the set of lowest order nucleon bubble diagrams are generated from cutting two nucleon and one pion line, i.e. in order to satisfy three-particle unitarity, we only have to renormalize the nucleons by the lowest order self energy diagram. Another method of improving the relativistic isobar model is to describe the Δ as an unstable particle which decays into a pion and nucleon. More specifically the bare Δ particle propagator is dressed by the lowest order πN bubble diagram (see Fig. 11). The form factor at the $\pi N\Delta$ vertex is chosen to be of the form

$$F_{N\Delta\pi} = f_{OBE}(k^2) g_{sc}((p-2k)^2). \tag{4.14}$$

Without the second factor g_{sc}, it was not possible to find a decent fit to the P_{33} phase shift. The function g_{sc} depends on the relative momentum between the pion and nucleon and it is taken to be on the mass shell $p_0 - 2k_0 = \sqrt{(\vec{p} - \vec{k})^2 + m_N^2} + \sqrt{k^2 + m_\pi^2}$. It can be interpreted, as in the Kloet-Silbar model, as a way of describing phenomenologically the πN scattering in the P_{33} channel. Assuming it is of the form

$$g_{sc}(p^2) = \left[\frac{\Lambda_{sc}^2}{\Lambda_{sc}^2 - p^2} \right]^{\frac{1}{4}} \tag{4.15}$$

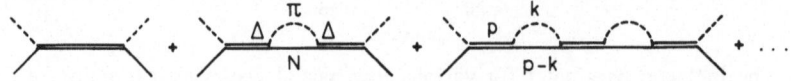

Fig. 11   Sum of bubble diagrams as a model for P_{33} πN scattering.

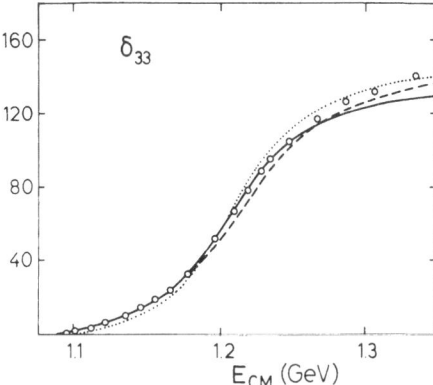

Fig. 12   The πN phase shift for the P_{33} channel using the Bransden-Moorhouse parameterization (----), the model as described here (-------) and in Ref. 44 ($\cdots\cdots$).
Experimental points are from Ref. 71.

the calculated phase shift is shown in Fig. 12 for $\Lambda_{sc}^2 = 0.85 m_N^2$ and $\dfrac{f_{N\Delta\pi}^2}{4\pi} = 0.32$ together
with the Bransden-Moorhouse parameterisation and the QP result of Ref. 44.

Although the physical ingredients in our relativistic isobar model are essentially the same as in the Kloet-Silbar calculations, our model has the basic advantage that the Pauli principle is satisfied. Moreover, it does not suffer from problems such as how to correct for the neglect of parts of diagrams (like the backward propagating pion in the Born term) and questions of undercounting the number of graphs as, for example, in the case of the NΔ box. On the other hand, the BS equation is in practice difficult to solve. In general we have to make a certain subselection of graphs. Here we have discussed the simplest case of the OBE model. Inclusion of one loop graphs such as the crossed two pion exchange graphs, although feasible, is already very hard. Another drawback of the BS equation should be mentioned. In the case that we would consider non equal mass scattering and we let one of the masses go to infinity, we essentially need all irreducible graphs to get the proper one body limit[59]. In particular, the ladder approximation does not suffice.

MESON THEORETICAL BASIS FOR DIRAC APPROACH

As a last topic I will discuss some calculations[60] which have been done using the results from the dynamical equations for the 2N system, as described in the previous section, to compute the Dirac optical potential. In the study of physical NN scattering, we have seen that the presence of the negative energy spinor states does have some effect but their neglect can be compensated for by slight modifications of the coupling parameters as long as we use pseudovector coupling for the pion. However, the precise nature of the NN interaction may affect the predictions profoundly for more complicated systems. In particular, whether the negative energy spinor state components are present or not in the description of proton nucleus scattering leads, at intermediate energies, to very different results in the polarization observables. This is precisely the result of the very successful Dirac optical approach[61-64].

The basic idea in the Dirac approach is that since nucleons are spin $\frac{1}{2}$ fermions, it may be more appropriate to describe the nucleons as Dirac particles, thereby emphasizing their relativistic nature. Within this scenario the Dirac equation has been used for the projectile proton together with some phenomenological forms for the optical potentials to describe various observables of proton scattering on spin zero nuclei with considerable success[61]. To provide a microscopic basis of the Dirac phenomenology, one may attempt, within the framework of multiple scattering theory, to describe the elastic scattering of a proton on a nucleus in terms of an optical potential[65]. At intermediate energies it is customary to calculate such a potential using a relativistic impulse approximation (see Fig. 13) in terms of the basic NN interaction. Corrections to these are assume or believed to be small. The resulting optical potential $U^{\rho,\rho'}$, which has matrix elements between all ρ spin states of the incoming and outgoing nucleon, is then used as input to a Dirac equation for the scattered nucleon. The Dirac equation has the form

$$[\vec{\alpha}\cdot\vec{p}+\beta m+U]\psi = E\psi \tag{5.1}$$

or more explicitly

$$(E_p-E+U^{++})\psi^+ + U^{+-}\psi^- = 0$$

$$(-E_p-E+U^{--})\psi^- + U^{-+}\psi^+ = 0 \tag{5.2}$$

where $\psi^{\pm}$ are the ρ spin state components of the Dirac wave function. Neglecting the ψ^- terms amounts to essentially a nonrelativistic description where pure kinematical effects of special relativity are taken into account. One can eliminate the ψ^- component to get

$$[E_p-E+U^{++}+U_{pair}]\psi^+ = 0 \tag{5.3}$$

with

$$U_{pair} = U^{+-}[E+E_p-U^{--}]^{-1}U^{-+}. \tag{5.4}$$

In actual calculations[64] one finds that the effect of U^{--} in Eq. 5.4 is small, and, as a result, the distinct difference between nonrelativistic and Dirac approach is the contribution from the Z-graph which is a ρ^2 density dependent correction to the nonrelativistic optical potential.

In practice, the one loop integral in the relativistic impulse approximation is not done explicitly, but one makes the so-called $t\rho$ approximation[66], in which leading order corrections due to Fermi motion vanish. A result of this approximation is the NN T-matrix becomes independent of the integration variable, so that the integral can be carried out involving only the one nucleon distribution function of the nucleus. It is clear that in order to determine U, in principle, complete information is needed of the NN interaction. Since the impulse approximation is calculated in the Breit frame of the proton-nucleus system, we have to know how t_{NN} transforms from the 2N cm system to the Breit frame. This is most conveniently done by expressing t_{NN} in terms of relativistic invariants in the Dirac space of the two nucleons. For physical on shell NN scattering five independent invariants are needed. The basic ansatz which has been made in the literature is to assume that the complete t_{NN} can be expressed in Dirac space as a linear combination of the five Fermi covariants[62].

$$t_{NN} = F_1 S + F_2 V + F_3 T + F_4 A + F_5 P \tag{5.5}$$

where

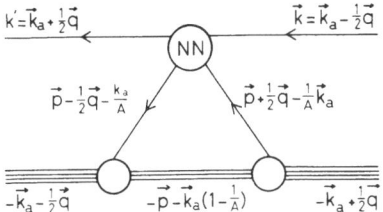

Fig. 13   The relativistic impulse approximation for nucleon nucleus scattering in the Breit frame.

$$S = 1, \; V = \gamma^{(1)} \gamma^{(2)}, \; T = \sigma_{\mu\nu}^{(1)} \sigma_{\mu\nu}^{(2)}, \; P = \gamma_5^{(1)} \gamma_5^{(2)} \text{ and } A = PV \; . \tag{5.6}$$

The scalar functions F_n can then be uniquely determined from NN data such as the phase shift parameters. The representation (5.5) is, however, ambiguous as has been realised by, among others, Adams and Bleszynski[67]. One can add any covariant term to it which has vanishing matrix elements between positive energy on shell states. For example,

$$t_{NN} = F_1 S + F_2 V + F_3 T + F_4 A + F_5 P + F_{P_v} [P_v - P - \tilde{S}(P_v - P)] \tag{5.7}$$

where $\tilde{S}$ is an operator interchanging the spinor indices of nucleons 1 and 2 and

$$P_v = \gamma_5^{(1)} (\not{p}_1 - \not{p}'_1) \gamma_5^{(2)} (\not{p}_2 - \not{p}'_2) \; . \tag{5.8}$$

The scalar functions F_n will be the same as found using Eq. 5.5, but now we have an additional unknown scalar function F_{P_v}. The reason for this ambiguity is that to know U, we in principle need the complete off shell 2N T-matrix, for which the five Fermi covariants do not form a complete set. Since, in general, the dimension of the Dirac space of two spin 1/2 fermions is given by 16, we have at most 256 linearly independent covariants to consider. Parity conservation cuts it down to 128. A further analysis based on time reversal invariance and charge symmetry shows that there are 56 independent covariants. A complete Lorentz invariant representation for off shell NN scattering can be formulated using the projection operators on the negative energy state of the i-th nucleon

$$\Lambda_i^{(-)} = \frac{-E_{p_i} \gamma_0^{(i)} + \vec{\gamma}^{(i)} \cdot \vec{p}_i + m_N}{2m_N} \; . \tag{5.9}$$

We may write

$$t_{NN} = \hat{F}^{11} + \Lambda_2^{(-)} \hat{F}^{12} + \hat{F}^{13} \Lambda_2^{(-)} + \Lambda_2^{(-)} \hat{F}^{14} \Lambda_2^{(-)}$$

$$+ \Lambda_1^{(-)} \left[\hat{F}^{21} + \Lambda_2^{(-)} \hat{F}^{22} + \hat{F}^{23} \Lambda_2^{(-)} + \Lambda_2^{(-)} \hat{F}^{24} \Lambda_2^{(-)} \right]$$

$$+ \left[\hat{F}^{31} + \Lambda_2^{(-)} \hat{F}^{32} + \hat{F}^{33} \Lambda_2^{(-)} + \Lambda_2^{(-)} \hat{F}^{34} \Lambda_2^{(-)} \right] \Lambda_1^{(-)}$$

$$+ \Lambda_1^{(-)} \left[\hat{F}^{41} + \Lambda_2^{(-)} \hat{F}^{42} + \hat{F}^{43} \Lambda_2^{(-)} + \Lambda_2^{(-)} \hat{F}^{44} \Lambda_2^{(-)} \right] \Lambda_1^{(-)} \; . \tag{5.10}$$

This representation has been chosen such that if we consider physical NN scattering only the operator $\hat{F}^{11}$ survives. There are 16 terms in Eq. 5.10 which are mutually independent. The operators $\hat{F}^{kl}$ can be expressed as a linear combination of 8 independent operators so that we have totally 128 scalar functions F_n^{kl}, in agreement with parity conservation

$$\hat{F}^{kl} = \sum_{n=1}^{8} K_N^{kl} F_n^{kl} \tag{5.11}$$

where for example the set of 8 covariants in the case of $\hat{F}^{11}$ is given by

$$K_n^{11} = \left\{ S,V,T,P,A,\gamma_2 Q_1 - \gamma_1 Q_2, P(\gamma_2 Q_1 - \gamma_1 Q_2), P(\gamma_2 Q_1 + \gamma_1 Q_2) \right\} \tag{5.12}$$

with $Q_i = (p_i + p'_i)/2m_N$. The covariants K_6, K_7, K_8 have been taken following Scandron and Jones[68]. For on shell scattering one can verify $F_6^{11} = F_7^{11} = F_8^{11} = 0$ which is in accordance with the fact that the five Fermi covariants form a complete set for on shell scattering.

From the above considerations it is clear that on shell scattering information is not sufficient to determine uniquely the full NN T-matrix. A relativistic dynamical model such as we have considered here is needed. From such a model we can construct uniquely all these 128 covariants, of which 56 of them are independent. In calculating all the functions F_n^{kl}, we have indeed verified that all the relations found from the symmetry arguments are numerically satisfied. This provides an excellent check on the calculations. Given these amplitudes the Dirac optical potential can be calculated. This has been done for the spin zero nucleus ^{40}Ca. To obtain a local optical potential it is necessary to approximate the energies $E(\vec{p}_n)$ occurring in the $\Lambda^{(-)}$ operators by m_N. After spin averaging over the struck nucleon in the nucleus and Fourier transformation in the momentum transfer, one obtains the optical potential. It is clear that the above approximation can be circumvented by doing a momentum space calculation. The optical potential U has 6 of the 8 allowed terms[69].

$$U_{DIRAC} = S(r) + \gamma_0 V(r) + \frac{1}{2} \left\{ \vec{\gamma} \vec{p}, C(r)/m_N \right\}_+ + i\vec{\alpha} \cdot \hat{r} T(r) + \left[S_{LS}(r) + \gamma_0 V_{LS}(r) \right] \vec{\sigma} \cdot \vec{L}. \tag{5.13}$$

Neglecting the terms T, S_{LS} and V_{LS}, which are of the order 1 MeV, the Dirac equation (5.1) with the potential U can be rewritten as

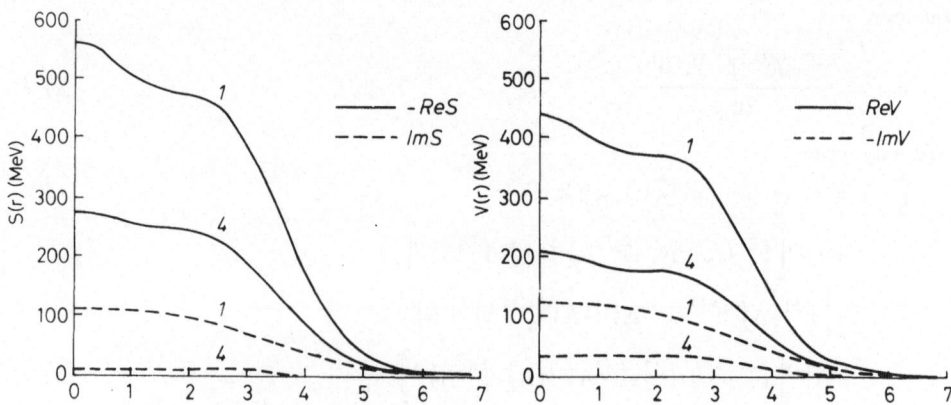

Fig. 14   Scalar and vector potentials for elastic proton ^{40}Ca scattering at 181 MeV T_{Lab}. Curves labeled 1 have been obtained using the NN representation with Fermi covariants, while those labeled 4 come from the microscopic model.

$$\left[-\gamma^0 E + m_N + \vec{\gamma} \cdot \vec{p} + S_{eff} + \gamma^0 V_{eff}\right] \tilde{\psi} = 0 \tag{5.14}$$

with

$$\tilde{\psi} = [1 + C(r)/m_N]^{\frac{1}{2}} \psi \tag{5.15}$$

and where the effective scalar and vector parts of the optical potential are given by

$$S_{eff}(r) = \frac{S(r) - C(r)}{1 + C(r)/m_N}$$

$$V_{eff}(r) = \frac{V(r) + EC(r)/m_N}{1 + C(r)/m_N} . \tag{5.16}$$

The resulting potentials are very different from the ones obtained using the ansatz (5.5). The results are shown in Fig. 14 for the case, that we restrict the struck particles to positive energy states. Similar results are found for the full calculation. The strengths of S and V are in general considerably smaller, of the order of 200 MeV, and there is relatively minor energy dependence, to be contrasted with the results using the Fermi covariants. Moreover, the strengths are in agreement with the relativistic nuclear matter calculations of the Brooklyn group[70]. In Fig. 15 we show the strengths of S_{eff} and V_{eff} for the case of nuclear matter using the dynamical model as a function of the lab energy. From this we see that S_{eff} and V_{eff} become abnormally large at low energies when the Fermi covariants are used, whereas the strengths obtained from the dynamical model are more reasonable. Although the shape and strength of the optical potentials are very different qualitatively, it is remarkable that similar predictions are found for the proton nucleus scattering observables.

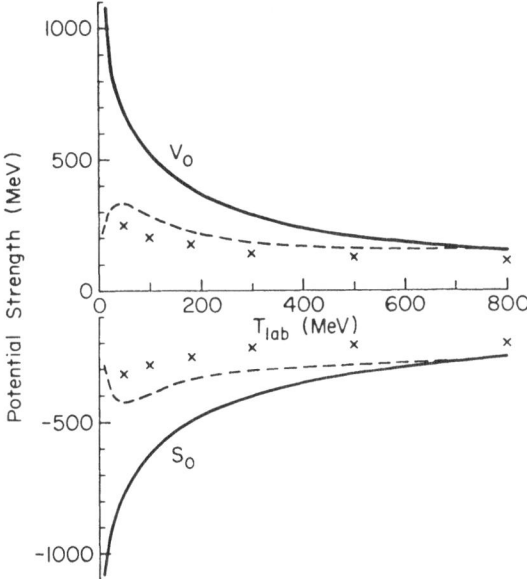

Fig. 15   Scalar and Vector potential strengths for nuclear matter based on the impulse approximation. Solid lines are obtained using the Fermi covariants. Dash lines are obtained by replacing the pseudo scalar term P in the Fermi covariants representation by the pseudo vector operator.

The strong energy dependence of the strength of the S and V found when we use the Fermi covariants ansatz suggests, in view of the presence of the pseudo scalar term P, that the pion contribution is in some way being favoured with an effective pseudo scalar coupling. Indeed if we simply calculate the pure OPE contribution, the same strong energy dependence is found in S and V. This indicates that if pair suppression should take place in NN scattering, the representation (5.5) is not reliable at least at lower energies. One may attempt to invent other simple representations which may be more reliable; for example, the one given in Eq. 5.7 when we assume that F_{P_v} is chosen such that the pseudo scalar term is replaced by a pseudo vector term. The result is shown in Fig. 13 by the dashed line. In so doing, the strength is more in accordance with the dynamical model calculations, whereas the observables of proton nucleus scattering are not closer to the corresponding results of the model. Although the obtained results of the relativistic effects in proton nucleus scattering are very encouraging and a remarkable agreement is found in the Dirac phenomenology with the experimental spin observables, it should be kept in mind that corrections of a different nature such as Pauli blocking effects and other medium corrections may also affect the proton nucleus scattering process.

REFERENCES

[1]   J. A. Tjon, Nucl. Phys. **A353**, 47C (1981).
[2]   A. C. Fonseca, Nucl. Phys. **A416** 421C (1984).
[3]   J. L. Friar, this summer school.
[4]   R. A. Amado, Phys. Rev. **132**, 485 (1963).
[5]   E. Alt, P. Grassberger and W. Sandhas, Phys. Rev. **C1**, 85 (1970).
[6]   P. A. M. Dirac, Rev. Mod. Phys. **17**, 195 (1945); ibid **21** 392 (1949).
[7]   F. Coester and W. Polyzou, Phys. Rev. **D26**, 1348 (1982).
[8]   I. Todorov, Constrained Hamiltonian approach to relativistic point particle dynamics, SISSA report, Trieste (1980).
[9]   E. E. Salpeter and H. A. Bethe, Phys. Rev. **84**, 1232 (1951).
[10] R. E. Cutkowski, J. Math. Phys. **1**, 429 (1960).
[11] For a review and many references on the BS equation, N. Nakanishi, Progr. Theor. Phys., suppl. **43**, 1 (1969).
[12] R. Blankenbecler and R. Sugar, Phys. Rev. **142**, 1051 (1966)
[13] A. A. Logunov and A. N. Tavkhelidze, Nuovo Cim. **29** 380 (1963).
[14] F. Gross, Phys. Rev. **186**, 1448 (1969).
[15] M. J. Zuilhof and J. A. Tjon, Phys. Rev. **C24**, 736 (1981).
[16] S. Weinberg, Phys. Rev. **150**, 1313 (1966).
[17] J. M. Namyslowski and H. J. Weber, Zeit. für Phys. **A295**, 219 (1980).
[18] L. L. Frankfurt and M. I. Strikman, Phys. Rept. **76**,215**(1981)**.
[19] B. L. G. Bakker, L. A. Kondratyuk and M. V. Terent'ev, Nucl. Phys. **B158**, 497 (1979).
[20] G. P. le Page and S. J. Brodsky, Phys. Rev. **D22**, 2157 (1980).
[21] M. Chemtob, Nucl. Phys. **A336** 299 (1980).
[22] A. D. Jackson, A. Landé and D. O. Riska, Phys. Lett. **55B**, 23 (1975).
[23] T. A. Tjon and M. J. Zuilhof, Phys. Lett.. **84B**, 31 (1979); Phys. Rev. **C22**, 2369 (1980).
[24] R. G. Arnold, C. E. Carlson and F. Gross, Phys. Rev. **C21**, 1426 (1980).
[25] J. J. Kubis, Phys. Rev. **D6**, 547 (1972).
[26] J. Fleischer and J. A. Tjon, Nucl. Phys. **B84**, 375 (1975); Phys. Rev. **D21**, 87 (1980).

[27] J. L. Gammel, M. T. Menzel and W. R. Wortman, Phys. Rev. **D3**, 2175 (1971).

[28] W. Strueve, Ch. Hadjuk and P. U. Sauer, Nucl. Phys. **A405**, 581 (1983).

[29] J. F. Mathiot and D. O. Riska, Phys. Lett. **133B**, 23 (1983).

[30] W. Jaus and W. S. Woolcock, Nucl. Phys. **A431**, 409 (1984).

[31] D. O. Riska, Isovector electromagnetic exchange currents and the nucleon-nucleon interaction, Helsinki preprint (1984).

[32] F. Zachariasen, Phys. Rev. **121**, 1851 (1961).

[33] K. Schwarz et al., Acta Phys. Austr. **53**, 191 (1981).

[34] G. Rupp and J. A. Tjon, to be published.

[35] M. J. Levine, J. Wright and J. A. Tjon, Phys. Rev. **154**, 1433 (1967); ibid **157**, 1416 (1967).

[36] G. C. Wick, Phys. Rev. **96**, 1124 (1954).

[37] K. L. Kowalski, Phys. Rev. Lett. **15**, 798 (1965).

[38] H. P. Noyes, Phys. Rev. Lett. **15**, 538 (1966).

[39] M. Jacob and G. C. Wick, Am. Phys. **7**, 404 (1959).

[40] A. Yokosawa, Phys. Rep. **64**, 47 (1980).

[41] N. Hoshizaki, Progr. Theor. Phys. **61**, 129 (1979).

[42] R. A. Arndt et al., Phys. Rev. **D28**, 97 (1983).

[43] P. Mulders, A. Aerts and J. de Swart, Phys. Rev. **D21**, 2653 (1980).

[44] W. M. Kloet and R. R. Silbar, Nucl. Phys. **A338**, 317 (1980); ibid **A364**, 346 (1981).

[45] I. R. Afnan and B. Blankleider, Phys. Rev. **C24**, 1572 (1981).

[46] M. Betz and T-S.H. Lee, Phys. Rev. **C23**, 375 (1981).

[47] M. Araki, Y. Koike and T. Ueda, Nucl. Phys. **A369**, 346 (1981); ibid **A389**, 605 (1982).

[48] M. Araki and T. Ueda, Nucl. Phys. **A379**, 449 (1982).

[49] A. S. Rinat and Y. Starkand, Nucl. Phys. **A397**, 381 (1983).

[50] J. A. Tjon and E. E. van Faassen, Phys. Lett. **120B**, 39 (1983); Phys. Rev. **C28**, 2354 (1983); ibid **30**, 285 (1984); to be published.

[51] T-S.H. Lee, Phys. Rev. **C29**, 195 (1984).

[52] A. Green and M. E. Sainio, J. Phys. **65**, 503 (1979); ibid **G8**, 1337 (1982).

[53] R. Aaron, R. D. Amado and J. E. Young, Phys. Rev. **174**, 2022 (1968).

[54] S. Morioka and I. R. Afnan, Phys. Rev. **C23**, 852 (1981).

[55] W. Rarita and J. Schwinger, Phys. Rev. **59**, 436 (1941).

[56] B. H. Bransden and R. G. Moorhouse, The Pion-Nucleon System, Princeton University Press, Princeton (1973).

[57] W. M. Kloet and J. A. Tjon, Phys. Rev. **C30**, 1653 (1984).

[58] W. M. Kloet and J. A. Tjon, Nucl. Phys. **A392**, 271 (1983).

[59] F. Gross, Phys. Rev. **C26**, 2203 (1984).

[60] J. A. Tjon and S. J. Wallace, Phys. Rev. Lett. **54**, 1357 (1985); Phys. Rev. **C32**, 267 (1985); ibid **C32**, 1667 (1985); to be published.

[61] B. C. Clarke et al., Phys. Rev. Lett. **50**, 1644 (1983); ibid **51**, 1809 (C); Phys. Rev. **C28**, 1421 (1983).

[62] J. A. McNeil, J. R. Shepard and S. J. Wallace, Phys. Rev. Lett. **50**, 1439 (1983); ibid **50**, 1443 (1983).

[63] J. A. McNeil, L. Ray and S. J. Wallace, Phys. Rev. **C27**, 2123 (1983).

[64] M. V. Hynes et al., Phys. Rev. Lett. **52**, 978 (1984); Phys. Rev. **C31**, 1438 (1985).

[65] A. K. Kerman, H. McManus and R. Thaler, Ann. Phys. **8**, 551 (1959).

[66] S. A. Gurvitz, J.-P. Dedonder and R. D. Amado, Phys. Rev. **C14**, 142 (1979).

[67] D. L. Adams and M. Bleszynski, Phys. Letter **136B**, 10 (1984).

[68]   M. D. Scandron and H. F. Jones, Phys. Rev. **173,** 1734 (1963).
[69]   L. S. Celenza and C. M. Shakin, Phys. Rev. **C28,** 1256 (1983).
[70]   H. R. Anastasio *et al.,* Phys. Rep. **100,** 327 (1983).
[71]   R. Koch and E. Pietarinen, Nucl. Phys. **A336,** 331 (1980).

NUCLEAR PHYSICS INVESTIGATIONS USING MEDIUM ENERGY COINCIDENCE EXPERIMENTS

George E. Walker[†]

Nuclear Theory Center and Physics Dept.
Indiana University, Bloomington, IN 47405
USA

I. INTRODUCTION

In the last decade, there has been considerable advantage in comparing information obtained in such reactions as (e,e'), (p,p') and (π,π'). As the field of intermediate energy physics moves into the era where high precision coincidence experiments such as (e,e'p), (e,e'π); (p,2p), (p,p'π); (π,π'p) and (π,2π) are performed, it seems useful to review the kinds of physics that will be investigated and the advantages of using electromagnetic and strong probes together. In general, the higher counting rates associated with strong projectiles make coincidence experiments with these probes somewhat easier. Thus, identifying interesting features first from such experiments may occur. On the other hand, studies of these features may depend on the use of the well understood but experimentally challenging electromagnetic probe coincidence experiments. We shall pursue this theme in some of the examples below. In the next section, we discuss possibilities using electromagnetic probe coincidence experiments. Following that, we present in sections III and IV some recent results using strong probes and some suggestions for future coincidence experiments. Finally, in the last section, we briefly summarize and suggest some joint studies using strong and electromagnetic exclusive reactions.

II. NUCLEAR PHYSICS WITH HIGH DUTY CYCLE ELECTROMAGNETIC MACHINES (≤1 GeV)

A. Overview of Possible Investigations

Several references[1] that were found particularly useful in reviewing this area are listed in the Reference section. Although this conference has a focus on "new vistas" in electronuclear physics, the emphasis herein will be on the important, familiar but *unrealized* opportunities allowed by medium energy (≤1 GeV) high duty cycle electron machines. Theoretical papers now almost twenty years old have discussed much of the potential of using (e,e'p) and (e,e'γ) to study nuclear physics but the lack of appropriate experimental facilities has impeded studies employing these reactions.

Basically, using electromagnetic probes one can map out the charge and current densities of nuclei. Because the electromagnetic interaction is weak and well known, the nuclear response is often straightforward to interpret. Some of the different aspects of nuclear structure that can be studied using coincidence experiments with electromagnetic probes are:

a) Nuclear giant resonances using (γ_T, γ) and $(e, e'x)$ where $x \equiv \gamma, N, \alpha$ or fission fragment. One can study the location, width and decay of resonances. Coincidence experiments allow one to eliminate the bremsstrahlung background, disentangle overlapping resonances, and study the coupling of resonances to other nuclear degrees of freedom.

b) Single nucleon degrees of freedom in nuclei using the same reactions (and with the same advantages) as mentioned in a). One can study the validity of proposed models of nuclear states using noncollective single nucleon degrees of freedom. The study of the *quasi-elastic nuclear response* falls in this category. For this process, it is useful to study such reactions as $(e, e'N)$, $(e, e'2N)$, $(e, e'\pi)$, $(e, e'\pi N)$, and $(\gamma, \pi N)$. The properties investigated in such reactions include single nucleon separation energies, momentum distributions, and shell model occupation probabilities. One can also obtain information on NN correlations and/or interactions in nuclei.

c) Nuclear fission using the reaction $(e, e'f)$. This reaction should allow additional studies of fission barriers (or the potential energy surface) and the dynamics leading to fission.

d) Non-nucleonic or sub-nucleonic degrees of freedom in nuclei using the reactions $(e, e'N)$, $(e, e'\pi)$, $(e, e'2N)$, $(e, e'\pi N)$, $(e, e'K^+)$, and $(e, e', \text{Heavy Meson})$. Using such reactions one may obtain information on the role of Δ's in nuclei including the Δ–nucleus interaction and relatedly, Δ properties in nuclei. Other topics, including the π–nucleus interaction, hypernuclei via $(e, e'K^+)$, nucleon and meson resonances in nuclei, and possible partial quark deconfinement (bag size) may be studied using some of these reactions. In planning for these coincidence studies, it is encouraging to note that the nuclear-nucleon quasi free response and the isobar production region tend to be nicely separated in the (q, ω) plane.

In addition to many aspects of nuclear structure, some of which are listed above, one may carry out reaction mechanism investigations using the $(e, e'x)$, $(e, e'2N)$, and $(e, e'\pi N)$ reactions. Some of the topics which might be studied include:

a) the validity of the factorization of the nuclear excitation and decay process;
b) the number of active target nucleons in the reaction (i.e., importance of two nucleon reaction mechanisms, including exchange currents);
c) the role of virtual non-nucleonic degrees of freedom in a given reaction;
d) the testing of assumptions regarding final state interactions of the strongly interacting particles observed in $(e, e'x)$ experiments.

Obviously high duty cycle, intense beam current and high *variable* energy are essential for studying many of the processes listed above. Trivially, sufficient energy is required to conserve energy and momentum in the excitation of nucleon resonances or the production of mesons. Thus, for *efficient* study of electroproduction of mesons and N*'s, electron beam energies of > 1 GeV are appropriate. Below one GeV (the region on which we concentrate) mainly pion production and $\Delta(1232)$ production studies are possible. Of course even if heavy particle production is not part of the coincidence study, substantial beam energy is an important consideration. For example, since the counting rate is often a practical limiting feature in the feasibility of experiments, it is important to have sufficient beam energy so that the desired momentum transfer to the nucleus can occur at forward angles (the Mott cross section increases approximately two orders of magnitude for $|\vec{q}| \sim 400$ MeV/c if one increases the electron incident energy from 200 MeV to 700 MeV). Moreover, it may be useful to remove nucleons from the nucleus so that the final nucleon has substantial kinetic energy. Since final state nucleon-nucleus interactions are expected to be a minimum near $T_p^{lab} \sim 200$ MeV, it is of interest to note that an incident electron must have an energy in excess of 1 GeV to produce a scattered (knocked out proton) with $T_p^{lab} \sim 200$ MeV for a forward electron scattering angle ($\sim 30°$) even for knockout from loosely bound shell-model states.

Basic Formulae for Coincidence Experiments

Consider experiments of the type (e,e′p). The basic geometry is shown in Fig. 1. The four-momentum of the (incident) [scattered] electron is denoted $(\vec{k}_1, E_1)$ $[\vec{k}_2, E_2]$. The square of the energy loss, $\omega \equiv E_1 - E_2$, and three momentum transfer, $\vec{q} \equiv \vec{k}_1 - \vec{k}_2$, may be combined in four vector notation as $q_\mu^2 = q^2 - \omega^2 (|\vec{q}| \equiv q)$. The momentum, energy, and scattering angle (relative to the $\vec{q}$ direction) of the detected particle, p, are denoted $\vec{p}$, E_p, and θ_p, respectively. The angle between the electron scattering plane and the $(\vec{p},\vec{q})$ plane (see Fig. 1) is given by ϕ_p.

The double coincidence cross section associated with the reaction shown in Fig. 2 can be obtained by using standard definitions and the Golden Rule. Assuming unpolarized particles (and final spins undetected). Born approximation (one photon exchange), and a local current coupling one obtains from the result[2]

$$\frac{d^4\sigma}{d\Omega_{k_2}dE_2 d\Omega_p dE_p} = \frac{2\alpha^2}{q_\mu^4} \frac{k_2}{k_1} \frac{pE_p}{M_T} \eta_{\mu\nu} W_{\mu\nu} \tag{1}$$

where α is the fine structure constant, M_T is the mass of the target and $\eta_{\mu\nu}$ is a tensor associated with the *electron* current and is given by

$$\eta_{\mu\nu} = 2E_1 E_2 \frac{1}{2} \sum_{\substack{electron \\ spins}} (\bar{u}(k_2)\gamma_\mu(k_1))(\bar{u}(k_2)\gamma_\nu u(k))^* \tag{2}$$

(see Ref. 3 for more details on the Dirac electron spinors, u). The tensor $W_{\mu\nu}$ results from the nuclear current and may be written in the form[2]

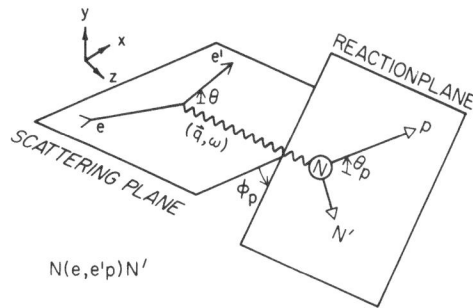

$$N(e,e'p)N'$$

Fig. 1   The exclusive (e,e′p) reaction using the notation adopted in the text.

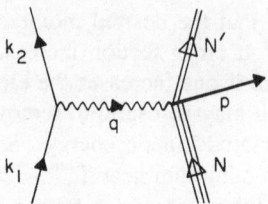

Fig. 2   The exclusive (e,e'p) process, assuming a virtual single photon exchange
          mechanism and using the notation adopted in Eq. 1.

$$W_{\mu\nu} = \delta_{\mu\nu}W_1 + \frac{N_\mu N_\nu}{M^2} W_2 + \frac{\frac{1}{2}(N_\mu p_\nu + p_\mu N_\nu)}{-p\cdot N} W_3 + \frac{p_\mu p_\nu}{m^2} W_4 . \tag{3}$$

In obtaining Eqs. 1-3, Lorentz invariance and electron current conservation has been
assumed[2]. The quantity N_μ denotes a component of the initial target four-momentum
($N_\mu^2 = -M^2$) and m is the nucleon mass. Note that a term of the form $\frac{1}{2}(N_\mu p_\mu - p_\mu N_\nu)$ is
not allowed in $W_{\mu\nu}$ in this case because $\eta_{\mu\nu}$ is symmetric. The $W_i(i = 1\rightarrow4)$ contain the
nuclear structure information and are, in general, functions of the Lorentz scalars
q_μ^2, $q\cdot N$, $q\cdot p$ and $p\cdot N$. In what follows, we present results appropriate for the laboratory
frame and use the notation and definitions given by DeForest[2]. It is convenient to rewrite
Eq. 1 using the following expressions appropriate for (e,e'p)

$$W_C = \frac{-q^2}{q_\mu^2} W_1 + \frac{q^4}{q_\mu^4} \left[W_2 + cW_3 + \frac{E_p^2}{m^2} c^2 W_4 \right] \tag{4a}$$

$$W_T = 2W_1 \tag{4b}$$

$$W_S = \frac{p^2}{m^2} W_4 \sin^2\theta_p \tag{4c}$$

$$W_I = \frac{-p}{E_p} \frac{q^2}{q_\mu^2} \left[W_3 + 2\frac{E_p^2}{m^2} cW_4 \right] \sin\theta_p \quad [c \equiv 1 - \left(\frac{\omega p}{qE_p} \right) \cos\theta_p]. \tag{4d}$$

Using Eqs. 4a-4d, Eq. 1 can be rewritten in the form

$$\frac{d^4\sigma}{d\Omega_{k_2}dE_2 d\Omega_p dE_p} = \frac{2\alpha^2}{q_\mu^4} \frac{k_2}{k_1} \frac{pE_p}{M_T}$$

$$\times [V_L(\theta)W_C + V_T(\theta)W_T + V_I(\theta,\phi_p)W_I + V_S(\theta,\phi_p)W_S] \tag{5}$$

where now the W's are functions of q, ω, p and θ_p. The V's are kinematic functions and
may be written in the form (neglecting the mass of the electron)

$$V_L = \frac{q_\mu^4}{q^4} 2k_1 k_2 \cos^2\theta/2 \tag{6a}$$

$$V_T = \frac{q_\mu^2}{2q^2} [2k_1 k_2(\cos^2\theta/2) + q^2] \tag{6b}$$

$$V_I = (E_1+E_2) \left[(q_\mu^2/2q^2)\, V_L\right]^{\frac{1}{2}} \cos\phi_p \tag{6c}$$

$$V_S = \frac{q_\mu^2}{2q^2} \left[4k_1 k_2 (\cos^2\frac{\theta}{2})\cos^2\phi_p + q^2\right] \tag{6d}$$

If the coincidence particle is not detected, the angle ϕ_p is integrated over and the term V_I vanishes while V_S becomes equal to V_T. Thus one sees that $V_L W_C$ yields the usual longitudinal contribution in inclusive (e,e'), while both V_T and V_S contribute to the familiar transverse form factor.

The term $V_L W_C$ contains the contribution from the Coulomb plus longitudinal spatial current, $V_T W_T$ is a contribution from the transverse current, $V_I W_I$ is an interference between a current proportional to the transverse (relative to $\vec{q}$) component of $\vec{p}$ and the Coulomb contribution, and $V_S W_S$ arises from the transverse current proportional to $\vec{p}$ just mentioned[2]. The four different W's can be separated by varying the electron scattering angle for fixed q and ω and the angle ϕ_p. Note that since cross section measurements for three values of $\cos\phi_p$ are required to separate the W's, it is necessary to carry out at least one non-coplanar experiment.

As mentioned earlier, (e,e'p) experiments, using the formalism above, can provide useful information on spectroscopic factors (occupation probabilities), single particle binding energies (shell model levels), and the momentum distribution of the initial bound nucleon. One can work in kinematic regions (away from the quasi-elastic peak) where nucleon-nucleon interactions play an important role in determining the coincidence cross section. It will be interesting to study (e,e'2N) reactions in this region in the future. Recent experimental results on (e,e'p) are discussed by DeWitt-Huberts[4] elsewhere in these proceedings.

If one has a polarized electron beam then, in the process $(\vec{e},e'p)$ an additional form factor, W_5, appears in the nuclear current tensor, $W_{\mu\nu}$. Such a term is allowed because $\eta_{\mu\nu}$, is no longer symmetric (due to the initial electron polarization). The additional term appears in Eq. 3 in the form

$$\frac{i}{2}\, W_5 \frac{(N_\mu p_\nu - p_\mu N_\nu)}{p\cdot N} . \tag{7}$$

In the laboratory frame where $N_\mu = M\delta_{\mu 0}$, this additional term enters the coincidence cross section in a term proportional to

$$\frac{q^2 p}{q_\mu^2 E_p}\, W_5 \sin\theta_p \sin\phi_p . \tag{8}$$

The W_5 contribution results from an interference between longitudinal and transverse pieces of the nuclear current. Note W_5 involves an imaginary part of a longitudinal-transverse interference whereas W_I is the real part of this interference[5]. The nuclear tensor, $W_{\mu\nu}$, has an effective coincidence nucleon polarization built into it because of final state interaction effects. The coincidence cross section as a function of ϕ_p, the angle the proton makes with the electron scattering plane, can be used in $(\vec{e},e'p)$ to distinguish between orbits having $j = l\pm\frac{1}{2}$ for the original shell model orbit of the ejected nucleon[6]. Note in a later discussion we shall see that similar studies are possible using the (p,2p) reaction. In this latter reaction, it turns out that the j signature of shell model orbits is quite pronounced even for unpolarized beam protons.

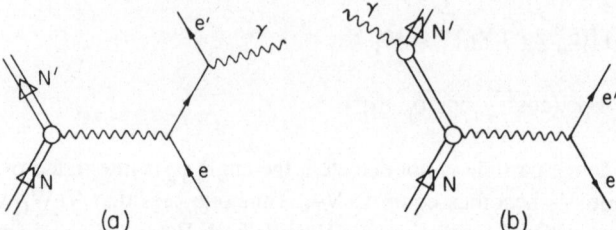

Fig. 3   The two graphs contributing to (e,e'γ). The electron bremsstrahlung graph (a) can, in general, interfere with the nuclear electroexcitation and photo-decay graph (b).

Eq. 5 can also be applied to study the decay of nuclear levels following electroexcitation using reactions such as (e,e'γ) or (e,e'p_0). The final coincidence particle γ or p_0 is detected when the intermediate nuclear resonance level decays. If a particle other than a photon is detected, the bremsstrahlung background is automatically eliminated. By studying the angular distribution of the coincidence decay products one can determine the appropriate nuclear currents and thus identify the resonance spin and parity or test assumptions concerning nuclear levels. These coincidence studies can be particularly useful in the case of overlapping resonances.

C. Examples of Coincidence Studies

As a first example of the potential of coincidence experiments for elucidating nuclear structure, consider a simple model due to Acker and Rose[7] using the reaction (e,e'γ). In lowest order, the cross section associated with this reaction comes from the contributions of the two diagrams in Fig. 3. Diagram 3a represents the traditional bremsstrahlung contribution to the reaction while the amplitude associated with diagram 3b leads to the cross section formula given in Eq. 1 (electroexcitation and subsequent γ decay of a nucleus). The cross section for the (e,e'γ) reaction given by the diagrams shown in Fig. 3, can be written as

$$\frac{d^3\sigma}{d\Omega_{k_2}d\omega d\Omega_k} = \frac{d^3\sigma_{brem} + d^3\sigma_{nuc} + d^3\sigma_I}{d\Omega_{k_2}d\omega d\Omega_k}. \tag{9}$$

More discussion and detailed formulae for each contribution to the cross section are given in Ref.7. Acker and Rose apply Eq. 9 to the following model. Assume a nuclear ground state with $J^\pi = 0^+$. Imagine a single nuclear excitation with unknown parity and spin I^π (to be determined by the (e,e'γ) experiment). Asssume the nuclear excitation is centered at an energy $\varepsilon_I = 20$ MeV relative to the ground state. The width, Γ, of the nuclear excitation is assumed to be Γ = .2 MeV. The excited nucleus is assumed to decay to the original nuclear ground state by emitting a photon of 20 MeV. (Note the bremsstrahlung related contributions to Eq. 9 are only important for observed photon energies that are very close to the difference in energy of the relevant nuclear energy levels associated with the process (here the relevant energy is $E_k \sim 20$ MeV)).

It may be instructive to note that the formula given by Acker and Rose for the nuclear excitation cross section [$d^3\sigma_{nuc}$ in Eq. 9] can be written schematically as

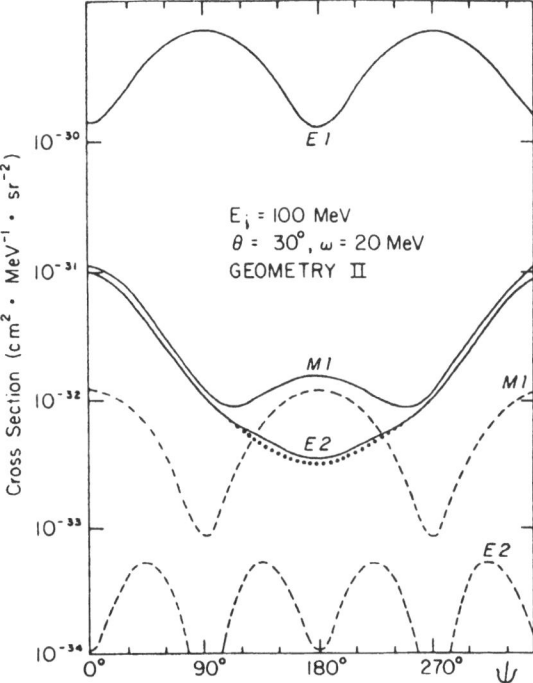

Fig. 4   The cross sections for the process (e,e'γ) assuming an E1, M2, or E2 intermediate excited resonance (see the Acker-Rose model discussed in this text). The dashed lines result from diagram (3b) alone while the dotted lines are given by diagram (3a) alone. Figure taken from Ref. 7.

$$\frac{d^3\sigma_{nuc}}{d\Omega_{k_2}d\omega d\Omega_k} = \frac{1}{(\varepsilon_I-\omega)^2 + \Gamma_I^2/4} \, (I|\gamma(k)|0)_T^2 \times$$

$$\times \{a(I|C(q)|0)^2 + b(I|T_\lambda(q)|0)^2 + c(I|C(q)|0)$$

$$\times (I|T_{el}(q)|0) + d\lambda(I|T_\lambda(q)|0)^2\}, \tag{10}$$

where $(I|\gamma(k)|0)_T^2$ is the square of the transition matrix element for decay of the level $|I)$ to the ground state $|0)$ via emission of a real photon of momentum k. The bracketed { } terms result from electroexcitation of the original nucleus to the excited state $|I)$. The symbol C represents the longitudinal transition operator, T_λ represents the transverse multipole transverse operators [$\lambda = +1(-1)$ represents $T_{el}(T_{mag})$]. The functions a→d depend on $\vec{q}\cdot\vec{k}$, $\vec{q}\times\vec{k}$, and $\vec{k}_1\times\vec{k}_2$. The angular dependence of Eq. 10 as a function of $\vec{k}$ will be significantly different depending on the multipolarity and parity of the excited nuclear level, and this fact can be used to determine I^π. One useful approach is to keep $\vec{k}_1$ and $\vec{k}_2$ fixed so $|\vec{q}|$ is held constant and vary the direction of $\vec{k}$. Acker and Rose[7] consider two possibilities: (a) varying $\vec{k}$ but keeping it in the plane of $\vec{k}_1$ and $\vec{k}_2$, and, (b) varying $\vec{k}$ so

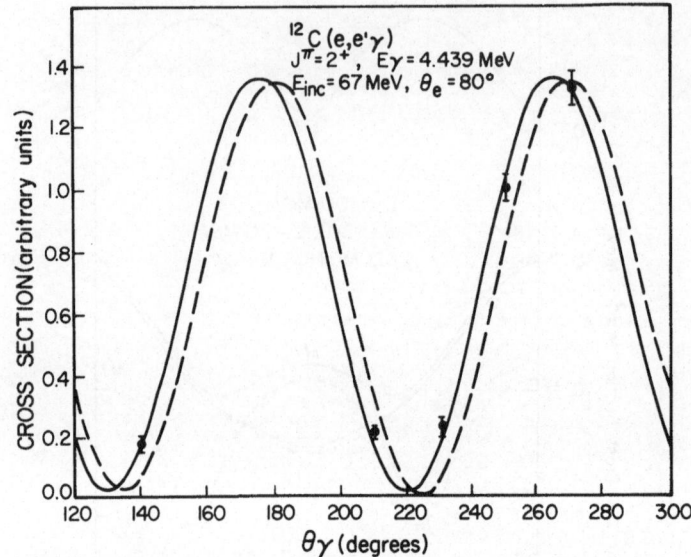

Fig. 5   The photon angular distribution for the ^{12}C(e,e'γ) reaction with an intermediate nuclear excitation of 4.439 MeV. The result indicates a relative negative sign between the longitudinal and transverse form factors. The figure is taken from Ref. 8.

that the plane $(\vec{k},\vec{q})$ is $\perp$ to the plane $(\vec{k}_1,\vec{k}_2)$. The results are shown in Fig. 4 assuming possibility (b) above. Note only low multipoles for the nuclear cross section are relevant since the momentum of the observed photon is low (~20 MeV). The predicted E1, M1, and E2 angular cross sections are distinctly different and thus could be used to determine I^π.

One can also use the (e,e'γ) reaction to obtain the relative sign of various contributions that do not interfere in the inclusive (e,e') cross section. An example has been given recently by Papanicolas et al.[8] who studied the ^{12}C(e,e'γ) reaction for excitation of the 2^+ (4.4 MeV) level. In this case the C2 multipole contributes to the longitudinal form factor $(F_L)^2$, while the E2 term contributes to the transverse from factor $(F_T)^2$, for the transition $0^+ \to 2^+$ (4.4 MeV) in (e,e'). From a Rosenbluth analysis one knows the ratio $(F_T)^2/(F_L)^2 = 5.8 \times 10^{-3}$. The Coulomb-transverse electric interference term in $d^3\sigma_{nuc}$ (see Eq. 10), causes a rotation of the 2^+ quadrupole pattern by 2.3° (see Fig. 5). The direction of the rotation of the quadrupole pattern is determined by the relative sign of F_T and F_L. For the example shown in Fig. 5, a relative opposite sign for F_T and F_L may be deduced from the existing experimental data[8].

The success of inclusive (e,e') experiments in determining nuclear multipole moments has been most dramatic for even-even nuclei where $J_{g.s.} = 0^+$. For even-odd nuclei, more than one multipole can contribute to the transition to a particular J^π final state, and, therefore, there is considerable interest in using the (e,e'γ) reaction to further study the multipole moments associated with these nuclei. Studies of this type should provide new information on core polarization in nuclei. P Coincidence experiments can, in principle, be quite useful in the studies of overlapping resonances. Consider a nucleus with a series of i overlapping resonances of width Γ_i whose angular momenta are denoted I_i. We denote the excitation energy of the center of resonance i by ω_i. We denote the initial angular momentum of the target as J_0 and the final angular momentum of the system (after decay of the intermediate resonances I_i) is denoted by J_f. The cross section for the nuclear decay contribution to the (e,e′x) overlapping resonance coincidence reaction can be written[9]

$$d\sigma^{J_i \to J_f} = (2\pi/v)\delta(E_1 - E_2 - E_x + E_{J_i} - E_{J_f})$$

$$\times \frac{1}{(2J_i+1)} \sum_{M_0 M_f M_x} \left| \sum_{I_i M_i} \frac{K_{J_f I_i}^{M_f M_i} A_{I_i J_0}^{M_i M_0}}{\omega - \omega_i + \frac{1}{2} i \Gamma_i^{tot}} \right|^2 \frac{d^3 k_2 d^3 p_x}{(2\pi)^6} \tag{11}$$

where M_x is the angular momentum along the quantization axis for the coincident particle, $A_{I_i J_0}^{M_i M_0}$ is the amplitude for electroexcitation from the ground state to the resonance I_i and $K_{J_f I_i}^{M_f M_i}$ is the amplitude for particle decay of the resonance I_i. If a heavy particle is detected, then there is no bremsstrahlung contribution to the coincidence cross section. If the final coincident particle is not detected then Eq. 11 reduces to a sum of squares of the amplitudes for electro-excitation of each of the overlapping resonances separately. One interesting point for the coincidence cross section here is not only that there can be interference between longitudinal and transverse contributions to the same resonance, but from Eq. 11 there can be interference between contributions to *different* overlapping resonances. Apparently there does not yet exist extensive studies using this feature. Kleppinger and Walecka[10] have studied the reaction $^{13}C(e,e′,p_0)^{11}B(3/2^-)$ assuming an intermediate giant dipole resonance excitation of 22.8 MeV. They make the standard assumption that the matrix elements in the reaction process can be factorized into, (a) an electromagnetic inelastic excitation form factor, (b) a Breit-Wigner line shape factor reflecting the width of the resonance, and, (c) an exit channel overlap factor for the final particles. Using additional experimental information and a Goldhaber-Teller model (with an empirical strength reduction factor) for the dipole resonance, predictions are made for the cross section as a function of the angle of the coincidence proton. Comparison between theory and the few experimental points suggests that there may exist another interfering resonance which should be included in the theoretical model[10]. This example suggests that such experiments can be used to identify and study overlapping resonances.

We have summarized the studies above because they represent some of the original motivations for carrying out coincidence experiments. Indeed there may be other, more attractive, techniques (such as using polarized electron inclusive experiments) for studying the interference of form factors. In fact if one has obtained detailed information on interference terms from other reactions, the low energy coincidence reactions discussed above may be used to test the reaction model for the excitation, coupling, and "subsequent" decay of nuclear states. For example, one could study the factorization assumption mentioned in the previous paragraph.

In addition to coincidence experiments involving detected photons and nucleons (or clusters of nucleons) an important class of reactions involves a final detected coincidence pion. The reaction (e,e′π) combined with (γ,π) and (p,p′π) may provide particularly useful information on the Δ-nucleon hole interaction and Δ propagation in nuclei. In theoretical studies of (e,e′π) on light nuclei, it has been emphasized that above 200 MeV electron energy loss, ω, diagrams such as that shown in Fig. 6 are important (in addition to the usual Born terms). For example, Tiator and Drechsel[11] find that in the region $2 < \omega < 400$ MeV cross sections can be altered typically by factors of ~2 due mainly to

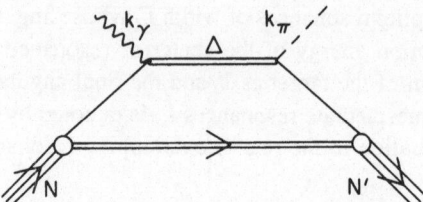

Fig. 6   A diagram resulting in an intermediate Δ contribution to electromagnetic pro-
          duction of pions from nuclei.

interference between Born terms and Δ(1236) formation amplitudes. More specifically,
of the four structure functions appearing in Eq. 5, only W_C is unaffected by inclusion of
Δ exchange current terms (the Δ formation amplitude yields a transverse contribution)[11].
Theoretical predictions associated with the ratio of π^+/π^- coincidence particles and a discussion of the role of Fermi motion in (e,e′π) are available[11], however, as yet there does
not exist sufficient experimental data to make detailed comparison between theory and
experiment. There is satisfactory agreement between theory and experiment for the reaction ^{3}He(e,^{3}H)e′π$^+$ for fixed incident electron energy (between 170-230 MeV) and fixed
triton angle as a function of triton energy in the region $4<T_{3_H}^{lab}<8$ MeV. Other lecturers
will discuss the electromagnetic pion production reaction in more detail. For our purposes it is just important to understand its importance for studying the formation and propagation of the Δ resonance. We discuss this again in later sections where we discuss
such coincidence reactions as (p,p′π) and (π,2π) and their relation to the (e,e′π) reaction
discussed in this section.

III. SELECTED INTERMEDIATE ENERGY (π,π′) AND (p,n) INVESTIGATIONS

Historically, strongly interacting projectiles have provided important information on
nuclear structure. Frequently this information has been complementary to information
available using electromagnetic probes. Recently medium energy (100 - 1000 MeV)
(π,π′), (p,p′) and (p,n) reactions have yielded additional tests of microscopic theories of
nuclear structure and reactions. We list a few examples below.

A. Studies of Isospin Mixing Via (π,π′)

The advent of medium energy charged pion beams has allowed one to use $(\pi^+,\pi^{+'})$
and $(\pi^-,\pi^{-'})$ reactions on nuclei to study nuclear isospin mixing. For example, consider
π^+ and π^- inelastic scattering on light self-conjugate nuclei leading to excited nuclear
states that are primarily particle-hole excitations. An isospin mixed particle-hole state
isospin wavefunction can be written, schematically, in the form

$$\alpha\{p\bar{p}+n\bar{n}\} + \beta\{p\bar{p}-n\bar{n}\}$$

(12)

       T=0   T=1

where the bar represents a proton ($\bar{p}$) or neutron ($\bar{n}$) hole state. The coefficients α and β
represent isospin admixture coefficients. We consider a situation where the spin-
configuration space wavefunction associated with the T=0 and T=1 states are essentially
identical. (For example, one might consider high spin stretched particle-hole states

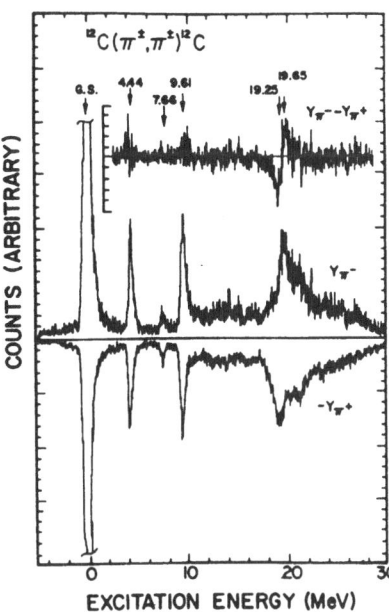

Fig. 7   Pion inelastic spectra for ^{12}C with $T_\pi^{lab} = 162$ MeV and a pion scattering of 70°. The difference of the π^+ and π^- spectra near 19 MeV may be useful in studies of isospin mixing. Figure taken from Ref. 12.

where $J = j_p + j_h$ such as the 4^-, T=0 and 1, 1d $5/2(1p\ 3/2)^{-1}$ states in ^{12}C or ^{16}O.) If the pure isospin states are not widely separated in energy, $\Delta E \leq 300$ KeV, the Coulomb interaction or other isospin symmetry breaking forces might be expected to mix the doublet of states. Comparison of the π^+ and π^- data associated with exciting these states can provide at least qualitative information on the degree of isospin mixing. If the final nuclear states possess essentially pure isospin, then the π^+ and π^- results should be identical (ignoring small Coulomb effects on the medium energy pion-nucleus optical potential). On the other hand, if there is isospin mixing (see Eq. 12), then one of the states will be predominantly $(p\bar{p})$ while the other isospin mixed member of the doublet will have a larger $(n\bar{n})$ component. The pion-nucleon interaction at intermediate energies is dominated by the $j = 3/2$, $t = 3/2$ resonance. Therefore, it follows that the π^+p (π^-n) interaction is stronger than the π^-p (π^+n) interaction. Thus, the isospin mixed state dominated by the $p\bar{p}$ amplitude $(n\bar{n}$ amplitude) will be relatively more strongly excited by the $(\pi^+,\pi^{+'})$ $(\pi^-,\pi^{'-})$ reaction. By subtracting the $(\pi^+,\pi^{+'})$ results from the $(\pi^-,\pi^{-'})$ data one should, therefore, be able to see isospin mixing effects if present. This is illustrated in Fig. 7. Of course, one must be careful to consider realistic nuclear structure models in interpreting the results but the possibility of studying isospin mixing is present. (Note that in (e,e') reactions leading to stretched configurations, $\Delta T = 1$ states dominate while $\Delta T = 0$ states are difficult to study because of the small isoscalar nucleon magnetic moment.)

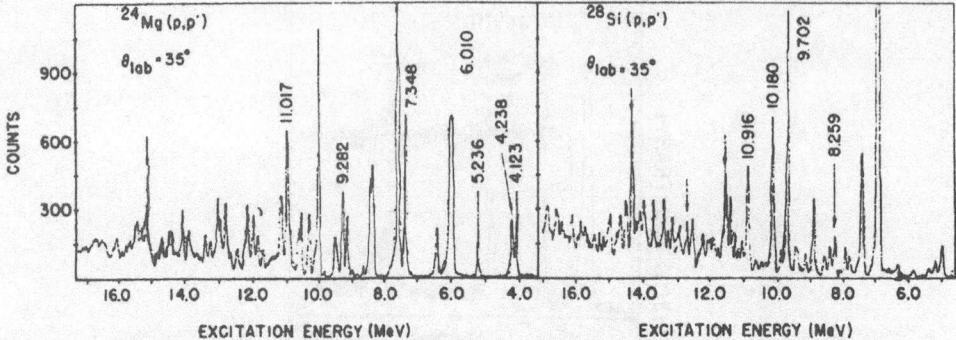

Fig. 8   Representative excitation spectra for 135 MeV (p,p') on sd shell nuclei. Figure
           taken from Ref. 13.

B. Basic Input for Interpreting (p,p') and (p,n) Reactions

    High quality intermediate energy proton inelastic scattering and charge exchange
high quality data now exist. In Figs. 8 and 9, we show some representative spectra for
(p,p'). Important features of the data include excellent energy resolution, variable
energy, and the availability of angular distribution data for cross sections and spin-
observables. The data is available in the region near 200 MeV. This is important
because in this general region the dominant piece of the nucleon-nucleon (NN) interaction has a minimum and, therefore, one might expect that corrections to the distorted
wave impulse approximation (DWIA) might be relatively smaller than, for example, at
lower energies. The energy dependence of the various central terms in the NN t-matrix
are shown in Fig. 10. Note that the largest term, t_0^c, is independent of spin and isospin
and has a minimum near $E_p = 300$ MeV. The next largest term, $t_{\sigma\tau}^c$, is responsible for the
excitation of Gamow-Teller type states. Note that the excitation functions shown in Figs.
8 and 9 are dominated by high-spin states due to the large momentum transfer associated

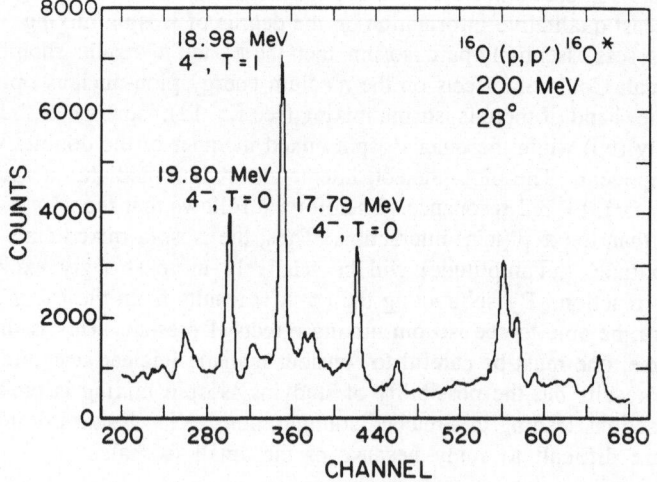

Fig. 9   Excitation spectrum for 200 MeV (p,p') on ^{16}O. Figure taken from Ref. 14.

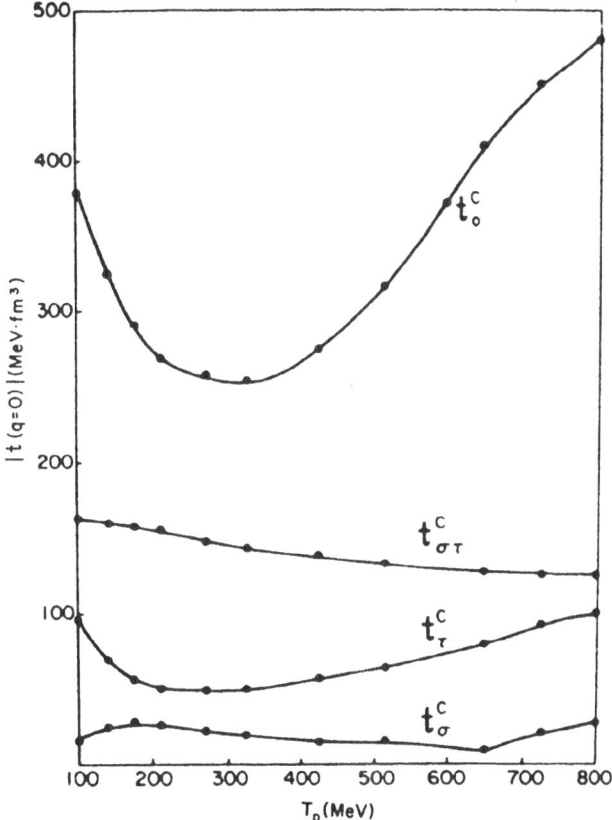

Fig. 10   Energy dependence of the magnitude of the various contributions to the central
N-N t-matrix used in standard DWIA analyses of (p,p′) and (p,n). Figure taken
from Ref. 15.

with the proton scattering angles shown. The $\Delta T = 1$ high spin states present in the spectrum are also seen in inelastic electron scattering. It is important to note that the states *predicted* to dominate the nuclear response in a RPA linear response prediction for (p,p′) and (p,n) do, in fact, dominate the intermediate energy strong probe *experimental* response functions even for momentum transfer of several hundred MeV/c.

One has been able to carry out very detailed theoretical studies of the (p,p′) nuclear response using the DWIA. The basic input is a parameterization of the on-shell nucleon-nucleon scattering amplitude M which is related to the transition matrix, t, via

$$t_{N_1N_2} = \frac{4\pi(\hbar c)}{E_{CM}} M_{N_1N_2}(E_{CM},\theta) \equiv \eta M_{N_1N_2} \tag{13}$$

where

$$M_{N_1N_2}(E_{CM},\theta) = A + B\vec{\sigma}_1 \cdot \hat{n}\vec{\sigma}_2 \cdot \hat{n} + C(\vec{\sigma}_1 + \vec{\sigma}_2) \cdot \hat{n} + E\vec{\sigma}_1 \cdot \vec{q}\vec{\sigma}_2 \cdot \vec{q} + F\vec{\sigma}_1 \cdot \vec{Q}\vec{\sigma}_2 \cdot \vec{Q}. \tag{14}$$

In Eqs. 14 A, B, C, E, F are functions of θ, E_{CM} and isospin. If $\vec{k}$ and $\vec{k}'$ are the initial and final momentum, respectively, in the center-of-mass frame, then the vectors $\vec{q}, \vec{Q}$ and n̂ can be written

$$\vec{q} = \vec{k} - \vec{k}',$$ (15a)

$$\vec{Q} = \vec{k} + \vec{k}',$$ (15b)

$$\hat{n} = \frac{\vec{k} \times \vec{k}'}{|\vec{k} \times \vec{k}'|}.$$ (15c)

Using spin singlet (P_S) and triplet (P_T) projection operators, the scattering amplitude, M, may be written in the form

$$M = A'P_S + B'P_T + C'(\vec{\sigma}_1 + \vec{\sigma}_2) \cdot \hat{n} + E'S_{12}(q) + F'S_{12}(Q)$$ (16)

where A', B', C', E', F' are functions of θ, E_{CM} and isospin and are related to A→F by the relations

$$A' = A - (B+E+F) \qquad E' = \frac{E-B}{3}$$

$$B' = A + \frac{(B+E+F)}{3} \qquad F' = \frac{F-B}{3}$$

$$C' = C.$$ (17)

One can transform M to an NN transition matrix, t, by using Eq. 13. It is useful to make the spin and the isospin dependence explicit and, therefore, t is written in the form

$$t = t_0^c + t_\sigma^c \vec{\sigma}_1 \cdot \vec{\sigma}_2 + t_\tau^c \vec{\tau}_1 \cdot \vec{\tau}_2 + t_{\sigma\tau}^c \vec{\sigma}_1 \cdot \vec{\sigma}_2 \vec{\tau}_1 \cdot \vec{\tau}_2$$
$$+ \left[t_0^{LS} + t_\tau^{LS} \vec{\tau}_1 \cdot \vec{\tau}_2 \right] \vec{L} \cdot \vec{S} + \left[t_0^T + t_\tau^T \vec{\tau}_1 \cdot \vec{\tau}_2 \right] S_{12}(q)$$ (18)

where the t_i are functions of the C.M. energy and the momentum transfer q. For direct terms, the NN central t matrix components having a subscript τ (σ) are isovector (spin-vector) operators in the nuclear target space for (N,N′) reactions on nuclei.

A topic of interest in studying nuclear currents is the investigation of nuclear longitudinal $(\vec{\sigma} \cdot \vec{q})$ and transverse $(\vec{\sigma} \times \vec{q})$ spin densities. It turns out that electrons and pions are sensitive to nuclear transverse spin densities. In a simple static model plane wave impulse approximation of the nucleon-nucleus inelastic response the differential cross section can be written:

$$\frac{d\sigma}{d\Omega} \propto |M_0 V^0|^2 + |M_1 V^t|^2 \quad \text{(natural parity excitations, } \Delta\pi = (-1)^J)$$ (19a)

$$\frac{d\sigma}{d\Omega} \propto |X_L V^l|^2 + |X_T V^t|^2 \quad \text{(unnatural parity excitations, } \Delta\pi = (-1)^{J+1})$$ (19b)

where M_0, M_1, X_L, and X_T are tensors defined in Ref.17. The longitudinal coupling, $\vec{\sigma} \cdot \vec{q}$, (transverse coupling, $\vec{\sigma} \times \vec{q}$) is denoted by $V^l (V^T)$. In terms of the coefficients given in Eq. 14, one obtains

$$V^0(q) = \eta(A^2 + C^2)^{\frac{1}{2}}$$ (20a)

$$V^l(q) = \eta E$$ (20b)

$$V^t(q) = \frac{\eta}{\sqrt{2}} (C^2 + B^2 + F^2)^{\frac{1}{2}}.$$ (20c)

To the extent that the transition operator range has some resemblance to the range of an underlying meson exchange potential, one can understand why V_τ^l (associated with pion

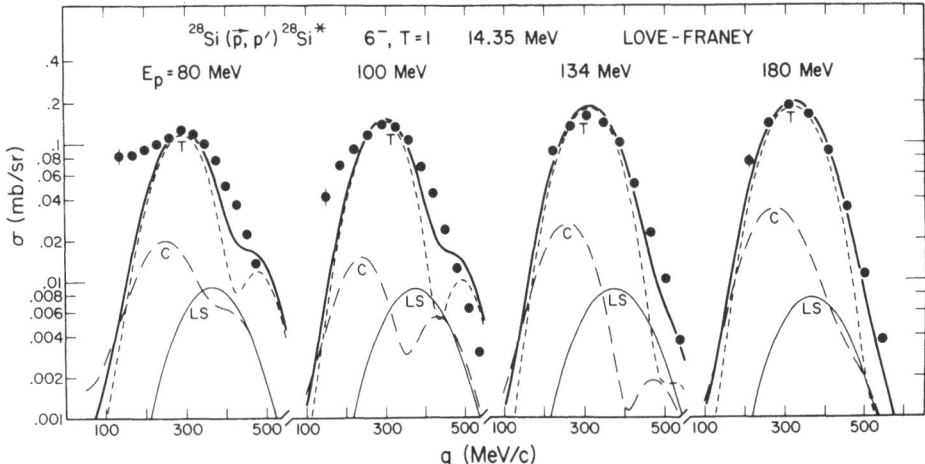

Fig. 11   Comparison of DWIA theory and experiment for ^{28}Si(p,p')^{28}Si leading to the "stretched" 6$^-$ T = 1 state at 14.35 MeV. The theoretical curves have been multiplied by 0.3. The Love-Franey NN interaction is discussed in Ref. 15. Figure taken from Ref. 18.

exchange) would be considerably longer ranged than V_τ^t (associated with $\rho-$ meson exchange).

It is useful to define polarization transfer variables D_{ij} where

$$D_{ij} = \sum_\mu Tr(M_\mu\sigma_i M_\mu^\dagger\sigma_j)/Tr(M_\mu M_\mu^\dagger) \qquad (21)$$

where M_μ and i, j subscripts refer to components in the $\hat{n},\hat{q},\hat{Q}$ system. One finds, for example,[17]

$$\sigma_0 D_{nn} = X_T^2(C^2+B^2-F^2)-X_L^2 E^2 \qquad (22a)$$

$$\sigma_0 D_{qQ} = -\sigma_0 D_{Qq} = 2X_T^2 Im\,(BC^*) \qquad (22b)$$

etc., where X_T and X_L are the transverse and longitudinal nuclear form factors associated with the nuclear matrix elements of $<\vec{\sigma}\times\vec{q}>$ and $<\vec{\sigma}\cdot\vec{q}>$, respectively. It is sometimes convenient to transform the D_{ij} to a set of quantities $D_{NN}, D_{LL}, D_{SS'}, D_{LS'}, D_{SL'}$ where $\vec{N} = \vec{k}_i \times \vec{k}_f$, $\vec{L}(\vec{L}')$ denotes a spin direction along $\vec{k}_i(\vec{k}_f)$ and $\vec{S} \equiv \vec{N} \times \vec{L}$, etc.. For completeness, we note that the polarization, P, along a particular direction is given by

$$P = \frac{1}{\sigma_0} \{\sigma_{++} + \sigma_{-+} - \sigma_{+-} - \sigma_{--}\}, \qquad (23)$$

while the expression for the analyzing power is

$$A = \frac{1}{\sigma_0} \{\sigma_{++} - \sigma_{-+} + \sigma_{+-} - \sigma_{--}\}, \qquad (24)$$

so that the difference, P-A, is given by

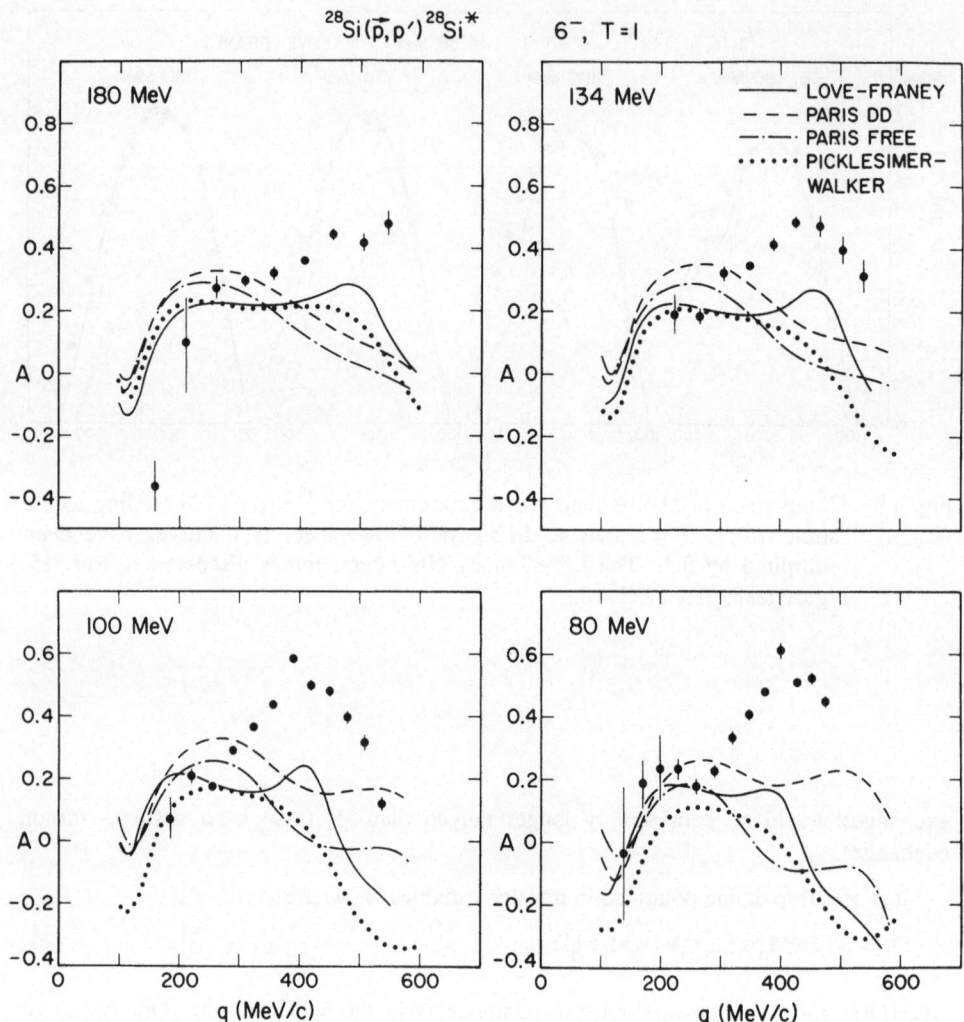

$$^{28}\text{Si}(\vec{p},p')^{28}\text{Si}^* \qquad 6^-, \, T=1$$

Fig. 12 Comparison between DWIA analyzing power predictions and experiment as a function of energy for the $^{28}\text{Si}(p,p')^{28}\text{Si}$ (6^-, T=1, 14.35 MeV) reaction. The theoretical curves utilize different effective NN transition operators and are referenced in Ref. 18 which is also the source of the figure.

$$P-A = -\frac{2}{\sigma_0} \{\sigma_{+-} - \sigma_{-+}\}. \tag{25}$$

C. Recent Experimental Results for (p,p′) and (p,n)

We shall show some experimental results for spin-observable quantities in what follows. Note that in a transition were a single orbital angular momentum transfer, ΔL, dominates that X_L becomes proportional to X_T and that under these conditions (in the

PWIA)

$$D_{NN}(\theta = 0^0) \quad \begin{cases} = 1 & \Delta J = \Delta L, \Delta S = 0 \text{ (IAS)} \\ = 0 \text{ for} & \Delta J = \Delta L, \Delta S = 1 \\ \leq -1/3 & \Delta J = \Delta L \pm 1 (GT) \end{cases} . \qquad (26)$$

We shall exploit the relations given in Eq. 26 to identify the Gamow-Teller (GT) resonance in a later discussion. We illustrate in Figs. 11-13, recent intermediate energy cross section and spin observable angular distribution (p,p′) data that has become available from the Indiana University Cyclotron Facility (IUCF)[18,19]. Points of interest include the fact that, (a) high quality data exists with which to confront existing structure and reaction models, (b) the cross section angular distribution data requires renormalization factors (sometimes similar to those required in inclusive inelastic electron scattering), and (c) the spin-observable data allows more stringent testing of the existing models. Note that none of the interactions shown give good agreement with the analyzing power data

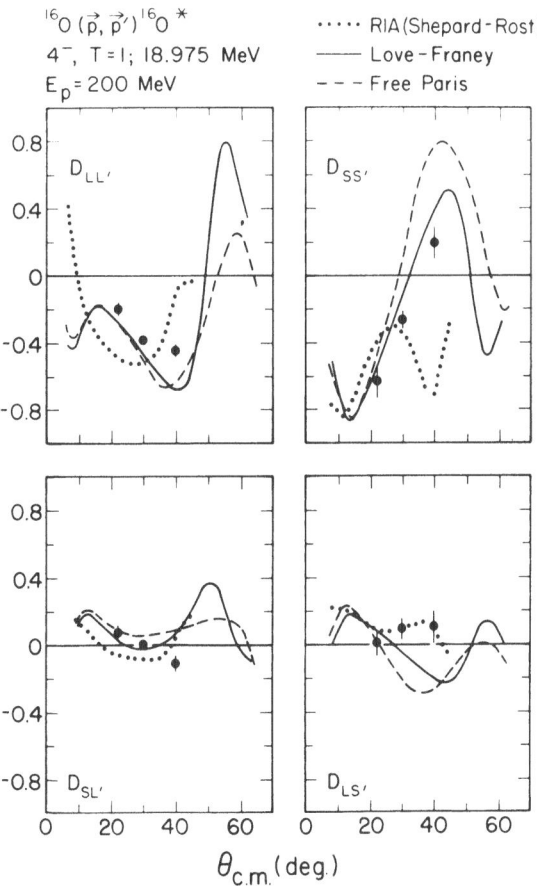

Fig. 13   Polarization transfer data and comparison with DWIA theory. The theoretical curves utilize different transition operators. The "RIA" calculation utilizes the relativistic impulse approximation without exchange and is discussed in Ref. 20. Figure taken from Ref. 14.

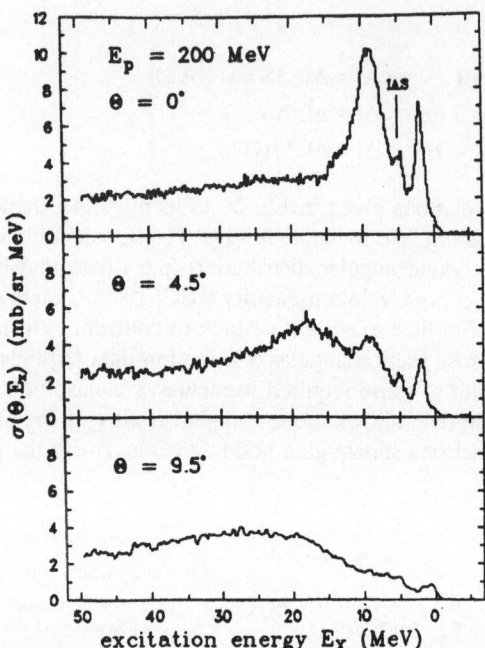

Fig. 14 ^{90}Zr(p,n) differential cross section spectra for $E_p = 200$ MeV showing the forward peaking of Gamow-Teller and isobaric analogue state (IAS) strength.
Figure taken from Ref. 24.

(see Fig. 12) beyond ~300 MeV/c. The polarization transfer variable $D_{LS'}$ shown in Fig.
13, is sensitive to several factors including non-local terms (exchange) and the *interference* between the central and tensor interactions[19]. This is in contrast to the usual (p,p′)
cross section angular distributions where, in a simple model the central and tensor terms
do not interfere. The situation is much like that for inclusive (e,e′) reactions where the
longitudinal and transverse form factors do not interfere while for (e,e′x) reactions, as
discussed in an earlier section, interference is possible.

The charge exchange reaction, (p,n), above 100 MeV has been an important area of
investigation in the last few years[21,22]. One of the main reasons (p,n) reactions have
received so much attention is that they provide an excellent way to study the distribution
of Gamow-Teller ($\sigma\tau$) strength in nuclei. More specifically, one would like to study the
Gamow-Teller matrix element, <GT>, defined by

$$<f|GT|i> = \frac{1}{(2J_i+1)^{1/2}} \ <f\| \sum_k \tau_k^\dagger \sigma_k \ \| i> \tag{27}$$

where the $\sum_k$ is a sum of the single nucleon operators $\sigma\tau$ over all nucleons in the nucleus.
One can show that, assuming nucleon degrees of freedom only, the total Gamow-Teller
strength summed over all final nuclear states $|f>$ should exceed $3(N-Z)$ where $N(Z)$ is the
number of neutrons (protons) in the target nucleus. Because the Gamow-Teller transition
involves $\Delta L=0$, it is important to measure the nuclear response near $q=0$ for such transitions to dominate the response (see Fig. 14). In order to study the matrix element given
by Eq. 27 effectively using the (p,n) charge exchange reaction, it is important to be well

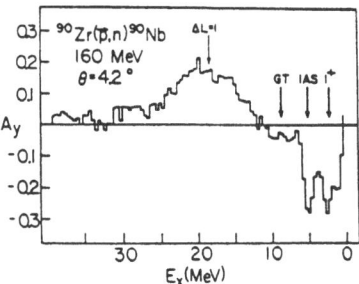

Fig. 15   The analyzing power spectrum for ^{90}Zr(p,n)^{90}Nb at $E_p = 160$ MeV and $\theta_{lab} = 4.2°$. Figure taken from Ref. 24.

above $T_p \approx 100$ MeV so that, (a) the $0°$ cross section gives a momentum transfer q which is essentially zero (note there is an energy Q value $\neq$ 0 for the charge exchange reaction so that $q \neq 0$ when $\theta = 0°$), and (b) the $t_{\sigma\tau}$ piece of the transition t matrix dominates over the t_τ term (see Fig. 10) -- otherwise, isobaric analogue transitions (IAS), excited via the t_τ term, may dominate the spectrum and make it difficult to extract Gamow-Teller strength. Of course, working in the region 200-300 MeV is also expected to minimize the effects of distortions and multistep corrections to the DWIA. The (p,n) reaction is normalized by studying states where the Gamow-Teller strength is known from beta decay weak interaction data. This allows one to calibrate effects associated with optical potential distortions and nuclear medium effects on the $t_{\sigma\tau}$ transition operator. An assumption is made that the normalization effects are not strongly state or nucleus dependent. In fact one finds that, for a wide range of nuclei only approximately 50-80% of the predicted strength is found[21,22]. Some of the reasons advanced for the discrepancy include, (a) Δ particle-nucleon hole effects so that an appreciable part of the strength is stripped out of the low energy excitation region into a region $\sim$ 300 MeV above the ground state, (b) effects of core polarization so that the wave functions used in making DWIA predictions are not sufficiently realistic, and (c) fragmentation of strength over a wide energy range (but still at relatively low energies <<300 MeV) so that the strength has remained undetected experimentally. Of course, the various mechanisms suggested are not mutually exclusive, and in fact, some combination of all three mechanisms may well play an important role in the final analysis. At the present time, the situation is not resolved, and thus it is particularly interesting to search for methods of detecting strength that does not appear in a resonance bump. In this regard, the characteristic signature (from simple models) of a value for D_{NN} at $0°$ of +1 for IAS type transitions and $\sim$ - 1/3 for GT transitions, is helpful[22-24]. Indeed from Figs. 15-17, one sees in the experimental analyzing power and D_{NN} data the possibilities for locating GT strength. Note that the simple model predictions yield qualitatively correct results for those situations where the resonance strength is already known to be of the GT or IAS type.

We note that in the studies of (p,p') and (p,n) briefly summarized above, input from electroweak interaction data has played an important role in pinning down nuclear structure factors and providing confidence that the DWIA (perhaps with density dependent interactions) is a useful analysis tool for analyzing medium energy proton inclusive reactions. An obvious question is how will this carry over into intermediate energy proton

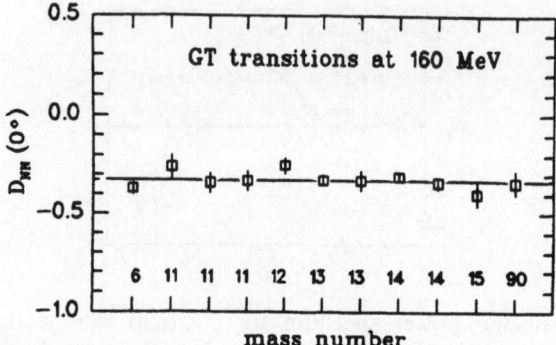

Fig. 16   The measured polarization transfer variable, $D_{NN}(0°)$, for known pure
          Gamow-Teller transitions compared with the value -1/3 predicted by theory.
          Figure taken from Ref. 24.

induced coincidence experiments such as (p,2p) and (p,p'π).

IV. COINCIDENCE STUDIES USING PROTON AND PION BEAMS

A. The (p,2p) Reaction

The (p,2p) reaction has historically yielded information on shell structure in nuclei.
On the one hand, counting rates are higher for this strong reaction than for the (e,e'p)
reaction. However, distortion effects and reaction mechanism uncertainities are more
important in (p,2p). In the following, we briefly discuss two intermediate energy applications of the (p,2p) reaction: one involving polarized protons, the other a study of reaction
mechanism vs. nuclear structure complications. First, we consider the ($\vec{p}$,2p) reaction. If
one assumes the momenta of the initial polarized proton (with polarization P_0) and two

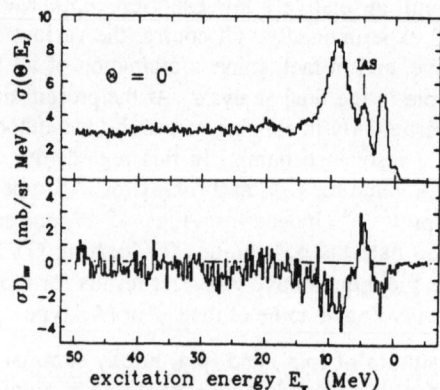

Fig. 17   The forward scattering differential cross section and polarization transfer cross
          section spectra for $^{90}Zr(p,n)^{90}Nb$ at $E_p = 160$ MeV. Figure taken from Ref. 24.

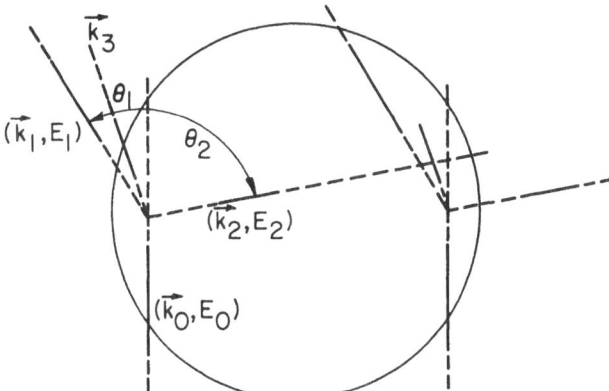

Fig. 18   Nuclear paths associated with the ejected nucleon in (p,2p) assuming asymptotic momenta and two different initial struck nucleon locations. Figure taken
from Ref. 25.

final protons (1,2) are coplanar, and adopts the factorized form for the DWIA then the
coincidence cross section can be written in the form[25]

$$\frac{d\sigma}{d\Omega_1 d\Omega_2 dE_1 dE_2} = \frac{4}{(\hbar c)^2} \frac{k_1 k_2 \bar{E}_0^2}{k_0 E_3} \frac{d\sigma}{d\Omega_{c.m.}} (T_0, \bar{\theta}, P_0, P_3)$$

$$\times \frac{1}{2J_A+1} \sum_f |g'_f(\vec{k}_3)|^2 \, \delta(E_1 + E_2 + E_{A-1} - E_0 - E_A) \qquad (28)$$

where 0, 1, 2, 3 refer to the projectile, two final energetic protons, and the initial bound
"knocked-out proton", respectively. The term $g'_f(k_3)$ is the initial distorted (because of
initial and final distortions) momentum distribution amplitude for the knocked-out
nucleon, $\dfrac{d\sigma}{d\Omega_{c.m.}}$ is the off-shell nucleon-nucleon center-of-mass cross section (for polarized initial nucleons but summed over final state spin-polarizations). The initial target
nucleon is effectively polarized as will be discussed below. In obtaining the factorized
form above, one assumes that distortion effects are not so great that one cannot use the
asymptotic momenta k_0, k_1, k_2 in defining $\dfrac{d\sigma}{d\Omega}$ (note $\vec{k}_0 + \vec{k}_3 = \vec{k}_1 + \vec{k}_2$). Such effects as the
Pauli principle on intermediate states and spin dependent terms in the optical potential
have also apparently been suppressed in writing Eq. 28 in the factorized form shown.

The validity of the factorization approximation has been studied[26] and will not be discussed here. It would be useful to write out the various terms in the transition amplitude
and see if there are some spin observables in ($\vec{p}, p\vec{p}$) where one might learn new information by specifically not using the cross section but using the non-factorized (p,p) amplitude keeping the rich spin dependence studied in the (p,p') reactions mentioned earlier.

It turns out that the effects of absorption (out of the elastic channel) and the strong
effective spin-orbit potential in the optical model can be used to advantage in (p,2p). The
basic idea is that the initial bound proton is effectively polarized (i.e. those protons that
participate in the observed (p,2p) reaction have a preferred spin direction). How this

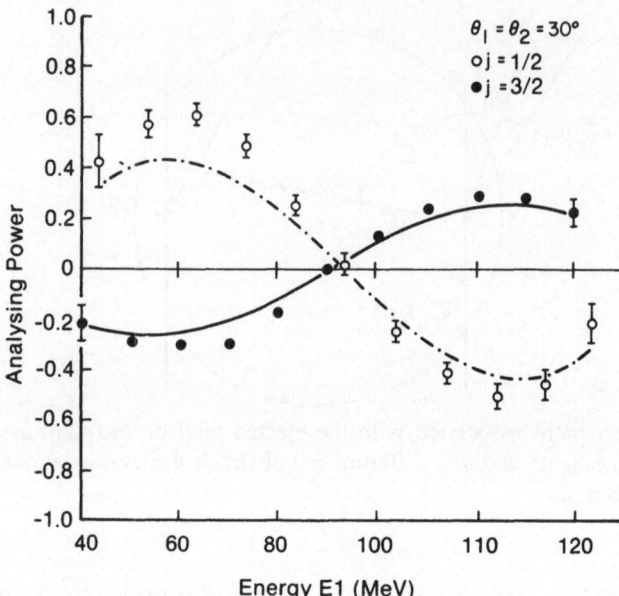

Fig. 19 Measured asymmetry and DWIA calculations for the process (p,2p) showing
the ability to distinguish $j = l \pm 1/2$ orbitals as discussed in the text. Figure
taken from Ref. 27.

occurs can be seen with reference to Fig. 18. Because the *nuclear* paths associated with
final nucleons 1 and 2 are longer on the left hand side of the figure, nuclear absorptive
effects cause the right hand side of the figure to give a larger contribution to the (p,2p)
reaction. (Use of the energy dependence of the mean free path could be used to enhance
this effect for certain kinematical conditions). Knowing $\vec{k}_0$, $\vec{k}_1$, and $\vec{k}_2$ allows one to
determine the direction of $\vec{k}_3$. Thus from the figure one can determine the direction of
$\vec{l}_3 = \vec{r}_3 \times \vec{k}_3$, which preferentially will contribute to the (p,2p) reaction. Now the proton-
proton interaction is spin-dependent so the direction of $\vec{s}_3$ will affect the predicted cross
section obtained from Eq. 28. Since the initial j-j coupled single nucleon states require
$\vec{j}_3 = \vec{l}_3 \pm \vec{s}_3$, the effective polarization of the bound nucleon for a given j orbit allows one to
distinguish (via the analyzing power angular distribution) between $j = l + \frac{1}{2}$ and $l - \frac{1}{2}$ by
using the $(\vec{p},2p)$ reaction[27]. An example is shown in Fig. 19.

The (p,2p) reaction can, in principle, also be used to investigate possible small components in the nuclear wavefunction. One example is the use of the $^{12}C(p,2p)^{11}B$ reaction to study possible small $1f^2$ components in the ^{12}C ground state wavefunction. The
basic idea is that the $\frac{1}{2}^-$(6.74 MeV) and $\frac{5}{2}^-$(4.45 MeV) states of ^{11}B are not expected to
be populated via a one step process in (p,2p) unless there are non-negligible $(1f)^2$ components in the ^{12}C g.s.. The (p,2p) reaction does lead to these ^{11}B excited states with
appreciable cross section[28]. The angular correlations for the two outgoing protons for the
4.45 MeV state apparently favour $l = 1$ (1p) knockout as opposed to $l = 3$ (1f) knockout[29].
The situation for the $\frac{7}{2}^-$(6.74 MeV) state is similar - although the presence of an
unresolved $\frac{1}{2}^+$ state (6.79 MeV) means that the angular distribution correlation experiments are less definitive here[29].

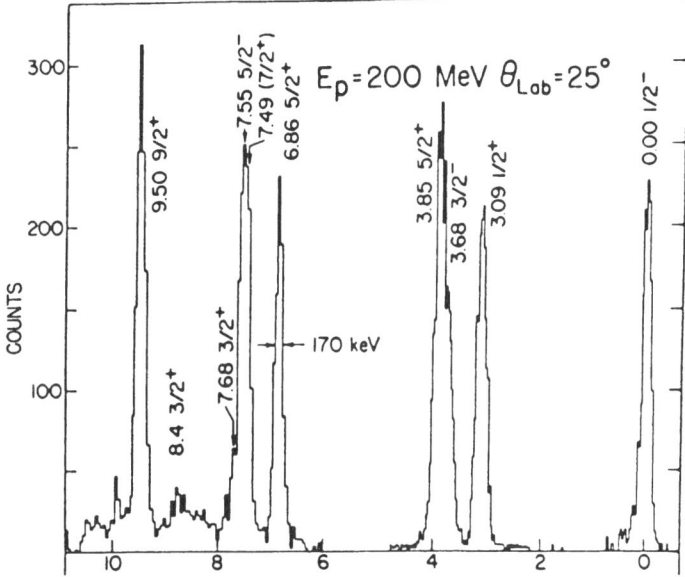

Fig. 20   An excitation spectrum for the ^{12}C(p,π^+)^{13}C reaction with 200 MeV incident protons showing the strong excitation both of states assumed to be of a single particle and two particle-one-hole nature. Figure taken from Ref. 32.

With regard to studies of this type, van der Steenhoven et al.[30], have recently concluded from studying the strength for exciting the 5/2 state discussed above that the (e,e′p) reaction should be a useful process for studying relatively small components of the nuclear wavefunction because two step processes induced by the strongly interacting exiting proton appear (experimentally) to be suppressed. One expects a cross-fertilization between (e,e′p) and (p,2p) associated with (p,2p) studies of (a) appropriateness of optical potentials, (b) reaction mechanism complications (processes induced by the exiting initial bound nucleon), (c) nuclear structure and (d) spin effects associated with the assumed nucleon-nucleon transition operators. High beam current, high energy, variable energy polarized beams and/or targets and, of course, high duty cycle machines are needed to fully use the complementary potential of the (e,e′p) and (p,2p) reactions.

B. Proton Induced Pion Production

In what follows we briefly discuss the (p,π) and (p,p′π) reactions and discuss some possible overlap with (e,e′π). There is considerable experimental data on the (p,$\pi^{\pm}$) reaction leading to bound or quasi-bound nuclear states in the proton projectile energy region $150 \le T_p \le 800$ MeV[31]. Both analyzing power and cross section angular distribution data is available. Proton induced pion production results in a large momentum transfer $q \ge 2k_F$(~550 MeV/c) to the nucleus. The reaction allows study of a process (pion emission) that plays an important role in binding the nucleus. It also has the possibility of providing wavefunction and/or reaction mechanism information of interest for other high-momentum transfer medium energy reactions such as (p,γ) and (e,e′π). Presently, there does not exist a theoretical approach that has been shown to yield quantitative agreement with the wide range of high quality data available. The situation is complicated by such effects as, (a) the importance of multistep processes because of the large

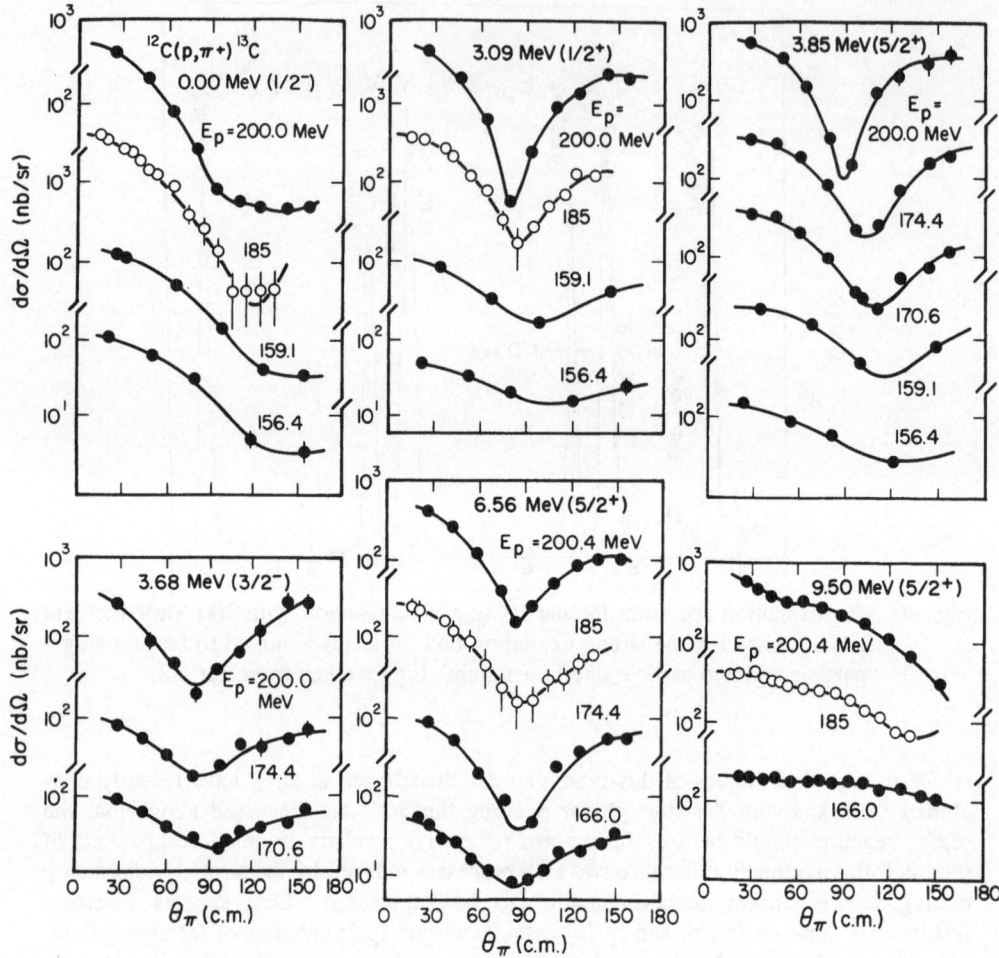

Fig. 21   Selected angular distributions for the $^{12}C(p,\pi^+)^{13}C$ reaction for several incident proton energies. Figure taken from Ref. 33.

momentum transfer to the nucleus-- (this may include an important two-nucleon mechanism involving a propagating intermediate Δ with medium corrections), (b) relativistic corrections and (c) nuclear structure, distorting potential, and vertex form factor uncertainties at high momentum transfer.

Experimentally, there has been emphasis on energies below ~ 250MeV and on targets of ^{90}Zr or lighter where the density of states is less and distortion effects are relatively reduced. The experimental results to date seem to have considerable lack of systematics (i.e., we have not yet recognized a pattern). In Figs. 20-23, we show some representative data[32-35]. In Figs. 20 and 21, typical excitation function and angular distributions are shown for $T_p \leq 200MeV$, $^{12}C(p,\pi^+)^{13}C$. The excitation spectrum is apparently composed of "single particle" states [g.s.($1p\frac{1}{2}$), 3.09 MeV($2s\frac{1}{2}$), and 3.85 MeV($d\frac{5}{2}$)] and two-particle one-hole (2p-1h) states [such as the 6.86 MeV $\frac{5}{2}^+$, and 9.5 MeV $\frac{9}{2}^+$ states]. Fig. 20 illustrates that the 2p-1h states can be as strongly excited as single particle states (depending on the nucleus). Fig. 21 includes several angular distributions exhibiting dips

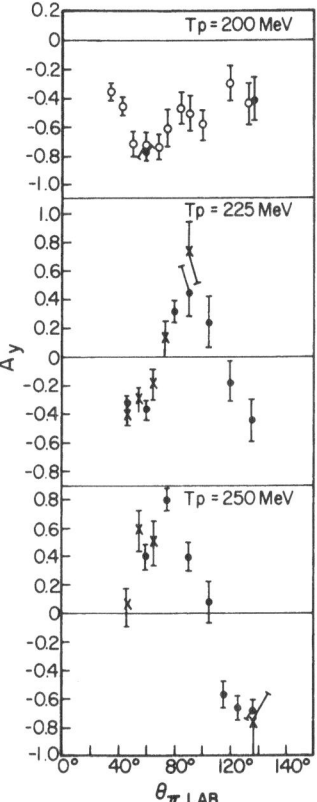

Fig. 22   The energy dependence of $A_y(\theta)$ for the reaction $^{12}C(\vec{p},\pi^+)^{13}C_{g.s.}$ for 200-250 MeV incident protons. Figure taken from Ref. 34.

near 90° (perhaps associated with a p wave, $\cos\theta$, dependence). The angular distributions for *some* single particle and 2p-1h states have the dip structure. The angular distribution associated with the analyzing power sometimes exhibits considerable energy dependence as shown in Fig. 22. The (p,π) reaction mechanism may, in fact, be several competing single nucleon and two-nucleon mechanisms. The (p,π⁻) reaction is believed to result essentially from a two-nucleon mechanism. Arguments based on a two nucleon model of the (p,π⁻) reaction have resulted in correct predictions of the j dependence of the sign of the analyzing power[35]. Recent studies of the (p,π⁻) reaction indicate a selectivity presumably associated with the excitation of high spin 2p-1h states (see Fig. 23)[36]. This is consistent with a two nucleon mechanism for a large momentum transfer process.

The first attempts to model the (p,π) reaction involved a DWBA single nucleon stripping model[31]. The fact that there is only one active nucleon in the model and the process requires evaluation of the final bound nucleon wavefunction at very high momentum transfer means there is considerable sensitivity to wavefunction and optical potential parameters. Apparently the observed strong excitation of 2p-1h states in (p,π⁺) and the observed (p,π⁻) data are not naturally included in this model because such processes naturally involve two active nucleons. In addition, it is known that single pion rescattering with an intermediate Δ formation is an important ingredient of theories that fit the

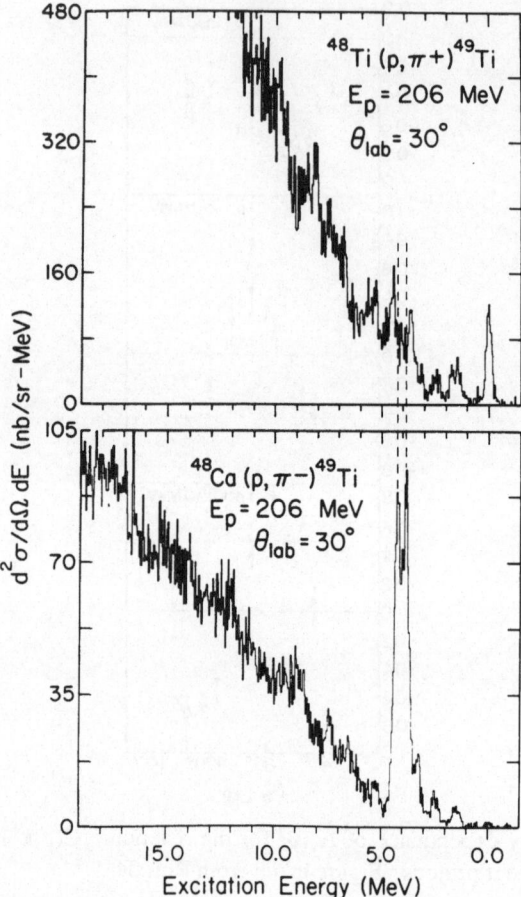

Fig. 23   An example of (p,π⁺) and (p,π⁻) showing the greater selectivity of the (p,π⁻)
reaction for the relative excitation of different states in the same final nucleus.
Figure taken from Ref. 36.

two body NN → dπ⁺ reaction. (Some of this effect can be included in an average way in
pion distortions). One more recent example of a single nucleon mechanism is the relativistic calculation of Cooper and Sherif[37]. It may be useful to supplement the two
nucleon mechanism discussed below with a *plane wave* (to avoid possible double counting) single nucleon-mechanism.

There have been several discussions of two-nucleon mechanism (TNM) models. In
what follows, we briefly summarize the approach we have taken[38]. We show in Fig. 24
some typical (resonant and non-resonant) diagrams associated with the TNM under discussion. The results we discuss are for a TNM mechanism incorporating an intermediate
propagating and interacting delta, with virtual pion and rho exchange including the
effects of realistic external proton and pion distortions. The details of the model, formulae, and calculational procedure are given in Ref. 38. The results for a study of the
^{12}C(p,π⁺)^{13}C(g.s.) transition ($T_p = 250$ MeV) indicate that *shapes* of angular distributions are not qualitatively changed by reasonable variations in the Δ−nucleus optical
potential, proton and pion optical potentials, and the choice of single particle orbitals

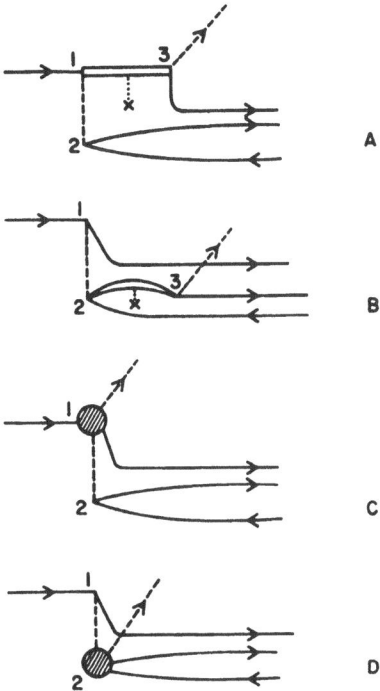

Fig. 24   Some diagrams appearing in a two nucleon mechanism of pion production. Diagram A (B) is referred to as a target (projectile) emission contribution resulting in intermediate Δ production. Diagrams C and D are non-resonant contributions.

(harmonic oscillator or Saxon-Woods). This relative insensitivity means that one can now carry out studies hopefully leading to a better understanding of the reaction mechanism(s) involved in pion production. Calculations to date have shown that it is important to include the propagation of the intermediate Δ and the energy transfer component in the intermediate pion propagator. The projectile emission piece (for the transition studied) dominates over the target emission term (see Fig. 25). This is of significance for comparative studies of $(e,e'\pi)$ and $(p,p'\pi)$ because it is the projectile emission isobar term in $(p,p'\pi)$ that is similar to the important intermediate Δ contribution to $(e,e'\pi)$ As an example, if one considers the projectile emission isobar term in $(p,p'\pi)$ and the isobar term in $(e,e'\pi)$ for the kinematic situation $T_e \sim T_p \sim 500\text{MeV}$ with an energy loss, ω, of ~ 200 MeV and $\theta_p^{scat} \approx \frac{1}{5}\theta_e^{scat} \approx 10°$ the energy-momentum parameters $(|\vec{q}|,\omega)$ of the exchanged boson resulting in Δ formation are similar. This region, where Δ effects should be important for the $(e,e'\pi)$ and $(p,p'\pi)$ reactions is an attractive arena for comparative studies. It would be instructive to have detailed coincidence studies of these processes in the continuum region where detailed nuclear structure considerations are less important and one can concentrate on reaction mechanism questions involving for example, Δ formation and propagation.

C. Pion Induced Nucleon Knockout and Meson Production

The final topic to be discussed involves the utility of pion induced coincidence studies such as, $(\pi,\pi'p)$ and $(\pi,2\pi)$. Of course, pion induced reactions resulting in heavy meson production or baryon resonances are exciting and practical possibilities for future research programs. These studies will provide information, in principle, on the formation and propagation of strongly interacting resonances in a many body environment and such investigations should be given a high priority. The actual observation of such reactions will depend on detection of the heavy meson or baryon resonance *decay* products and subsequent analysis using a Dalitz plot. Thus, these studies are intimately related to studies such as $(\pi,\pi'p)$ and $(\pi,2\pi)$.

Recently there has been interest in large energy loss pion inelastic and charge exchange experiments and the validity of using the DWIA in the associated theoretical interpretation. Of particular interest has been the role of the nuclear medium in altering the πN interaction and how this might be expected to affect inclusive quasi-elastic and charge exchange data. An understanding of the important modifications of the medium on the resonant part of the pion-nucleon interaction is apparently crucial for studying such processes as (π,π') and $(\pi,\pi p)$. Considerable progress along these lines has resulted from development and application of the isobar-hole model[39]. Thies[40] has studied the effects of medium modifications of the Δ resonance on inelastic pion scattering in the resonance region. He finds that effects treated in the isobar-hole model (including binding, Pauli blocking, and true absorption) significantly reduce the effective πN t matrix below and near the resonance and increase it above the resonance. The isobar-hole model, which includes these effects, yields superior fits to the quasi-elastic pion scattering data[41] (compared to a lowest order DWIA theory which neglects medium corrections). Karapiperis and Moniz[42] have pointed out that medium modifications of the pion-nucleon interaction must be taken into account before one can begin to use π^+/π^- ratios or differences to study nuclear isospin mixing (see earlier example of (π,π') results and Fig. 7).

Exclusive reactions[43,44] such as (π^+,π^+p) and (π^-,π^-p) where one can focus on the quasi-free reaction, identify specific final states, and compare relatively strong (π^+p) and relatively weak (π^-p) reactions and have been especially useful in isolating medium effects and validating some of the main predictions of the isobar-hole model. For example[44], strong deviations from the free ratio of $(\pi^+,\pi^+p)/(\pi^-,\pi^-p)$ are seen in the forward cross section data for 1p shell removal in $^{16}O(\pi,\pi'p)^{15}N$. Near the maximum of the π^+ cross section, significant enhancements are obtained while ratios smaller than the free result are obtained away from the π^+ cross section maximum. Deviations from the simple free Δ dominance model are expected from Δ–hole model results[44,45]. There are additional effects not included in first order (medium modified) models. For example a second order process where the intermediate Δ interacts with another nucleon could result in that nucleon's ejection. The additional $(T=1,\Delta-N)$ interaction appears to provide an important interference mechanism for understanding the exciting new and anticipated exclusive $(\pi,\pi N)$ data.

Several groups[46–50] have discussed the importance of the $(\pi,2\pi)$ reaction. There is already some information on $(\pi,2\pi)$ cross sections and one can expect additional data in the next few years. Using the distorted wave impulse approximation and the local density approximation, Cohen and Eisenberg[47] have emphasized the sensitivity of the $(\pi,2\pi)$ results to the nuclear spin-isospin strength distribution and to g', the Migdal spin-isospin parameter. In their model two nucleon contributions are estimated to be of the order of a few percent[47]. A more exact treatment[48] of the elementary $\pi N \to \pi\pi N$ amplitude results

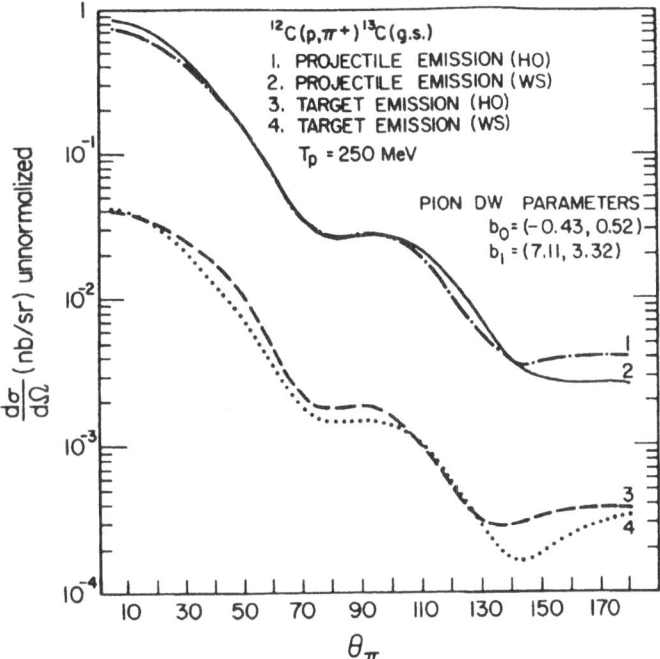

Fig. 25   The differential cross section contribution arising from projectile and target emission diagrams assuming harmonic oscillator (HO) or Saxon-Woods (SW) single nucleon bound orbitals. Full proton and pion distortions have been included in this TNM calculation. Figure taken from Ref. 38.

in a large increase in the total $(\pi,2\pi)$ cross section on protons ($\sim$ a factor of 4) compared to the Cohen-Eisenberg model. Enhancement effects (associated with g') would result in a further increase in the predicted cross section. An additional contribution to the $(\pi,2\pi)$ reaction involves pion absorption leading to an intermediate Δ which by a subsequent $\Delta N \rightarrow \Delta\Delta$ interaction leads to double Δ formation[49]. The $\Delta\Delta$ term has a 2π decay mode (via $2\Delta 2N \rightarrow 4N2\pi$). The ratio $R \equiv \sigma(\pi^-,\pi^+\pi^-)/\sigma(\pi^-,\pi^-\pi^-)$ is apparently quite sensitive to the inclusion of the 2Δ mechanism. For example, in the energy region $T_\pi \sim 300\text{--}400$ MeV it is found that R is decreased by more than a factor of two by inclusion of the $\Delta\Delta$ mechanism[50]. Recently, Oset and Vincente-Vacas[51] have studied $(\pi,2\pi)$ reactions on nuclei not using some approximations made by other authors. They find that precritical enhancements are unimportant but that there does result roughly a factor of two enhancement in the cross section due to the different relation between energy and momentum for pions in a many-body environment. It seems that the $(\pi,2\pi)$ reaction mechanism is not simple. It may be true that the reaction is *sensitive* to a variety of interesting effects associated with virtual intermediate Δ's or pions propagating in the nuclear medium. Whether the reaction will be useful in providing definitive information with regard to these effects is uncertain. It is clear that experimental data is badly needed, for a variety of nuclei and kinematic conditions, and for the possible initial and final pion charge states. The next generation of theoretical calculations, in addition to building on the theoretical results only briefly discussed above, should include Δ–nucleus interactions in the Δ formation piece of the calculation. Connections between theoretical studies of this reaction and $(p,p'\pi)$ and $(\pi,\pi'p)$ in terms of vertex functions and medium-modified propagators should be exploited. Information from these exclusive reactions should provide important constraints on the role of two nucleon (Δ formation) transition operators in inclusive experiments such as (p,p') and (π,π').

V. SUMMARY

There has been considerable advantage in previous joint studies of inclusive reactions involving electromagnetic and strong probes, such as (e,e'), (p,p'), and (π,π'). One expects this to continue as detailed studies of nuclei are performed using exclusive reactions such as (e,e'x), (p,p'x) and (π,π'x) [x=γ,π,p, etc.]. In these lectures we first discussed some possibilities available using ≤ 1 GeV high duty cycle electromagnetic facilities. Some of the interesting areas include studies of overlapping resonances, multipole transitions in J$\neq 0$ nuclei, coupling of resonances, to other nuclear degrees of freedom, and the formation and propagation of nucleon resonances and heavy mesons in nuclei. Although the electromagnetic interaction is weak and well-known there exist important strong interaction components in the coincidence experiments discussed. This introduction of strong interactions as an *important* contributor to the reaction mechanism is both an opportunity and a challenge. It means there is the possibility of learning important new physics but that the uncertainities and complications of strong interactions must be faced. This certainly motivates an overlap with strong probe exclusive experiments (now not only for exchange of deduced nuclear structure information but to treat certain parts of the reaction mechanism consistently in different reactions). We have included a discussion of some recent strong probe inclusive studies to show the quality of data and degree of success of theoretical models currently employed. We have also briefly discussed studies either recently reported or anticipated associated with the (p,2p), (p,p'π), (π,π'p) and ($\pi,2\pi$) reactions. In many of these studies isobar formation and propagation in the nuclear environment plays an important role in the reaction mechanism. It will be very interesting to see if combined studies can yield a consistent picture at the level of non-relativistic hadrons for these diverse exclusive reactions. The possibilities have *not* been discussed for studying the role of, (a) quarks and gluons, and (b) relativistic field theory in describing the results of the exclusive experiments considered in these lectures. Certainly these experiments may offer possibilities for such exciting studies. However, it seems probable that strong evidence for non-trivial relativistic or quark effects will be revealed only after *consistent* studies involving non-relativistic hadronic reaction theory has been shown to be inadequate for understanding the anticipated high quality exclusive data.

† Work supported in part by the National Science Foundation.

REFERENCES

[1] S. Frullani and J. Mougey, Ad. Nucl. Phys. **14,** eds. J. W. Negele and E. Vogt (Plenum, N.Y. 1984); *Nuclear Physics Research with a 750 MeV Cascade Microtron,* Univ. of Illinois Report (1982); *Nuclear Physics Research with a 288 MeV Cascade Microtron,* Univ. of Illinois Report (1984).

[2] T. DeForest, Jr., Ann. Phys. (NY) **45,** 365 (1967).

[3] J. D. Bjorken and S. D. Drell, *Relativistic Quantum Mechanics, McGraw-Hill (NY), 1964.*

[4] P. K. A. DeWitt Huberts, 1985 NATO Advanced Study Institute, Banff, Alberta, Canada (to be published).

[5]   T. W. Donnelly, 1985 NATO Advanced Study Institute, Banff, Alberta, Canada (to be published).

[6]   S. Boffi, C. Giuesti, and F. D. Pacati, Nucl. Phys. **A435,** 697 (1985).

[7]   H. L. Acker and M. E. Rose, Ann. Phys. (NY) **44,** 336 (1967).

[8]   C. N. Papanicolas *et al.,* Phys. Rev. Lett. **54,** 26 (1985).

[9]   D. Drechsel and H. Uberall, Phys. Rev. **181,** 1383 (1969).

[10] W. E. Kleppinger and J. D. Walecka, Ann. Phys. (NY) **146,** 349 (1983).

[11] L. Tiator and D. Drechsel, Nucl. Phys. **A360,** 208 (1981).

[12] C. L. Morris *et al.,* Phys. Lett. **86B,** 31 (1979).

[13] G. S. Adams *et al.,* Phys. Rev. Lett. **38,** 1387 (1977).

[14] C. Olmer, *Antinucleon and Nucleon-Nucleus Interactions,* eds. G. E. Walker *et al.,* (Plenum, NY 1985).

[15] W. G. Love and M. A. Franey, Phys. Rev. **C24,** 1073 (1981); (E) **C27,** 438 (1983); W. G. Love, A. Klein and M. A. Franey, *Antinucleon and Nucleon-Nucleus Interactions,* eds. G. E. Walker *et al.,* (Plenum, NY 1985).

[16] J. M. Moss *Spin Excitations in Nuclei,* eds. F. Petrovitch *et al.,* **355,** (Plenum, NY 1984)

[17] J. M. Moss, Phys. Rev. **C26,** 727 (1982).

[18] C. Olmer, *et al.,* Phys. Rev. **C29,** 361 (1984).

[19] C. Olmer, (private communication).

[20] J. Shepard, Proc. LAMPF Workshop on Dirac Approaches to Nucl. Phys., Los Alamos, NM (1985).

[21] C. Gaarde, J. S. Larsen and J. Rapaport, *Spin Excitations in Nuclei,* eds. F. Petrovitch *et al.,* **65,** (Plenum, NY 1984).

[22] C. D. Goodman and S. D. Bloom, *Spin Excitations in Nuclei,* eds. F. Petrovitch *et al.,* **143,** (Plenum, NY 1984).

[23] T. N. Taddeucci *et al.,* Phys. Rev. Lett. **52,** 1960 (1984).

[24] T. N. Taddeucci, *Antinucleon and Nucleon-Nucleus Interactions,* eds. G. E. Walker *et al.,* (Plenum, NY 1985).

[25] G. Jacob, Th. A. J. Maris, C. Schneider and M. R. Teodoro, Nucl. Phys. **A257,** 517 (1976).

[26] P. G. Roos *et al.,* Phys. Rev. Lett. **40,** 1439 (1978).

[27] P. Kitching *et al.,* Nucl. Phys. **A340,** 423 (1980).

[28] H. G. Pugh *et al.,* Phys. Rev. **155,** 1054 (1967).

[29] D. W. Devins *et al.,* Aust. J. Phys. **32,** 323 (1979).

[30] G. Van der Steenhoven *et al.,* (NIKHEF-K preprint-1985); private communication, P. K. A. deWitt Huberts.

[31] Some recent reviews of the experimental and theoretical situation include D. F. Measday and G. A. Miller, Ann. Rev. Nucl. Part. Sci. **29,** 121 (1979); B. Hoistad, Adv. Nucl. Phys. **11,** 135 (1979); H. W. Fearing, Prog. Part. Nucl. Phys. **7,** 113 (1981); *Pion Production and Absorption in Nuclei - 1981,* AIP Conf. Proc. No. 79, edited by R. D. Bent (AIP, NY 1982); G. E. Walker, Comm. on Nucl. Part. Phys. **A11,** 155 (1983).

[32] F. Soga *et al.,* Phys Rev. **C22,** 1348 (1980).

[33] F. Soga *et al.,* Phys. Rev. **C24,** 570 (1981).

[34] G. J. Lolos *et al.,* Phys. Rev. **C25,** 1086 (1982).

[35] W. W. Jacobs *et al.,* Phys. Rev. Lett. **49,** 855 (1982).

[36] S. E. Vigdor *et al.,* Nucl. Phys. **A396,** 61c (1983).

[37] E. D. Cooper and H. S. Sherif, Phys. Rev. **C26,** 3024 (1982).

[38] M. J. Iqbal and G. E. Walker, Phys. Rev. **C32,** 556 (1985).

[39] M. Hirata, F. Lenz and K. Yazaki, Ann. of Phys. **108,** 116 (1977); M. Hirata *et al.,* Ann. of Phys. **120,** 205 (1979); Y. Horikawa, M. Thies and F. Lenz, Nucl.

Phys. **A345,** 386 (1980); F. Lenz *Spin Excitations in Nuclei,* eds. F. Petrovitch *et al.,* 267 (Plenum, NY 1984).

[40] M. Thies, Nucl. Phys. **A382,** 434 (1982).

[41] M. Baumgartner *et al.,* Phys. Lett. **112B,** 35 (1982); C. H. Q. Ingram *et al.,* Phys. Rev. **C27,** 1578 (1983).

[42] T. Karapiperis and E. J. Moniz, Phys. Lett. **148B,** 253 (1984).

[43] E. Piasetzky *et al.,* Phys. Rev. **C25,** 2687 (1982).

[44] G. S. Kyle *et al.,* Phys. Rev. Lett. **52,** 974 (1984).

[45] M. Hirata, F. Lenz and M. Thies, Phys. Rev. **C28,** 785 (1983).

[46] R. Rockmore, Phys. Rev. **C11,** 1953 (1975).

[47] J. Cohen and J. M. Eisenberg, Nucl. Phys. **A395,** 389 (1983).

[48] R. S. Bhalerao *et al.,* Phys. Rev. **C30,** 224 (1984).

[49] G. E. Brown, Phys. Lett. **118B,** 39 (1982); B. Schwesinger *et al.,* Phys. Lett. **132B,** 269 (1983).

[50] L. C. Liu (private communication).

[51] E. Oset and M. J. Vicente-Vacas, Univ. of Ill. Preprint 170, (1985).